Vienna Circle Institute Library

Volume 14

Editorial Address
Wiener Kreis Gesellschaft
Universitätscampus, Hof 1, Eingang 1.2
Spitalgasse 2-4, A–1090 Wien, Austria
Tel.: +431/4277 46504 (international) or 01/4277 46504 (national)
Email: vcs@univie.ac.at
Homepage: https://vcs.univie.ac.at/

This peer-reviewed series includes monographs and edited volumes, which complement the format of the related series "Vienna Circle Institute Yearbook" and "Veröffentlichungen des Institut Wiener Kreis", both published with Springer Nature. The books mainly deal with individual members of the Vienna Circle, the entire Schlick-Circle as a collective, its adherents and critics, as well as with related topics of Logical Empiricism and its periphery in historical and philosophical perspective. Specifically, they are based on so far unpublished primary sources and feature also forgotten and marginalized issues as significant contributions to the most recent research in these fields.

Roberto Gronda • Adam Tamas Tuboly
Editors

Ernest Nagel on Science and Philosophy

Volume I: Biographical Materials and Correspondence with C. G. Hempel

Springer

Vienna Circle Institute Library

Editors
Roberto Gronda
Dipartimento di Civiltà e Forme del Sapere
Università di Pisa
Pisa, Italy

Adam Tamas Tuboly
Institute of Philosophy
ELTE Research Centre for the Humanities
Budapest, Hungary

ISSN 1571-3083
Vienna Circle Institute Library
ISBN 978-3-031-99457-9 ISBN 978-3-031-99458-6 (eBook)
https://doi.org/10.1007/978-3-031-99458-6

The correspondence in Chapter 4 is published by the permission of The Archives of Scientific Philosophy, Archives & Special Collections, University of Pittsburgh Library System.

This Springer imprint is published by the registered company Springer Nature Switzerland AG
The registered company address is: Gewerbestrasse 11, 6330 Cham, Switzerland

Preface

This volume was in the making for several years, although it was put aside occasionally. Nevertheless, seeing the renewed interest in Ernest Nagel's life and work, and in Carl G. Hempel's philosophy as well, we decided to finish the editorial work, including as many references and materials as possible to make the volume useful for scholars who would like to engage with the philosophical and institutional history of twentieth-century philosophy of science.

This is the first volume of our work that focuses on Ernest Nagel's ideas on science and philosophy. The second volume, tentatively titled "Selection of (Un)-published Materials," is under preparation (scheduled for 2026) and will include a larger and broader selection of Nagel's unpublished talks and lectures, and also many of his rare, hardly accessible, or simply forgotten papers, booklets, and reviews. Our aim with these two volumes is not just to call attention to Nagel's life and work, but also to show the depth and comprehensiveness the philosophy of science and analytic philosophy had in the twentieth century—and still could have—through the figure of Nagel.

We are truly grateful to Professor Friedrich Stadler for his encouragement and interest in the volumes. We were supported in the last few years by Yvonne Nagel, who also contributed a longer biography to the first volume, and shared her materials with us many years ago. Alexander and Sidney Nagel gave us permission to use these materials, and provided incessant support and interest in their father's work. Miranda and Peter Andrew Hempel also supported us and gave permission to work on their father's life and philosophy and to publish our results.

Our most sincere gratitude goes to the staff of the Archives of Scientific Philosophy at Pittsburgh University, where the Nagel–Hempel correspondence is hosted. We are truly indebted to their work, and we hope that our volume will call the attention of many readers to the Archive and their materials without which this volume would have been impossible.

For the relevant correspondence of Hempel and Nagel see:

- Ernest Nagel Papers, 1925–1985, ASP.2020.01, Archives of Scientific Philosophy, Archives & Special Collections, University of Pittsburgh Library System, Box 2, Folders 52–57, Box 3, Folders 1–4.
- Carl Gustav Hempel Papers, 1903–1997, ASP.1999.01, Archives of Scientific Philosophy, Archives & Special Collections, University of Pittsburgh Library System, Box 28, Folders 1 and 2.

The photos in Chap. 2 came from the personal collection of Yvonne Nagel, and we reproduce them with the permission of Yvonne, Alexander, and Sidney Nagel.

Pisa, Italy/Budapest, Hungary
July 2025

Roberto Gronda
Adam Tamas Tuboly

Contents

About the Editors

Roberto Gronda is Associate Professor in Philosophy of Science at the University of Pisa. He works on the history of American philosophy, especially on pragmatism, and the philosophy of scientific expertise. His latest works include *Dewey's Philosophy of Science* (Springer, 2020) and *Pragmatism and Philosophy of Science: Historical and Contemporary Perspectives* (ed. with G. Marchetti and M. Janack, 2026, Routledge).

Adam Tamas Tuboly is research group leader of the MTA Lendulet Values and Science Research Group at the ELTE Research Centre for the Humanities, Budapest and senior research fellow at the Department of Behavioural Sciences at the Medical School, University of Pécs, Hungary. He works on the history of philosophy of science, especially on logical empiricism. His latest works include *Interpreting Carnap* (ed. with A. Richardson, CUP, 2024), *Otto Neurath in Britain* (with Ch. Burke, CUP 2025), and *The Routledge Handbook of the History of Philosophy of Science After Kant* (ed. with F. Padovani, 2026).

Chapter 1
In Focus: Ernest Nagel and American Philosophy Between Pragmatism and Logical Empiricism

Roberto Gronda and Adam Tamas Tuboly

Abstract This chapter introduces the latest inquiries about logical empiricism and situates this volume in the context. After that, we discuss Nagel's place and role in the history of scientific philosophy and analyzes his relationship to pragmatism and logical empiricism.

Keywords Ernest Nagel · Pragmatism · John Dewey · Logical empiricism · History of philosophy of science

1.1 Dancing Around Logical Empiricism

HOPOS: The International Society for the History of Philosophy of Science Association is usually divided—at least at its congresses—into two major subgroups, namely "Before Kant" and "After Kant". At these occasions, the "After Kant" sessions are often dominated by talks about the various (historical) aspects of logical empiricism. Even though the term—"logical empiricism"—is taken broadly and thus enabling one to research and discuss a vast quantity of materials and perspectives from the 1920s to the 1970s, Rudolf Carnap and Otto Neurath remain in focus in these discussions, alongside Moritz Schlick and Hans Reichenbach.

This is understandable for various reasons. There are many stories to be told and arguments to be made about how in the 1970s, 80s, and 90s the industry of the Vienna-Circle-scholarship got off the ground and dominated the still-forming

R. Gronda
University of Pisa, Pisa, Italy
e-mail: roberto.gronda@unipi.it

A. T. Tuboly (✉)
MTA Lendület Values and Science Research Group, ELTE Research Centre for the Humanities, Budapest, Hungary

Department of Behavioural Sciences, Medical School, University of Pécs, Pécs, Hungary

R. Gronda, A. T. Tuboly (eds.), *Ernest Nagel on Science and Philosophy*, Vienna Circle Institute Library 14, https://doi.org/10.1007/978-3-031-99458-6_1

HOPOS community from the early 1990s onwards.[1] From a philosophical point of view, however, Carnap, Neurath, Schlick, and Reichenbach were the figures who influenced parts of the philosophical landscape and the general agenda of logical empiricism in both Vienna and Berlin. As Carnap and Reichenbach emigrated to the United States in the mid- and late 1930s, they had numerous students who followed their doctrines and kept their spirit alive, thus their type of logical empiricism came to dominate the United States from Chicago to Los Angeles.

Almost everybody who studied philosophy of science in the twentieth century had to go through some texts of Carnap and Reichenbach; yet Karl Menger, Hans Hahn, and Philipp Frank rarely came up in the relevant curricula after the 1960s. Thus, if we want to understand our own positions in the field, if we are aiming to reconstruct and see why we are studying *these* problems within a *particular* framework, then we must study and reconsider the philosophers who had major impacts on the field.

From a historical point of view, it is not necessary for Carnap, Neurath, Schlick and Reichenbach to dominate the scene. Already by the 1970s, a partial reinterpretation and reconsideration of the Vienna Circle had taken place, mainly due to Robert S. Cohen, who worked in close collaboration with the widows, friends, and colleagues of the Circle members to translate and publish their most important writings in the *Vienna Circle Collection* series (first by Kluwer, later by Springer). The selected writings of Schlick, Reichenbach, the first major English translations of Neurath, were supplemented by the volumes of Karl Menger, Hans Hahn, Felix Kaufmann, Philipp Frank, Eino Kaila, Béla Juhos, Friedrich Waismann, and even of Josef Schächter. But without detailed studies of the secondary literature that could have contextualized and explained some of these esoteric works to contemporary audiences, they fell into the void like water drops into a fountain.[2]

As time went on, people started to recognize that the Vienna Circle was a much bigger endeavor and a slow move from the core to the periphery became visible. Important, though largely forgotten people, such as Herbert Feigl, Carl G. Hempel, Friedrich Waismann, and recently Philipp Frank began to receive scholarly

[1] There is neither the need nor the space here to repeat all the major landmarks of scholarship, which has been done primarily by Jordi Cat, Michael Friedman, Rudolf Haller, Elisabeth Nemeth, George Reisch, Alan Richardson, Friedrich Stadler,Thomas Uebel, and many others. For a comprehensive overview see Stadler (2015) and for a more recent, personal narrative of the events, see Sarkar (2024, esp. Chap. 9).

[2] While Carnap has never fallen from the canon and thus his works required less reinterpretation to influence philosophers (although, of course, the actual reinterpretation of his early German works contributed much to the renewed and *refined* interest that the field has now experienced for decades), Otto Neurath was quickly forgotten after his death in 1945. Decades-long work, first of Rudolf Haller, then independently by Friedrich Stadler, Elisabeth Nemeth, and Thomas Uebel was required to put Neurath's original works under right and relevant light. Nothing like that has happened to most of the other members of the Circle until recently.

attention.[3] As a result, their already translated and available works earned new and informed interpretations and contextualization, and their *actual roles* within the movement were now emphasized against a more detailed and supported philosophical and socio-cultural background. The Vienna Circle returned to what it was once, a huge *circle of scholars* with different backgrounds and interests.

Besides rediscovering the *full circle*, so to speak, messengers and promoters of logical empiricism were reconsidered as well to understand how the caricature of the movement (Verhaegh 2024) was born and became established within such a limited time. The influential role of A. J. Ayer in Great Britain, and Charles Morris' and Ernest Nagel's place in the United States gained not only new historical perspectives, but also a systematic canvass on their favored topics such as ethics or reduction.[4]

Scholarship on logical empiricism in general is being transformed for various reasons. Firstly, as noted above, the topics and subjects are constantly changing, and it is hard to tell at the moment which layer of the movement will be the last one that *genuinely* had *any* kind of justifiable influence on the history and philosophy of empiricism/analytic philosophy/scientific philosophy.[5] Secondly, the methods of philosophy and history of philosophy are also changing. Though archival work was always one of the basics and foundations of HOPOS-work, due to the increasing digitalization of recent years (that saw a renewed and exponential increase during the COVID-19 pandemics), archives became more democratized. That is, a broader range of scholars gained access to those materials that were previously restricted to very few people, i.e., to those who were either living close to the relevant facilities or had the financial and institutional background to visit those archives across the globe. In that sense, the age of digitalization that has finally reached scientific archives has also brought about the democratization of the history of philosophy.

With the opening of the archives, new types of materials became available on a grand scale: personal and professional correspondence, diaries, manuscripts, personal libraries, lecture notes, and various further official and personal documents. One typical example might be the Carnap-Neurath correspondence (that was intended to have been published in the 1970s) and the diaries of Carnap—these materials are having a significant impact on how Carnap and Neurath are currently

[3]On Feigl see Neuber (2025); on Hempel see the essays in Fetzer (2000) and Dewulf (2018, 2022); on Waismann see Makovec and Shapiro (2019); on Frank see Reisch and Tuboly (2021) and Uebel (2003, 2011).

[4]On Ayer see Tuboly (2021); on reduction in Nagel see recently Chen (2025) and references there.

[5]Note also, however, that it is not just the figures and scholars who are constantly changing, but also the general narrative in which the same old figures have been reinterpreted. Recently, Alan Richardson (2023) has attempted to transform and change our understanding of what the Circle was up to without changing much the core figures that have dominated the field for decades. That work is being analyzed in a special issue of *Asian Journal of Philosophy*: https://link.springer.com/collections/gebcgjijig.

being studied.[6] Further examples include the recent publications on Kurt Gödel's life and work based on previously unknown archival materials, the publication of Philipp Frank's last unfinished book manuscript, and Josef Schächter's newly available Viennese writings: all these projects mark new paths for our understanding of the movement of logical empiricism and the rise of scientific philosophy.[7] It is true that most of these materials were already used in the 1970s and 80s (especially by Haller, Stadler, Uebel, Dahms, etc.), but now they are available to more people. Perhaps the interest of the community has also changed: as various biographies have recently appeared (see Edmonds 2020; Sigmund 2017; Misak 2020; Burke and Tuboly 2025), they somehow show that the archives are no longer primarily used to find forgotten philosophical points to use in an argument, but rather to get a clearer and more transparent cultural and personal background and context of the given Vienna Circle member.

This volume contributes to these two strands of scholarship. Namely it concerns a figure, Ernest Nagel, who was treated for a long time as a secondary logical empiricist next to Carnap and Reichenbach, and who received less attention than Neurath or even Waismann.[8] Furthermore, it concerns some secondary materials, namely a selection of Nagel's correspondence and his personal biography.

1.2 Who Was Ernest Nagel?

As the second chapter of this volume is a major biography of Nagel (written by Yvonne Nagel), a full reconstruction of his life and work is not required here. We shall concentrate only on a few key moments (for the life and works of C.G. Hempel, another prominent figure in this volume, see Chap. 3).

Nagel was born on 16 November 1901 in Nové Mesto, Moravia, then part of the Austro-Hungarian Empire. His family emigrated to the United States in 1911 and became an American citizen in 1919. He lived in New York for most of his life: he first studied at the City College of New York (CCNY), where he obtained a Bachelor of Science degree in 1923. At the CCNY, he studied with Morris R. Cohen, who was to have a lasting influence on his philosophical views (Suppes 1994). He then moved to Columbia University to receive a master's degree (1925) and a PhD in philosophy

---

[6]For the Neurath-Carnap correspondence, see Cat and Tuboly (2019, 512–685) and Damböck et al. (2024a); for Carnap diaries, see Damböck (2021, 2022); the result of using these materials is to be found in Damböck and Wolters (2021), Damböck et al. (2024b), and Richardson and Tuboly (2024).

[7]For Gödel, see Lethen (2021) and von Plato (2025), Frank (2021), and Heinrich-Ramharter and Stadler (2025) on Schächter.

[8]Besides articles on Nagel (especially on his theory and idea of reduction), general and systematic exceptions are the edited volume Neuber and Tuboly (2022) and the special issue of Berofsky and Levi (2012). Of course, Nagel's only *Festschrift* (Morgenbesser et al. 1969) provides the best traditional clue to assess his contemporary significance.

(Nagel 1930), with a dissertation on the logic of measurement, some extracts of which were published as separate articles.[9]

At that time, Columbia was characterized by a distinctive form of naturalism, which combined Aristotelian and pragmatist elements: Frederick Woodbridge and John Dewey were the most recognizable figures in the two camps (Jewett 2011; Verhaegh 2025c; see also Anton 2005). Nagel's naturalism emerged from these sources and remained a central feature of his philosophy during his entire career. Later in an interview, Nagel admitted that at Columbia, besides Woodbridge, "the other person who, of course, certainly had an influence on me was John Dewey" (Nagel and Nunn 2022, 294).[10]

After receiving his PhD, he began to teach—mainly logic, both introductory and advanced, and history and philosophy of science—at Columbia University, where he remained until 1970, with only a one-year break in 1966–67, when he accepted a position at Rockefeller University. During his career, he also served as President of the American Philosophical Association Eastern Division for the year 1954–55 and was elected to the National Academy of Sciences in 1977.

Nagel published consistently throughout his life. From the time he was a graduate student, he developed the habit of writing numerous reviews of books and articles, first as a way of earning some very much-needed money and then as a forced strategy for keeping up with the literature. As Yvonne Nagel has correctly pointed out (see Chap. 2 in this volume), the philosophical relevance of many of these reviews is demonstrated by her father-in-law's decision to republish them in some of his later books.

Among his monographs, three are particularly noteworthy. Firstly, the book he wrote with his teacher and mentor Morris R. Cohen, *An Introduction to Logic and Scientific Method* (Cohen and Nagel 1934), which was to become a widely read text in the field. Secondly, the short monograph *Principles of the Theory of Probability* that Nagel wrote for the *International Encyclopedia of Unified Science* (edited by Rudolf Carnap, Charles Morris, and Otto Neurath) and which was a culmination of a series of articles that he had published in the 1930s (Nagel 1939). Finally, in 1961 Nagel published his masterpiece, *The Structure of Science: Problems in the Logic of Scientific Explanation* (Nagel 1961), which reconstructed many of the debates—about scientific explanation, the nature of theories, theory reduction—that had taken place over the previous three decades (see further Chap. 3 here).

---

[9] Chapter 2 on "Measurement" was published in Erkenntnis (Nagel 1931), chapter 4 on "Nature and Convention" was published in a revised and extracted form in *The Journal of Philosophy* (Nagel 1929a), and one might get a sense of what was going on in chapter 5 on "Logic, Intuition, and Consistency" from an article on "Intuition, Consistency, and the Excluded Middle" published again in *The Journal of Philosophy* (1929b). We intend to reproduce the whole dissertation in Gronda and Tuboly (2026).

[10] In the same vein, May Brodbeck (1961, 998) wrote in her review of Nagel's *The Structure of Science*, Nagel was unable to leave behind instrumentalism in favor of a full-blooded realism because "John Dewey was an early, strong, and, as must now be seen, ineradicable influence on his thought."

The Structure of Science was Nagel's last major philosophical work. He continued to work and publish regularly after its publication, but mainly in the form of articles and essays (that were collected in his *Teleology Revisited and Other Essays in the Philosophy and History of Science*, see Nagel 1979b). In 1977 he was invited to give the third series of John Dewey Lectures at Columbia University. He died of pneumonia on 20 September 1985 at the age of 83.

1.3 Nagel's Place in the History of Scientific Philosophy

In his recent volume on European intellectual migration, Sander Verhaegh has summarized Nagel's importance as a bridge-builder:

> Nagel, for instance, is a perfect example of someone who still engaged with a broad range of schools and traditions in the 1930s. He ... was a student of the pragmatist John Dewey ..., played an important role in the development of the Columbia school of naturalism ... as well as the reception of logical empiricism in North America ... but was also involved in debates with phenomenologists ..., Horkheimer's Frankfurt School in exile ...., and Leo Strauss .... Ironically, Nagel was one of the scholars who helped coin the phrase 'analytic philosophy' ... and would eventually become one of America's best-known philosophers of science in the analytic tradition. (Verhaegh 2025b, 16)

Nagel was not a typical narrow-minded logician and philosopher of science who was unable to reflect on traditions external to his own canon. For scholarly and perhaps for existential reasons as well (to write solicited reviews as a mean to supplement his salaries), Nagel was not only well-read in every form of twentieth-century intellectual history, but also, he engaged with the most important figures and occasionally even contributed to foundational and on-going debates. Only the list of Nagel's published and unpublished writings could do sufficient justice to such a claim (for that, see Gronda and Tuboly 2026), but even his correspondence with Carl G. Hempel (Chap. 4 here), and his intellectual biography (written by Yvonne Nagel, Chap. 2 here) present enough evidence to pique interest in Nagel's life and achievements. Below we consider only two of Nagel's major reference points, namely pragmatism (Sect. 1.3.1) and logical empiricism (Sect. 1.3.2).

1.3.1 Nagel and Pragmatism

It would be wrong, however, to think that the activity of bridge-building that Nagel was so keen on was not fraught with difficulties. Indeed, despite Nagel's insistence on the theoretical continuity between pragmatism and logical empiricism (Nagel 1941/1954), their integration was far from straightforward.

In particular, we argue that Nagel's relationship with pragmatism is so complex and multifaceted that it cannot be evaluated along a single axis. As we will try to show, Nagel's use of pragmatism should not be understood solely in terms of his original and technical interpretation of pragmatist themes—not to mention in terms of the biographical circumstances of his encounters with pragmatism. The historical context in which Nagel's appropriation of pragmatist themes took place must also be taken into account. In this context—which was characterized by the process of transformation of American philosophy of science as a result of the influence exerted on it by the philosophers who had immigrated to the United States from Europe (see the chapters in Verhaegh 2025a)—the use of, and reference to, pragmatism had a cultural and political import that transcended the strictly philosophical implications of the adoption of specific pragmatist theses. As we will argue, the use of, and reference to, pragmatism also had an intended *performative* effect, aimed at promoting a particular image of what philosophy of science should look like at a time when the status of the discipline was under discussion.

We will therefore examine Nagel's relationship to pragmatism under three different headings: (a) his personal encounters with pragmatists; (b) his theoretical appropriation of pragmatist theses; (c) his 'political' use of pragmatism. We deal with points (a) and (b) in the present section, leaving point (c) to the next. By distinguishing three different levels of analysis, we hope to provide a better and richer overview of how Nagel positioned himself in the context of the growing discipline of philosophy of science.

Before considering Nagel's relationship to pragmatism, it may be useful to say a few words about classical American pragmatism as a movement. Pragmatism is commonly presented as a distinctive tradition in American philosophy—sometimes as *the* most distinctive contribution to philosophy (Fisch 1942/1986, 6)—along with the various forms of realism, naturalism, and idealism that dominated the philosophical debate in the first three decades of the twentieth century (see, among the others, Kuklick 2001; see also Misak 2013). Contrary to what is often assumed, however, classical pragmatism was not a single, unified movement, but rather a family of thinly connected theoretical positions that were held together by personal relationships between their proponents and by their agreement on some general theses, such as the fallibilistic nature of human knowledge and the intrinsic relation between action and knowledge (Haack 2024).

Charles S. Peirce famously wrote that he preferred to drop the word pragmatism and label his own philosophical views 'pragmaticism', a name "ugly enough to be safe from kidnappers" (Peirce 1905/1935, 166; CP5.414). There is good reason to believe that Peirce was not chastising James with this harsh remark, but some other thinkers who, in his view, had gone too far in their efforts to take pragmatist insights out of the scientific laboratory in which they were developed; in any case, it seems safe to say that Peirce was not happy with the direction pragmatism was taking in the first decade of the twentieth century. Peirce also disliked Dewey's work on logic: he felt that Dewey had developed a "natural history of thought" that lacked the rigor of a scientific discipline (Hickman 1986). On his part, Dewey decided not to use the word "pragmatism" to label his logical views as formulated in his *Logic: The Theory of Inquiry*, as he believed that "so much misunderstanding and relatively futile controversy have gathered about the word that it seemed advisable to avoid its use" (Dewey 1938/2008, 4).

This was the unity of pragmatism. Accordingly, a finer grid of interpretation is needed to assess Nagel's relationship to the various forms of pragmatist philosophy available in his time than that which is provided by the broad category of "American pragmatism".[11] With this in mind, a good place to start is to look more closely at how Nagel came to be exposed to pragmatist views during his education.

Nagel studied with Morris R. Cohen at CCNY and then moved to Columbia, which was a stronghold of pragmatist naturalism at that time (Verhaegh 2022; Suppes 1994, 258–259). Morris was a realist who absorbed the teachings of the great Harvard philosophers—Josiah Royce and William James—of whom he was a doctoral student. He developed a relational and contextual form of realism based on the view of things as elements in a network (Cohen 1914), thus giving birth to an original philosophical outlook which, according to one interpreter, crystallized and exemplified what Nagel considered to be "America's most significant contribution to philosophic intelligence" (Nagel 1947/1954, 53; see Cahoone 2016, 2017; Schliesser 2022; Suppes 1994).

Although not sympathetic to pragmatism, Cohen was influential in establishing Peirce as a leading figure in American philosophy. He seems to have played a role in arranging for the publication of Peirce's *Collected Papers*, but more importantly he contributed to the rediscovery of Peirce's philosophy by editing the influential collection of Peirce's essays *Chance, Love, and Logic: Philosophical Essays*—published in 1923, the year Nagel graduated from CCNY (Peirce 1923). Nagel's acquaintance with Peirce's philosophy can be thus reasonably attributed to Cohen's influence and mediation.

At Columbia, Nagel was exposed to a different kind of pragmatism, more naturalistic and socially oriented than that of Peirce. Nagel attended only one course with Dewey on *The Types of Logical Theory*, which was a course for graduate students that Dewey taught at Columbia in 1927–1928 (see Nagel and Nunn 2022, 269; Johnston 2020). We know that Nagel was not impressed by Dewey as a teacher: he found him quite confusing and difficult to understand, as he seemed to use his classes as an occasion to think about the problems he was concerned with. As Nagel recalled later in an interview:

> He was a very difficult lecturer. You didn't know what on earth he was talking about.... almost as bad as his writing. I still recall the way he would come in with a piece of paper no larger than the palm of his hand on which he had something written. And then he would talk, and you had the sense of a man who didn't have a finished thesis to teach but he was thinking through a problem, and it was terribly impressive. And every once in a while he would stop and go to the window and look out; he was still thinking about this. I remember some of us

[11] Not to mention the figure and role of Charles Morris who was also an important bridge-builder between pragmatism and logical empiricism (see Morris 1937). Although Nagel acknowledged the importance of Morris' work in merging together pragmatism and logical empiricism, he was more skeptical of his semiotics and the embrace of semantic analysis presented in his texts (Nagel 1938d). Over time, Morris became rejected by most of the participants in these debates (including major pragmatists such as Dewey and Arthur Bentley and logical empiricists such as Neurath), and his character is still highly underappreciated in the history of twentieth century philosophy of science (cf. Reisch 2005, Chap. 16).

used to keep awake by taking notes, because he was so slow and he drawled. (Nagel and Nunn 2022, 269)

Nagel was also negatively biased towards Dewey as a result of the low regard in which he was held by Cohen (Nagel and Nunn 2022, 269 and 307). However, some of Nagel's views—for example, his attention to the relationship between science and common sense (Nagel 1961, Chap. 1) and his emphasis on what he called "contextual naturalism"—have a distinctive Deweyan flavor and can be read as a sign of the lasting impression that Dewey left on his philosophy.

When it comes to addressing the question of the influence of pragmatism, broadly conceived, on Nagel's philosophy of science, it is useful to distinguish between the adoption of general pragmatist ideas and the acceptance of specific theses that are peculiar to one of its versions. The former category includes the idea of fallibilism and the rejection of what Dewey called "the quest for certainty" (Dewey 1929/2008; see Nagel 1968/1979), as well as a deep respect and attention to the procedures and results of the empirical sciences. It also includes the partial nature of verification that Nagel embraces in his early article "Verifiability, Truth, and Verification", which seems to be in strong continuity with the pragmatist conception of doubt as grounded in specific reasons, as opposed to the Cartesian view of doubt as a universal and unrestricted activity of questioning every bit of received knowledge (Nagel 1934a; see also Peirce 1868/1984). From this perspective, pragmatism is consistent with the understanding of contemporary science as a self-correcting enterprise (Nagel 1959/1979).

Much more philosophically interesting is the theoretical debt Nagel owed to certain pragmatist theses, the most relevant example being his early adoption of an account of probability as truth-frequency largely inspired by Peirce (Nagel 1933; Galavotti 2022). In his article, "A Frequency Theory of Probability", Nagel defines probability as "the *relative frequency* of the effectiveness of a certain type of inference" (Nagel 1933, 537). In order to clarify the nature of probable reasoning necessary to support the definition of probability as truth-frequency, Nagel refers to Peirce's notion of a leading principle of an inference. A leading principle is a general habit or rule of inference which guides a class of arguments (Peirce 1880/1933, 17; CP 3.164); it is therefore different from the premises of an argument. As Nagel (1933, 537) puts it, "[a] leading principle can be defined as a proposition formulating a class of inferences." He then argues that a leading principle "denotes a relation whose domain is a class of propositions p and whose converse domain is a class of propositions q such that q is inferable from p in virtue of this principle" (Nagel 1933, 537). The probability that a proposition (or set of propositions) q is true, relative to a leading principle from which it is inferred from a proposition (or set of propositions) p, turns out to be "the limiting value of the relative frequency with which the principle leads to true conclusions from premises which are true" (Nagel 1933, 538). In the light of all this, it seems no exaggeration to describe Nagel's early account of probability as thoroughly Peircean.

Nagel's relationship to Deweyan pragmatism is more difficult to assess. Nagel's knowledge of Dewey's philosophy was first-hand and first-rate, particularly with regard to Dewey's logical theory, as they discussed the manuscript of Dewey's 1938 *Logic* (Hickman 2022, 207).[12] In addition, Nagel wrote an article with Dewey and Sydney Hook entitled "Are Naturalists Materialists?" (Dewey et al. 1945). We know from a letter sent to Hook that Dewey was "happy to subscribe" to what Nagel had written in reply to Sheldon's article "A Critique of Naturalism" (Sheldon 1945; Dewey 2008, John Dewey to Sidney Hook, (13128) 1945.07.21). There were only two changes that Dewey thought it useful to make: first, he wanted to emphasize more the strong connection between the naturalism they advocated and the advance of contemporary science; second, he objected that Nagel had conceded too much to the idea of the supposed private character of mental events. Overall, however, many of the categories which Nagel had relied on in the text were thoroughly Deweyan, especially the distinction between experience and knowledge.

Such a use of Deweyan categories is by no means an exception on Nagel's part. For example, to cite just one notable example, Nagel's account of the relationship between the "objects of gross experience" and the objects of science, as formulated in the article "The Logic of Reduction in the Sciences" (Nagel 1935a), owes much to Dewey, even from a terminological point of view. Indeed, it was Dewey who, in *Experience and Nature*, had spoken of the "primacy and ultimacy of gross experience" (Dewey 1925/2008, 24). A similar remark can be made about Nagel's use of the formula "generic traits [or characters] of existence", which was one of the key notions of Dewey's naturalistic metaphysics and which Nagel often referred to in his analysis—which changed significantly over time—of the issue of the connection between logic and ontology (Nagel 1930, ii; Nagel 1935/1956).[13]

The use of Deweyan concepts and terms, however, belongs to an early phase of Nagel's philosophical development. Due to Nagel's exposure to logical empiricism, and as a consequence of the refinement of the philosophical debate in the US, the influence of pragmatism on his philosophy began to fade. This is true not only of Dewey's influence, but also of Peirce's. Indeed, even in the field of the philosophy of probability, where his debt to Peircean pragmatism was more explicit, Nagel moved away rather quickly from his early truth-frequency account towards a less idiosyncratic position (Galavotti 2022).

When asked by Remmel Nunn, in a 1983 interview that was part of the *Columbia Center for Oral History*, to reconstruct his relationship to classical pragmatism and American philosophy, Nagel expressed his opinion that neither Cohen nor Dewey's

---

[12] As a renowned expert on Dewey's works, Nagel was asked to write the introduction to the new edition of *Logic: The Theory of Inquiry*, to be published as the twelfth volume of Dewey's *Later Works*. But again, Nagel had his problems: "I re-read the book and it was a very difficult task for me to go through that; it's not very well written [. . .]. So I was really in a great quandary. I didn't want to just make this a paean of praise; at the same time I have a great deal of piety towards Dewey. So, this was sort of a struggle between being honest and being pious" (Nagel and Nunn 2022, 308). For the introduction, see Nagel (1986).

[13] To assess the similarities and differences between Dewey's and Nagel's philosophies (of science), one must engage first with Dewey's philosophy of science first. As a start, see Gronda (2020) and Brown (2012).

work had stood the test of time and were far less significant than he had believed. Speaking of Dewey, Nagel remarked that his greatest limit was that he "didn't work out the details of the theory that he held" (Nagel and Nunn 2022, 308). Dewey's philosophy could well provide some good insights, Nagel argued, but it lacked the detailed treatment of the issues he touched upon. In this respect, Dewey's philosophy was disappointing. Nagel thus was not entirely a pragmatist. But he certainly gained a lot from the pragmatist tradition.

1.3.2 Nagel and Logical Empiricism

Nagel was no logical empiricist either. He was not educated or socialized as one, but he was one of the most ardent and loudest defenders of the movement in the States. After reading their works, Nagel came to know most members personally in 1934–35 when he was in Europe on a Guggenheim Fellowship. The details of the trip are described by Yvonne Nagel (see Chap. 2); thus, it is sufficient here to note that Nagel attended the *Eighth International Congress of Philosophy in Prague* (and the Circle's *Vorkonferenz* a few days before the international gathering, see Stadler 2015, 166–171). There the "most interesting and vital philosophers present" were the positivists (Nagel 1934b, 591) He repeatedly emphasized in his congress-report that the sessions on logic and science that were filled with logical positivists were the best, the most comprehensible, and to the point.

After visiting Vienna, Cambridge, and many cities where positivists and akin philosophers resided, Nagel returned to the States and wrote down his experiences, making the term "analytic philosophy" famous once and for all (Nagel 1936a, 1936b). In New York he started to publish reviews of the positivists' works and wrote papers about their scientific philosophy. One of Nagel's reasons for running basically a campaign for logical empiricism was his disappointment that much of contemporary American philosophy lacked precision and clarity. The encounter with logical empiricists represented "salutary stimuli" to an entire generation of American students of philosophy since the "vigorous and technically precise manner" with which logical empiricists stated their philosophical thesis forced them to "re-examine assumptions [they] had accepted as a matter of course" (Nagel 1956, xii).

One goal that Nagel was seeking was to develop an appropriate philosophical position for himself as an American philosopher dealing with a foreign tradition of greater sophistication. In his early writings Nagel often insisted on the affinities between logical empiricism and Peirce's empiricism. The most relevant example of this conciliatory approach is the important article "Charles S. Peirce, Pioneer of Modern Empiricism" which was read at the *Fifth International Congress for the Unity of Science*, held at Harvard University in 1939, and later published in "Philosophy of Science" in 1940 (see Nagel 1940/1954). In arguing that attention must be paid to Peirce's empiricism—Nagel wrote that "no account of the development of contemporary empiricism is adequate which neglects the writings and the influence of Charles Peirce"—he explicitly placed the two traditions if not on a par, at least in strong continuity. And by saying that "[i]t is therefore a happy sign that so many of the central ideas of the present movement have been independently developed on both sides of the Atlantic", Nagel was simultaneously signaling that no difficulties were to be expected about the compatibility of American and European empiricism and emphasizing the fact that American philosophers represented an original and independent voice in the philosophical debate.

As is well known, the question concerning the relationship between pragmatism and logical empiricism was hotly debated at the time. When Charles Morris wrote about the implementation of European logical empiricism in the States (again, in the context of the *Fifth International Congress for the Unity of Science* at Harvard in 1939), he noted that while American philosophers have written about the logical analysis of the sciences, it was "doubtful" whether these writings were equal "in either quantity or quality to the achievements of other thinkers", especially of those of European logical empiricists (Morris 1938, 26). The issue of the lack of precision of classical American philosophers was not one that Nagel avoided. In dealing with Peirce's pragmatic maxim, he was ready to admit that "Peirce's own formulation [. . .] leaves much to be desired in the way of explicitness and clarity" and that "more recent formulations, such as those by Professor Carnap and others, have the same general intent but superior precision" (Nagel 1940/1954, 93).[14]

For his part, Morris was optimistic that the new European philosophy would find fertile ground in the United States:

> [C]ertainly the main lines of contemporary development (Peirce notwithstanding) have not been directed from these shores; that they may be so directed in the future is partly due to American capacity to learn from others and to the presence in our midst of many thinkers whose lives were first lived elsewhere. (Morris 1938, 26–27)

There is little doubt that Nagel would have agreed wholeheartedly with these words of Morris; logical empiricism was offering something new, something exciting and precise, and because of the various similarities between the two traditions (engagement with experience and science), it would not be difficult for American philosophers (or at least many of them) to accept the new developments.

[14]It must be noted, though, that verification attracted the attention of American philosophers early on, and C. I. Lewis wrote a highly critical but stimulating piece on the relationship between experience and meaning as early as 1934. This was the paper that prompted Schlick's (1936) response and refinement of verification, and provided a good occasion for Carnap's (1936, 1937) first papers on the liberation of verification (dating back to 1931), which also marked his way to American publicity. Nagel himself devoted a paper to this question in 1934, with the telling title "Verifiability, Truth, and Verification" (Nagel 1934a).

In the beginning, Nagel was enthusiastic to help and promote logical empiricism, and even overcompensated his occasional critical remarks, as he noted to Neurath in a letter:[15]

> I have been asked to give a lecture before a group of students and teachers of mathematics on the Unity of Science Movement. The talk is scheduled for about two weeks hence, and is planned to be mimeographed and distributed. I look forward to it, and expect to have a good time being a propaganda agent. Toward the end of February I will also give a lecture at Cooper Union, which will be broadcast, the subject be Logical Empiricism. And some time this winter I am having a semi-popular article published in a semi-popular magazine on the same subject.[16]

Nagel ended his list of activities by asking Neurath, "Don't you think I am a pretty active missionary on behalf of a good cause?" He certainly wanted to be one.

The semi-popular paper mentioned was "The Fight for Clarity: Logical Empiricism" that appeared in *The American Scholar*. In that paper, Nagel presented logical empiricism within a traditional philosophical setting: just as Aristotle was already discussing "the methodological assumptions of science", logical empiricism "serve[d] as a cathartic to men's beliefs and ideas" by clarifying ideas. "In other words," says Nagel (1938b, 47), "according to this conception of philosophy its aim is to perfect principles of logic and method and to exercise them upon the various statements occurring in everyday discourse and the sciences." Thus, philosophy does not discover empirical matters of the world (the sciences are doing that) or the supernatural realm (that belongs to a wrong-headed metaphysics), but philosophy is a clarification process to arrive at the most reasonable and rational beliefs.

But Nagel was never just a missionary, or blind follower of the movement, a propagandist of positivism, or an uncritical defender of the European movement that was about to emigrate to the States. He was critical of various aspects of the Vienna Circle. Already in his congress-report, he noted that the pragmatist Charles Morris' biggest result was to supplement the positivists "purely formal studies on the nature of symbols by considering the social and biological contexts in which the latter function" (1934b, 592).

Morris himself offered this view, not as a criticism, but as a chance for engagement between the Americans and the Europeans. After discussing (as quoted above) what American philosophers could learn from European logical empiricists, he went on to argue that in "the reverse situation" when one is concerned about the "scientific treatment of socio-humanistic subject matters [...] our thinkers seem to me to have

[15] Nagel mentioned his review of the first three volumes of the *International Encyclopedia of Unified Science*, where he criticized Neurath's repetitive and "rambling essay" that did not make clear what does "unity" actually mean; and called Niels Bohr and Bertrand Russell's papers "incidental" (Nagel 1938a, 691), which would have been better left out of the volume. He asked Neurath in a letter, "I also said many complimentary things. Do you think you can forgive me?" (Nagel to Otto Neurath, November 2, 1938 (Otto Neurath Nachlass, Wiener Kreis Archiv, Rijksarchief in Noord-Holland, Haarlem, the Netherlands, item, 275).

[16] Ernest Nagel to Otto Neurath, November 2, 1938 (Otto Neurath Nachlass, Wiener Kreis Archiv, Rijksarchief in Noord-Holland, Haarlem, the Netherlands, item, 275).

made decisive contributions and to be in a strategic position of leadership" (Morris 1938, 27). Since psychological and social sciences, along with the theories of values, have been especially developed in the States, American philosophers, especially the pragmatists, were in a privileged position to assess the socio-humanistic background of science scientifically.[17] "Thus pragmatism has for decades been preparing the ground for the integration of the socio-humanistic studies with the wider corpus of scientific knowledge and procedure" (Morris 1938, 28).

Morris went on—and again, based on his writings, Nagel would have agreed with this—to say that logical empiricism and its unity of science movement needed to be integrated with pragmatism:

> Such integration is important if the unity of science program is not to remain a torso without head; it is humanly important because of its cultural ramifications and implications. An exact understanding of the nature of mind, value, art, and moral behavior should be an incentive in furthering creative cultural forces and in releasing them from the subservience to science which a scientifically minded age has seemed, wrongly, to impose [. . .]. (Morris 1938, 28)

To enable such an integration, you need more than symbolic logic, although that played its part too; Nagel thought that logical empiricism did not always function ideally in that regard. Right at the beginning, when he reviewed the *Gesammelte Aufsätze* of Schlick that appeared shortly after his death in 1936, Nagel wrote in agreement that "Schlick was very sensitive to the cultural and intellectual significance of the history of philosophy, and was less inclined than some of his confrères to throw aside as useless verbiage the writings of 'the great thinkers'" (1938c, 248). While, in general, Nagel was much closer to Carnap than to Schlick, he recognized Schlick's special knowledge of the history of philosophy and how he used it in his philosophy.[18]

Nagel emphasized the role and importance of history already in his impressions of European philosophy:

> it seems to me that a better knowledge of the history which they condemn would have saved many of the analytical philosophers from serious error; for, it will be seen, the latter frequently discuss the traditional problems however disguised they may be by a different terminology. Moreover, the historical approach, when wisely cultivated, can frequently produce the same kind of intellectual catharsis and dissolution of pseudo-problems as does the analytic method. (Nagel 1936a, 7)

Logical empiricists counted amongst analytic philosophers and surely some of them were among those whose mistakes originated from their ahistorical attitude. In

[17] It is no coincidence that almost all of the monographs of the *International Encyclopedia of Unified Science* dealing with socio-humanistic subject matters—with the notable exception of Neurath's *Foundations of Social Sciences* (1944)—were assigned to American scholars: Dewey's *Theory of Valuation* (1939) and Abraham Edel's *Science and the Structure of Ethics* (1961) were the only two published monographs, but three other monographs—only planned but never completed—on the sociology of science, the interpretation and judgment of art, and humanities and unified science were assigned to Louis Wirth, Meyer Schapiro, and Abraham Kaplan, respectively. On this point, see Morris (1960) and Reisch (1995).

[18] On Schlick as a more traditional philosopher, see Tuboly (2022) and Ferrari (2023).

chastising European logical empiricists for their lack of historical knowledge, Nagel was signaling a difference of approach: he was very interested in the history of science from an early stage, as it is testified by his 1939 paper on "impossible numbers" (Nagel 1935/1979) and his eighty-page long essay on the formation and history of formal logic within geometry (Nagel 1939/1979). Later, Nagel wrote extensively on the "origin of dialectic and syllogism in Greek philosophy", "the rise of modern science and its institutional career", "the scientific revolution", and other topics, although all of them remain in the archives (see Gronda and Tuboly 2026). But Nagel's extensive reviewing activity, and even his correspondence with Hempel, shows the breadth and depth of his historical knowledge and interests.

Furthermore, Nagel was very critical of the overtly formalistic tendencies that were adopted by many members of the movement after the 1940s. While Nagel (1935b) reviewed Carnap's *Logische Syntax der Sprache* (1934) favorably (although itself a highly formalistic work), he was highly dismissive of the semantic and methodological works that Carnap published in the 1940s and 1950s (see, for example, Nagel 1940/1954, 1942). We argue that, in Nagel's hands, pragmatism acted as a corrective to the hyperformalist tendencies that were present in the logical empiricist movement. Some of his references to the pragmatist tradition should therefore be read as part of a set of argumentative tools aimed at defending an alternative view of the philosophy of science.

Such a strategic use of pragmatism is nowhere more evident than in Nagel's *Introduction* to *Teleology Revisited and Other Essays in the Philosophy and History of Science*. Nagel's aim in these introductory pages is to address some of the criticisms raised by the so-called "new philosophy of science" against the "alleged ahistorical character of the 'orthodox approach'", and particularly the idea that "the observational evidence for a scientific theory can be assessed by using the rules of a formal calculus" (Nagel 1979, 3). Contrary to what the new philosophers of science believe, Nagel argued, the orthodox approach was not wholeheartedly committed to an ahistorical and purely formalistic account of scientific activity. The latter was Carnap's understanding of philosophy of science, but it was not shared by "most of the older generation of writers on the subject (for example, C. S. Peirce, Josiah Royce, John Dewey, M. R. Cohen, or P. W. Bridgman)", as well as by "some of Carnap's fellow logical positivists (such as Philipp Frank or Richard von Mises)."[19] As Nagel pointed out:

> Unlike Carnap, none of these thinkers subscribed to an ahistorical evaluation of the evidence for a scientific theory; and none of them identified the rationality of science with the use of exclusively formal canons for assessing claims to knowledge. It is misleading to ascribe to all representatives of the "orthodox approach" to the philosophy of science the beliefs that are idiosyncratic of what at best is a relatively small subset of that group of thinkers. (Nagel 1979b, 3)

---

[19]Thomas Uebel (2015) has argued that a certain form of pragmatism (American pragmatism mediated through European lenses) had a substantial influence on the so-called members of the "First Vienna Circle", such as Philipp Frank and Otto Neurath.

Nagel did not subscribe to the "ahistorical evaluation" and when he made his concrete criticism of Carnap's inductive program in the 1963 Schilpp-volume (see Nagel 1963), his relationship with Carnap indeed suffered a major blow (a recurring topic in the correspondence with Hempel, see Chap. 4 here). As he wrote to Hempel,

> I had great difficulty in doing that paper, in large measure because of my rather negative appraisal of C[arnap]'s inductive logic; and I regret that my worst fears have apparently been realized. My paper is of course full of very complimentary and genuinely felt comments; but it is true that if I am anyway near correct in some of my criticisms, the paper is hardly the sort of thing one contributes to what is essentially a Festschrift. I have not heard from Carnap for years, and I regret more than I can say that I have apparently hurt him. (See letter 106 in Chap. 4)

However, Carnap's gentle character and appreciation of scientific and scholarly skepticism and criticism won at the end of the day:

> On the whole, I am relieved that Carnap was not as displeased as I feared he would be – after all, I rejected pretty much in toto the work on which he has spent a larger fraction of his life than on anything previous; and he would have to be inhuman not to show some signs of disappointment with me. It was a difficult piece for me to write, just for this reason. (See letter 119 in Chap. 4)

But in line with Morris, Nagel often emphasized the behavioral and practical amendments that pragmatism offered to the logical analysis of science. In fact, Nagel occasionally emphasized some theoretical advantages of Peirce's approach to those of logical empiricists. Peirce's insistence, for example, on the behavioral dimension of meaning was conducive to a more concrete analysis of the meaning of scientific concepts, which could be clarified by looking at "the complex of relevant practices involved in their use", as well as to a sounder and more scientific account of knowledge, the latter being conceived of as a fallibilistic, objective, and cooperative activity of problem solving. As Nagel put it, "the contemporary insistence upon methodological principles as resolutions and rules of language, fails in its objective to exhibit the logical articulation of science if it neglects the practical compulsions involved in the construction and manipulation of technological and laboratory apparatus".

In this reconstruction, pragmatists and American philosophers of the older generation are invoked as proponents of a more appropriate conception of what philosophy of science should look like. Once again, Nagel's concern was to find a *via media* between a purely formal approach to the philosophy of science and an equally unacceptable non-epistemic and essentially irrationalist account.

Significantly, thus, the view with which Nagel expressed sympathy, and which he believed could overcome the shortcomings of the two approaches he found unsatisfactory, was pragmatism. It is worth quoting the passage in its entirety:

> According to this view, it is a fundamental error to define the rationality of a pattern of thought (or more generally of a mode of behavior) in terms of supposedly self-evident *a priori* principles. On the contrary, on this view a mode of behavior is rational if it regularly (or for the most part) achieves the objectives for the attainment of which the behavior is undertaken. In consequence, if the aim of scientific inquiry is to obtain satisfactory answers to the questions (or problems) that initiated the inquiry, the rationality of science consists in

> its use of procedures (whether symbolical or overtly experimental) which have been shown to be capable of doing just that. There is no antecedent guarantee that a proposed rule for the conduct of inquiry, if adopted, will achieve the aim of the undertaking, however "inherently rational" the rule may seem. Just as the effective rules or modes of procedure for constructing enduring chairs and tables can be ascertained only by reflecting upon the practice of cabinet makers, so the effectiveness of a proposed method for achieving the objectives of inquiry can be established only by reflecting upon the historical fortunes of that method (Nagel 1979b, 3–4).

Even at this late point of Nagel's philosophical career, it is the pragmatist notion of inquiry (or conduct of inquiry), with its contextual and temporal character and its capacity for self-evaluation and self-correction, that is said to provide a stable ground for the philosophy of science that could avoid reducing the latter to either a formal calculus or a "sociological" inquiry into the non-epistemic reasons that led a group of scientists to accept a particular theory.

Because of these, Thomas Mormann (2022) has characterized Nagel's relation to logical empiricism as a process "from cautious enthusiasm" (showing already the signs of his criticism of the ahistorical character of the movement) to "profound disenchantment", recognizing that logical empiricism did not live up to most of its promises. But it does not mean that Nagel would have been hostile to the movement or its figures. He helped Hempel, Frank and others to become acclimatized in the States, and he even joined Frank's Institute for the Unity of Science for years. There Nagel organized conferences (for example about sociology of science and W.V.O. Quine's "Two Dogmas" paper), directed essay contests, and helped Frank to push logical empiricism in less formalistic ways (see Reisch and Tuboly 2021). Nagel also served as referee for Frank's book manuscript on philosophy of science (see Frank 1957), especially because of his less formalistic tendencies and broad view of philosophy of science (Reisch and Tuboly n.d.). The manuscript was too long, and its last part remained unpublished for some time (and appeared later as Frank 2021), while those parts that got published became known as Frank's "text-book" and were criticized, for example, by Hilary Putnam (1958) for that broad-minded and metaphysics-engaging character that Nagel appreciated (see Tuboly 2017).

During all these changes, logical empiricism was always a reference-point for Nagel, and he was happy to make the movement international and rebuild it anew after Schlick's death and the emigration of its members. As soon as the *Vienna Circle* had died in Europe, the international unified science movement took off and became the representative of the "systematic house-cleansing of traditional philosophic speculations" (Nagel 1938b, 49). Just as Peirce wanted to make our ideas clear, logical empiricism and Nagel were interested in clarification and the mental health of society:

> One of its functions, that of serving as a disinfectant to the thinking of men, would alone justify the continuance and spread of the movement. It rests its case not on an appeal to authority or the emotional needs of men but on an appeal to a persistent effort to think clearly. The movement is an important arm in the interminable warfare against obscurantism and for clarity. (Nagel 1938b, 59)

With all this medical language, Nagel saw the purifying, curative, and therapeutic work that had to be done in society to remedy all the socio-political and cultural blows rationality suffered in the interwar years. Irrationality and mistrust in science were the diseases, and the critical-rational thinking of logical empiricism was the cure. For this role he took, Nagel was appreciated by many; as A. J. Ayer (1961, 197) wrote about him, "[Nagel] has been a relentless enemy of philosophical *Schwärmerei*, and has proved himself adept at puncturing pretentious theories, both in the philosophy of science and in other fields."

1.4 On the Volume

The volume has two major parts. Chapter 2 is a long intellectual biography of Ernest Nagel, written by his daughter-in-law, Yvonne Nagel, a retired professor of mathematics at the University of Wisconsin–Madison. The paper goes through the most important achievements and events of Nagel's life from his childhood immigration till his last days and reconstructs a fascinating life from personal and scholarly letters. The correspondence Yvonne Nagel is using here was kept by the family for decades, and she has been preparing Nagel's cultural journey through the lens of the letters for years. It contains many interesting remarks for the general reader, but many of the long-quoted letters also speak to hardcore Nagel-scholars.

The second major part of the volume consists in the Nagel and Carl G. Hempel correspondence (Chap. 4) that lasted six decades and is preceded here by an editorial introduction (Chap. 3). The letters are reproduced entirely without a selection, and they represent some of the major developments of scientific and analytic philosophy in the long twentieth century.

We have decided to present the Nagel-Hempel correspondence for various reasons. Both Nagel and Hempel are important defenders and propagators of logical empiricism in the United States; though both of them were emigrants, one of them was educated and socialized in the United States and was greatly influenced by pragmatism, the other was an educated in Europe, who became naturalized quickly and published his most important results as an American philosopher. In a certain sense, though, Nagel and Hempel could be seen as minor figures from the viewpoint of the usual historiographies of logical empiricism. They are rarely placed alongside Carnap, Schlick, Neurath, and Reichenbach (see Sect. 1.1 above). But in one sense they did the most to propagate and acclimatize the movement in the States. Although Reichenbach had a lot of students who became influential analytic and scientific philosophers (like Hilary Putnam or Wesley Salmon), Nagel and Hempel were teachers of generations who carried the torch for the movement in the background for decades (like Isaac Levi; see Sarkar 2024).[20] When Carnap died in 1970, the

[20] Further students of Nagel include Henry E. Kyburg, Judith Jarvis Thomson, Kenneth Schaffner, and Patrick Suppes. Hempel's students include, among others, Adolf Grünbaum, John Earman, Richard Jeffrey, Jerrold Katz, Gerald Massey, and Lawrence Sklar.

most influential positivist from the original Viennese circle passed away.[21] Although Nagel participated on only a few occasions in 1935, and Hempel did the same in 1929, both knew enough of the original European discussions and mentality to be counted amongst the inner circle.

They also lived through the most intense times of institutionalization (Dewulf 2020), when major journals and departments were established, when scientific philosophy was transformed from a socially engaged cultural movement into a professional and technical enterprise, and both shaped and reshaped the public and scholarly face of philosophy of science as a new discipline.

This volume is, however, not a critical edition, but it stems basically from our interest and enthusiasm for the correspondence of Nagel, and we hope that others will see it as engaging as we did when we first encountered the materials presented here.

References

Anton, John Peter. 2005. *American Naturalism and Greek Philosophy*. Amherst: Humanity Books.

Ayer, A. J. 1961. Review of Nagel's The Structure of Science. *Scientific American* 204 (6): 197–203.

Berofsky, Bernard, and Isaac Levi. 2012. Aspects of Explanation, Theory, and Uncertainty: Essays in Honor of Ernest Nagel: Special issue. *The Journal of Philosophy* 109 (8/9).

Broadbeck, May. 1961. Structure of Science. Review of Nagel (1961). *Science* 134 (3484): 997–999.

Brown, Matthew J. 2012. John Dewey's Logic of Science. *HOPOS: The Journal of the International Society for the History of Philosophy of Science* 2 (2): 258–306.

Burke, Christopher, and Adam Tamas Tuboly. 2025. *Otto Neurath in Britain*. Cambridge: Cambridge University Press.

Cahoone, Lawrence. 2016. American Realism, Objective Relativism, Columbia Naturalism, and Justus Buchler. *Transaction of the Charles S. Peirce Society* 52 (3): 416–434.

Cahoone, Lawrence. 2017. The Metaphysics of Morris R. Cohen: From Realism to Objective Relativism. *Journal of the History of Ideas* 78 (3): 449–471.

Carnap, Rudolf. 1934. *Der logische Syntax der Sprache*. Wien: Springer.

Carnap, Rudolf. 1936. Testability and Meaning. *Philosophy of Science* 3 (4): 419–471.

Carnap, Rudolf. 1937. Testability and Meaning – Continued. *Philosophy and Science* 4 (1): 1–40.

Cat, Jordi, and Adam Tamas Tuboly, eds. 2019. *Neurath Reconsidered: New Sources and Perspectives*. Cham: Springer.

Chen, Bohang. 2025. Nagelian Reduction and Approximation. *European Journal for Philosophy of Science* 15: 7.

[21] Feigl lived even longer than Nagel (Feigl died in 1988 and Nagel in 1985). However, although Feigl was actually one of the founders of the original discussion group around Schlick with Friedrich Waismann, our understanding of his role in the history of scientific philosophy in general, and logical empiricism in particular, is still very underdeveloped. Fortunately, scholars such as Michael Bennett McNulty and Alan Richardson have recently taken up Feigl in conference talks, and Matthias Neuber has published extensively about Feigl (see Neuber 2011, 2017, 2018, 2022, 2025).

Cohen, Morris R. 1914. Qualities, Relations, and Things. *The Journal of Philosophy, Psychology and Scientific Methods* 11 (23): 617–627.
Cohen, Morris R., and Ernest Nagel. 1934. *An Introduction to Logic and Scientific Method.* Harcourt, Brace & World.
Damböck, Christian, ed. 2021. *Rudolf Carnap: Tagebücher, Band 1: 1908–1919.* Hamburg: Felix Meiner Verlag.
Damböck, Christian, ed. 2022. *Rudolf Carnap: Tagebücher, Band 2: 1920–1935.* Hamburg: Felix Meiner Verlag.
Damböck, Christian, and Gereon Wolters, eds. 2021. *Der junge Carnap in historischem Kontext: 1918–1935/Young Carnap in an Historical Context: 1918–1935.* Cham: Springer.
Damböck, Christian, Johannes Friedl, and Ulf Höfer, eds. 2024a. *Rudolf Carnap/Otto Neurath: Briefwechsel.* Hamburg: Felix Meiner Verlag.
Damböck, Christian, Johannes Friedl, and Ulf Höfer, eds. 2024b. *Ways of the scientific world-conception: Rudolf Carnap and Otto Neurath.* Leiden/Boston: Brill.
Dewey, John. 1925/2008. Experience and Nature. In *John Dewey: The Later Works*, ed. Jo Ann Boydston, vol. 1. Carbondale: Southern Illinois University Press.
Dewey, John. 1929/2008. The Quest for Certainty. In *John Dewey: The Later Works*, ed. Jo Ann Boydston, vol. 4. Carbondale: Southern Illinois University Press.
Dewey, John. 1938/2008. Logic: The Theory of Inquiry. In *John Dewey: The Later Works*, ed. Jo Ann Boydston, vol. 12. Carbondale: Southern Illinois University Press.
Dewey, John. 1939/2008. Theory of Valuation. In *John Dewey: The Later Works*, ed. Jo Ann Boydston, vol. 13, 189–251. Carbondale: Southern Illinois University Press.
Dewey, John. 2008. *The Correspondence of John Dewey, 1871–1952 (I–III). Volume 3: 1940–1952.* Electronic Edition.
Dewey, John, Sidney Hook, and Ernest Nagel. 1945. Are Naturalists Materialists? *The Journal of Philosophy* 42 (19): 515–530.
Dewulf, Fons. 2018. Revisiting Hempel's Contribution to Philosophy of History. *Journal of the History of Ideas* 79 (3): 285–406.
Dewulf, Fons. 2020. The Institutional Stabilization of Philosophy of Science and its Withdrawal from Social Concerns after the Second World War. *British Journal for the History of Philosophy* 29 (5): 935–953.
Dewulf, Fons. 2022. The Emergence of Scientific Explanation as a Problem for Philosophy of Science: Aristotle, Nagel and Hempel. In *Ernest Nagel: Philosophy of Science and the Fight for Clarity*, ed. Matthias Neuber and Adam Tamas Tuboly, 67–87. Cham: Springer.
Edel, Abraham. 1961. *Science and the Structure of Ethics.* Chicago: The University of Chicago Press.
Edmonds, David. 2020. *The Murder of Professor Schlick: The Rise and Fall of the Vienna Circle.* Princeton: Princeton.
Ferrari, Massimo. 2023. After the Tractatus: Schlick and Wittgenstein on Ethics. In *Wittgenstein and the Vienna Circle: 100 Years After the Tractatus Logico-Philosophicus*, ed. Friedrich Stadler, 127–160. Cham: Springer.
Fetzer, James, ed. 2000. *Science, Explanation, and Rationality: Aspects of the Philosophy of Carl G. Hempel.* Oxford: Oxford University Press.
Fisch, Max Harold. 1942/1986. Justice Holmes, the Prediction Theory of Law, and Pragmatism. In *Peirce, Semeiotic and Pragmatism. Essays by Max H. Fisch*, ed. Kenneth Laine Ketner and Christian J. W. Kloesel, 6–18. Bloomington: Indiana University Press.
Frank, Philipp. 1957. *Philosophy of Science: The Link Between Science and Philosophy.* Englewood Cliffs: Prentice-Hall.
Frank, Philipp. 2021. *The Humanistic Background of Science.* Eds. George Reisch and Adam Tamas Tuboly. Albany: SUNY Press.
Galavotti, Maria Carla. 2022. On Nagel's Truth-Frequency Theory of Probability. In *Ernest Nagel: Philosophy of Science and the Fight for Clarity*, ed. Matthias Neuber and Adam Tamas Tuboly, 173–188. Cham: Springer.

Gronda, Roberto. 2020. *Dewey's Philosophy of Science*. Cham: Springer.
Gronda, Roberto, and Adam Tamas Tuboly, eds. 2026. *Ernest Nagel on Science and Philosophy. Volume 2. Selection of (Un)published Writings*. Cham: Springer, forthcoming.
Haack, Susan. 2024. Ugly Enough to be Safe from Kidnappers: 'Pragmatism', 'Pragmaticism', and the Ethics of Terminology. *Transactions of the Charles S. Peirce Society* 60 (1): 1–22.
Heinrich-Ramharter, Esther, and Friedrich Stadler, eds. 2025. *Josef Schächter's Philosophical Writings and Documents in the Context of the Vienna Circle: Before the Second World War*. Cham: Springer.
Hickman, Larry. 1986. Why Peirce Didn't Like Dewey's Logic. *Southwest Philosophy Review* 3: 178–189.
Hickman, Larry. 2022. Dewey's Logic: Reviews and Criticisms. *Análisis. Revista de investigación filosófica* 9 (2): 189–212.
Jewett, Andrew. 2011. Canonizing Dewey: Naturalism, Logical Empiricism, and the Idea of American Philosophy. *Modern Intellectual History* 8 (1): 91–125.
Johnston, James Scott. 2020. *John Dewey's Later Logical Theory*. Albany: SUNY Press.
Kuklick, Bruce. 2001. *A History of Philosophy in America, 1720–2000*. Oxford: Oxford University Press.
Lethen, Tim. 2021. *Gespräche, Vorträge, Séancen: Kurt Gödels Wiener Protokolle 1937/38*. Cham: Springer.
Lewis, Clarence I. 1934. Experience and Meaning. *The Philosophical Review* 43: 125–146.
Makovec, Dejan, and Stewart Shapiro, eds. 2019. *Friedrich Waismann: The Open Texture of Analytic Philosophy*. Basingstoke: Palgrave Macmillan.
Misak, Cheryl. 2013. *The American Pragmatists*. Oxford: Oxford University Press.
Misak, Cheryl. 2020. *Frank Ramsey: A Sheer Excess of Powers*. Oxford: Oxford University Press.
Morgenbesser, Sidney, Patrick Suppes, and Morton White, eds. 1969. *Philosophy, Science, and Method: Essays in Honor of Ernest Nagel*. New York: St. Martin's Press.
Mormann, Thomas. 2022. From Cautious Enthusiasm to Profound Disenchantment: Ernest Nagel and Carnapian Logical Empiricism. In *Ernest Nagel: Philosophy of Science and the Fight for Clarity*, ed. Matthias Neuber and Adam Tamas Tuboly, 89–108. Cham: Springer.
Morris, Charles. 1937. *Logical Positivism, Pragmatism and Scientific Empiricism*. Paris: Hermann et Cie.
Morris, Charles. 1938. The Unity of Science Movement and the United States. *Synthese* 3 (12): 25–29.
Morris, Charles. 1960. On the History of the International Encyclopedia of Unified Science. *Synthese* 12 (4): 517–521.
Nagel, Ernest. 1929a. Nature and Convention. *The Journal of Philosophy* 26 (7): 169–182.
Nagel, Ernest. 1929b. Intuition, Consistency, and the Excluded Middle. *The Journal of Philosophy* 26 (18): 477–489.
Nagel, Ernest. 1930. *On the Logic of Measurement*. Dissertation, The University of Columbia.
Nagel, Ernest. 1931. Measurement. *Erkenntnis* 2: 313–335.
Nagel, Ernest. 1933. A Frequency Theory of Probability. *The Journal of Philosophy* 30 (20): 533–554.
Nagel, Ernest. 1934a. Verifiability, Truth, and Verification. *The Journal of Philosophy* 31 (6): 141–148.
Nagel, Ernest. 1934b. The Eighth International Congress of Philosophy. *The Journal of Philosophy* 31 (22): 589–601.
Nagel, Ernest. 1935/1956. Notes Toward a Naturalistic Conception of Logic. In *Logic Without Metaphysics and Other Essays in the Philosophy of Science*. Glencoe: The Free Press, 39–53.
Nagel, Ernest. 1935/1979. 'Impossible Numbers': A Chapter in the History of Modern Logic. In Nagel (1979a), 166–194.
Nagel, Ernest. 1935a. The Logic of Reduction in the Sciences. *Erkenntnis* 5: 46–52.
Nagel, Ernest. 1935b. Review of Carnap's Logische Syntax der Sprache. *Journal of Philosophy* 32: 49–52.

Nagel, Ernest. 1936a. Impressions and Appraisals of Analytic Philosophy in Europe I. *The Journal of Philosophy* 33 (1): 5–24.
Nagel, Ernest. 1936b. Impressions and Appraisals of Analytic Philosophy in Europe II. *The Journal of Philosophy* 33 (2): 29–53.
Nagel, Ernest. 1938a. Review of IEUS. *The Journal of Philosophy* 35 (25): 689–693.
Nagel, Ernest. 1938b. The Fight for Clarity: Logical Empiricism. *The American Scholar* 8 (1): 45–59.
Nagel, Ernest. 1938c. Review of Schlick's Gesammelte Aufästze 1926–1936. *The Journal of Philosophy* 35 (9): 246–248.
Nagel, Ernest. 1938d. Review of Morris (1937). *The Journal of Philosophy* 35 (5): 133–134.
Nagel, Ernest. 1939. *Principles of the Theory of Probability*. Chicago: The University of Chicago Press.
Nagel, Ernest. 1939/1979. The Formation of Modern Conceptions of Formal Logic in the Development of Geometry. In Nagel (1979a), 195–259.
Nagel, Ernest. 1940/1954. Charles S. Peirce, Pioneer of Modern Empiricism. In *Sovereign Reason and Other Studies in the Philosophy of Science*, 89–100. Glencoe: Free Press.
Nagel, Ernest. 1941/1954. Recent Philosophies of Science. In *Sovereign Reason and Other Studies in the Philosophy of Science*, 36–49. Glencoe: Free Press.
Nagel, Ernest. 1942. Review of Carnap's Introduction to Semantics. *Journal of Philosophy* 39 (17): 468–473.
Nagel, Ernest. 1947/1954. Philosophy and the American Temper. In *Sovereign Reason and Other Studies in the Philosophy of Science*, 50–57. Glencoe: Free Press.
Nagel, Ernest. 1956. Introduction. In *Logic Without Metaphysics and Other Essays in the Philosophy of Science*, ix–xviii. Glencoe: The Free Press.
Nagel, Ernest. 1959/1979. Modern Science in Philosophical Perspective. In Nagel (1979a), 7–28.
Nagel, Ernest. 1961. *The Structure of Science: Problems in the Logic of Scientific Explanation*. New York: Harcourt, Brace, & World.
Nagel, Ernest. 1963. Carnap's Theory of Induction. In *The Philosophy of Rudolf Carnap*, ed. Paul A. Schilpp, 785–825. LaSalle: Open Court.
Nagel, Ernest. 1968/1979. The Quest for Uncertainty. In Nagel (1979a), 64–83.
Nagel, Ernest. 1979a. *Teleology Revisited and Other Essays in the Philosophy and History of Science*. New York: Columbia University Press.
Nagel, Ernest. 1979b. Introduction. In Nagel (1979a), 1–6.
Nagel, Ernest. 1986. Introduction. In *The Later Works of John Dewey*, ed. Jo Ann Boydston, vol. 12, ix–xxvii. Southern Illinois University Press.
Nagel, Ernest, and Remmel Nunn. 2022. Interview with Ernest Nagel by Remmel Nunn. In *Ernest Nagel: Philosophy of Science and the Fight for Clarity*, ed. Matthias Neuber and Adam Tamas Tuboly, 258–314. Cham: Springer.
Neuber, Matthias. 2011. Feigl's 'Scientific Realism'. *Philosophy of Science* 78: 165–183.
Neuber, Matthias. 2017. Feigl, Sellars, and the Idea of a 'Pure Pragmatics'. In *Logical Empiricism and Pragmatism*, ed. Sami Pihlström, Friedrich Stadler, and Niels Weidtmann, 125–137. Cham: Springer.
Neuber, Matthias. 2018. *Der Realismus im logischen Empirismus. Eine Studie zur Geschichte der Wissenschaftsphilosophie*. Cham: Springer.
Neuber, Matthias. 2022. Herbert Feigl. In Edward N. Zalta and Uri Uri Nodelman (eds.) *Stanford Encyclopedia of Philosophy*, https://plato.stanford.edu/archives/win2022/entries/feigl/
Neuber, Matthias. 2025. Herbert Feigl on the Idea of a 'Scientific Humanism'. In *American Philosophy and the Intellectual Migration: Pragmatism, Logical Empiricism, Phenomenology, Critical Theory*, ed. Sander Verhaegh, 191–209. Berlin: De Gruyter.
Neuber, Matthias, and Adam Tamas Tuboly, eds. 2022. *Ernest Nagel: Philosophy of Science and the Fight for Clarity*. Cham: Springer.
Neurath, Otto. 1944. *Foundations of the Social Sciences*. Chicago: The University of Chicago Press.

Peirce, Charles S. 1868/1984. Some Consequences of Four Incapacities. In *Writings of Charles S. Peirce. A Chronological Edition, Volume 2 1867–1871*, ed. Edward C. Moore, 211–242. Bloomington: Indiana University Press.

Peirce, Charles S. 1880/1933. On the Algebra of Logic. *American Journal of Mathematics* 3 (1): 15–57. Reprinted in *The Collected Papers of Charles Sanders Peirce, Volume 3 and 4: Exact Logic (Published Papers) and The Simplest Mathematics*, eds. Charles Hartshorne and Paul Weiss. Harvard: Harvard University Press, 3.164.

Peirce, Charles S. 1905/1935. What Pragmatism Is. *The Monist* 15 (2): 161–181. Reprinted in *The Collected Papers of Charles Sanders Peirce, Volumes 5 and 6: Pragmatism and Pragmaticism and Scientific Metaphysics Mathematics*, eds. Charles Hartshorne and Paul Weiss. Harvard: Harvard University Press, 5.414.

Peirce, Charles Sanders. 1923. *Chance, Love, and Logic. Philosophical Essays*. New York: Harcourt, Brace, & Company.

Putnam, Hilary. 1958. Review of Philipp Frank's Philosophy of Science. *Science* 127 (3301): 750–751.

Reisch, George. 1995. *A History of the "International Encyclopedia of Unified Science"*. Dissertation, The University of Chicago.

Reisch, George. 2005. *How the Cold War Transformed Philosophy of Science: To the Icy Slopes of Logic*. New York: Cambridge University Press.

Reisch, George, and Adam Tamas Tuboly. 2021. Philipp Frank: A Crusader for Scientific Philosophy. In *Philipp Frank, The Humanistic Background of Science*, ed. George Reisch and Adam Tamas Tuboly, 1–67. Albany: SUNY Press.

Reisch, George, and Adam Tamas Tuboly. n.d. *Humanistic Philosophy of Science: Philipp Frank's Life Between Science and the Humanities*. Unpublished manuscript.

Richardson, Alan. 2023. *Logical Empiricism as Scientific Philosophy*. Cambridge University Press.

Richardson, Alan, and Adam Tamas Tuboly, eds. 2024. *Interpreting Carnap: Critical Essays*. Cambridge: Cambridge University Press.

Sarkar, Sahotra. 2024. *Vienna Circle: The Story of Logical Empiricism*. New York/London: Routledge.

Schlick, Moritz. 1936. Meaning and Verification. *Philosophical Review* 45: 339–369.

Schliesser, Eric. 2022. Philosophy of Science as First Philosophy: The Liberal Polemics of Ernest Nagel. In *Ernest Nagel: Philosophy of Science and the Fight for Clarity*, ed. Matthias Neuber and Adam Tamas Tuboly, 43–65. Cham: Springer.

Sheldon, William H. 1945. Critique of Naturalism. *The Journal of Philosophy* 42 (10): 253–270.

Sigmund, Karl. 2017. *Exact Thinking in Demented Times: The Vienna Circle and the Epic Quest for the Foundations of Science*. New York: Basic Books.

Stadler, Friedrich. 2015. *The Vienna Circle: Studies in the Origins, Development, and Influence of Logical Empiricism*. Cham: Springer.

Suppes, Patrick. 1994. Ernest Nagel 1901–1985. In *Biographical Memoirs, Volume 65. National Academy of Sciences*. Washington, DC: National Academics Press.

Tuboly, Adam Tamas. 2017. Philipp Frank's Decline and the Crisis of Logical Empiricism. *Studies in East European Thought* 69 (3): 257–276.

Tuboly, Adam Tamas, ed. 2021. *The Historical and Philosophical Significance of Ayer's Language, Truth and Logic*. Basingstoke: Palgrave Macmillan.

Tuboly, Adam Tamas. 2022. Making Logical Positivism Less Logical: The Case of Schlick and von Mises. In *The Socio-Ethical Dimension of Knowledge: The Mission of Logical Empiricism*, ed. Christian Damböck and Adam Tamas Tuboly, 105–125. Cham: Springer.

Uebel, Thomas. 2003. Philipp Frank's History of the Vienna Circle: A Programmatic Retrospective. In *Logical Empiricism in North America*, ed. Richard Hardcastle and Alan Richardson, 149–169. Minneapolis: University of Minnesota Press.

Uebel, Thomas. 2011. Beyond the Formalist Criterion of Cognitive Significance: Philipp Frank's Later Antimetaphysics. *HOPOS: The Journal of the International Society for the History of Philosophy of Science* 1 (1): 47–72.

Uebel, Thomas. 2015. American Pragmatism and the Vienna Circle: The Early Years. *Journal for the History of Analytical Philosophy* 3 (3): 1–35.

Verhaegh, Sander. 2022. Nagel's Philosophical Development. In *Ernest Nagel: Philosophy of Science and the Fight for Clarity*, ed. Matthias Neuber and Adam Tamas Tuboly, 43–65. Cham: Springer.

Verhaegh, Sander. 2024. Logical Positivism: The History of 'Caricature'. *Isis* 115 (1): 46–64.

Verhaegh, Sander, ed. 2025a. *American Philosophy and the Intellectual Migration: Pragmatism, Logical Empiricism, Phenomenology, Critical Theory*. Berlin: De Gruyter.

Verhaegh, Sander. 2025b. Introduction: American Philosophy and the Intellectual Migration. In *American Philosophy and the Intellectual Migration: Pragmatism, Logical Empiricism, Phenomenology, Critical Theory*, ed. Sander Verhaegh, 1–28. Berlin: De Gruyter.

Verhaegh, Sander. 2025c. Columbia Naturalism and the Analytic Turn: Eclipse or Synthesis? In *American Philosophy and the Intellectual Migration: Pragmatism, Logical Empiricism, Phenomenology, Critical Theory*, ed. Sander Verhaegh, 77–99. Berlin: De Gruyter.

von Plato, Jan, ed. 2025. *Mathematical Logic in Vienna*. Cham: Springer.

Chapter 2
Ernest Nagel: A Biography Told Through Letters

Yvonne Nagel

Contents

Abstract This chapter surveys the life and work of Ernest Nagel through his personal correspondence with family, friends, and colleagues.

Keywords Ernest Nagel · L. Susan Stebbing · Sidney Hook · Pragmatism · History of philosophy of science · Rudolf Carnap

2.1 Personal Introduction

I first met Ernest Nagel and his wife Edith in 1970 when they came to Madison, WI, to visit their son Alex Nagel. The following summer, Alex invited me to the family's summer home in Vermont. At the time Ernest was about 70 years old and Edith a

Y. Nagel (✉)
Department of Mathematics, University of Wisconsin-Madison, Madison, WI, USA
e-mail: ynagel@math.wisc.edu

R. Gronda, A. T. Tuboly (eds.), *Ernest Nagel on Science and Philosophy*, Vienna Circle Institute Library 14, https://doi.org/10.1007/978-3-031-99458-6_2

decade younger. Since we were both 26 at the time, they seemed very old to me. They lived in an apartment owned by Columbia University most of the year but spent their summers in Vermont. Like several of their academic friends such as Sidney Hook and Meyer Shapiro, they had bought summer homes in Vermont during the Depression when farms were cheap and anyone who had a job (as they did) was relatively rich. After Alex and I married in 1973, we saw them often.

Ernest and Edith rarely spoke about their younger days. I knew that they had both been the children of immigrants, had both earned PhDs (he in philosophy and she in physics) and taught at various universities in New York. When I met them, he had retired from Columbia University but still participated in some seminars. She taught Physics at City College, retiring around 1975. I met some of their close friends during visits to Vermont and New York. The Hooks owned the house just up the road from their place in Vermont, so we saw Sidney and Ann every summer. I also met the art historian Meyer Shapiro and the mathematician Paul Smith. In New York I met the physicist I. I. Rabi and the sociologist Sigmund Diamond. When we spent the academic year 1975–76 in Princeton, we met the philosopher Carl G. (Peter) Hempel. It was clear that they had all been friends for many years, but they rarely talked about their early years.

After Ernest passed away in 1985 and Edith in 1988, we stored two boxes of his correspondence in our basement. Alex and his brother Sidney had given most of Ernest's papers to Columbia University, but these boxes seemed to contain more personal letters. These boxes moved with us, silent witnesses to another era. Finally in 2008 a letter from Susan Goulian, Sidney Hook's daughter, prompted me to open the boxes and read their contents. Ernest had written to Sidney in 1980 asking about his recent illness. In the letter he said:

> ....The two of us, each in his own way, have lived through hard times before, and I am confident that we shall do so again. I would fly out to see you, were it not for the conviction that this would accomplish little if anything and that I would simply be in the way. But if you think otherwise, just say the word! ...

I hoped to find a letter from Sidney to Ernest to send to Susan. However, as I read through the letters, I found myself transported to another era. I understood better what he meant by *hard times.* Although I had planned to read just one letter, I ended up scanning more than 1000 letters and transcribing about 550 of them and I wrote this biography to try to understand what I discovered about Ernest and his generation of scholars.

Although I am neither a philosopher nor a historian, I found what I discovered about Ernest and his correspondents fascinating and not just because of the family relationship. The letters were mostly written during the years 1925 through 1950, which was an interesting time historically because of WWII and because for Ernest it corresponded to that period in his life when he became an academic. There are several themes in this biography.

Ernest and many of his friends belonged to the great wave of immigrants who came to the U.S. between 1880 and 1920. Most of his high school and college friends were the children of poor working-class immigrants. Their parents often had little education. How was it that their children earned advanced degrees and became academics at places like Columbia University and NYU? One answer of course is that they worked very hard. Ernest and his close friend Sidney Hook taught High School or evening classes while they were graduate students at Columbia University. However, they were also helped by mentors and scholarships.

Ernest and many of his closest friends were raised as Jews, but not all of the people he knew were Jewish. His wife Edith, for example, was the daughter of Swedish Lutheran immigrants. Many of them later lost their faith. Several of them became communists or communist sympathizers.

One of the most important events in Ernest's life was the year (1934–35) that he spent in Europe on a Guggenheim Fellowship. During that year he made a point of visiting the people whose philosophical ideas most interested him. These included people like Rudolf Carnap, Hans Reichenbach, Alfred Tarski, Susan Stebbing and Otto Neurath. After he returned home, the Nazi invasions of several European countries sent a flood of refugees to England and the U.S. Many of them passed through New York. Ernest helped many of them, as well as his own relatives from Czechoslovakia, with the all-important affidavits and visas. He also helped them get settled in the U.S. and sometimes found jobs for them.

Finally, and most importantly I found the letters themselves fascinating both because of the ideas and events they described and because of the way the letter writers expressed themselves. They wrote extremely well—often describing recent events and their reactions to them. The freshness of the writing makes their pain or joy more vivid. One cannot fail to be moved by reading Susan Stebbings' descriptions of the devastation wrought by wartime bombing of London, for example.

I hope that the format of the biography—letters interspersed with commentary—will give the reader this same sense of immediacy and discovery and will help him or her understand these people and their times better.

2.2 Beginnings (1901–1919)

> None of us are self-made men and those who think they are, are generally no credit to their makers. The language in which our thinking moves, the ideals to which we are attuned in the formative years of our childhood, our habits, occupations, and pastimes, even our gestures, facial expressions and intonations, are so largely the social products of generations of teaching, that no man can understand himself and his limitations unless he understands his heritage; and it is very difficult to understand one's heritage, or anything else, unless one approaches it with a certain amount of sympathy. (Cohen 1949, 229)

On May 9, 1911, the Kaiser Wilhelm II docked at Ellis Island after a long sea voyage, which had started in Bremen. Among the second-class passengers who disembarked were a mother from Hungary and her two children—a nine-year-old boy named Ernö and his sister Iren, age 4. They were part of the huge wave of immigrants from Europe—especially Eastern Europe—who came to America between 1880 and 1920. They didn't know what to expect. The father of this family,

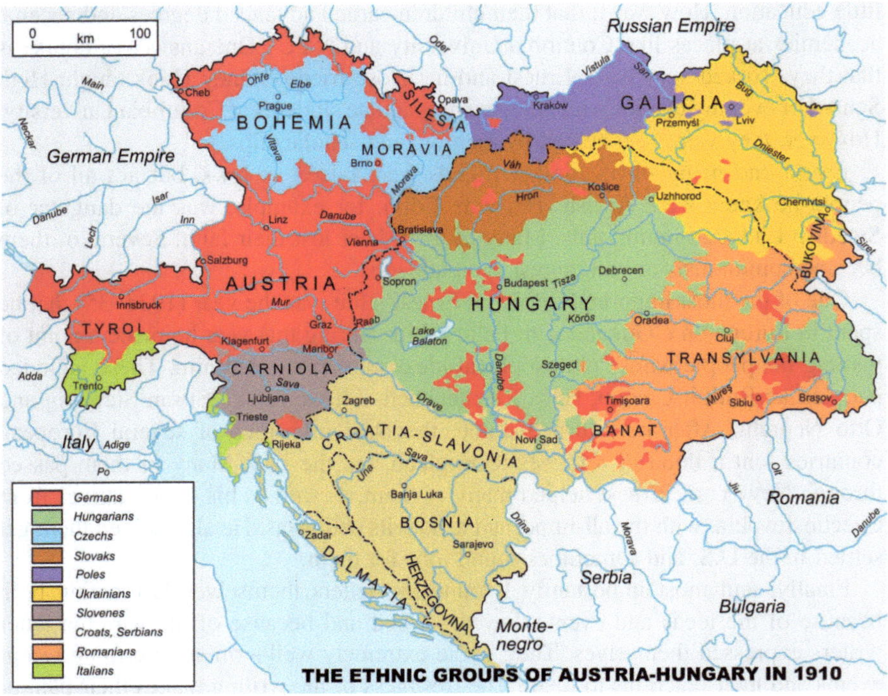

Fig. 2.1 The Ethnic groups of Austria-Hungary in 1910. (Copyright: Wikimedia Commons)

Isidore Nagel, had disappeared from the family home in Nové Mesto nad Vahom several months ago. After months of silence, his wife, Frida (née Weisz), received a letter asking them to meet him in New York and so they had come. Ernö had caught chicken pox on the voyage and was frightened that he would be sent back alone to Hungary. However, he passed the physical exams and they were reunited with their father and settled down to a new existence in New York. The contrast between the life they left and the one they were about to enter was great. The Nagels were German-speaking Orthodox Jews who lived in a predominately Slovak area of the ethnically diverse, polyglot Austro-Hungarian Empire. The map below shows the country the year before the Nagels left (see Fig. 2.1).

Nové Mesto is a small town (its population in the 2001 census was about 21,000) located about 100 miles northeast of Vienna on the Vah River. According to Wikipedia, Hungarian Jews have existed there since at least the eleventh century. After struggling against discrimination throughout the Middle Ages, the Jewish community grew to be 5% of Hungary's population by the early twentieth century and was prominent in areas of science, art and business. Members of the Nagel and Weisz families worked as small shopkeepers, just like Ernest's father Isidore.

What was life like then? Below is a postcard Isidore Nagel sent to Frida Weisz in 1900 just before they were married (Fig. 2.2). He wrote in German, but the town is

Fig. 2.2 Postcard of Isidore Nagel. (Private Collection of Yvonne Nagel)

named Vág-Újhely which is the Hungarian version of the Slovak Nové Mesto nad Vahom ("New Town on the Vah").

Ernö's parents, Isidore Nagel and Frida Weisz, both came from the smaller, nearby town of Vrbové. Isidore was born in 1870 and Frida in 1879. They were married in 1900 and their first child, Ernö, was born on November 16, 1901. They had a second son named Robert and a daughter Iren who was born on January 1, 1907. Robert died before the family emigrated to America in 1911.

We know very little about Frida's family and even less about Isidore's. According to family stories, the Nagels were more observant Jews than the Weisz's and so they considered the marriage a poor match. However, Frida's family may have been more prosperous than Isidore's (although they were not wealthy). Here is how Ernest described Vrbové when he visited it in 1934 (letter to his fiancée Edith Haggstrom dated September 18, 1934):

> The week in Vrbové was another week wasted, but it couldn't be helped. Fortunately, the weather is ideal, so that I walked a bit. The village was more primitive than it seemed in retrospect. But my relatives are very religious - at least they conform to all the details of outward manifestation with great zeal; I shocked those on my father's side especially with my ignorance in such matters and downright incorrigibility. The situation is not without pathos, however. Religious observances are the only events which are something more than humdrum, and which lend a shade of meaning to very barren lives. And on the other hand, the young are caught in the iron hand of tradition, and cannot help themselves. My cousin is in a sorry plight - she is pretty, intelligent, a good Hausfrau and has a zest for life. But she is made miserable because she is still unmarried and her brother, two years older, and eager to [be] married himself but unable to until the girl [his sister] is married, doesn't speak to her and goes out of his way to fasten more closely traditional bonds upon her. Without a large dowry marriage is impossible for a female. And everywhere family and financial considerations rank foremost - it is simply impossible to acquire an education out of keeping with

Fig. 2.3 Picture of Isidore and Frida Nagel. (Private Collection of Yvonne Nagel)

> one's birth and social position. Had we not immigrated to America I would probably be a small shopkeeper in some obscure village. That would doubtless have been a small loss to the world, but it would have been the loss of the world to me. (Nagel 2020, Box 2, Folder 40)

Below is a picture of Isidore and Frida (Fig. 2.3) taken when they were married in 1900 and a picture of Ernö around age 3 (Fig. 2.4). Like his father he had protruding ears.

2.3 Immigrant Life (1911–1919)

In New York, Isidore had opened another small shop at 1502 2nd Ave. selling *Ladies and Gent's furnishings* (that is hats, gloves and other small items). The family lived in various apartments near the store. The addresses they lived at include: 1808 2nd Ave., and 233 East 79th St. Once a year, they would have their photograph taken in a photography studio. They also exchanged postcards and pictures with relatives in Hungary, later Czechoslovakia. Although they arrived knowing no English, Ernest and Irene thrived in the public schools. Ernest won several awards for scholarship including a medal for German. After graduating from middle school in 1914, he attended the Lutheran High School. Because his father's business skills had not

Fig. 2.4 The young Ernö Nagel. (Private Collection of Yvonne Nagel)

improved in their new home, Ernest sometimes had to drop out of school to earn money to support the family. Here is how he described this stage of his life in the Oral History interviews conducted at his home in 1983. These interviews were part of the Columbia University Oral history project. The interviewer was Remmel Nunn:

> **Nagel:** Well, as I was remarking, I had some difficulties when I went to high school because of the financial condition of my family. And I left school once for about a half a year and eventually returned. Then a similar crisis arose a year after that, and my father asked me to go out and earn some money. The money couldn't have been very large, because I remember I served as an office boy, and the salary there was $5 a week. Five dollars meant a good deal in those days, but still it wasn't a magnificent salary either.
>
> Now, the second time around when I think I went to see the vice-principal to say, "I have to leave school", I forget what the laws were - why I was able to do this - but apparently one could at that time. And this time both the vice-principal and a teacher in history whom I knew made a very serious effort to find me some work after school hours, which they did. I served as a page in one of the public libraries. That was I think something like $ 20 a week. You put books in their places and that kind of thing.
>
> **Nunn:** What year was this? Do you recall exactly what year it would have been?
>
> **Nagel:** It probably was in 1920, something like that. I finished elementary school in 1914, and I went to the Lutheran high school, and I was there I think for four years. So it must have been around 1918, 1919 that I started City College. (Neuber and Tuboly 2022, 259)

Fig. 2.5 Nagel family circa 1919. (Private Collection of Yvonne Nagel)

In fact, the Nagel family (Fig. 2.5) struggled financially. Ernest's father lost his small shop and had to take a job as a shipping clerk for a record store. Ernest augmented the family income with his job at the library. He also saved some money by roller skating instead of taking the bus or subway.

Another great influence on Ernest's cultural and social development was the Settlement house movement. The 79th St. Settlement house which Ernest belonged to was a branch of the Henry St. Settlement. It was located at 232 79th St. which was close to the Nagel apartment. We can get a glimpse of how it contributed to his development through his correspondence with a Miss Elizabeth Whitmarsh and the director Jean Hanson. He corresponded with Miss Whitmarsh from 1917 through 1922 and even mentioned her in a letter he wrote to his fiancée Edith Haggstrom in 1934. Ernest acknowledged the importance of this settlement house in his own life in a letter he wrote to Columbia Dean Harry Carman in 1945. After thanking the Dean for sending him a copy of his (Carman's) Wayne University address on education, he wrote:

> There is one matter upon which you briefly touch, however, which raises in me a doubt. You suggest by your citation from someone unnamed (p. 9) that a student admitted to college come from a home in which ideas and things of the spirit are familiar and customary. I wonder, for I cannot help thinking of myself. My parents were immigrants, and though they were decent people by any standards, their own background and their circumstances gave them no access to things of the spirit other than that which a moderately orthodox religious upbringing provided. Ideas and things of the spirit, such as I have managed to acquire, I have

Fig. 2.6 Elizabeth Whitmarsh at Smith College. (Private Collection of Yvonne Nagel)

> acquired in the main outside the home -- in high-school, at the settlement house to which I belonged but above all in college and university. Perhaps, in the future, boys and girls applying for college entrance will all come from homes in which parents themselves will have had a liberal education. But in any event I would hate to deny to anyone the opportunities of a college education on the ground that the home from which he comes did not antecedently supply the stimulus for a liberal culture. (Nagel 2020, Box 1, Folder 30)

Elizabeth Whitmarsh taught in the Department of Spoken English (i.e. Speech and Drama) at Smith College in Northampton, MA (Fig. 2.6). During the summer of 1917 she worked at the 79th St. Settlement House where she met Ernest and several other young people. She was especially interested in plays, music and books. From their correspondence it seems that she would often discuss books or politics or other matters with Ernest and she invited him to attend concerts, operas and plays (some written by her friends). She started a glee club at the Settlement House and encouraged Ernest in his studies and violin playing. In the fall of 1917, she returned to Smith College but continued to write to Ernest and some of the other boys at the Settlement House. Later, in 1918–19, she moved to Pittsburgh to help care for her ailing mother but continued to write to Ernest. We have a picture of her and several of her letters to him. Unfortunately, we have none of his letters to her.

Miss Whitmarsh was obviously impressed by Ernest's abilities. She was constantly admonishing him to set his sights high and improve himself in various ways (intellectually, culturally and even athletically). Here is an excerpt from one of her letters to him (Nagel 2020, Box 5, Folder 44):

Pittsburgh, Pa. June 1919

I wanted so much to see you and talk over some plans for your coming College days. I still have a desire for you to go to Yale or Harvard or Columbia. I understand Yale has so many ways in which scholarships can be secured and opportunities can be had for working your way through. There is always a big advantage for leaving your own city and having to make adjustments to a new environment. Will you write to the Secretary of Yale University, New Haven, Conn. for such information? I was talking to one of the professors at Smith concerning a course such as you desire and she suggested that an all round course including natural sciences and literary and language courses would best prepare you for specializing. But not to specialize until your graduate course.

I know how you hate the Athletic life, but please try to give attention to it. It makes for sane, wholesome and normal thinking and living. The biggest and finest creature in the world is an all round normal man or woman...

Yours Sincerely,
E.P. Whitmarsh

Despite her advice, he ended up taking a job she considered "beneath him" after graduating from DeWitt Clinton High School in 1919 with High Honors in History. The job was probably a continuation of his job as a page at a library. He also went to City College where tuition was free, and he could save money by living at home. It is possible that he also chose not to leave New York because his family needed him. She sent this letter from Pittsburgh on August 25, 1919 (Nagel 2020, Box 5, Folder 44).

Dear Ernest

... This has been a trying summer for you especially since the work you are doing is so foreign to your tastes. I can't understand why you accepted work of this nature. Did you secure it through the High School?

All over the country there is now a movement for vocational guidance and bureaus are being established for the purpose of finding the right work for the right person. We have in our office a department through phily(?) & all boys under twenty one years of age can secure suitable positions. Hasn't New York an office of this kind? I see you need me there to scold you for being willing to work at this sort of thing. ...

E.P. Whitmarsh

Despite two absences from school, Ernest graduated on June 26, 1919, from DeWitt Clinton High School at 59th St. and Tenth Ave. with general honors and special honors in history. The address on Miss Whitmarsh's letter indicates that the family may have moved to the West side of Manhattan by then.

2.4 Higher Education (1919–1930)

Attending New York's City College was a crucial turning point in Ernest's life. The intellectual stimulation he received from Professors like Morris Cohen and fellow students such as Sidney Hook (future philosopher), Bob Wolff (future physicist), and William (Willie) Finkel (later an English Professor at CCNY and husband of Ernest's sister Irene) was incalculable. He graduated with a bachelor's degree in 1923. Morris Raphael Cohen (1880–1947) had a great influence on Ernest as well as many other students who attended City College in the Thirties. Here is how Ernest described Cohen's teaching methods (in his 1983 Oral Interview):

> **Nunn:** Wasn't Professor Cohen, Morris Cohen, quite a fiery debater and lecturer?
>
> **Nagel:** No, no, no, he wasn't fiery at all. He was sarcastic; he didn't suffer fools gladly. And he had a great wit and a great presence of mind. I mean: I don't know whether you've heard any of these stories about him. This course I think I mentioned in the philosophy of civilization, in which one part was a critique of various areas of social progress. And at the end of the semester one student spoke up and said, "Professor Cohen, you've destroyed all our illusions about these various theories, but you haven't replaced them with anything positive." And Cohen came back as quick as a flash and said, "Look, it's not recorded that Hercules had any other job except to clean the Aegean stables." (Neuber and Tuboly 2022, 278–279)

Ernest Nagel and Sidney Hook both received their bachelor's degrees from CCNY in 1923. They were accepted into the Graduate program in Philosophy at Columbia but could only attend classes in the late afternoon because they had to work. They each taught during the day and evening in order to support themselves and pay for tuition at Columbia. Ernest earned a master's degree in mathematics from Columbia in 1925, which qualified him to teach mathematics in High School. In the evening, he taught adults at an evening high school. He could only take classes at Columbia in the late afternoon.

Ernest became the main breadwinner in his family when his father died of a heart attack at work on October 17, 1926. The death was probably unexpected – Isidore was 56 years old. He and Frida had sent postcards to their daughter Irene from Coney Island where they were vacationing in July, 1926. At the time, Irene was 19. She was at Fleishmanns, NY in the Catskills where she may have been working as a camp counselor. During the school year, she attended City College and planned to work as a high school math teacher. Ernest described his difficulties as a student and breadwinner and the lack of opportunities available to him in the 1983 Oral Interview:

> **Nunn:** So you didn't really plan to teach high school as a career?
>
> **Nagel** One couldn't very well do that in those days. You had to have a certain number of years of experience before you did. And I taught elementary school for a year and a half, and I took the examination to teach with a high school license, and I taught mathematics in high school.
>
> Now, it was rather a busy, strenuous life in a sense. I taught high school, and then I went on with this job of being a page. And then when you could squeeze in some graduate courses, you did that. So I had at least three responsibilities.
>
> **Nunn:** Did you have to do a great deal of traveling back and forth? Were you in very disparate parts of the city?
>
> **Nagel** Well, I worked in several public libraries. Some were way downtown on the east side. Eventually I worked in the library on East 79th Street, and so that was quite easy. But it was this constant running around from one place to another. And so in a sense I didn't get the most out of my college education. I was involved, of course, in these other things. Well, let's see: I taught mathematics in the high schools until ... it must have been 1928. Meanwhile, my father died, so I was the breadwinner of the family - that sort of thing. And while I continued taking graduate courses - first in mathematics until I got the master's, and then I transferred to the Department of Philosophy -, there was no hope (at least I didn't

have any hopes) that I would ever be able to do any college or university teaching. One took graduate courses because you wanted to, not because it offered any way of ....

Nunn: Why was there no hope?

Nagel Well, for a number of reasons. In the first place, the demand for people teaching in a college was not very great. Secondly, the number of places where somebody with my background could teach was very small.

For example, I was brought up as an Orthodox Jew. In fact, that was the situation. And in those days, although the Department of Philosophy was free of this kind of outlook, many of the departments of Columbia were strongly ethnic-oriented; and to go out of town would have been even more difficult, partly because I couldn't very well leave. I had a sister who was going through college and my mother, so I had to be around to run the place, so to speak. (Neuber and Tuboly 2022, 259–260)

At Columbia, Ernest met Meyer Schapiro who became a close friend of both himself and Sidney Hook. Meyer, who later became a prominent professor of Art History at Columbia University, was born in Lithuania. He immigrated to the U.S. with his family in 1907 and grew up in a Jewish neighborhood of Brownsville in Brooklyn. Like Sidney Hook, he also attended Boy's High School in Brooklyn. Like Ernest, he also found a second home at a Settlement House. In his case, this was the Hebrew Settlement House in Brownsville where he was the only child in an evening class taught by the painter and sculptor John Sloan. In 1920, at the age of 16, he graduated from high school. In September of that year, he entered Columbia College as the holder of both a Pulitzer scholarship and a Regents scholarship. He was a prodigy who studied mathematics, philosophy and art history at Columbia. Although he painted and drew all his life, he felt he did not have the talent to become an artist. Instead, he pursued a doctorate in art history at Columbia, doing much of his research in Europe, especially in France. He received his doctorate in 1929. He later became an influential art critic and historian, renowned for his lectures, his promotion of modern abstract art and artists and Marxist and Socialist theories.[1]

Sidney became the student and later the protégé of John Dewey, who was already famous as the leading philosopher of his day. As Sidney explains in his Autobiography, *Out of Step,* he initially set himself up as the "official opposition" to Dewey's philosophy of *Pragmatism* by interrupting him several times during his course with questions which reflected the metaphysical and logical viewpoints of Morris Cohen and Bertrand Russell, the heroes of his undergraduate years. However, when he sat down at the end of the year to write a definitive refutation of Dewey's arguments, he found, instead, that what he wrote confirmed them. Both Ernest and Sidney agreed that Dewey was not an inspiring lecturer. As Ernest said in his 1983 Oral Interview:

Nunn: Someone that you had mentioned in passing, John Dewey, one of the most famous of the Columbia philosophers of that era: what were your impressions of Dewey when you were studying at Columbia?

Nagel Well, while I was still teaching (I think it was) in the high school in New York I had only one course with Dewey – I couldn't tell you of any other. This was this course in Types of Logical Theory, and I think that course used to meet from three to four twice a week, something of that kind. So I could just make it from teaching in the school system.

[1] See the obituary of Meyer Schapiro; http://writing.upenn.edu/~afilreis/50s/schapiro-obit.html.

> He was a very difficult lecturer. You didn't know what on earth he was talking about . . . almost as bad as his writing. I still recall the way he would come in with a piece of paper no larger than the palm of his hand on which he had something written. And then he would talk, and you had the sense of a man who didn't have a finished thesis to teach but he was thinking through a problem, and it was terribly impressive. And every once in a while he would stop and go to the window and look out; he was still thinking about this.
>
> I remember some of us used to keep awake by taking notes, because he was so slow and he drawled.
>
> I remember there was a sequence of his courses – maybe two or maybe three of them - in which he talked about the meaning of "this." Finally at the end of this he said, "Well, it's clearer to me now than it was before." (laughs)
>
> But I never got to know him very well. One of the reasons was that he had the appearance so much of a sage that I felt I didn't have anything to say to him.
>
> Another reason I suppose - I'm not sure: much as I respected him and learned so much from his views, he had an attitude which wasn't entirely congenial to the young. He was very much of an activist, you know, and his thinking was preparation for acting. And while up to a point this sounded fine, I was more of a contemplative kind, and I think I had the prejudices of Morris Cohen. So I didn't quite feel that I was able to talk to him at any length. He once did ask me to look at one chapter he had written for his book on logic, and I made some comments, and he was very kind and appreciative and so on. (Neuber and Tuboly 2022, 269)

Sidney, however, was attracted by Dewey's activism. Dewey in turn found Sidney's brashness appealing. He was also flattered that Sidney was moving philosophically in his direction. Dewey invited Sidney to dinner at his house (this was very unusual as Professors and graduate students normally did not socialize). Sidney became very close to Dewey, working with him on his manuscripts and his public activities on behalf of various causes. Dewey may have helped Sidney win a University Fellowship during his last year as a graduate student, 1926–27. Sidney had already begun to publish while he was a graduate student and this work together with the Fellowship made it possible for him to complete his PhD in 1927. Sidney became invaluable to Dewey who regarded him as his successor. As Sidney wrote in his autobiography

> Once when Dewey had made some kind remarks about what I had written – something not unusual – I observed: "I don't know why you should make such a fuss about it. All I was doing was restating what you have said." I have always treasured his reply: "Well, Sidney, when I read what you write about me, I understand myself better." (Hook 1987, 91)

Sidney and Ernest remained close friends until Ernest died in 1985. However, they were temperamentally very different. Sidney described this in his autobiography:

> There were about a half dozen other individuals, each interesting in his or her own way, who loosely belonged to the group and felt, however, vaguely, a sense of concern for and loyalty to it. Among its central figures was Ernest Nagel. He was the keenest intellect among us, and yet at the same time the most modest. Everyone was aware that he had an unfailing eye for quality. Although he never criticized many things said and done, he praised nothing that fell short of his high standards. Despite this, he was more tolerant than the rest of us toward differing points of view. He was soft-spoken and retiring, but perhaps by way of contrast with the others, the members of the group felt he would always be on the side of fairness and justice. No one was surprised, except Nagel himself, when his merit was recognized in the academy. It was as if we had known it all the time.
>
> Ernest became and remained my best friend until his recent death. In his memory I hoped to write something of permanent interest, but since he was at the center of our circle of friends in those days, it is appropriate to write however briefly something about him here. I

am proud that I was the first to recognize and cherish, behind his shy and quiet demeanor, his extraordinary intellectual gifts long before any of his teachers and colleagues did. I dedicated my first book, *The Metaphysics of Pragmatism,* to him with an esteem and an affection that deepened with the years and was never affected by any serious difference between us, although on many matters we did not simultaneously see eye to eye.

He was less impulsive and much more reflective than I. We had no secrets from each other, although I suspect he was much more aware of my failings than he made apparent. He never praised anything of mine or of anyone else's that fell short of his high critical standards, but his inherent kindness would often temper the winds of his criticism. Unfailingly liberal in social outlook, he was genuinely sympathetic for many years to some of my social, political and educational activities except during my period as a purveyor of Marxism in its several varieties, which he regarded as inadequate, based on the available scientific evidence.

In intellectual and social but not metaphysical outlook, he was closer to our teacher, Morris Cohen, than I, except in two important respects. He was constitutionally incapable of hurting anyone in the process of teaching and learning, and he suffered from a modesty so deep that it was sometimes disabling. His models were not his contemporaries but the great figures in the fields of his manifold interests. Although temperamentally different, we complemented each other. However, I could never infect him, try as I might, with the buoyancy, hope and cheerfulness that would crop up in me even in times of personal distress. He enjoyed my zest and *joi de vivre* – so he said – but he never seemed himself to possess them. Yet he radiated a feeling of quiet serenity that was, I fear, never really a reflection of his true feelings. (Hook 1987, 78–79)

Unlike Sidney, who finished his degree in 1927 and Meyer Schapiro who graduated in 1929, Ernest was not able to complete work on his dissertation until 1930. His father had died in 1926, leaving him with the financial burden of supporting his family. Although Sidney urged him to apply for a University Fellowship, like the one he had won, Ernest refused to do so, saying that others were better qualified and more needy than he. However, Ernest's life changed dramatically in 1928. In February he became romantically involved with a young woman named Miriam Bidner. He mentions her in this letter to Sidney sent from Fleischmanns, NY in the Catskills. Apparently, he brought her home to live with his family. His mother and sister did not object. They lived together in a Common Law marriage for 4 years. Miriam Bidner was born in Russia in 1908 and immigrated to the U.S. in 1920 with her family. Like others in their social circle, she was interested in social issues and politics. She was also a Communist.

Sept 4, 1928

Dear Sidney,

My mother, Miriam, and I are spending a week or so in these hills. The weather now is glorious, the air crisp and invigorating, the trees all hues of red. But the boarding house season has been the worst ever, and the Bartos family has been worrying itself sick. They are breaking up house for the winter, sending the children to relatives and friends, while the parents hire themselves out in order to pay back bills. Bartos's temper now knows very little restraint, everybody ducks when he approaches, and if Klink weren't present to jolly the guests back to good humor, very few would remain I'm sure. (Nagel 2020, Box 3, Folder 10)

The other change affected his entire career. He received financial help from his mentor, Morris Cohen. Here is how he described it to Remmel Nunn:

Nunn: Did you go to Columbia on a fellowship?

Nagel: No, you see, in those days the idea that a graduate student would be supported by the government was fantastic. Nobody would ever think of that. And so both I and friends of mine who were following a parallel path took graduate courses for which you paid yourself by doing some work outside. Things have changed - I know that. You see, I've always felt that Columbia should never institute a rule which made it impossible for people to take part time graduate work, which would have excluded many people in my generation. I don't know what the rules are at present. Probably they require a larger number of hours of attendance than I was able to give. My recollection is that I took about six points a semester. In those days it wasn't very expensive, because I think they charged $ 12 a point. That, of course, was eventually changed.

So I continued this sort of many-dimensional existence for quite a number of years until I finally took the qualifying examination in the Department of Philosophy, the qualifying examination for being a candidate for a doctor's degree. These things have changed completely, but it may be of interest. At that time what you were required . . . the examination that you had to pass consisted of the following:

First of all, you had to take a written examination in the history of philosophy for two days. I never had any courses in the history of philosophy at Columbia. Of course, Woodbridge was teaching the courses. He taught them at 9 o'clock in the morning, however, when I was busy with other things.

Nunn: But you had taken history of philosophy courses at City College.

Nagel Yes, the qualifying examination consisted of that. If you passed that, then you could present yourself for a two-hour oral examination before the Department. At that time the requirements were that the oral examination consisted of answering questions on four philosophers in the history of philosophy. Two of them had to be ancients, and two of them had to be moderns. And my choice was Plato and Plotinus for the ancients and Spinoza and Charles Peirce for the moderns.

Now, I'd like to just mention something that was really critical in my own life. The chairman of the committee was John Coss. I don't know whether the name is familiar to you. John Coss was the executive officer of the Department. He was also director of the summer session, and he was really the kingpin in many ways. The real power behind him was Woodbridge. I don't think that bothered him, but Woodbridge did.

Well, toward the end of the examination, Coss asked me (this is very vivid in my memory) why did I pick these four. I said well, all of them represent what I thought was a strong, rationalistic outlook on the world and that I'd learned to respect this approach very highly from my old teacher at City College, Morris Cohen.

Well, Irwin Edman was present at that time, and he was a good friend of Cohen; and the next time he saw Cohen, he told Cohen about this. I suppose I must have done fairly well in the oral examination, because Cohen suddenly became interested in me; and he was able to get me a fellowship/scholarship, which enabled me to give up teaching while I tried to do the dissertation for the degree; and he did that for two years. I think the stipend on that fellowship was something like $ 2000 a year, which again, of course, is absurd in these days, but $ 2000 was an awful lot of money then. (Neuber and Tuboly 2022, 260)

Morris Cohen also took a personal interest in Ernest. He wrote to him in June 1928 from the MacDowell Colony in Peterborough, New Hampshire, to invite him to visit his family in Lake Placid:

Dear Ernest:

Mrs. Cohen, Felix and I would very much like to have you come up and stay with us at our house in Lake Placid from July 15 to August 15. You can go out with the boys mountain climbing, swimming or canoeing, talk philosophy to me, or read and write in your own room. But the air in Lake Placid is lovely and quiet and I think that you will like it. I very much hope that your plans will not interfere so as to prevent your acceptance. If you have committed yourself to teach in the summer my urgent advice to you is to get a substitute. It will be much better if you get a rest and begin on your doctor's dissertation.

Talk it over with your mother and let me know as soon as you can. Trusting that you are all well, I am, with best wishes,

Cordially yours,
Morris R. Cohen

1928 was also an important year for Sidney Hook. He had finished his dissertation in 1927 under John Dewey and applied for and won a Guggenheim Fellowship. So, in June, 1928 he and Carrie sailed for Europe on the SS. Hamburg where he planned to visit several German Universities. Their correspondence during that year revealed a lot about themselves.

The 1928–29 school year started off well for Ernest. He had a new romantic interest, and the scholarship Cohen had arranged for him freed him from financial responsibilities for a year. He looked forward to not teaching but worried that if he dropped the Evening High School job, they would not rehire him later (Nagel 2020, Box 3, Folder 10):

Sept. 13, 1928

Dear Sidney,

. . . We returned to school this week, and I find the place scarcely bearable. The fireworks will begin Monday when I announce to the principal my leaving. Irene has a teacher-in-training position there, and she will probably bear the brunt of her ire. I had a card from the principal of the N.Y. Evening H.S. this week to fill a vacancy there. I went to see him, asked him what would happen if I declined now. He assured me I would remain at the head of the list, and that I should come to see him in January. I will take an appointment without fail in Feb. - I almost took it now, but decided the distraction and travel was too much. Besides, I have been looking forward to long uninterrupted evenings at home for part of the year anyhow, and for five months I will indulge myself. . .

Ernest and Sidney wrote many letters to each other during 1928–29 when Sidney was in Europe on his Guggenheim Fellowship. Ernest relished the freedom which the Cohen scholarship brought him, but his self-doubt and tendency toward depression rapidly brought him down. As the semester wore on, Ernest's feelings of unworthiness and responsibility for the money Cohen had obtained for him developed into a serious depression. By November, he had a breakdown. He escaped to the Valeria Rest Home to recuperate and wrote to Sidney (Nagel 2020, Box 3, Folder 10):

November 14, 1928

Dear Sidney,

I wanted to reply to you immediately upon receipt of your article, but as you see I have no control over my fortunes. I was firmly convinced I was going insane, found myself chattering to myself like a monkey, gesturing to the empty air, incapable of putting two thoughts together, or reading a book continuously. Of course, everybody whom I see became very much worried, my condition even got to Cohen's ears, he came to see me last week, and under the combined pressure of everyone who had a finger to lift I made this hegira. I need not tell you how cheap and small I feel, nor how ashamed I am to write you at all. You surely know, Sidney, what you have been to me during the past few years, how heavily I draw upon your generosity and strength. I had hoped some day to justify your confidence and to show myself not incapable of sharing in your friendship on terms of relative equality - not of

Fig. 2.7 Portrait of Sidney Hook and Ernest Nagel. (Private Collection of Yvonne Nagel)

course in range, or acumen, but at least by cultivating a small patch acquire a sufficiently authoritative voice to be heard for reasons other than courtesy. (Never mind the medley of metaphors) I now feel so damnably inferior to everyone I know, that I am puzzled they still are taken in - I got thru these years palming off a fake on everyone, and it is really impudence for me to write you at all. I know you will say this is a temporary state of mind, and will pity my frazzled nerves - but pity is a rotten basis for any healthy relationship or any claim upon the attention of others.

I expect to be here for two weeks - the place is very beautiful, the food good, the room comfortable, and the price amazingly reasonable - $2.50 per day. It is not a Jewish crowd and I expect to make few acquaintances. But I am trying to give myself a chance to recover, away from home. At moments I realize I must be driving my mother frantic with anxiety...

After reading Ernest's letter, Sidney wrote back to reassure him (Fig. 2.7). In this remarkable letter, he also states what Ernest's friendship means to him and shows insight into his own personality (Nagel 2020, Box 3, Folder 10).

Berlin, Nov. 27, 1928

Dear Ernest,

Nerves - Ernest - nerves! But you can't fool me! You are no more on the verge of a breakdown than I am - and the proof is that your sense of humor is still as gleaming and effective as ever. But if you stop writing me - I shall either take the first boat back or write a thousand-word cable collect.

You probably have swung back to normalcy by now. The whole incident reminds me of those two crucial years in Cohen's life and of his decision, once and forever, to shed his sensitive cloak and sharpen his claws. It would be unjust to suspect, in face of the thoroughness of his transformation, that he never was sensitive at all - that his delicacy was only expediency. But at any rate, the toughness was good for him - it gave him peace and tinged his outlook with a pinch of resignation until his bones set. Now that you know what panic, fear and Selbst-zerfleischung [self-flagellation] are - you can consider yourself hardened, free from all the petty terrors and stings which come from a hypertrophic imagination or from the actual uncertainties of fortune. Before long, you yourself will see the situation in perspective and smilingly dismiss the year as the thorny transition to self assertion.

I can only guess at the depths of your feeling, for in similar situations I react in a more superficial, yet healthier, fashion. The harder I'm hit by anything, the stronger my comeback; the more abysmal my failure, the more thorough my renunciation; the keener the frustration, the intenser my belief in the futility of striving. It is a vulgar process of compensation but it seems to be operated by an automatic mechanism - and it succeeds marvelously in turning the edge of defeat and removing the smart of failure. If I weren't emotionally constructed that way - I'd go to the dogs in no time, for all my life I've been grasping for things outside my reach. I cannot devote myself to one task without entertaining notions of tackling another at the same time; I cannot fasten myself to one place without wishing to be somewhere else; I cannot look into the eyes of one dainty precious dove without thinking of the whole flock impatiently flapping their wings elsewhere. But I really am not a brazen rascal or an ambitious fraud as many people think and as I myself often think - my interests are really genuine and intense, I am like a piece of magnesium ribbon in violent reaction all the time - to everything and everyone. I can't understand it but I can make my emotions obey my will (as when I am shameless) and bend my will to my reason (when I am humble and submissive) and then discover that my reason has been harnessed all the time to my emotions. What a pretty puzzle I am! If I were born into the REPUBLIC, I am sure Plato would have killed me - or I him.

If it makes you any happier to believe that you have drawn upon my generosity and strength go right ahead but much as I should like to swallow the pious fiction I really can't. I can easily show that the reverse is true, point to the many times you have clarified my thinking, set me right with a phrase, warned me with a word, etc. I could recall now the occasion on which I offended the bunch with one of the few honest confessions I ever made to them namely that you were the only person from whom I have learned and gained in the circle of my friends. But I won't dwell upon that. It is not half as important as what I owe you emotionally.

I could never bring myself to say this openly for I was afraid of that suspicious little twinkle in your eye and your half-shy, humorous chuckle. But just as Carrie has improved my character, you have purified it. I have found myself capable in your presence of thinking thoughts, dreaming of things, resolving upon acts - which would never come to me under any other conditions. I who am shameless before everyone else found myself concealing all of my mistakes, immoralities and callousness before you. I reconstructed my personality during the last six years of our friendship - or rather developed an aspect it had never had - a saving aspect. I don't know how to explain it. It wasn't worship. It wasn't merely ideal society. There was a trace of spirituality in it. Don't smile, but I used to feel that I had washed my soul clean at times. And when I strove to analyze it I could only say – "It's his intelligence, gentleness and goodness". But it was more than that. If it hadn't been for you I believe that my character would have gone to the devil - my enthusiasms would have been merely natural animal exuberance, and my intelligence a flexible instrument of deep cunning. You can't pooh-pooh it - dear Ernest - I've FELT it and FEEL it.

I don't dare reread what I've written - my face would burn and I would destroy this letter - I feel as if I have opened the most secret chambers of my heart to the light. But you will believe and understand.

Sidney

Sidney traveled freely in Germany during 1928–29. Like his mentor, John Dewey, he was interested in the political movements of the time. In 1929 Sidney received an invitation to visit the Marx-Engels Institute in Moscow from its director, David Riazonov. He wrote to the Guggenheim Foundation asking for an extension of his fellowship, taking great delight in the thought of having a capitalist organization finance his trip to a Communist country. He wrote to Ernest (Nagel 2020, Box 3, Folder 11):

January 13, 1929

To show you what a scoundrel I am, I shall confide a secret: I am trying to chicane the Guggenheim people into extending my tenure for two months so that I can go to Moscow on capitalist money. [David] Ryazanov, the director of the Marx-Engels Institute, has invited me to continue my researches there. When I got your letter I was just finishing a seventy page tentative report with which to bait my application for renewal...

His request was granted and he and Carrie spent the summer in Moscow. Ernest recovered from his depression, but was not able to finish his dissertation in a year. (during the second semester he resumed his evening job). However, Cohen came to his rescue again (Nagel 2020, Box 3, Folder 11; cf. (Fig. 2.8):

Fig. 2.8 Portrait of Morris R. Cohen. (Private Collection of Yvonne Nagel)

March 25, 1929

Dear Sidney,

.... I have had a surprising experience last week Tuesday night (last week). I cut evening school to hear Weyl. I expected to meet Cohen there but he didn't appear. But next morning he called me up whether I wouldn't call on him that evening (I haven't told him of my evening job). I went with the expectation of having a repetition of the Wheelwright affair. He was alone and we spent (or rather I) a pleasant two hours gossiping and talking philosophy. Suddenly he asked me what my plans were for next year - I told him I saw nothing but going back to school [the evening high school registration is so poor and decreasing that I do not think it will last much beyond this year]. He waved that aside and said that he could force me into the college [City College] this year but that I wouldn't find that pleasant with Overstreet conceiving the function of philosophers to be delivering Unitarian sermons. Further, that not having a degree my salary would be small, and that it would be better to wait til next year when I'd have it, and when there would be a great demand for logic. He thought he could raise another $2500. I raised my hands in horror, told him what an unhappy year I've led and how little I've done. He belittled my objections, and while he did not decide what he would do, he assured me the idea of going back to high school was "no plan." What do you think, Sidney? Would it be downright knavery if I lived on other men's toil for another year? I have got used to some leisure and can work now with considerable pleasure - but what guarantees have I that I'll turn out anything, next year or ever? I haven't faith enough in myself to say "boo" to a shadow...

Ernest

And here is Sidney's reaction. In this letter he also reports that although the extension of his Guggenheim came through, the offer from Chicago did not (Nagel 2020, Box 3, Folder 11).

Berlin, April 5, '29

Dear Ernest,

Your letter came a few minutes ago and the reason for my haste in answering it is that I fear you will wave another $2500 away out of a mistaken sense of obligation. For heavens sake! Whose money is it anyhow? This is not naiveté on my part but honest indignation. Why a fur coat for a rich man's mistress costs more than what your stipend amounts to! The people who fork their money over to Cohen are not depriving themselves of anything and get a kick out of their patronage of the arts and sciences. And since there is no question here of competition, you cannot raise the lame excuse of possibly not meriting an award. Look how free you are - you don't have to play at being pious or politically orthodox, and as for your work - what you'll do, you'll do, and what you can't you won't. There's a lot of sense in a tautology sometimes....

My fellowship has been renewed for two months and I shall leave for Moscow sometime in June. At the end of this month I shall take a trip down the Rhine visiting various universities for about 3 or 4 weeks. I return to Berlin at the end of May...

Moore of Chicago wrote me a touchingly apologetic letter explaining why the offer he had spoken so much about never came. I am grateful that I don't have to decide now between New York and a better job. Plans for the Symposium are going strong and Wheelwright has asked me whether I think you will be a future contributor. I am to do a book review for the first issue - if I do it. But my chief task as a contributing editor (or member of the editorial board) is to induce Dewey and Cohen to fork over manuscripts.

Sidney

Sidney returned to the Philosophy Department at NYU. He and Carrie settled into an apartment in Brooklyn and on September 30, 1930, they became the parents of John Bertrand Hook, named after John Dewey (Fig. 2.9) and Bertrand Russell, Sidney's

Fig. 2.9 Portrait of John Dewey. (Private Collection of Yvonne Nagel)

philosophical heroes at the time. Sidney was very busy with his academic work and political activism.[2]

He was active in the AWP (American Workers' Party) headed by A. J. Muste. At times he considered leaving academia to work full-time with the AWP and on a magazine of political commentary. He debated the meaning of Marxism with Max Eastman who had also studied with Dewey at Columbia. He was invited to Yaddo, an artistic and intellectual colony near Saratoga Springs, New York, to lecture on Marxism.

Even though he had now become a father, Sidney kept up his romantic liaisons. By 1931 he was deeply involved with Ann Zinken, a young woman who had attended some of his classes at N.Y.U. His neglect of his family and parental duties brought him into conflict with Carrie.

Ernest started writing reviews and articles for various philosophical journals while he was a graduate student. He did this partly to earn money (he could earn a small amount for each article), but also to keep abreast of developments in his field. His ability to remember everything he read made him a valuable resource for his friends. In several letters, one of his friends might say "I am trying to find an article by so-and-so about such and such" and he would come up with the exact reference.

[2]On the Hook-Nagel letters, see further Verhaegh (2025), and also Verhaegh (2022) on Nagel's letters to Hook about his European journey.

He kept a small notebook with a list of everything related to his scholarly writings. It shows that his first published review appeared in the *Journal of Philosophy* in 1927. When he died in 1985 there were 365 entries in the notebook, not counting an article which appeared posthumously. The entries included articles, reviews, books, and translations and reviews of his works.

Ernest finished his PhD in 1930 with a dissertation *On the Logic of Measurement* which he published privately (a requirement at that time).

2.5 Postdoctoral Years (1930–1934)

After Ernest finished his PhD in 1930, he held several jobs to earn money. Morris Cohen helped him get a one-year appointment at City College. The next year (1931–32), John Coss, the executive officer of the Columbia philosophy department offered him a position as an instructor. He accepted the position even though there was no guarantee of tenure. Since it did not pay very much, he also taught a class at City College. In 1931, Sidney, Ernest and Felix Cohen (Morris Cohen's son) organized a Festschrift to celebrate Morris Cohen's 30 years of teaching at City College. Later, Cohen proposed that Ernest collaborate with him on an introductory book on logic. They worked on this for several years, finally publishing it in 1934. The book, titled *An Introduction to Logic and Scientific Method*, was largely Ernest's work.

In 1931, Columbia hired L. Susan Stebbing (a philosopher from Bedford College in England with connections to the Vienna Circle) as a visiting professor. Ernest described his first meeting with her in the 1983 Oral Interview with Remmel Nunn:

> **Nagel:** She was teaching in one of the colleges of the University of London. I knew her because the year before that she had come to Columbia. Woodbridge was going to be away on a year's leave as a Roosevelt professor in Berlin. And so in order to replace him they got Susan Stebbing, and that was my first experience in meeting an English philosopher. She was a very determined lady.
>
> This will be typical. When she came was when I started teaching a graduate seminar at Columbia, and this was supposed to be a graduate seminar in logic or something like that. And I was asked whether I would share this seminar with her, and I said, "By all means." And she really dominated that seminar, and I barely opened my mouth for fear of the lady, you know. I remember sometimes I would make a comment, and then she would turn to me: "Now, don't be a fool, Nagel," and this sort of thing obviously didn't inspire me to say more.
>
> I wasn't reconciled to that until in December of that year when there was a meeting of the Philosophy Association. It was at Harvard. I attended that, and there was a joint discussion on something or other – I forget now what – by Susan Stebbing and Whitehead.
>
> **Nunn**: Alfred North Whitehead?
>
> **Nagel**: Yes. At one point Whitehead made some remark and then she said to him: "Don't be a fool, Whitehead." (laughter) (Neuber and Tuboly 2022, 265)

Susan was a very interesting person. She and Ernest became very good friends. Although they only met for two periods: fall, 1931 and spring and summer of 1935, their friendship continued through the letters they wrote until Susan's death in 1943.

Fig. 2.10 Portrait of L. Susan Stebbing. (Private Collection of Yvonne Nagel)

Fortunately, her letters to him survive. This correspondence attests to the depth of their friendship. Her full name was Lizzie Susan Stebbing, although she used L. Susan Stebbing for formal letters and just Susan for informal ones (Fig. 2.10). She was born on December 2, 1885, in North Finchly, Middlesex, the youngest of six children. She studied at Girton College, Cambridge, graduating in 1908.

In 1914, She took an MA degree at King's College London University with a thesis on "Pragmatism and French Voluntarism". She was a pupil of William Ernest Johnson but was most influenced by G. E. Moore. She was a part-time lecturer at Kings from 1913 to 1915, and a frequent contributor to meetings of the Aristotelian Society where she met Russell and Alfred North Whitehead, and renewed her acquaintance with G. E. Moore. In 1915 she became part-owner of a school for girls in Hampstead (the Kingsley Lodge School) with her sister Helen and two friends, Hilda Gavin and Vivian Shepherd, where she taught history part-time. She was involved in the running of this school for some years and lived there most of the time. In 1915 she also became a part-time Lecturer at Bedford College, University of London; in 1920, a full-time Lecturer and in 1933 she became Professor there. In 1930 she published *A Modern Introduction to Logic*, an introductory logic textbook. She was president of the *Aristotelian Society* from 1933 to 1934. She was promoted to Professor at Bedford College in 1933, becoming the first woman to hold a chair in philosophy in Britain.

During the Fall semester of 1931, Susan, Ernest, Ann Zinken (Fig. 2.11) and a "Mr. Edel" would often meet for discussions. Susan recalled these afternoons fondly. She became good friends with Ernest and Ann, inviting them to visit England when they were in Europe in 1934. At the time, Ann was an undergraduate studying

Fig. 2.11 Portrait of Ann Zinken. (Private Collection of Yvonne Nagel)

philosophy and "Mr. Edel" might have been Abraham Edel, who was a 23-year-old graduate student in philosophy at Columbia at that time.

Although Ernest and Sidney enjoyed professional success during the years 1930–34, their personal lives were in turmoil. Sidney and Carrie separated in 1930, shortly after their son John Bertrand Hook was born. Sidney was deeply involved with Ann Zinken, an undergraduate philosophy student whom he had hired to do secretarial work. Miriam Bidner left Ernest in September 1932. The reasons are not clear. He mentions two facts: that she no longer wanted them to live with his mother and that she wanted a child. He was focused on his career at that point and thought that if he gave in to Miriam, he would have to give up his dream of teaching philosophy in a university. She said that she no longer felt he needed her. He was devastated when she left. However, he continued to support her financially and also encouraged her to complete work on her degree in economics at NYU. About a year after she left, he started dating Edith Haggstrom, a young woman who had graduated from Barnard College in 1933 and was starting to do graduate work in Physics at Columbia. They shared a love of books and conversation.

Edith Alexandra Haggstrom was the daughter of two Swedish immigrants, Berthe Carlsson and Alexander Haggstrom. They met in New York and were married on November 9, 1907. Alexander was a carpenter from Bygdó, Sweden who did rough work such as sanding floors. His brother, Olaf Albin Haggstrom, who was 5 years older than Alexander, was already in New York when he arrived. Albin had married an American citizen, Ida Albertina Sabini, in 1900. Berthe came from Göteborg, which is located on the southwest coast of Sweden. The harbor in Göteborg was the main point of departure for Swedish immigrants to the United States. They had three children. Their first child, a daughter named Edith Alexandra, died when she was 6 weeks old. When their second daughter was born on July 28, 1911, they gave her the same name. They returned to Sweden after this Edith was born and probably lived in Göteborg, near (or with) Berthe's parents. Their son Gustave (Gus) was born in Sweden on March 30, 1914. Bertha returned to New York on November 21, 1914, with the two children but without Alexander. He may have returned separately, or he may not have accompanied them to Sweden. In the 1915 census the family lived on 148th St. in the Bronx with three Swedish boarders. Alexander was listed as head of household and Berthe as his wife. By the 1920 Census, Bertha was listed as head of household and Alexander no longer lived with them.

Two unfortunate things happened while Edith was growing up. Her father was unable to work and may have been institutionalized for alcoholism.[3] In any case, Bertha was forced to support the family with her wages as a maid and raise their children by herself. In 1933 she worked for a Mrs. Vanderbilt. The other tragedy was that Edith was stricken with polio which left her right leg weaker and a little shorter than her left. Although she was able to walk without a crutch as an adult, she did limp slightly.

Both Edith and her brother Gus were ambitious. Edith studied mathematics at Barnard College where she became intrigued by the Schrödinger equation. She then went on to study Physics at Columbia University. Gus attended Stuyvesant High School in Manhattan. After graduating he spent some years "bumming around", attending night classes at CCNY, associating with left-wing radicals and boxing. After a few years of this he managed to enter Columbia College with Edith's help. He graduated in 1941 with an M.D. Degree and served as a Medical Officer in the Navy in the Pacific during World War II.

In 1933, Ernest went to Yaddo in Saratoga Springs, to work on the logic book with Morris Cohen. Here are excerpts from some letters Edith wrote to Ernest in 1933 (Nagel 2020, Box 2, Folder 39; cf. Fig. 2.12):

> June 30, 1933
>
> Dear Ernest, . . .
>
> Saratoga Springs sounds much more peaceful than I imagined it to be. You ought to be able to enjoy yourself there. I hope Professor Cohen will be lazy so that you may be so too. Thank you for thinking of me on a horse and not in the ridiculous posture of falling off. Speaking of Baltimore, I realize that I must have been very naive to think of it as being as exciting as I did then. But then again, I'm sure I would be just as excited over being at Saratoga Springs now as I was at being in Baltimore. . .
>
> Yours sincerely
>
> Edith Haggstrom

[3] Ellis Island records show that a Gustaf Alexander Haggstrom arrived in New York from Sweden on November 3, 1921. His wife is listed as Bertha Haggstrom on 148th St. but his residence is given as "In Hospital" with a handwritten note "escaped from hospital".

Fig. 2.12 Portrait of Edith Haggstrom. (Private Collection of Yvonne Nagel)

August 29, 1933

Dearest,. . .

Now, perhaps, I have loved you with my ardor, and shall turn to an account of what I did today. This morning I read the final part of Cities of the Plain which as you said was a very striking scene. There is some rather beautiful and sympathetic writing coming directly after it. I then went in for some house-cleaning and washing and cooked supper. This evening a girl whom I hadn't seen for a long time, who graduated from Barnard the year before I did dropped in. We talked mathematics and physics. She wants to know something about hydrodynamics and asked me whether I knew any references. I wonder whether you know the names of the standard good books. . .

With all my love,

Edith

September 6, 1933

Dearest,

This letter, I'm afraid will not be very interesting for I am very tired just now. As usual, I arranged things very stupidly and instead of resting last night, I went to see "Reunion in Vienna" at the Allenton. This morning I started out early to see about that job which as I feared turned out to be a fake. I went to see Prof. Rabi[4] this afternoon. He could not suggest my doing anything just now though he sent me to the astronomy department.

At the end of the discussion he said that this was a man's world, that a woman had to be three times as good as a man to fill the same job and wound up looking very woe-begone

[4]Refers to I.I. (Isidor Isaac) Rabi. He was the son of Galician Jewish Immigrants. He received his PhD in physics from Columbia University in 1927 and spent 2 years in Europe studying with Niels Bohr, Werner Heisenberg, Wolfgang Pauli and Otto Stern before returning to Columbia to teach. During World War II he worked on the atomic bomb. He won the Nobel Prize in Physics in 1944.

announcing "I have a daughter and I'm very sorry." I'm glad I spoke to him for he was pleasant and I didn't feel I was doing a rather ungraceful thing.

Sweetheart, I hope you won't mind if I intrude with my advice into your affairs with Cohen. After you had left, Tuesday morning, I gathered the impression that Sidney feared that things were heading for a fight between you and Cohen. He seemed to think that it would be a mistake to let that happen. I have worried about it too, and have two things to say. Firstly, Cohen may be of help to you in the future. Secondly, if such a thing did happen I think that it would upset you and add an extra weight to a burden of worries. You have a lot of work to do both before school starts and afterwards. Please, darling, don't be annoyed with me (or Sidney) either for gossiping, intruding or for exaggerating the situation. I should perhaps have waited until I saw you again before saying this. Anyhow, I hope that the book has gotten along a little bit this week and that you can see your way to the end of it.

Anne and I have made plans to come out on the eleven o'clock train Saturday morning. I'll be so happy to see you again. It is lovely weather, even in the city and should be glorious out there.

With all my love,
Edith

Ernest also applied for and received a Guggenheim Fellowship for the following (1934–35) academic year. Since he and Edith planned to marry, he hired a lawyer to secure a Mexican divorce from Miriam. His mother, Frida, also visited Europe in 1934. She left in June, before he did. The purpose of her visit was to settle the estate of her mother Julia Weisz, née Justh, who owned several properties in Vrbové. She stayed until October 1934, visiting her father Ignatz Weisz in Vrbove, her brother Isidore Weisz and his wife Anna and daughter Liesl in Vienna and her brother Ernesto Weisz in Fiume, Italy (now Rijeka, Croatia, both names mean "river" in their respective languages). There were also other relatives in Vrbove such as the Menheims. The Menheims owned a large store in Vrbové and had a son named Laci who had been a childhood friend of Ernest's. Frida and Ernest would sometimes meet in Vienna at the home of her brother or in Vrbove.

Frida and Ernest used Isidore Nagel's naturalization papers when they applied for passports. Frida sailed for Europe first, arriving at Le Havre on June 11, 1934. Ernest taught summer school at Columbia so that he would have enough money for his trip. Edith remained in New York and worked to earn money so that she could join him later. Their plan was that Edith would join him in the Spring, and they would marry in Europe. Ernest finally left for Europe on August 17th. He described the first half of his 1934–35 trip to Europe in his 1983 Oral Interview:

. . .while I was at Columbia I applied for a Guggenheim. This was in '33 I think or '34. I got the Guggenheim Fellowship in '34, maybe in the spring of '34, and the next fall I went abroad. There was an International Congress of Philosophy in I think early September of '34. This was an International Congress in Prague. I attended that, and I wrote a report of the Congress for the Journal of Philosophy.

I had sort of a grand tour of Europe. I went to Prague first, and I had some relatives at that time in Vienna, and I stayed in Vienna for about a month, maybe two; and I got to know some of the people in Vienna. For example, I got to know Schlick; I got to know Waismann; and also people in other departments - I mean people who were members of the Vienna circle.

Then I went on to Poland. I went to both Warsaw and what at that time was known as Lwow. It's no longer part of Poland, because Russia took it. The center of logic, at least in

> East Europe in those years, was Warsaw. I got to know Lukasieicz and some others. And also in Lwow there was a most interesting man by the name of Leon Chwistek. He was a kind of polymath. I mean he made an interesting contribution to logic. He was a painter. He was a literary critic. And he was strongly left-oriented in his political views. That was really a wonderful experience, to meet a person like that. (Neuber and Tuboly 2022, 263)

We can see that one of his main objectives was to meet members of the Vienna Circle and the philosophers associated with it. We shall describe the Vienna Circle and related philosophical groups in the next section.

2.6 Fall, 1934: The Vienna Circle

The philosophers whom Ernest wanted to meet in Europe shared his interests in logic and in formulating a method of philosophical analysis which could be applied to the study of natural systems such as language, mathematics and the social or empirical sciences. Having read scholarly articles by them, this was his first opportunity to meet them in person. Many of them belonged to the Vienna Circle or had extensive contacts with its members, so we shall pause to briefly sketch the history of this group and introduce some of its members. Some of them later emigrated to America and became close friends or associates of Ernest Nagel.

It started with a group of young academics from various disciplines who met during the years 1907–1912 for discussions in Viennese coffeehouses. This "first Vienna Circle" included the physicist Philipp Frank, the mathematician Hans Hahn and the philosopher and economist Otto Neurath. Those discussions ended in 1912 when Frank moved to Prague to assume the chair left vacant by Albert Einstein. In 1922, Moritz Schlick was hired (with the support of Hahn) to teach philosophy at the University of Vienna. Schlick organized informal discussions with the mathematicians. In 1924, two of his students, Herbert Feigl and Friedrich Waismann, asked him to hold regular evening discussions. In 1925–26 they studied recent work in the foundations of mathematics by Gottlob Frege, Bertrand Russell, and Ludwig Wittgenstein. By now the group also included the mathematician Olga Hahn (she was the blind wife of Otto Neurath and sister of Hans Hahn), Felix Kaufmann, a lecturer in the philosophy of jurisprudence, the mathematician Kurt Reidemeister and the philosopher Victor Kraft as well as Feigl and Waismann.

In 1924–25, Schlick and Hahn considered two candidates for the position of lecturer in philosophy at the University of Vienna. These were Hans Reichenbach and Rudolf Carnap (Fig. 2.13), who had similar backgrounds in mathematics, physics and philosophy. They were also good friends, having met at a conference in 1923. Carnap had been a student of Gottlob Frege in Jena while Reichenbach was at that time primarily a philosopher of physics. Feigl (1969/1981, 61) later wrote that they selected Carnap because Hahn, who was a great admirer of Russell, "was convinced that Carnap would carry out in detail what was presented merely as a program in some of Russell's epistemological writings."

Fig. 2.13 Portrait of Rudolf Carnap. (Private Collection of Yvonne Nagel)

In 1926 Reichenbach became a professor of physics at the University of Berlin, with strong recommendations from Albert Einstein, Max Planck and Max von Laue. Carnap and Reichenbach continued their correspondence on philosophical issues and in 1929 they became co-editors of *Erkenntnis* (Knowledge). Reichenbach founded the Berlin Circle, a discussion group like the Vienna Circle. The mathematician Kurt Grelling and the philosopher Walter Dubislav were also founding members. The Berlin group later included Reichenbach's students Carl Gustav Hempel and Olaf Helmer. Hempel did not like his given names and therefore asked that his friends call him "Peter", the name used by his first wife Eva. He later became a close friend of Ernest's, so we shall refer to him as Peter Hempel in the rest of this book. There were strong connections between the Berlin and Vienna Circles because of the friendships between Carnap and Reichenbach and between Richard von Mises and Philipp Frank.

In the 1930s there were many other scientists, mathematicians and philosophers in Europe who enjoyed discussing scientific and philosophical ideas. Notable people included the Dutch mathematician Luitzen Egbertus Jan Brouwer (founder of the Intuitionist school), the Viennese philosopher Karl Popper, the Polish logician Alfred Tarski (Fig. 2.14) and Ludwig Wittgenstein (who was from Vienna).

There were also many contacts between this group and English philosophers. Carnap wrote to Russell while he was a student because he could not afford to buy a copy of the *Principia Mathematica.* Russell then copied out large sections of the book by hand and mailed them to Carnap. Susan Stebbing invited Carnap to visit

Fig. 2.14 Alfred Tarski with his son Jan. (Private Collection of Yvonne Nagel)

England in 1934. Susan was a point of contact between the Vienna and Berlin Circles and two British groups, the *Aristotelian Society* and the *Mind Society*.

Kurt Reidemeister suggested that they read and discuss at length Wittgenstein's book, *Tractatus Logico-Philosophicus* which had been published in 1921. Hans Hahn prepared them for this task by offering an extracurricular evening course in which he discussed the major ideas of the *Principia Mathematica* of Alfred North Whitehead and Bertrand Russell, which had been published in 1910. Wittgenstein had initially gone to England to study with Russell, but soon afterwards, Russell became a great admirer of Wittgenstein, regarding him as a genius.

There were several contacts between European and American philosophers in the 1930s. Schlick had an American wife and was fluent in English. Feigl said that some of the conversations at Schlick's house were in English. In 1929 Schlick was a visiting professor at Stanford University for a semester. Paul A. Schilpp served as his assistant. Schilpp, who was German, later became well-known for his Library of Living Philosophers, which he started at the urging of Philipp Frank. Herbert Feigl spent 9 months at Harvard University on an International Rockefeller Research Fellowship in 1930. In the same year, he emigrated to the U.S. with his wife, taking a position in the philosophy department of the University of Iowa.

Several American philosophers also visited Europe. Two Americans with interests similar to Ernest's visited Europe in the 1930s. They were Willard van Orman Quine, who traveled to Europe on a Sheldon Traveling Fellowship in 1932 after earning his PhD in philosophy at Harvard University and Charles Morris whose sabbatical trip to Europe in 1934 paralleled Ernest's. Quine had met Bertrand Russell while he was working on his PhD at Harvard during 1930–32. In 1932–33 he studied with Alfred Tarski in Warsaw for 6 weeks and with Carnap in Prague. As he wrote in his Homage to Carnap:

Carnap was my greatest teacher. I got to him in Prague 38 years ago, just a few months after I had finished my formal studies and received my PhD. I was very much his disciple for six years. In later years his views went on evolving and so did mine, in divergent ways. But even where we disagreed he was still setting the theme; the line of my thought was largely determined by problems that I felt his position presented ... Carnap and Ina were a happy pair. He was 41, she even younger. Along with their intense productivity there was an almost gay informality. If you combine strong intellectual stimulation, easy laughter, and warm friendliness, you have an unbeatable recipe for good company; and such were the Carnaps. (Quine 1970, xxiii, xxiv)

Charles Morris was an Associate Professor of Philosophy at the University of Chicago from 1931 to 1947. He spent the summer and fall of 1934 in Europe on a sabbatical. He also attended the International Congress in Prague in September 1934 and visited Vienna where he participated in discussions of the Vienna Circle. He was sympathetic to the logical positivist movement.

Ernest first met many of these philosophers at the *Eighth International Congress on Philosophy* which met in Prague in early September 1934. He attended as a participant (he delivered a paper on the last day) and as a reporter. He had agreed to write a report on the Congress for the *Journal of Philosophy*. Two things impressed him about the philosophers he met. First, they came from a wide variety of disciplines—mathematics, physics, law, medicine and sociology as well as professional philosophy. Second, they pursued their interest in philosophy without much hope of an academic career. This was partly because of economic conditions but primarily because of religious and cultural discrimination, particularly discrimination against Jews. Several of them pursued their philosophic interests as *privatdozents* (unpaid lecturers) but made their living by teaching in high school (Alfred Tarski, Karl Popper) or other work (Felix Kaufmann worked for a Dutch oil company). Some of them like Waismann and Edgar Zilsel taught at Adult Education Centers, one of the biggest achievements of Red Vienna.

Ernest wrote to Sidney Hook about some of the people he met at the conference. Sidney had also met Reichenbach during his Guggenheim year 1928–29 and had been impressed by him.

...I am very glad I came, since I got to know in a preliminary way a number of people. Reichenbach is a jolly fellow and he has been very kind to me. I am almost certain to visit him in Istanbul, though I suppose it is not altogether necessary - his book on probab. is coming out in October in Holland. He is interested in getting it into English, and I shall write Harcourt Brace and also Cohen asking them to consider it. He is not happy in Turkey, his classes are very elementary and he must instruct by having a native translate his lectures sentence by sentence into Turkish. But materially he is quite well off and likes living in Constantinople. I think I'll be able to hit it off with him in an excellent way. He inquired about you, and everything I told him interested him. Carnap is a splendid specimen, *streng wissenschaftlich* and the soul of courtesy. I had only one chance to talk with him but expect to renew acquaintanceship when I return to Prague on my way south from Poland. For my plans at present are these: stay in Vienna until the beginning of October, then go to Warsaw and perhaps Lwow for at most six weeks (I met Lukasiewicz and Tarski and have told them I shall visit them) After that I'll return to Prague, for how long I don't yet know. It is a beautiful city *und das Leben ist sehr billig*.

Although many points of view were expressed at the Congress, Ernest characterized the *Wiener Kreis* by its "anti-metaphysical" tendency (by which he meant propositions which are not amenable to direct experience) and the central role played by the theory of meaning.

Ernest's Guggenheim visit to the continent coincided with the waning years of the *Wiener Kreis*. Hitler's rise to power as the Chancellor of Germany in January, 1933 and anti-Semitism there led to the dismissal of many Jewish professors from German universities. Hans Reichenbach and Richard von Mises lost their jobs at the University of Berlin and moved to Istanbul to teach. Reichenbach's students Olaf Helmer and Peter Hempel also left Berlin. Hempel left because his fiancée, Eva Ahrends, was Jewish. Helmer moved to England where he worked on a second PhD on Russell's paradox under Susan Stebbing. Hempel moved to Brussels, Belgium, to work with Paul Oppenheim, an independently wealthy industrialist with strong interests in philosophy. Albert Einstein was visiting the United States when Hitler came to power and decided to stay. He settled in Princeton, N.J.

There was also political turmoil in Austria. In March 1933, Chancellor Engelbert Dollfuss suspended parliament and banned the Austrian Nazi party, assuming dictatorial powers. There was an uprising after he suppressed the socialist movement in February 1934. Dollfuss was assassinated as part of a foiled coup attempt by Nazi agents in July 1934. During this time, Otto Neurath (Fig. 2.15), a socialist, was in Moscow to establish an ISOTYPE museum. Uncertain whether to return to Vienna, he spent a month in Prague visiting Carnap and Frank. Deciding that it was too risky

Fig. 2.15 Portrait of the young Otto Neurath. (Wikimedia Commons)

to go to Vienna, he fled to the Hague with his wife Olga Hahn and a small group of assistants including Marie Reidemeister, the sister of Kurt Reidemeister. There he set up his Mundaneum Institute for visual education and continued his organizing efforts for the Unity of Science movement. This movement was in many ways the successor to the Vienna Circle. Hans Hahn died. Many members of the movement who were Jewish or communist or both were anxious about their futures and made plans to emigrate to England or America. Herbert Feigl, who had had a Rockefeller fellowship to study at Harvard in 1930, had already emigrated to the U.S. in 1931 because he realized he had no future in Austria. Rudolf Carnap made plans for a lecture tour in the U.S. in 1936 during which he hoped to find a more permanent job. Friedrich Waismann would emigrate to England later. In fact, many of the people whom Ernest met during 1934–34 in Europe later emigrated to America and, as we shall see, Ernest helped them.

After attending the Congress, Ernest returned briefly to Vienna, but the University was closed until the end of October, so he spent a month in Poland visiting people in Warsaw and Lwow (Lemburg). He met with the mathematicians Jan Łukasiewicz, Alfred Tarski and Adolf Lindenbaum. Tarski, who was born in 1901, the same year as Ernest, had been the youngest person ever to earn a doctorate at Warsaw University. He completed his doctorate in 1923 and in the same year, changed his surname from Teitelbaum to Tarski and his religion from Judaism to Catholicism. Nevertheless, antisemitism was very strong in Poland and he could not obtain a University position. He served as Łukasiewicz's assistant but supported himself by teaching mathematics at a Warsaw secondary school. He married a Catholic woman, Maria Witkowska who also taught at the same school. By the time Ernest met him, Tarski had written several important books and papers. Carnap had invited Tarski to give some lectures in Vienna where he met Kurt Gödel. Ernest's weekly letters to Sidney Hook and his fiancee Edith Haggstrom were filled with news about his meetings with these people (Nagel 2020, Box 3, Folder 17).

> to Sidney from Ernest (from Vienna) September 9, 1934
>
> I have had no chance to see anyone yet. The University here won't open till the end of October, and I haven't even got the addresses of some of the people I ought to go and see. I've made a date with Felix Kaufmann for next week. He interests me very much - you may remember . . .he lectures on jurisprudence, is interested in theoretical economics, gave his first allegiance to Husserl (still is a kind of phenomenologist) and was a pupil of Kelsen. Of course he is Jewish - I met him in Prague, and is only a Privat Dozent. He was about to be made a professor some time ago, when a special law (for his benefit) was passed excluding him. He makes his living as a director of some kind of an oil company.

Kaufmann was sort of a peripheral member of the Vienna Circle. These remarks show that even at this early date, Ernest had wide philosophical interests. He was interested not only in science, but also in law, economics and, as we will see later, in sociology (Nagel 2020, Box 3, Folder 17).

> To Edith from Ernest (from Vienna) October 9, 1934
>
> Yesterday I paid another visit to a man I stated in my project I wanted to meet - a Dr. Friedrich Waissmann. He has been writing a book (Logik, Sprache und Philosophie) for eight years, without it's being finished yet. It will be in large measure an exposition of

Wittgenstein, but it seems that between starting to write the book and the present, Wittgenstein has been changing his own position in a chameleon like fashion. I spent two hours in Waissmann's company, not learning much except that he has changed his position, and discussing probability, logic, and verification. There is only one other man in Vienna I shall try to meet, (I already spoke to him at Prague) a Professor Schlick, but probably I won't be able to until I return from Poland.

To Sidney from Ernest (from Warsaw) October 17, 1934

This is a pretty difficult place to be in - German is rarely spoken, the climate is unpleasant and I have an awful time trying to keep warm. I have seen Lukasiewicz, Tarski and Lindenbaum for a brief space each, and to-morrow I have a date for a longer session with the former. On the whole, I think it was silly for me to have come here. I've been reading some of the stuff - very elegant, very formal, but nothing I would not have got by grubbing in libraries elsewhere, especially home. I asked Tarski some questions, but he is frankly not interested in "large" issues, and boasts he is a mathematician, not a philosopher.

To Edith from Ernest (from Warsaw) October 30, 1934

But I have met some friendly people last week, and they have made life sort of bearable. To be particular, I met a young lady and her fiancé. She was in England for eight months on a fellowship, so insists on talking English in order to practice. He is a mathematician, she a philosopher with special interest in probability and induction. They have a common friend, a chap whose specialty is theoretical (topological) mechanics. Both of the men make philosophy their hobby and are great admirers of Campbell and Bridgman.[5] So I feel quite at home in the group. Also, they all enjoy music very much, ...

The young lady was Janina Hosiasson, a philosopher who had just returned from England. She had studied with Bertrand Russell among others, and had earned a PhD there in 1930. She worked as a high school math teacher and also as a translator (from English to Polish). Her fiance was Adolf Lindenbaum, who was also a friend of Tarski's. Ms. Hosiasson later wrote to Ernest asking him for help with translating some terms in books by Russell and Dewey (Nagel 2020, Box 3, Folder 18).

To Edith from Ernest (from Lwow) November 9, 1934

That very same day I met the two professors for whose sake I went to Lwow - Ajdukiewicz and Chwistek (Last night I slept very little also, and this letter no doubt shows the effects of it. But I feel I must write to you before the day is done). Both were very friendly, and the latter is a charming character. I had dinner at his house and later tea. He is a typical European Gelehrter [learned man]: is a literary and social critic, flirts with communism (it is illegal in Poland), and the "dean" of the Polish futurist painters, and has the chair of logic. He is fifty, has a sixteen year old daughter and a wife who looks thirty five. In parting, he gave me a small painting of his and a snap-shot of himself; his daughter who also paints, also gave me a sample of her work. It was all very charming, and we shall have some colorful pictures on our walls. I had the feeling, however, both in the case of Chwistek and others, that they feel themselves neglected, and that they would like to enjoy a great American reputation; and hence that their friendliness to me was in a measure motivated by self-interest

To Sidney from Ernest (from Lwow) November 11, 1934

That same day I met Ajdukiewicz (he has two articles in the present volume of Erkenntnis) and Chwistek. Both were very friendly, and the latter a very interesting and

[5]Percy Bridgman (1882–1961) was an American physicist and philosopher who taught at Harvard University. He won the Nobel Prize in physics in 1946.

Fig. 2.16 Portrait of Leon Chwistek. (Private Collection of Yvonne Nagel)

> likeable character. As I think I already wrote you, he is a painter, is recognized as a spokesman for the futurist group, writes on literary and social criticism, and has pronounced communist sympathies. The first thing he wanted to talk about was politics. He is a little naive, . . . and apparently sees social questions in their crudest outlines only. He, like most of the people I've met thus far, has an obvious hankering after an international reputation and felt very flattered that someone should come all the way from New York to see him. . . . The condition of the Jews in Poland is hopeless, many of the younger ones (I've been told 40%) are Communists, many are Zionists who hope to be able to get to Palestine, and in the academic circles it is just impossible to become more than a Gymnasium-lehrer, with perhaps a Dozentship. I can understand why so many hug Jewish culture closer to themselves. Chwistek told me (he is not a Jew) that most of his students are Jewish, and that it is to him the saddest sight to see the zeal of their scientific effort in spite of unspeakable poverty.

In 1929, Leon Chwistek (Fig. 2.16) and Tarski were the two chief candidates for the professorship at Lwow. Chwistek won, partly because of a letter Russell wrote in which he said that he (Chwistek) was a very able man, but that he had not heard of Tarski.

Ernest was moved by the plight of the Jews in Poland. As he later (November 28, 1934) wrote to Sidney (Nagel 2020, Box 3, Folder 18):

> the Poles I met . . . take philosophy to be a serious business. They regard logic to be the analysis of linguistic forms, just as does Carnap here,. . . The saddest thing to me in Poland is the plight of the Jewish students, and it is surprising what a consciousness of being a Jew one develops in that place. They have practically no future in the universities, even though they are frequently the outstanding scholars. Both Lukasiewicz in Warsaw and Chwistek in Lwow commented upon the, to them, hardly explicable zeal of their students, almost all of whom are Jewish, who continue to study in spite of abject poverty and without any hope of social recognition.

After leaving Lwow, Ernest went to Prague where he visited the physicist Philipp Frank and the philosopher Rudolf Carnap, whom he had first met at the Congress there in September. Both taught at the German University in Prague. While Ernest was in Poland, Carnap and his wife Ina had visited England where he had been invited to give three lectures. Susan Stebbing found them a very attractive couple as did Ernest. He was impressed by their warmth and friendliness. He wrote to Edith about them:

> I have seen Carnap and his wife this week, and am supposed to visit them again to-morrow. I wrote him last Saturday, and to my amazement he came with his wife to see me next Monday – climbing the five flights of stairs to my room. He is an exquisite person, the soul of sweetness, and a splendid mind. She too is charming and very friendly. They are married only two years or so, she is a former pupil of his, and he has been divorced from his first wife. (He must be around 45 and she is 30 – she told me so herself) So you see in some ways I find much in common with him. He has been in England for some lectures, and Miss Stebbing wrote me how much she was taken with both of them. He plans to go to America next year if a lecture tour can be arranged, and I hope then you will get to know both of them.

And also to Sidney Hook:

> I have been seeing Carnap and his wife. Both are excellent people, very friendly and unconventional, and he impresses me the more I see of him. He is preoccupied with very formal analyses - syntax - but I think it is first-rate stuff even though it leaves much unanswered. I am half convinced that our frequent to-do with the "ontological basis" of formal logic is a mistake, and that the "necessity" we worry about is a conventional and linguistic one. But I know that for the moment I am much taken with the neatness of formulation with which he gets rid of some skeletons, and sadly easily return to my former rut. Certainly Carnap would regard as outside his domain the "derivation" of formal characters from existence and I'm not sure what is left of the problem when it gets formulated in his "formal mode" of speaking rather than the "rational-mode". I do think his Logische Syntax is an important book, and that it supplies "us" with plenty of ammunition against many directions. But I so long you were here with me to discuss the things, entre nous and with him, and I feel cheated out of much in having you so many thousands of miles away.
>
> Carnap is a Soc. Democ. but is disgusted with their performance and has given up hope for the moment. He cannot join the C[ommunist] P[arty] because of their intolerance and international policies. He feels drawn to Trotsky, but cannot swallow his own (Trotsky's) petty politics and carping criticism of the internal affairs in Russia. I think he would enjoy reading your things - so if you can spare them send them to him Pod Homolkou 146, Prague XVII. He remembers your "Impressions of Germ. Philos." with pleasure. He has lectured in England last month, apparently with some success, and is negotiating with Duggan's Institute for a lecture tour in America next year. He would be glad to get a call from an Amer. Univ. - Prague is not very congenial to him, and he thinks everything is going to blow up here very soon. So he is trying to master English with considerable success, and is friendly to me because of the chance he has to practice it. All this is of course unofficial; and he has asked me not to write about his plans, especially because Duggan's Institute insists that everything be arranged through them. He is very frank about his private affairs - he has been divorced, and remarried, this time to a very charming former student, with whom he has lived for a few years "outside the law." He thought I might have some prejudices on the subject, and added that the latter was too "strong" for even the European professors!
>
> Ernest

Duggan's Institute refers to the Institute for International Education. It was founded by Stephen P. Duggan who was a professor of diplomatic history and later the history of education at CCNY. He founded the institute in 1919 together with Nobel laureates Elihu Root and Nicholas Murray Butler. Root and Butler received the

Nobel Peace Prize in 1912 and 1931, respectively. It was created to promote peace through international educational exchanges. During WWII, the IIE established a Committee to Aid Displaced German Scholars. Edward R. Murrow was its assistant director at this time.

Susan Stebbing was also very friendly. He had given her (and several others) copies of the Logic book he had written with Cohen. She, in fact, had been asked to review it for *Mind*, the Journal of the *Mind Society* whose editor was G. E. Moore. Her review was generally favorable, but she was scrupulously honest. She had written a book on logic in 1930. Here are excerpts from two of her letters to him (Nagel 2020, Box 5, Folder 11):

> August 28, 1934
>
> Dear Dr. Nagel,. . .
>
> I wrote a review of your book on Logic, & expect it will appear in the next number of Mind. I was interested in speculating as to which parts had been written by you & which by Prof. Cohen – or was it a joint production throughout? I hardly think so. However, I did not let my speculations appear in my review, as to authorship. I like the book, & think it very good. I don't like the chapter on Critical Evaluation, especially what is said about art & about poetry. I wish that had not been put in. It seemed to me amateurish. I hope you don't mind my saying so. I also didn't think your treatment of history was any better than my own, & mine is so bad I wish it had never been written. I like your collection of exercises. . .
>
> Susan Stebbing

In the following letter, she mentioned the Carnaps' visit and invited Ernest to also visit England (Nagel 2020, Box 5, Folder 11).

> October 14, 1934
>
> Dear Dr. Nagel,
>
> I was ever so glad to have your letter. I've had no time to write sooner, & not time now to write as long a letter as I meant to do. Ever since Oct. 3rd I have been in a great rush. I have been seeing a lot of Prof. Carnap & his wife, & like them both very much. I think he is a fine personality, & I have learnt much from discussing with him. His lectures here were excellent, & were very much appreciated. . .
>
> I wonder whether there will be any chance of your being in England the first week of July next year. Our Joint Session (the equivalent of your Yale meeting I attended in 1931) will be from July 5–8, & this next year it meets at Bedford College. Moore will be there, & many others. I do so wish that you could be there too. If you do find it possible to come, I shall be glad if you could be my guest. But I do not ask you now, formally; I only mention it – so that you could bear it in mind, in case you could arrange to be in England then. . .
>
> Yours ever,
>
> Susan Stebbing

After visiting Prague, Ernest returned to Vienna, where he attended the bi-weekly meetings of the Vienna Circle. He found that the members of the Circle had split into two groups—a conservative wing headed by Schlick and Waismann which adhered to the ideas of Russell and Wittgenstein and the radical wing headed by Carnap. He wrote up his impressions of European Philosophy in two long articles titled "Impressions and Appraisals of Analytic Philosophy in Europe I and II" which appeared in successive issues of the Journal of Philosophy in January 1936.

2.7 Guggenheim Year, View from New York, Spring, 1935

While Ernest was traveling in Eastern Europe to meet and talk to the philosophers whose interests he shared, he corresponded regularly with family and friends back in the States. He wrote every week (and sometimes more often when he was lonely) to Sidney Hook and Edith Haggstrom. He also wrote to his sister Irene Finkel and her husband Willie, who were taking care of details of his divorce from Miriam and marriage to Edith. He wrote to Morris R. Cohen, his co-author on the Logic Book and their publisher Harcourt. He also wrote to John Coss, the Columbia professor who was in charge of teaching appointments in the Philosophy Department and to Professor Herbert Schneider who handled correspondence for the *Journal of Philosophy*.

Before leaving for Europe, Ernest was worried that his Guggenheim fellowship would not cover all his expenses. He had taught summer school at Columbia to earn extra money, and he borrowed $500 from Sidney, intending to give at least $400 of that amount to Miriam. In the end he gave all of it to Miriam. Morris Cohen generously offered to give Ernest all of the profits from the first year of sales of their logic book. While Ernest initially demurred, he accepted it in the end, but was determined to repay the debt later.

Oct. 26/34

Dear Ernest:

Enclosed is the check from the publishers of the Logic Book. The second installment will reach you a little after Christmas. When you are financially on easy street we shall begin to discuss the question of how to divide the royalties. For the present it is not in order.

You will be glad to learn that not only is Willie [Morris's son] independent now, drawing a salary at Iowa State University, but Leonora [his daughter] is instructor of French at Smith College.

There is no news otherwise. There is a new row at the College here, but that is nothing really new. I am planning a paper on the Statistical View of the Universe and one on Legal Philosophy. With regular classes and the New School, and the accumulating duties of the Conference on Jewish Relations that is about as much as I can hope to achieve before New Years...

Cordially,
Morris R. Cohen

Ernest spent about a month in Vienna before visiting Poland and Prague. He returned to Vienna after leaving Prague. He almost felt at home there because he was fluent in German and because he had family there. His mother's father, Ignatz Weisz, lived nearby in Vrbove and his uncle Isidore Weisz lived in Vienna. He also had an uncle who lived in Fiume on the Adriatic coast near Venice. His father's family also lived in Vrbove. They were more Orthodox than his mother's family. The story he told about them is that they first welcomed him, offering him fruit, but then they took the fruit away because he did not know the blessing for fruit. His mother's family was more cordial. Both of her brothers had married non-Jewish women and were happy about his impending marriage to Edith. He visited his uncle Isidore Weisz and his wife Anna in Vienna once or twice a week. They would invite him for dinner or to concerts or musicals. Anna owned a hat shop. Her wealthy customers would often give her tickets to shows which she shared with him. Their twelve-year-old daughter Liesl was good at baking cakes. He could live quite cheaply in a flat without running water. He even considered spending the winter in Vienna. These excerpts from his letters describe his life there (Nagel 2020, Box 3, Folder 17).

Prague, Sat. Sept 8, 1934

Dear Sidney. . .

The newspapers here talk of much excitement in the U.S. on account of the textile strike. But the news is very inadequate, and I miss very much the complete accounts of N.Y. papers. All the papers I've seen here are strongly anti Nazi, and in Vienna too everywhere I've been the sentiment is against them. The Wiener Kreis is pretty self-conscious politically, and Neurath, who is a burly fellow and a bit of a clown, always takes care to point out the theological, reactionary tendency of every other school of philosophy.

He described his living arrangements to Edith:

Vienna, Sept 18, 1934

Dearest,

I am living in a Pension, two blocks from the University, and just about as far from the famous RING. It is an excellent place, and for the first time since I left New York I feel almost at home. I have a large room, about twice the size of the one at home, and it is very light and airy. The one drawback is that it has no running water - all the rooms which have it happen to be occupied; so instead there are two enormous porcelain basins with equally enormous water jugs on a stand, with a slop pail underneath. But I have access to a bath anytime, and that is something extremely rare, as you will find out when you come to Europe. And to cap it all, it is very reasonable: room with three good meals is eleven shillings plus ten percent for tips per day, which means about seventeen dollars a week. If I could remain indefinitely in Vienna, there would be no question of your sailing for these parts within three months. But I will have to start moving again within a month, - travelling is not cheap, and I don't know what lies ahead of me in Poland.

. . .although I barely know the city, I will hate to leave Vienna very much. The poverty in some places strikes you in the face. But there is a refinement, dignity and politeness everywhere and the splendors though of a bygone age are to me all the more appealing. I have not yet been anywhere, but perhaps next week, if my report gets done and if they play a Mozart opera, I'll try to get in. To night I'm going to a movie (kino) with my mother and her brother. She is leaving for Fiume Thursday, and will probably be back before I leave.

Edith had graduated from Barnard College with a degree in physics. She wanted to continue her studies in physics and earn a doctorate, but in 1934 she took a job with social services to earn money for travel. She hated the job which mostly involved delivering checks to welfare recipients and keeping track of payments. The constant walking about was difficult for someone like her who had polio as a child. Encouraged by Ernest who had worked two jobs while in graduate school, she tried to arrange her schedule so that she could finish by 4:00 p.m. and take evening classes. However, she was too tired at the end of the day to do this. She tried to keep up her skills and interest in physics by reading books on statistics, eigenvalues and quantum mechanics. In October, she applied for a lab job at Johns Hopkins University and traveled to Baltimore for an interview. Ernest was relieved when she did not get the job as he missed her and was counting on her arriving in January instead of waiting until April (Nagel 2020, Box 2, Folder 39).

August 23, 1933
Dearest,
This will be a very short note for I have not much time and not very much to tell you. I worked until 6:30 this evening getting out a case that I thought was important. Nevertheless this has been the most pleasant day I have had so far. First, I slept enough last night, secondly I had a pleasant lunch with a new friend, and thirdly, I had more investigation and less delivering of checks to do today which means I didn't have to walk so much. The work will not have the same fascination for me as it seems to have for other people. I don't like the idea of social service, that is, the service part of it, where you try to help the wretched in a kindly and wise way. Personally I think that most social service workers are as fascinated by the morbid details they come across as their clients are interested in murders and the daily news. My reaction to the general scene is that all you can do is to try to snap up the mechanical procedure of giving relief by punctuality etc., or by lending aid to some political movement that would burn up this louse infected area. The people too, are like vermin.

I'm going to try to arrange my work so that I can be finished at four o'clock and take some courses at Columbia. I hope I won't be physically too tired to do that. It will be necessary for a proper balance to study a little. . .
Love, Edith

She also chided Ernest for constantly comparing himself (unfavorably) to others and feeling unworthy. In the following letter she tells him that she did not get the job in Baltimore (Nagel 2020, Box 2, Folder 41):

[To Dr. Ernest Nagel Warsaw, Poland] October 17, 1934
Dearest,
I was thoroughly disgusted with your recent letters to me and Irene. What is the matter with you? Your condemnation of your ability and papers are always so flatly contradicted by other people's reports, that if your modesty is real it is slightly insane. I know that you are lonely, dear, and that you are often depressed but you don't have to get so emotional over your work. In a way you honor your work by making it your heart and soul, in another you downgrade it by not realizing that philosophy has to be greater than any of your contributions to it. You will never be able to say the final word about probability or anything else. You should enjoy approbation when you get it but a mind of your stature is made childish when it is so much aware of other people "holding their noses" Why do you have to be coddled? I heard from Sidney, Edel and others that Montague thought your paper a fine work. What more is expected of a report of a congress than careful analysis and good plain English? I would wager anything that it contains the additional luxury of a little brilliant satire, and if you had to rest on the laurels of your book look at Dewey's review of it. I can tell you are worried about Stebbing's review. If it is stern you will be in the dumps, if good it will be of no importance to you. May I suggest that whatever she says it may have bearing on the subject matter of logic rather than on the brilliance of Ernest Nagel and Morris Cohen. I dislike making such obvious and high sounding statements because they are repetitious. Nor can they be an answer to the enigma that is you for you always remain the same. Forgive me if I seem cross sweetheart, and don't be angry with me though I am angry at you.

Sweetheart, you must be having an awful time writing letters to everyone you know here. Irene shows me all of hers and Sidney showed me his last and taken all together your correspondence is immense. You are so regular and so faithful. I find it difficult to write at stated times. There are times when I feel tired and worn out and not my real self at all. . . .

I received a letter from Willie Cohen this week from Iowa. It is a very simple letter, very much like himself. He says he does nothing all day but work in the lab from 9 in the morning to 10 at night and is very happy doing it. He is using the Caeserium experiment for his doctorate. I did not get the job at Hopkins. They took an older person with more experience. I am afraid you will never have a chance to think of me as a physicist. Now, however, that the next four months are spread out in a known fashion before me, I may be able to settle down to something. . . .

Have the recent events in Europe had any greater significance to you who are nearer them than to us here who read the Times? There has been an awful lot of talk about it being another Sarajevo and I was worried over you. My first reaction was one of spite at Alexander[6] but sensibly, I don't suppose it will make any difference to the mass of people in Yugoslavia and will only upset international relationships further. And Spain has been terrible. Everything turns out as badly as the worst villain could hope.

Edith

P.S. . . . Anyway, dearest, I want to talk to you , not write to you and that's why it's all so awkward.

Meanwhile, Willie and Irene Finkel worked with Ernest's lawyer Daniel Bronstein to make sure that all was in order for Ernest and Edith to marry. They were planning to marry in Europe, perhaps in Italy, as Ernest had an uncle in Fiume. Willy Finkel had worked hard to get the necessary papers, but there were difficulties as he wrote to Ernest just before Edith left for Europe (Nagel 2020, Box 2, Folder 15).

January 12, 1935

Dear Ernest,

I need not tell you how happy I am that your distress of last week, which was ours, has been displaced by joy. Now that we are all breathing easier, perhaps some historical details may be of interest.

Getting the Italian authentication followed the same lines as getting the papers themselves. Bronstein had said the matter would take two days. It took almost three weeks. He kept putting me off from day to day – now the papers were in the hands of the other lawyer, now they were in the Italian consulate, now they had been returned to the other lawyer, now they were on the way up to his office and could I come for them. I came – they hadn't come up – he didn't know why – they would be up over the week-end surely. I telephoned before I chased down town. The faker pretended to call the other lawyer and let me listen in to one end of a telephone conversation which he was obviously having with nobody at all. Since I had prepared for these delays by leaving considerable leeway, I remained rather complacent in the face of all these monkeyshines. Eventually the papers came. With considerable relief, I told Edith it was all over but the celebration.

About two days later your cable came. Needless to say, I was very much taken aback. Bronstein had never said anything about residence requirements; he had, in fact, told me that all you had to do to get married was to take the authenticated papers to any court in Italy or France. I called him at his home Friday night (your cable arrived in the morning, but I did not learn of it until about six in the evening). His reactions to my news were those of a fool: "There must be some mistake." When I had impressed him with the fact that there was no mistake: "Gee, your friend sure has had trouble all along the line"; "He's in a pickle, all right" etc. I told him that condolences would not suffice (he apparently did not realize the seriousness of the situation) and that I was interested in finding a way out. Despite his usual glib talk, he proved to be completely ignorant of the legal requirements which govern marriage in foreign countries. I left with his promise to look up the law, the following morning, and told him I would call at his office the following day (Sat.) at noon, to learn what he had found out.

Before I went to keep my appointment with him, I visited the Italian consulate. The secretary there told me that Italy did not recognize divorces of any kind, and that remarriage

[6]King Alexander of Yugoslavia was assassinated on October 9, 1934, in Marseilles while on a state visit to France.

of a divorcee was impossible in Italy. He added, that if divorcees were married in another country, Italy would recognize the parties as man and wife. (I am at a loss as to how to reconcile this information with that which you sent us about marriage in Italy. I may add that he was very emphatic in what he said.)

This news was sad, but was dependable and consequently more satisfactory than Bronstein's "crap" of the evening before. When I got to his office, I told him what I had learned, and asked him what he knew. He said he had consulted the standard book on the subject, but could find in it nothing about residence requirements. He gave me this information in the over familiar tone which usually meant that the matter would be permitted to drag indefinitely, for no good reason. At that I got quite angry and blamed it all on him - I said that it was his fault that you had not been able to get married before you left, that now he had stranded you in Europe, that you could not go to Europe every week, etc. I then told him that it was obviously his business to get you out of the predicament in which he had placed you - and that I expected him to get you out of it.

I asked him point blank whether he had ever had a client who had gotten a Mexican divorce and then remarried in Europe. He replied in the negative. I then told him to get in touch with a lawyer who had had such experience - for I wanted reliable facts - not just talk. He then called another lawyer (in my presence), and explained the situation to him. The latter said: (1) that you could get married in France after ten days residence, and (2) that since the girl was under 25 she would have to have her parents' consent. On the basis of this information I promised to have Edith come to his office the following afternoon to sign the necessary papers which would then be signed by her mother and then returned to Bronstein for notarization by his office and by the French consulate. Under these circumstances it was necessary to tell Edith about the news, which I did assuring her that things would turn out all right.

Despite Bronstein's assurance, and my reassurance of Edith - I had little peace of mind, for I had lost all faith in Bronstein. Accordingly I went to various consulates so that I could find out the requirements from official sources. On Monday, therefore, I called on the Czech Consulate - the legal advisor was gone for the day (though my visit was during the office hours indicated on the door). The Austrian office told me I needed a lawyer and gave me two addresses; the French consulate informed me that the divorce was O.K., but that the papers had to be authenticated by the French Consulate in Mexico; the residence requirement was one month; it was necessary to publish banns in France two days before the wedding. I told her of the predicament you were in. She was very helpful. The banns might be published here - her office would authenticate them and post them on the bulletin board of the French Consulate - a procedure which would be accepted in lieu of publication in France. Residence might be started immediately by you alone. Finally she suggested that I had better send the lady herself. I said I should, and thanked her heartily for her cheering news.

My last visit was to the Spanish Consulate - Divorce O.K. - but authentication by consulate in Mexico was required; no residence requirement.

To put the rest of the story briefly. Edith went to the French consulate, also to the Swiss & Czech. As a result of the information obtained by all of us, the banns were posted at the French Consulate and the divorce papers were sent back to Mexico for reauthentication by the French, Spanish and Swiss Consulates, there. All of the papers will be in Edith's hands when she sails (we hope) so that there will be no obstacle in the way of your getting married in each of three different countries if you like. (You might then write a paper on a comparative study of marriage ceremonies, etc.)

All in all, I should say that despite the perturbation you experienced, you ought to consider yourself a pretty lucky fellow. If things had not happened the way they have, you would have been unable to get married in Europe (in all likelihood). For the authentication of the foreign consulates in New York City - which Bronstein had thought sufficient - are absolutely useless.

I shall write you about your other affairs within a few days. I close this letter with the hope that your life with Edith may fulfill your fondest wishes.

With love,

Willie

Sidney Hook's domestic life was even more chaotic. Although he was separated from Carrie, she was still bitter and they argued about Johnnie. He had been living with Ann Zinken for a while but they quarreled frequently. Also, Ann started seeing another man (Leon Edel) and contemplated moving to Chicago (Nagel 2020, Box 3, Folder 17).

Sept. 3, 1934

Dear Ernest,

. . . I am much more myself than I was when I wrote you last but things are in a mess and they are sure to get messier still. After having begged me with tears streaming down her face to take her back, I said yes to Ann on the condition that she would leave N.Y. for a year, and if we both feel like it after she had spent a celibate year, we would get married and settle down. Well, no sooner had I said yes than she became frightened at what the action would involve. It became clear to me that she had tied herself up much more seriously than she realized or even I. She had brought this fellow to her house and introduced him to her parents as her husband. And for a few days he lived there as such. His whole family had practically adopted her as his wife & the fellow seems to be passionately in love with her and is still in the honeymoon phase of his emotions. If she quits him cold he may go to pieces. Ann confesses only to a "great fondness" for him but she feels, I have reason to believe, more deeply than that.

I think she was taken aback by my willingness to resume the status quo with a difference and although she professes to feel the same way towards me as in the past I doubt it very much. I guess I still love her - my torment would not have been so great if I did not. I have the strength, however, to cut her out of my life or to wait until the auspicious moment arises when she can make a graceful and hurried exit out of her present liaison. This last is her plan. But it costs me more hurt than the final and irrevocable break would. At least I think so but I cannot be sure. I can stand the days but the nights are full of horror for me. I don't believe I have had a sound night's sleep since I came back from Maine. And in my case that means a lot. Sometimes I think that whoever is controlling my life, it is not I and so I can't tell how things are going to turn out. But the odds are about even that Ann's plan will fail, (i.e., I shall make it fail altho it is a cinch that the longer she lives with this fellow the more difficult it will be for a person of her nature to break away) I shall report faithfully all developments. I fear greatly, however, that I weary you with this indecent exhibition of my weaknesses. . .

Love in all languages,

Sidney

Sidney's professional life was hectic. He was the chair, responsible for scheduling classes for the professors in the philosophy department at NYU. He had also become involved in political battles (Nagel 2020, Box 3, Folder 17).

Sept 15th

Dear Ernest:

. . . After I left Rosenthal's I went to Rochester where I had a glorious time and got paid for it, too. The thing was in the nature of a personal triumph and if I didn't know what a weak sister I really am I might develop a swelled head as a result of it. I stayed two days and on the second day read three (3) papers (!). In the morning the stuff on the Imp. of a Point of View, in the afternoon, at special request, the gush about my Exp. Nat. for Kallen's volume, and in the evening something which I composed hurriedly on the Curriculum of a Liberal Arts

College. What particularly impressed me about the group I spoke to was their appearance, voices, and rather infantile - politically speaking - obsession with the problem of violence. They are terribly pinkish, swear by the coming revolution and really do not know what it is all about. At the first crack they will faint or run back to God - who by the way was a rather unpopular chap with this Nat Council on Religion. I was also impressed with the lack of a critical point of view (except for some of them). But they can be useful - very much so after the revolution.

I took a mid-night train out of Rochester with their hearty farewell songs to me still ringing in my ears and got to Pittsburgh the next morning in time for the Active Workers Conference of the A[merican].W[orkers].P[arty]. It was held 15 miles out from Pittsburgh at a special national meeting of the P.O.C. I had been called to consider the question of unity with the C.L.A. (Trotskyist) I stayed 3 days there and had many battles. Meetings lasted until dawn. Hardly anyone slept. The field workers reported an event throughout the country. I met new types of people - wonderful raw material for a rev[olutionary] party. Well, my political strategy worked out well. Against great odds, psychological obstacles, and material difficulties, Burnham & I got the POC unanimously to accept recommendations which makes unity now a matter of a few months. It was a great victory - it will either kill the AWP or make something out of it. Meanwhile we have been hammering the Trotskyists hard to drive their sectarianism out of them & put them into a position where the marriage will be fruitful.

The political scene on the left is rapidly changing and to my genuine dismay a group of people . . . are looking to me for political leadership - a leadership which I cannot give them unless I throw everything over & devote myself to the series of fluid messes which are called political problems. I am looking forward to unity as affording a convenient opportunity for me to retire gracefully to the Vorhallen of philosophy. I really am very tired of playing political chess but I have to see certain moves through.

Things are starting off at school with a bang. The Y[oung].C[ommunist].L[eague] by a rump meeting called when school was still on a vacation got the NSL chapter to issue a leaflet demanding that students be permitted to register for courses even if they can't pay fees - which means nobody would pay & the school go into bankruptcy or at best reduce faculty salaries & personnel. The leaflet is as insulting & provocative as anything can be; "demands" that the Chancellor & Dean "appear" before the Evening Council Session (captured by the NSL) to answer etc etc. They are picketing & calling upon students not to register until this & other demands (some of them good, most of them bad) have been granted. Crazy as ever - but they can do mischief - even if they hurt the students which they are sure to do. Burnham who is faculty advisor of the NSL fell all over himself "explaining" to the administration that this was done behind his back. (I had warned him last year that the Y.C.L. would put him in the spot) I have been standing off - realizing that nothing I can say can affect the attitude of the emotionally poisoned YCL'ers to make a stink for the smell of it. Ultimately it will be bad for the whole student movement.

I'm starting school not much rested and with very little zest left. I feel old & played out, not suffering but rather torpid & indifferent to things. Part of it is due to a rather prolonged round of philandering in which I have engaged in order to forget Ann - It failed of course. She is more in my thoughts than ever. And so here I am - a monkey's uncle - waiting for something to turn up. She still swears she loves me.

How I wish I were away from here and with you in Europe. . . .

I will write soon again

Sidney

One consequence of Sidney's activities was that his political actions were reported in the press. For example, this article appeared in the New American under the headline "College Communism":

The press carried reports a few days ago of the merger at a meeting in New York City of two groups, "so radical that even Communists have refused to have anything to do with them," into an organization called the Workers Party of the United States.

The merging bodies were the Communist League in America and the American Workers Party.

Acknowledging frankly that their political creed was based on an intention to overthrow the American Government by force, the leaders of the new body declared they hoped to foment a revolution in America in a few years.

To our amazement, we find two professors of New York University mentioned as prominent in the new organization.

Professor James Burnham of the Faculty of New York University is reported to have been elected editor to the "New Internationale," a monthly magazine the group intends to publish.

Dr. Sidney Hook, another New York University professor, is said to have been elected a national organizer. The report in question concludes with the statement that leaders of the new organization declared that they would be guided by the revolutionary theories of Trotsky, as opposed to the Socialist theories of Stalin...

Because he had told her she had to leave New York, Ann left for Europe in late October 1934. Sidney and her father paid for her trip. She first visited Susan Stebbing with whom she had corresponded since they met in the fall of 1931. She also traveled with Edith and Ernest on at least two occasions and seems to have enjoyed her *wanderjahr*. Sidney on the other hand, was miserable (Nagel 2020, Box 3, Folder 17).

October 17, 1934

Dear Ernest:-

This is the first free moment I have had to answer your long letter of October 3rd. School, meetings, preparations for Ann's departure & send off parties have literally consumed every moment of time. Monday night I gave a little party to Ann - Willie, Irene, Edith,

Edel & his wife were there. We drank your health and we all got gloriously tipsy...

I am not as active politically as I was last spring but meetings still take time & the demands upon me have increased under the phrase "responsibility to those who look to you for leadership." So you see, I am trying to follow your advice - moderately. Yet - there is little time left for philosophy. The curriculum matters & administrative duties & cabal take more time than I thought they would: - & as you predicted so uncannily affairs with Ann represent a series of crises not conducive to work. It is I now who am the passive factor in the situation & she the active. She promised me that she would not see this other fellow - but she is. She told me she had made it clear to him she was returning to me but what she actually told him is that she would go abroad in order to make up her mind about him. Her word is worthless now on anything because she doesn't know whose she is. I have a premonition that she won't return to me – & if she does unless she makes a clear break with him it will be another hell – only worse. The irony of it! Now that I am resolved to settle down for good - have met her family, dropped all social concealment, made plans for the future etc – she is putting on a Carrie – no Hortmann stunt for me.

Write me when you expect to be in England. The only clear advantage I have over this other chap is that before she returns you will be able to see her and talk to her. Things may be settled before then one way or the other but I shall keep you informed.

Sidney

Ernest was eager for Edith to come to Europe, but he was worried that they might not be able to find a place to stay together before they married. As with Miriam, he only told a few close friends about her. He did not tell many of his professional contacts such as Morris Cohen about either Miriam or Edith. In Europe the Carnaps and Susan Stebbing knew about Edith. Ina Carnap gave him this advice (Nagel 2020, Box 1, Folder 32):

Pod Homolkou 146, January 23, 1935

Dear Dr. Nagel, Your reasons for a marriage in hurry are not very illuminating to me. Or better, I think you could go around the difficulties in a better way. I do not think, that hotels really ask whether you are married or not, I do not think, they trouble about. I remember, when the Hempel's came to Prague for the Congress, I took the room for them and asked the proprietor of the hotel whether there are difficulties for getting one room, if people are not married. And he laughed very much about this idea, saying that he has no moral-control over his guests. But it may be, that in Italy or France you have such difficulties. Why then not shorten your stay in these moral countries and return to more hospitable fields? Sometimes you can avoid the difficulties by taking a private-room, not in a hotel. In England, no hotel will ask for your passport (and besides that, two single rooms are not more expensive than one with two beds). In Tyrol – where we hope to meet you – you will not be asked for your passport. And finally Czechoslovakia is a place, where you may live quite untroubled. If you do not marry in USA there is another difficulty (besides that they ask for a longer residence): you have to pay a lot for the translation of your papers; and there are much more papers necessary than you can ever dream! There are also difficulties if you and Edith have different religious confessions. When we married, Carnap was without any c., and I was still a Roman-Catholic. And I could not marry him before I left the Church. The rules are different in the countries. But I expect, it will be horrid everywhere. I am addressing this letter to Vienna but I am hoping that you are far away in a French sea-port, holding your lady in your arms!

With warm greeting from us both
ina[7] and R. Carnap

Edith arrived in Villefranche on January 27, 1935. They discovered too late that Italy did not allow divorced persons to remarry (and would not until the 1970s). They spent about a month in Nice (Fig. 2.17), France, to satisfy the residency requirements, staying in a Pensione recommended by Ernest's aunt Anna. While they waited for the Mexican divorce papers to arrive, they entertained themselves by going to movies and operas, taking drives in the country and learning French (with the help of an elderly Frenchwoman who had taken a liking to Edith). However, weeks passed without word from the lawyer or Willie. Meanwhile, Ernest fretted about the lost time, the expense and the lack of good cigarettes and coffee.

Sidney's divorce from Carrie was finalized in early February, but he had not heard from Ann for 3 weeks and wrote to Ernest (Nagel 2020, Box 3, Folder 19):

2/23/35

Dear Ernest,

When I wrote you last I thought that the settlement papers with Carrie had been satisfactorily drawn. Well a lot has happened in the interim & only now are these papers being prepared for signing... The terms are hard & if I lose my job only Carrie's good will can keep me out of jail. But I couldn't wait any longer & I knew Ann would never understand further delay. Und im Grunde genommen es ist mir ganz egal. I have been so many different kinds of a fool in life that it makes little difference if I'm a fool in some more respects...

[7] Ina always wrote her name with lower case letters and called her husband Carnap—never Rudolf.

Fig. 2.17 Ernest and Edith in Nice, France. (Private Collection of Yvonne Nagel)

> Ann writes me she has been borrowing money (or intends to) from you. Please Ernest don't give her any. Both her father & I have been writing now for some definite word as to where she is in order to cable her funds. So in case she writes just tell her to cable home & in 24 hours she will have all she needs. I can't make out from her letters what keeps her busy in Paris. But her goings-on seem awfully sumptuous.
>
> Your mother just telephoned to say that your last note just received mentions the fact of Ann's arrival at Nice. Hurrah! How happy I am! I feel she is in safe hands…
>
> I don't know how much of a burden it will be upon you but if it is at all possible I am hoping that Ann will spend the next 6 weeks with you & Edith - or near you - especially if you go to Italy - as you ought - you are so near, you know. I hope my divorce goes through by the middle of April, the latest, & to have Ann home by May 1st. Tell her to be sweet & patient with me when she returns - I shall certainly need it …
>
> Sidney

Toward the end of February, Ann visited them in Nice. She spent her *wanderjahr* visiting England (she saw the British philosopher Susan Stebbing, for example), France and Italy. Ernest wrote to his mother:

> Ann brought a camera with her and we took pictures of each other. They came out quite well, and my moustache shows clearly. But unfortunately we mislaid them, and though we searched high and low we can't find them again. If we do we shall send some to you, and if not we shall snap others.

Ernest and Edith left Nice at the end of February without marrying (though they would tell people that they had married in Nice). They spent a couple of weeks in

Fig. 2.18 Nagel in Rome, 1935. (Private Collection of Yvonne Nagel)

Rome (Fig. 2.18) with Ann, traveled the length of Italy to Sicily, then visited Ernest's uncle in Fiume on the way back. His next letter to Sidney came from Rome (Nagel 2020, Box 3, Folder 19):

Rome, March 7, 1935

Dear Sidney,

We are a week in Rome, the girls have been exploring the treasures of antiquity, and I have been staying in my room trying desperately hard to do a little reading and sketching an outline for the philosophy of mathematics. I feel sick at heart at the year I have been wasting, - such opportunity to work won't ever come again; but then I have always been a wastrel and am simply running true to color...

I have written to Santayana, i.e. to his London agents, asking for permission to see him, and also to [Federigo] Enriques. Neither has yet answered, I expect the latter will, but am quite resigned to be snubbed again by the former. When we shall be in Florence I will try to see C.A. Strong, who lives in Fiesole; and that will be all the Italian contacts I will try to make. We expect to be out of Italy before the end of April, and then to England via Paris, and Holland. Italy is a very expensive country, and before you have finished paying for the bare essentials you have spent twenty dollars a week per person. But we are in it, and will have to stay. We shall make a short trip to Naples and Sicily (Ann will precede us by about two weeks), for the railroad fares are very cheap with the fifty percent reduction.

Ann told me that she wrote you she borrowed some money from me your strenger Verbot notwithstanding. There really was no help for it, she had no money left, and it is no inconvenience to me. So don't be hard on her. I've been acting as a sort of general discourager of spending money, and if you get reports of my bearishness, please put it down to my taking on the office of the holder of (your) purse strings.

Fig. 2.19 Nagel at Columbia University. (Private Collection of Yvonne Nagel)

It is easy and tempting to spend money, and I think one has to have something of the miser in one's bones to refrain. I will let Ann have as much as she needs, but you may be sure that there will be enough left for me. . .

Ernest

Despite Ernest's worries, his stay in Rome was one of the high points of his Guggenheim year abroad (Fig. 2.19). He discussed philosophy with several eminent scholars, but left the sightseeing to Edith and Ann (Fig. 2.20). The following letter is unusually upbeat (Nagel 2020, Box 3, Folder 19).

Rome, March 14, 1935

Dear Sidney,

This morning Ann left for Naples, Paestum, and Sicily. She will be gone about ten days, and we will see her again on her return to Rome.

I think she is in good spirits, a little tired of Europe, and just waiting for you to hoist the flag so that she can return to the roost. I see no reason why you should be worrying about her in the least. And I hope that in about two months time there will be nothing else to say except that "they lived happily ever after."

I have been rubbing shoulders with very eminent people this week. Enriques replied, asking me to call. I spent an hour with him, he was very pleasant and genial, and the next morning he called for me at the hotel to take me for a walk. It was raining torrents, but we went out anyway, and talked philosophy and got wet at the same time. It was very amusing. He walked under a huge umbrella with the rain dripping off the ribs down my neck, and stopped every twenty feet to be able to gesticulate more easily. Until I got into the swing of his frequent stopping I was ahead of him half the time and had to stop and retrace my steps to hear him. He talks a wretched German mixed with Italian and French, but we managed to

Fig. 2.20 Ann Zinken and Edith in Nice. (Private Collection of Yvonne Nagel)

understand each other. There is not much weight to him philosophically, although I did not have to meet him to find that out. But he is at least familiar with the important literature, and has a refreshing dislike of scholasticism, which he thinks is very dangerous, and a suspicion of formalism, which is for him the gateway to the medieval age. He invited me to have supper last night, and there I met some famous people: the mathematician G[ino]. Fano (I don't know whether you know about him: I read some of his things when I was doing higher geometry), and [Tullio] Levi-Civita. The first is a charming old gent, and the latter is a tiny man with an enormous voice and a wife twice his size. Unfortunately, it wasn't as exciting as it sounds, for they talked Italian most of the time and I was able to get the drift only occasionally. But none of the company present were specially interested in foundation problems, and Levi-Civita replied to some question of mine that he doesn't understand what the issue is between the different schools.

I also had a reply from Santayana, and am to have lunch with him to-morrow. I felt a little excited at first, but feel dread now, and hope I'll have wit enough to open my mouth. I will write you all about it in my next...

This week I had a letter from Harcourt [publisher of the Introduction to Logic book he wrote with Cohen] with a check for royalties amounting to four-hundred-sixty dollars, for sales up to the end of December. The check is an advance sent at Cohen's initiative, for royalties are not distributed until next October. It is very decent both of Cohen and the publisher; and I suspect that C. must know that I am not alone in Europe and that I therefore need money. I have sent the check back to Willie to deposit for me, so that I will be able to split it with C. when I get back. The sales of the book surprise me; but I do not think they will continue at this rate for another year. In any case, I will have a few dollars to my name when I come home, and it will make the business of paying off debts a trifle easier...

With love, as ever,

Ernest

Ernest's lunch with Santayana was quite memorable. He recalled it 48 years later in the Oral Interview:

> **Nagel** I also met somebody that I was not sure I would ever see in the flesh, and that was [George] Santayana. He was living in Rome at that time in a hotel. If I may just interrupt this sequence, I read Santayana when I was in college in a course that Cohen gave called The Philosophy of Civilization. It was a year [long] course, and the first semester consisted of a survey of various theories, a psychological theory, an economic theory and so on. He went through a whole bunch of them, and of course he indicated what the limitations were. The second semester was devoted to reading Santayana's Life of Reason, and that I must say was an unforgettable experience for me. Here you had a man who had been all over the universe, so to speak; and yet his points of view on different sorts of things as they were absorbed by one mind: I can't really say how much I appreciated that.
>
> I only met him once - in Rome. I didn't know where he lived. I wrote to his agent in London, and eventually I heard from him in Rome, and he invited me to have lunch with him, and he was really a wonderful host. At that time his assistant or secretary was Daniel Corey. I don't know whether that name is familiar to you. He was there. We had lunch in a very elegant restaurant. Santayana ordered what you would have to eat and then told you how to eat it and so on. And after the lunch Corey left, and then Santayana took me for a walk on the Capitoline Hill in Rome, and he talked the way he wrote.
>
> He had in his hand a book which surprised me. This was Heidegger's Sein and Zeit. And I said, "I was very much surprised that you should read that book, which seems to me so incompatible with your general point of view." And he said, "Oh, no, no. You must understand this book. You must interpret it." And he began to interpret Heidegger, and it turned out to be Santayana's philosophy, you see. (chuckles) But it was really wonderful. (Neuber and Tuboly 2022, 264)

After leaving Italy, Ernest and Edith traveled to Paris, then Holland to meet Brouwer and his followers. Unfortunately, Brouwer was in Sweden. However, they were warmly welcomed by Otto Neurath, one of the leaders of the Vienna Circle. They then went to England. Susan Stebbing had invited them to Cambridge. Ernest had met her in 1931 when she visited Columbia for a year to teach. Ernest wrote of Neurath (Nagel 2020, Box 4, Folder 40):

> April, 1935
>
> We had a fairly pleasant time In Holland. I met Neurath again at The Hague. He is the embodiment of Viennese courtesy and put himself out to be friendly. His Mundaneum Institute is now located there. You probably know he invented a method of visual statistics which had been widely used in Vienna before the February affair, and has been also adopted in Russia. He has been there several times to inaugurate the system.[8]

Meanwhile, Ann Zinken had returned to New York and Sidney (Fig. 2.21). Sidney's plan to banish her so that he could forget her had backfired. As he wrote to Ernest (Nagel 2020, Box 3, Folder 19):

> Monday May 27th (1935)
>
> Dear Ernest:-
>
> I am writing this letter in the shade of a wormy apple tree on Jim Rorty's farm in Connecticut. On Saturday May 25th Ann and I got married at Greenwich - and for the last forty eight hours we haven't had a single quarrel. And so the last page of an old story is

---

[8] Neurath was a socialist who had worked for the government of "Red Vienna".

Fig. 2.21 Sidney and Ann Hook in 1934. (Private Collection of Yvonne Nagel)

turned and a new volume begun. It will probably be an interesting book and I wish I could tell what will be in it.

If you & Edith haven't already attended to matters in Paris - set your mind completely at rest. The thing is ridiculously easy in Connecticut - no questions asked, no documents required, everybody pleasant, the town lives on it. And after the mumbo jumbo legal ceremony is performed don't ask the smiling justice of the peace what his fee is for he replies "There is no fee but the customary gratuity is \$10." Just hand him in a sealed envelope two single bills. I was a dope & it cost me five dollars.

The procedure is as follows: Take an early morning train any weekday to Greenwich (1 hour). Walk to Town Hall and ask for a license Fee \$1. (5 min it is). Take return train (1 hour). After 5 calendar days return to Greenwich & get hitched in the same time. It's even simpler than it sounds.

Ann came home to discover that she was foolish not to have brought more stuff into the country than she did. Particularly liquor stores which are admitted duty free up to the legal amount. But if I were you I wouldn't be bothered with liquor & such tripe.

Willie told me about the raise for next year - and I hope you are properly jubilant as we all are. That means a promotion for next year.

Now where are your gloomy forebodings? - Prof. Nagel - A Columbia Don you shall be! The situation with M. Adler has gone from bad to worse. Dewey reports that he has been informed that the Catholic hierarchy is trying to get Adler to take orders.

I can think of no one more fitted to give the law seminar with Patterson than you. And I simply refuse to take seriously your lamentations. What do you know about the law? Well, not much but considerably more than Dewey, Adler & McKeon who gave this course before. Shame on you for your silly qualms.

Overstreet wrote in answer to my letter of inquiry re Ginsberg that he simply does not recall any such fellow & so cannot tell me anything but that his failure to recall should not be held against Ginsberg. He ended with a fervent plea to give Littman a job. Well, whatever else can be said about Overstreet he is a swell guy!

One touch of sad news. I don't remember whether I wrote you about Gail Kennedy. His salary has been raised & he has been giving two years notice to find another job. Lamprecht (?)'s behind it - presumably.

More later,
Sidney
p.s. Ann sends her love.

Ernest and Edith spent 5 weeks in England—from late May to mid-July. While they were in England Willie sent them a cablegram with the single sentence:

UNIVERSITY SALARY THREE THOUSAND CONGRATULATIONS LOVE WILLIE

This was a big raise from what he had previously earned. Unfortunately, it did not come with a promotion. He also heard that he could continue to teach at City College.

Cambridge was a bit disappointing. Ernest had hoped to attend Wittgenstein's lectures because the members of the Vienna Circle had spent much time studying and discussing Wittgenstein's philosophy. Here is how he recalled this time during the Oral Interview in 1983:

> **Nagel:** Then I tried to listen in on Wittgenstein, and I asked Moore, was it possible for me to do this. And he said, "Oh, of course. You'll have to ask his permission, Wittgenstein's permission." So I went to see him in his rooms (they were in Whewell Court, as I remember, in Cambridge) and asked whether he would permit me to go to his courses. And his first response was no. He said he doesn't like philosophical tourists. He was very impressive, had very piercing eyes, a big shock of black hair. Then he added: "You know, you wouldn't understand what I'm saying anyhow." So I made a terrible mistake. I said, "You know, I spent some time in Vienna and I went to some of the lectures of Schlick and Waismann and so I think I have some preparation for this." And he became really angry. He said, "You don't think that this is to be an adequate preparation to understand what I'm saying!" Well, I still persisted, and finally he said to me, "You know, whenever I lecture I suffer; and if I see a stranger, I suffer even more. Now would you want to see me suffer even more?" And what can you say to that?
>
> **Remmel Nunn**: So you didn't ever get to hear Wittgenstein.
>
> **Nagel:** So I was out. I couldn't do that. But I think I profited by the experience I had in Cambridge. It was a curious place in those days, because Wittgenstein was the only thing that students talked about. Moore was no longer top dog and neither were the other people...
>
> I got to know something of Wittgenstein's views because there was something called "The Blue Book," which was a sort of under-the-counter circulation. I don't know who really took the notes on this course, but I was able to borrow a copy of that; and you got the sense of what Wittgenstein was about, at least thought you had a sense of it. Eventually there was also a "Brown Book."[9] Both of them have been published since Wittgenstein's death...
>
> Maybe I can tell you another Wittgenstein story. One day Mrs. Wisdom invited myself and my wife for a late Sunday breakfast, and so we went; and when we arrived, among the other people that were there was Wittgenstein. And I thought, "Gee, this is wonderful. I'm here with the real McCoy."
>
> Well, Mrs. Wisdom asked us if we wanted tea or coffee, and Wittgenstein said coffee, and he began to glow about the stimulus he gets from coffee. And I thought this was just the foothills of a conversation, and so I really put my foot into it. There was a psychologist here at Barnard College who performed an experiment on some different people. To some of them he gave coffee, but they didn't know it was coffee. Others he didn't give coffee, but he told them it was coffee. And he found that people responded to their expectations. Since Wittgenstein was talking about it, I told him this story. And my God, that spoiled the whole morning! He became furiously angry, and the rest of the time there was a denunciation of experimental psychology. (Neuber and Tuboly 2022, 266–267, 276)

[9] The *Blue and Brown Books* were dictated by Wittgenstein to students Francis Skinner and Alice Ambrose. Alice Ambrose was an American studying in England. She later taught philosophy at Smith College.

In the Thirties, it was considered scandalous for unmarried couples to live or travel together. Ernest was worried that people might find out that he and Edith were not yet married. However, Ann provided a cover story for them. She wrote to them (Nagel 2020, Box 3, Folder 20):

> June 26, 1935
> I met Mrs. Cohen at the L.I.D. camp - I mentioned the fact that I met you abroad, & she immediately asked whether you were married. When I said yes, she asked whether it was in the winter or spring - also when Edith came. I said that you were married in Nice & that I acted as unofficial bridesmaid - why I went to that length I don't know. But that's the dope.
> I said you were looking well and she seemed genuinely pleased. She mentioned the fact that you didn't send Cohen an announcement of your marriage - I pointed out that Sidney & I didn't send out announcements either & glided over that embarrassment ...
> For Edith - Housekeeping is no joke - my fingers are peeling from so much dish washing & I'm worn out trying to put things in order. Its too bad that you have to return to a worst scramble of annoying things ...However, you'll have lots of people helping you.

On August 15, 1935, following Sidney's advice, Edith and Ernest took the train to Greenwich Connecticut where they were married by a justice of the peace (Nagel 2020, Box 3, Folder 20):

> We were up at Greenwich yesterday, and were home again by noon. We had a woman judge who seemed to have a sense of humor and warned us that the ceremony would be short. It really was. We had nothing to say except "I do", and both of us felt that we didn't get enough for the two dollars. Well, that's over.

2.8 Academic Immigration (1935–1938)

The years 1935–40 were good ones for the Nagels. After the turmoil of the previous 5 years, both Ernest and Sidney were able to settle down to more stable lives. Ernest had a secure job as a lecturer at Columbia University and also taught a class at City College. His income was enough to enable him to pay off his debts and rent an apartment for himself and Edith separate from his mother's. (This had been a point of conflict between himself and Miriam). Edith continued to work on her PhD in physics at Columbia. They were even able to save a little money.

In May 1935, shortly after his marriage to Ann Zinken, Sidney wrote to Ernest (Nagel 2020, Box 3, Folder 20):

> 7/13/35
> Dear Ernest:
> This time I really was remiss in letting almost 2 weeks go by before writing. But what two weeks they were! Full of all sorts of things only some of which I can enumerate. I am buying a farm in Vermont! I can hear you gasp. I have given a $10 deposit on a $1600 place which is 8 hours riding distance from the city. It is really a question of having a place for the summer and not week-ends. I won't know whether I have been wise or foolish until this time next year but at any rate at the worst can sell without too much of a loss. Shapiro has a place 15 miles away (a mere stone's throw as distances are reckoned in Vermont). It is pretty primitive but I think worth the money. I am planning to live there summers and do some writing and you & Edith can consider yourselves half partners in the place...

Fig. 2.22 Drawing of Hook. (Private Collection of Yvonne Nagel)

> And now, prepare yourself for a further shock – which you may accept as deliberately devised proof [of] my serious desire to turn over a new leaf, settle down, & some day struggle back to the green pastures of philosophy – ... I am on my way (1 month) of becoming a father to a second child. Its a great secret yet. It will mean more burdens but in the long run more happiness - I hope.
> Sidney

The Nagels enjoyed Vermont immensely. Edith, who had thought she would never see a place as lovely as Italy, was smitten. The following year, armed with the $850 they had managed to save from Ernest's salary they looked for a place of their own. They ended up buying the small house down the road from the Hooks. It wiped out their savings. However, this purchase gave them lasting pleasure. They became part of a community of academics from New York and Boston who spent their summers in Southern Vermont. The group included many of Ernest's old City College friends as well as other academics: Sidney Hook, the art historian Meyer Schapiro, the mathematician Paul Smith, Ernest's brother-in-law Willie Finkel (who now taught in the Speech department at City College). They formed an informal *Vermont Circle.*

Meyer and Lillian Schapiro were frequent guests at the Nagel and Hook homes in Vermont. Meyer was an amateur artist. He made these charcoal sketches of Sidney (Fig. 2.22) and Ernest (Fig. 2.23) on one of these visits.

His professional life flourished. Richard McKeon, who had taken a leave from Columbia in 1934 to go to the University of Chicago, accepted an offer to stay there permanently as Professor of Greek philosophy. Ernest was asked to take over

Fig. 2.23 Drawing of Nagel. (Private Collection of Yvonne Nagel)

McKeon's course in logic. He started a discussion group modeled on those he had seen in Europe such as the Vienna Circle and the Aristotelian Society. He invited people with diverse backgrounds to participate. The group, which did not consist exclusively of Columbia academics, would meet in its members' homes for discussions. One member would read a paper, and the others would discuss it. Here is a sample notice from 1939 (attached to the abstract of a paper with the title "Mathematical Systems").

> The next meeting of the New York Philosophical Circle will be held at 8 p.m. Saturday, February 25, 1939, in the apartment of V.J. McGill 227 East 57th St., New York City. Mr. J.C.C. McKinsey will read a paper the abstract of which is enclosed.

Ernest was becoming better known as a philosopher. He served on the executive council of the Association for Symbolic Logic (ASL). He was a consulting editor for their journal and, later when Cooper Harold Langford retired as editor in 1939, became a co-editor (along with Alonzo Church). By 1940, he was also on the executive committee of the Eastern Division of the American Philosophical Society. His salary was reasonable, but he was still employed as an instructor in Philosophy on a year-to-year contract, although he taught and advised graduate students. He had held this position since 1931, the year after he got his PhD. Other colleges were interested in hiring him (Nagel 2020, Box 2, Folder 3).

Brown University
Providence, Rhode Island
Department of Philosophy
Jan. - 26/37
Dear Dr. Nagel,
A few days ago I was asked to recommend someone for a position at Oberlin. They want someone with considerable training in mathematical logic and natural science. The position is to be either an instructorship or an Asst. Professorship, and the salary between $2000 and $4000.
It has long been a matter for wonder to us that so able a man as yourself should not have a higher rank than instructor. It does not seem right; and I therefore took the liberty of recommending you for the position if they desire a man of more experience, and more of a logic specialist, than one of our own recent graduates, whom I also recommended.
I cannot conceive that Columbia would let you go unless they have quite lost their senses; and I do not know whether the position would interest you even at the top figure, for the Columbia salaries are I believe pretty good – But I thought it would not do any harm. If they should approach you, you will know what is up.
Cordially yours
J. Ducasse

Ernest was finally promoted to Assistant Professor in April 1937.

Morris R. Cohen planned to retire from City College at the end of the Spring, 1938 semester. The College then made a handsome offer to Ernest. In his 1983 Oral Interview he described it like this: "A real problem arose because Cohen was going to retire from City College, and then he asked me whether I would want to come back to the College full time. This was a real problem because I realized how important it was to teach at an institution where you had a chance to teach graduate students." Ernest consulted his friends about the offer before deciding to stay at Columbia. Susan Stebbing and Rudolf Carnap had different opinions about what he should have done (Nagel 2020, Box 5, Folder 15).

February 4, 1938
Dear Ernest,
...What I want to say is – that I could almost wish you'd accepted the Associate Professorship at $ 5800 at C.C. But I can't of course estimate the intellectual aspects, & if it is better at Columbia, then I'm sure you are wise to wait. Let it be known how much C.C. wants you! I'm not at all surprised they do. It is true you are 'promising' - & it is true you've not had time to shew all the promise, or rather, to sculpt [?] it. But you are always overworked & that is why I want you to have a bigger salary. You are, of course, still young as philosophers go, & seem young to me, who am fifty, & so I hope you won't despair.
Susan

These two letters show the Carnaps' response to the offer (Nagel 2020, Box 1, Folder 34):

February 3, 1937
We were very pleased to get your letter and first of all we want to tell you our most cordial congratulations because of the offer from City College although it is not yet official. This is of course a hard problem. But I hope if the offer becomes official Columbia will be induced to promote you and give you a more or less official promise for further promotion in the future so that you can easily find the decision to turn down this job which really involves rather much work as it seems.

March 5, 1937

We think that your decision not to take Cohen's place was entirely right if you think that you have some chance for promotion at Columbia. And sooner or later the time for it has to come We hope: soon.

In the end, he decided to stay at Columbia because he wanted to be able to teach and advise graduate students in Philosophy. He was granted tenure later that year, becoming an Associate Professor in 1939 and a full Professor in 1946.

American philosophy was also enriched by the many academics who were fleeing the Continent because of the increasing aggressiveness of Nazi Germany. Already in 1934, several of them were searching for jobs in the U.S. At the September, 1934, meeting in Prague, Hans Reichenbach had told Ernest that although his salary in Istanbul allowed him to live comfortably there, he would not mind moving to England or America. Ernest wrote to Sidney and Morris Cohen asking for help in finding him a position. In 1935, Sidney was chair of the philosophy department at NYU. When a professor went on leave, he found that he could offer a one-year appointment to a philosopher. He asked Ernest whether he should offer it to Reichenbach or Carnap. In the end, he chose Reichenbach because he thought he would be more likely to take it. Sidney solicited letters for Reichenbach—including this (translated) one from Albert Einstein:

Princeton N.J., 31. January 1935.

Dear Sir:

I have followed the activity of Professor Reichenbach as a scientist and a teacher from the very outset of his career with great interest, and at the time was one of those responsible for his receiving a chair at the University of Berlin. His chief field of activity is the conceptual analysis of the foundations of the natural sciences, chiefly physics, in which field he has made significant contributions quantitative and qualitative. As teacher he brings to his subject a clear and suggestive approach, well calculated to educate young people to critical thinking.

It would give me great pleasure to learn that he has been selected for the post and I am convinced that in such a case you will experience no disappointment.

(signed) A. Einstein

Einstein himself had been visiting the U.S. when Hitler came to power in 1933 and decided not to return to Berlin. Since he was well-known, he was able to get a position at the Institute for Advanced Studies in Princeton, N.J.

Sidney offered the one-year appointment to him, but Reichenbach, who had signed a 5 years contract with the University in Istanbul, was unable to break his contract (Nagel 2020, Box 4, Folder 64):

2.8.35. Smyrna

on board S.S. Hermes

Dear Dr. Nagel, . . .

It is indeed a great pity that I could not realize Sidney Hook's kind invitation. But the Turkish government insisted on its right to have students instructed by me on the theories of Descartes and Kant, instead of allowing me to go to New York exposing my own theories. One year later, I would have come, because my contract then may be given notice to by myself; but unfortunately New York University could not postpone.

You speak of my return to Berlin – I hope that the probability of getting a second invitation to U.S.A. is greater than that of my return to Berlin, for the Hitler government is

Fig. 2.24 Rudolf and Ina Carnap. (Private Collection of Yvonne Nagel)

> still in a very strong position. I am sure that it will fall down one day; but historical developments march with larger steps than the lives of men.
>
> Yours very sincerely
> Hans Reichenbach

Rudolf Carnap was also interested in finding a job in the U.S. He had contracted with the International Institute for Education (also sometimes called Duggan's Institute after Stephen Duggan, one of it's founders) to arrange a lecture tour in the spring of 1936. Ernest wrote to Professor Coss asking him whether Columbia could invite Carnap to lecture or teach. Carnap forwarded it to the IIE since they were making the arrangements. Unfortunately, these plans fell through since the IIE wrote that there was not sufficient interest in lectures in philosophy. Later, though, Carnap received an invitation to attend the Harvard Tercentenary (August 31—September 18, 1936) to participate in a symposium on the topic "Factors Affecting Human Behavior" and read a paper in Logic and also an invitation to visit the University of Chicago for 3 months (January through March, 1936). The latter was arranged by Charles Morris whom Ernest had met at the Prague meeting in September, 1934.

Carnap wrote to Harvard asking whether he could also teach during the 1936 summer session (Fig. 2.24). Ernest helped him draft the letters (Nagel 2020, Box 1, Folder 32):

July 15, 1935

Dear Dr. Nagel,

Carnap is very busy with some preparations for the Paris-Congress, so you get only a few lines from me. Many thanks for your letter of June 3 and your card of June 21. In the meantime we have been in Vienna and now we are here in Pistyan which is a famous rheuma-spa. It was my idea of coming here with the intention of polishing up Carnap's health for America. Perhaps you remember that he had some pains in his shoulders.

Best thanks also for the fine letter you have prepared as to be Carnap's answer to the Harvard-invitation and also for your suggestions about the courses. What a pity that you leave for America and my prophecy that we won't see you again in Europe has been so true. I should prefer to be a bad prophet but to see you again before you leave. It would have been fine to discuss with you the content of the courses, but as this is not possible, your suggestions are very valuable. What a pity that we cannot spend this August together in the Alps! Feigl is coming also (takes part in the first days of the Congress at Paris) and then returns to Iowa. Woodgers are coming (I expect that you have made their acquaintance in the meantime – how did you like them?), and only Nagel and his lady won't come! . . .

Yours

Ina Carnap

with cordial greetings

R. Carnap

Shortly after returning to the U.S. Ernest received letters from Olaf Helmer and Carl G. Hempel. They had been students of Reichenbach in Berlin, but had left Germany in 1933 after Hitler came to power. Susan Stebbing had introduced Helmer to Ernest while Ernest was in Cambridge. Hempel had moved to Brussels, Belgium in order to work with (and for) Paul Oppenheim, a wealthy industrialist who had serious interests in philosophy.

Here is Helmer's letter (Nagel 2020, Box 2, Folder 51):

September 15, 1936

Dear Dr. Nagel,

I would like to ask your advice on a matter that I have been considering for some time. I hope you will forgive me for troubling you with my affairs. I am doing so because your advice would be the most useful, and I should be very grateful, if you could help me.

I have just finished my two years' research work under Susan Stebbing by taking the PhD. Degree with Bertrand Russell as the second examiner. I am now trying very hard to find some sort of position in this country, but the chances in this respect are exceedingly little for a foreigner. The Universities naturally tend to employ Englishmen in preference to foreigners, and a job as a school teacher – though much more easily obtainable – does not appeal to me very much, and besides, the labour permit is nearly always refused for an occupation of this kind.

So far my study here has been financed by a small grant (£ 100 a year) from the Academic Assistance Council. At present, this grant runs up to October, but it may possibly be extended for another year, if I do not find any employment by the time of its expiration.

In this case – and now I am coming to my main question – I was thinking of using this money for going to the States and looking there for an academic position. Do you think that it would be wise to do that, or is the risk of not finding anything within one year too great? You may be able to advise me more easily, if you know my special subjects. I took my German degree in Mathematics and the English one in Logic. Within Mathematics, I am especially interested in Pure Mathematics, that is, in Higher Algebra, Theory of Numbers, Theory of Functions, Theory of Aggregates, Logical Foundations of Mathematics; and within Logic, in the whole field of modern logic, that is, in Logistics, Many-valued Logic, Axiomatics, Logical Syntax, Semantics.

If, on the whole, you consider it expedient for me to go to the States, may I ask you to answer, if possible, the following special questions.

(I) Which University town would you advise me to go to in view of my subjects? (I am fairly sure that I should be able to get any introductions that were necessary.)
(II) What approximately is the minimum amount of money possible to live on a week, inclusive of everything?
(III) I am afraid that even with the greatest economy I should be unable to get along for a year with my grant of £ 100 only. Do you think that there is any chance of getting a sort of au pair job at one of the colleges? If so, what do you think is the best thing for me to do to find out about a job of this kind?
(IV) Another question, which is perhaps the most difficult one to answer, is this. I am engaged to be married, and if it is possible, I should like to get married before I go to America and take my wife with me. In that case, she would be able to take a small amount of money with her, say 1000 Dollars. Under normal circumstances, the mere idea of marriage in my present conditions must seem almost absurd. But I see that, for instance, in England the cost of living for a married couple who cater for themselves need hardly be more than that for one person who has to pay boarding-house and restaurant prices. For this reason I feel that the slight financial advantage of going alone is not worth a long separation from my fiancée. On the other hand, I do not know whether the chance of getting a post or grant by the end of the year is strong enough to justify our both using all our small capital. Of course, I can hardly expect you to give me any definite advice on a problem like this, but I should very much like to hear the opinion on this point of somebody who knows the American conditions well.

Again please forgive me for troubling you with these personal problems.

Thank you for sending me your articles on Philosophy in Europe. I have read them with great pleasure and found them very interesting.
Please give my compliments to your wife.
Yours sincerely,
Olaf Helmer

Helmer married his fiancée, Elaine Holding and came to the U.S. in November, 1936. He was able to survive by giving lectures at various Universities while searching for a more permanent job. However, he had a difficult time.

Peter Hempel was not planning to emigrate in 1936. He wrote to Ernest from Brussels (Nagel 2020, Box 2, Folder 52):

Brussels den 14.XI.1936
Dear Dr. Nagel, ...
Dr. Oppenheim and I were very pleased to hear that you had a favorable view of our little book. We both thank you from afar, that you were kind enough to review it. We are looking forward to reading your review. You would make me very happy and the subject interests me greatly as before....
Your Carl G. Hempel

Karl Popper wrote three letters to Ernest in 1936. Popper came from an upper middle class Viennese family which had converted from Judaism to Lutheranism before he was born. As a young man he was interested in Marxism and worked for a while as a laborer and cabinetmaker. He eventually earned a doctorate in psychology at the University of Vienna in 1930 but was only able to find a position teaching mathematics and physics at a secondary school. In 1935 and 1936 he visited England, trying to find a position there. Susan Stebbing mentioned that the Academic Assistance Council there was trying to help him in a letter she wrote to Ernest in November 1936. Schlick's murder on June 22, 1936, clearly worried Popper. He wrote to Ernest for help (in German) on August 4, 1936. Ernest responded quickly, advising him to learn English. Popper then wrote him two more letters (in English).

6th September, 1936

8, Neukraeftengasse,

Vienna, XIII.

Dear Dr. Nagel,

I was very happy when I read your kind letter, and I feel so thankful towards you because you answered it so promptly. I think I am rather capable to estimate this fact being myself at least such bad a correspondent as you are. I beg you not to feel yourself obliged to answer this letter with a similar promptness. I am convinced by your kind letter that I may count on you and on your friendly assistance, and if I don't hear from you for some time I shall know that there is nothing of importance to say at the moment and that I shall hear from you again as soon as you have tell me any news concerning my plans [sic].

Your assistance and advice means more to me than even a great influence! (Intending to follow your advice[10] I am writing this letter in a language which I hope is not too far away from what you would call English!)

I am convinced that my coming to America without plenty of money to look there for a position will not be a very pleasant undertaking. But I don't see any other possibility. To stay here has become impossible for me. Under other causes I have not the least intention to fight under the command of Herr Hitler, nor to enter an army controlled by him even in this so called 'times of peace'. (The 'outbreak of peace' between Germany and Austria – as we use to call the agreement of the 11th of July – has made the situation more sincere [sic] than it was ever before.) All this made it clear to me that I must not think too much about the difficulties of an emigration. Better to think about its possibilities only.

My main hope is that I have allowance to refer to the recommand-ations [sic] of authorities like Einstein, G.E. Moore, Niels Bohr, and Bertrand Russell.

As to your question regarding the date of my voyage I can only say that I intend to go as soon as I have got the American visa. At the moment the problem how to get the visa is still extremely difficult. I cannot declare that I don't intend to stay in America or to look there for a position, and therefore I cannot get the tourist's visa. As the Austrian's immigration quota is not exhaustet [sic], I could get an immigration visa quite easily but I would need for it two things: (1) some Invitations of Universities to deliver some lectures (because I have to prove my statement, that this is the occupation I shall practise [sic] in America), and (2) a so called 'affidavit' – a declaration which I fear is very difficult to get.

I understand from a friend that there is still an other form of visa which is called 'professor's visa', and which can be obtained without such difficulties; but only, if one has either got an academic position in America or in Europe, or at least an invitation as a 'guest-professor' for a certain time. (It may be not called 'guest-professor' but 'visiting-professor').

These difficulties are at the moment my main problems.

Niels Bohr has promised to write about me to people in America, in case that I know somebody to whom to write would be useful. But I am so ignorant of American circumstances that I have not the slightest idea to whom to apply. Informations [sic] in this direction would be very valuable for me.

Let me please tell you an other time how kind and friendly your letter has been and how thankful I feel about it!

With my kindest regards,

Yours very sincerely,

(signed) Karl Popper

---

[10] I practice as much English as I can.

His second letter shows that he contacted the Rockefeller Foundation and that he was worried about getting an affidavit. Many scholars planning to emigrate at that time had similar concerns. They tried to arrange visiting positions or lecture tours through the IIE or Rockefeller foundation and they asked influential people for help in getting affidavits.

6th September, 1936
8, Neukraeftengasse,
Vienna, XIII.
Dear Doktor Nagel,
please don't be frightened about getting a new letter from me so soon. But as I have to ask you a question and to tell you an idea, I think it is better to do so at once.

The question: I had a talk in Copenhagen with Mr. Warren Weaver, Director for the Natural Sciences, the Rockefeller Foundation, 49 West 49th Street, New York. He sayed [sic] to me that he was rather interested in my case having heard about me in London, and that he is thinking about possibilities to help me. The Rockefeller Foundation not being in the situation of giving fellowships to philosophers or to physicists he sayed that he would think about getting a lectureship for me or something of that kind. I have not had the occasion fully to inform him about my sizuation [sic] and about my plans. He is not even aware of my relations to Bohr, nor does he know, that Prof. Hans Thierring, (Ordinarius of Theor. Physics at the Vienna University) made attempts, about one year before, to get me a Rockefeller fellowship (from the European of the Foundation) but failed as no fellowships in physics are to be held any more.

I did not hear any more from Mr. Weaver. Do you think that I shall write to him? And if so, do you know something I could suggest him in my letter?

The idea: As far as I know it is a habit in America that lecturers and Professors go from time to time as visiting professors to other colleges or to Europe, and so on. Do you not think that such a case could be comperatively [sic] easy occasion of getting a temporary lectureship as a substitute?

Please don't answer this letter if you don't think it to be of importance! I hope you will not feel to be troubled by me too much!

With many thanks,
Yours sincerely,
(signed) Karl Popper

Unfortunately, despite the help of many people, including Susan Stebbing and Ernest Nagel, Karl Popper was not able to get an academic position in either the U.K. or the U.S. However, in 1937 he was able to get a position in New Zealand, where he became lecturer in philosophy at Canterbury University College of the University of New Zealand in Christchurch. He became quite famous after writing his influential book *The Open Society and Its Enemies.*

Rudolf Carnap and his wife Ina arrived in New York in December 1935. He had agreed to teach at the University of Chicago for 3 months (January–March 1936). Ina did not like Chicago, finding it dark and noisy even though people were very friendly towards them. Of the people they met there, they liked Charles Morris and his wife Gertrude best. Morris had arranged Carnap's visit to Chicago. They met Morris Cohen who was teaching at Chicago that year. Carnap complained that he had to spend all his time preparing lectures and did not have time for his own research. However, he enjoyed meeting the professors in the Philosophy department as well as some from other departments. Many of them came to his lectures and participated in discussions. However, Ina did not feel comfortable in Chicago society. She wrote to Edith and Ernest on January 19, 1936 (Nagel 2020, Box 1, Folder 33):

> People here are very friendly towards us; we like still best of all of them Mr. Morris and his wife. For my taste all these visits and invitations cost a lot of time which could be spent more useful. But especially as long as Carnap is for sale it seems unavoidable. Last Saturday was a dinner given by the department, McKeons, Carnaps and R.F. Perry from Harvard were the honor-guests. For our taste it was rather formally. I even had to get a new evening dress because my European seemed not to be formal enough. I had my place between Perry from Cambridge – who is a very nice and amusing fellow – and McKeon – with whom I did not know what to speak. What a shame that so much power is given in the hands of a single man. So far, things for Carnap are not near to a decision; but Morris thinks, Carnap has quite good chances. There are a few men outside of the philosophical department interested in Carnaps work; perhaps that's of some help. . . .
>
> Carnap feels quite well and in his right element. I have a little more European egg-shells on me and it takes me longer time to get used to the lively life. But you know, I have to be a support for Carnap and therefore I have to be brave.

He and Ina learned to drive and bought a car so that they could drive to the Universities where he had been invited to give lectures. These included Iowa (where Herbert Feigl taught), Urbana, Michigan, Cornell, Columbia and Princeton among others. He also planned to teach two courses during the six-week summer session at Harvard (July–August 1936) and to attend the Harvard Tercentennial in September 1936. In Spring, 1936, he actively searched for a more permanent academic position. His search was rewarded with a handsome offer from Princeton University. In March 1936, Princeton offered him a professorship with an annual salary of $5000 and 10 hours of teaching per week, mostly undergraduate courses. However, Carnap did not accept it immediately, wanting to see how Chicago would respond. When Chicago offered him a tenured professorship with an annual salary of $5500 and only 6 hours of teaching per week, mostly graduate courses, he accepted it (Creath 1990, 207, 211).

After a vacation in Skyland, Virginia, the Carnaps returned to Cambridge, Massachusetts so that he could teach two classes during the summer session. They stayed in a house which Naomi Quine had arranged for them to rent at 7 Francis St. (Creath 1990, 236). They invited Ernest to live with them for Carnap's 6 weeks long course at Harvard and participate in the discussions. (He did not attend the lectures, because he would have been required to pay). Herbert Feigl was also there. Edith, however, stayed in New York because she wanted to attend summer classes taught by Enrico Fermi at Columbia University. The worsening political situation in Europe brought many prominent academics to America. Here are excerpts from Edith's letters to Ernest during this period. She felt inspired by Fermi's lectures but rebuffed by the other, mostly male, students (Nagel 2020, Box 2, Folder 44).

> July 33, 1936
>
> Dearest, . . .
>
> I'm a little depressed this afternoon. Sometime I feel a need or desire to discuss physics or some of the things that came up in Fermi's class with some one. The boys all seem to talk to each other freely. I approached Lloyd Matz with a question to-day and though he answered me there was little interest on his part to go on. When you get the feeling that you are doing your work in a complete vacuum and that you are no nearer maturity in your

subject than you were years ago it is a little depressing. You know what I'm talking about, don't you dear. Fellowship in work is rather important, I think, because if you have no discussion you can never tell how completely you understand a thing.

However, when I get home and am alone by myself for a while my interest and ambition in physics for its own sake picks up again. . . .

Edith

August 8, 1936

Dearest, . . .

Darling yesterday and the day before I spent almost all day and evening in the lab because the boy was getting ready to make a run and since you weren't home I thought I might as well stick around as long as I could. Last night I came home at 10:30. . . .

I'm a little worried now about the hours they keep at the lab. I hope I'll be able to arrange it so that it will be a little regular. Torry almost sleeps in the lab and they think nothing of working all through the night. It is a slow business getting your hand into the various techniques. The set ups are very delicate and there is always the problem of wanting to do things oneself to get the practice and being afraid to do them because one hasn't the experience to handle them. If there aren't many runs next week I'm going to try to do some glass blowing. It's interesting and awfully important. The machines are really made by the boys themselves with all the glass tubes, leads, pumps [then] blown and sealed in the apparatus by their own hands. It will take me a long time to be able to do anything like that. Then I also have to get used to really handling the various machines like the thermo-coupler, vacuum tube electrometer and the like which I've only read about and forgotten about. . .

There is a girl here for Fermi's course who was a student of Sidney's at N.Y.U. It seems to me all N.Y.U. girls are of the same stamp. Somehow they all seem to have the same mode of making themselves attractive. One of the boys told her that I was married to you and so she introduced Sidney's name. Evidently she too fell under Sidney's spell for she likes him very much.

I spoke to Fermi yesterday after class. If Fermi seems charming in class he is a thousand times more charming outside. I can't describe it but he is really lovely and at the same time crystal clear in his ideas and helpful in trying to teach you. He works in the lab here all day and night, I understand, on a neutron problem which requires a lot of liquid nitrogen which he can't get in Italy. As soon as he finishes he'll return to Rome.

There's a good deal for me to learn about the theoretical sides of molecular beams. All these things join up together and so I shall read both Fermi's paper and some papers on molecular beams at the same time. I don't know yet whether there is an exam in Fermi's course. . . .

Edith

Ernest found the discussions in Carnap's seminar stimulating. However, Ina, who had suffered hunger during World War I, made him eat everything on his plate. After he returned to New York, he sent them a box of candy to thank them. Ina thanked him but said that since Carnap had the good principle never to touch candy, he had just provided her with a lifetime supply. When they told the Carnaps that they had just bought a house in Vermont, she responded:

Will you visit your farm? (Our best congratulations, you capitalists!) Carnap is authorizing me to offer you that you may always borrow money from him if you want to. He would be very glad to do it. I don't think that you have done a very curious thing in buying it. The only question which arises within me is: who ever will work on the 24 acres? Neither you nor your friends look very much like farmers.

After they returned to Chicago, Ina started to work on a doctorate in sociology. She reported that Carnap complained sometimes that his secretary was often too busy to type his letters and reports.

On March 5, 1937, they wrote to both Ernest and "Van" Quine about Carnap's new research seminar and the fact that he would be allowed to hire some assistants. Carnap wanted advice about who to hire (Nagel 2020, Box 4, Folder 55):

March 5, 1937

Dear Edith and Ernest, . . .

Immediate reason for this letter: The Department of Philosophy at Chicago is planning a Research Seminar under Carnap's supervision with a few assistants (salaries between 1000 and 1500 dollars). You know, how delighted we would be if we could have you here. But unfortunately there is no chance for higher positions in the near future. (And you would not come to Chicago anyway). Carnap is to make a list of candidates for these assistantships. He wonders whether you would recommend Gruen and/or Wohlstetter for such a job. What about their abilities of doing research work and of stimulating discussions in the seminar (this and maybe a little teaching besides would be their job). In Harvard they were rather silent (like Ernest himself). Or could you recommend somebody else? Carnap has in mind of organizing the seminar on a postgraduate level, the knowledge of the elements of the fields is presupposed (Formal logic; Syntax, Semantics; Foundations of Mathematics; Logic of the Empirical Sciences – this is the plan for three quarters). We do not yet know how the president will behave on the whole question and how much money he will grant. The Department plans of having the assistants for one year only and then others, 1 to 4 men; McKeon thinks, how many the president will grant will depend upon his liking or disliking of the suggested men. Carnap would love to have Hempel and Helmer (Tarski, Waismann are also in the background but their English will not be sufficient at the time), he is considering Weinberg (he has not read Weinbergs' book; do you think he would be worthwhile? As you say, he is a lonely soul and deserves encouragement; and this one year might give him a start somewhere else). But it would not be wise to suggest too many Europeans at one time. Will you help Carnap and give your opinion about possible candidates? And as far as you can without much trouble and looking up people will you answer the following questions about them? : age; studies; where and when; degrees; marital status; publications; teaching experience; position and salary at present, if any (it seems to me as if I had read somewhere that Gruen is teaching at the City College). Carnap has to present his list next Friday. Could you answer via airmail – no need to write a long letter!

Quine, who had a wife and child and a very low salary as an instructor wrote back that he wished the offer were for a permanent position and a salary of $3500 (Creath 1990, 232–234).

In the end, however, Carnap was only allowed to hire two assistants at $1000 each for only 1 year. He hired Hempel and Helmer and so they went to Chicago for the 1937–38 academic year (Nagel 2020, Box 1, Folder 34).

March 14, 1937

Dear Edith and Ernest,

Thanks for your letter of March 8th and the nightletter which arrived today. In the mean time the big Research Seminar Plan has somewhat collapsed because McKeon reported that in their last Meeting the Board of Trustees have ordered that all expenses have to be cut down as far as possible, which means no new positions should be created etc. Since McKeon and the President are quite in favor of the plan of the Seminar, McKeon said to Carnap, it might be that in a few weeks things may look different and that all hope is not yet lost for a realization on a smaller basis, i.e. in the best outcome two assistantships with $1000 dollars salary a year each. You can imagine Carnap's disappointment, especially since he had nursed the hope that Hempel would be invited for a better paid post. Carnap's order of candidates was finally: Hempel, Helmer, Weinberg, McKinsey (whom Quine had praised enormously). I don't think, Hempel would accept a $1000 dollar job, but would rather wait for an improved situation. If you tell Gruen about the situation, you may add that there is some chance that the situation will be better another year if the first year should be a success (and if the first year will be realized at all). Since Gruen has a job at present, and since the assistantships are for one year only, he does not lose a thing by waiting. . . .

I do not manage very well to do all the work I pretend I am doing. At least poor Carnap sometimes complains softly that he had a more efficient secretary before she started to become a schoolgirl. Sometimes I neglect his correspondence, and very often I am not able to type on manuscripts, especially if papers are due at the school. And I cannot do it without working in the night. But what I am doing is nothing in comparison with Edith's work in the Lab until midnight! And don't forget that we live in an Apartment Hotel where service is included. I have only to do the shopping, cooking and washing dishes. And of course, washing small laundry and darning Carnaps' socks. But if the rush is too big, then I prefer to buy new socks for him and to preserve the ones which need darning in a special drawer. You can believe me that this drawer always stays plumbful. . . .

With love to both of you,
---> from R.C. Too!
Yours,
ina

Olaf Helmer and Peter Hempel accepted the offers even though $1000 was barely enough to cover their expenses. Helmer, who had been searching for work since October 1936 without success, returned to Europe for the summer. He fretted over the meager salary, but Hempel made the best of the situation (Fig. 2.25):

Fig. 2.25 Peter and Eva Hempel with Joseph H. Woodger. (Private Collection of Yvonne Nagel)

Chicago, Nov. 14, 1937

Dear Ernest, . . .

We get more and more accustomed to Chicago and like living here much better than in the first week. We got acquainted with several very nice people, and everybody we met so far has been very friendly with us.

Olaf and I enjoy our activity at the University very much. Carnap's lecture course on logistic which I regularly attend seems to me a masterpiece from every point of view, and I am learning a good deal from it. The seminar is, of course, of a particularly high interest; it would be so fine to have you here for it.

Besides, Helmer and I are glad to lead the discussion group in connexion with the seminar. A few weeks ago, another group which has been given the title "Scientific Method Group" has been founded on the suggestion of Mindell's friend Passin and a second student, Krader; the attendants are students of different fields who are interested in scientific philosophy and in the rejection of metaphysics (particularly in the Adlerian form). These people are very nice and interested; unfortunately, some of them overestimate their degree of logical insight and training and are a little too eager to make a criticism of metaphysics which at the bottom is still very naïve. Helmer and I try to do our best in order to inform and train the people a little and to prevent the discussion from taking a merely negative and critical aspect. In the first meeting, Carnap came himself and gave a talk on empirical science and metaphysics; and now we are discussing some questions of the logic and methodology of physics. The next point will be a short analysis of mathematics and logic, and then we shall try to apply the points of view thus obtained to different other branches of empirical science, beginning perhaps with psychology. In the last meeting, a student of Adler's was introduced; he seems to be a nice and reasonable man, and we hope that his arguments will be stimulating in the discussions. . . .

Now I have to trouble you again with a question. Carnap and Olaf suggested me to visit, after the Princeton meeting, some more eastern universities in order to get personally acquainted with logicians and philosophers; they think this might also be helpful for finding a job. So I shall try to meet Prof. Hull in Yale and perhaps Prof. Köhler whom I know from Berlin and who is now at Swarthmore College. So you think there are still other people whom I ought to see? If so, it would be very kind of you to let me know; I then should write them beforehand and ask whether I may come and see them in the first days of January. . .

Very cordially, Peter [See letter 7 in this volume's Chap. 4]

Since their duties as Carnap's assistants were light—they mostly participated in discussions—they spent much of their time traveling to various Universities to give talks, hoping to get a job.

For the 1938–39 academic year, Olaf Helmer got a position teaching mathematics at the University of Illinois at Urbana but Hempel was unable to get a position in the U.S. and so he returned to Brussels for the fall to work with Paul Oppenheim and take care of personal matters. He wrote Ernest several letters from there. In one, dated September 19, 1938, he mentions that Tarski had visited and told him about a paper of Mazurkiewicz which he though very important to the field of the axiomatization of probability. He wrote to Ernest from Brussels:

Eva is going to stay in Brussels; but I have to see my mother and to talk many questions over with Eva's parents. Her father was kept in a concentration camp for 3 ½ weeks; he was released a few days ago. Since he can no longer earn his living in Germany, he is forced to emigrate; he has applied for a visa for the US, but probably he would have to wait 2 to 4 years before his case could be decided; so he is considering emigration to Australia. For the time being we are trying to obtain for him a permit for a temporary stay in England where he might wait until the Australian visa is granted, which may last half a year or so; but it is very difficult and it lasts long to get it (i.e. the English permit). His family would have to wait in Germany which will be possible as his wife is "Aryan". At least during his intermediary stay in England, we shall have to support him, and after that, we shall have to contribute for a while to the means of living of his family; this obligation adds to my strong personal inclination a material urge towards finding a position in America which offers certain chances of stability.

Besides, Grelling has now definitely emigrated from Germany, whereas his wife, who is "Aryan" is staying with the children in Germany for economical reasons. Now and then, she comes to see him. But both of them, who are very likeable people, are now, so to speak, waiting anxiously for our definite departure, because then G's financial status with Oppenheim will improve, and his wife might leave Germany for good too.

Yesterday, we received the good news that my mother has been allowed by the Devisenstelle to present us with the steamship tickets; this is why we shall cross on a German boat, the Bremen, which is due in New York on February 2. Please do not trouble to come to the pier; we are no longer greenhorns in New York, you know; and we shall ring you up when we are in.

We intend to stay in New York for some time; Eva will even try to find some job for herself and settle down when I am traveling around. I should be very glad to meet the members of your discussion group and whomever you think it might be good for me to meet for scientific or for practical reasons. I shall also try to contact Kaufmann whom I know from Vienna and whom we met last in New York a day before sailing back to Europe. - My further program has not yet been fixed; it will depend for a good deal on the suggestions and eventual introductions I might receive in New York...

Apart from the terribly depressing political events of the last weeks and months, we are feeling quite well here. The presence of Grelling whom Oppenheim invited, upon my suggestion, to become his collaborator after our departure in 1937, is very stimulating for me, and also personally he is very sympathetic, and we are having a really enjoyable time with him. Whereas Grelling and Oppenheim are continuing their studies on the concept of Gestalt, I am still busy with an analysis of confirmation and disconfirmation I began with Oppenheim already in 1937. We have been getting on quite nicely,...

Yours, Peter [See Letter 15 in this volume's Chap. 4]

Peter and Eva returned to Chicago in 1938. She got a job taking care of the children of a Professor Adams at the University (possibly James Luther Adams). He continued to work on his research. Olaf Helmer found Urbana tolerable, but his wife Eileen found it impossible. She wrote to the Nagels (Nagel 2020, Box 2, Folder 50):

Urbana, Illinois

April 30, 1939

Dear Nagels both:

I've been meaning to write to you for something like the last eight months, but the part of it that I spent in England last year was taken up chiefly in having my tonsils removed in the most painful way at present on the market, probably an English Patent, and I was too wholeheartedly absorbed by the enormity of my sufferings to be an agreeable correspondent. Troubles never come singly, and the next one was Urbana. I got as far as starting several letters to you, all in somewhat of a Hecuba-After-The-Seige-of-Troy manner, but somehow, seeing the history of Urbana down in black and white, just sent up my temperature, so I abandoned the whole project until such time as I should have become better adjusted to the environment – which, incidentally, God forbid.

The Twin cities (dwarf) boast a better grouser than I can ever hope to be. Bill Barrett.[11] When he feels particularly bad, he comes round to us, and we all grouse together, then Bill

[11] William Barrett (1913–1992) had received his PhD in philosophy from Columbia University before teaching at Urbana. He wrote several letters to Ernest complaining about the philosophy department there. He later taught philosophy at NYU (from 1950–1979).

goes home and says he'll write to Nagel about it. You've probably heard the worst from that source, and believe me, it's no exaggeration – it couldn't be. In giving my version of the situation, I just don't know where to begin. There's such a wealth of material. I might start with the Tuesday tea club, or would you prefer me to get right down to the Sewing Circle? The Book and Drama Club is our really highbrow institution. The chief inducement that was offered to members was that there was no need for them to read. This was just as well, as I doubt if many can read, anyhow. Well, I went to one of those meetings. It was presided over by a lady who might have, and probably did step straight out of a Helen Hopkinson drawing. She reviewed a book for us called And Tell of Time. She didn't attempt to embark on anything as destructive as criticism, but contented herself by giving a a synopsis of the story, and saying that it was a very good book because it was written by a faculty wife like ourselves, the lady who wrote it was a clever lady because the title came from Euripedes, it said so on the dust wrapper, and finally, because it was a best seller. Then we all had tea. One need never feel lonely in a place like this. In fact, it takes a rather highly developed technique to dodge the faculty wives whose mission in life seems to be the hunting down and bringing to heel of other faculty wives. One woman here who seems to share my feelings about this round of gaiety has solved her problem by announcing that she is a Communist. This will probably cost her husband his job, so I'm still racking my brains to think up something that is, at the same time, undesirable and respectable. So far, I've been saying that I'm so sorry, but I can't go to the sewing circle because I'm expecting a cable from Europe to call me to my mother's sick bed. An old one, but what can I do?

From Olaf's point of view, the place might be worse, but not much. At least he is not so isolated as Bill, because the Math Dept is large, about 60 all told, and among these are some who seem to be worthwhile, and with whom he can have discussions. The work itself is pretty lousy. He has a 12–15 hour schedule, freshman and sophomore, and very few of the kids have got beyond the stage of counting coloured beads. Few of them will ever get beyond it. The system is pretty sickening, because it sets great store by a college education, compels the poor little devils to sweat their guts out for grades, and ruins all chance of any sort of academic standard. Olaf says that the thing that make him uneasy in his mind is that these young morons from the engineering school will be building bridges one of these days. A thought like that would sober anyone up. Most of the students simply don't have a chance to prosper even if they have the capacity. They're terribly poor, and in order to get through school, they have to work like hogs, and are grossly underpaid. But let it not be said that here in Illinois we lag behind the beau monde of Harvard. There, they swallow goldfish, Here we swallow live rats. I'm trying to track down the hero of this last exploit, because our kitchen is overrun with mice, and there seems to be no other way of disposing of them.

The Art Department is a gem. Word got round that I was an artist, and I was accordingly called upon and asked if I did portraits or landscaped, and in either case, would I care to join the faculty wives sketching class under the direction of Mr. Roysher of Cleveland. It was beyond me. I couldn't trust myself. I said that I used to do a little sketching now and then, but that since my marriage I'd been too busy to take it up.

The advantages of this place (let's be fair) are that we have an extremely pleasant man as head of department, Coble, that we are not badgered to go to church, and that there are a number of really nice and intelligent people here, and that we can go to Chicago often, and do. Chicago once seemed to me to be on about a par with Manchester. Now it seems like Paris pepped up…

We were in Chicago a fortnight ago, and saw quite a lot of the Hempels, and heard all their news. They have an apartment there now, quite a pleasant one, and will possibly stay there during the summer as Carnap will be there, although this is not quite certain. Eva has a job, not an ideal one, but if it works out not too badly, she thinks that she'll keep it even if Peter finds something for this fall. The job is looking after children, and is from 8 o'clock in the morning til 6:30 in the evening, and is in her locality, but I expect that you've heard from her by this time. The outlook for Peter doesn't seem too good at the moment, in fact it's rather depressing. I do wish that he could find something. We haven't heard from them since we came back here, and I've been wondering if anything came of the possibility of the Cincinnati job. Bill Barrett was at the Missouri meeting, and did his best to introduce Peter to anyone he knew, but it was rather a case of the halt leading the blind. Carnap didn't go to that meeting which was a pity, nor did Morris, which was a greater pity, but Perry did, which was probably the greatest pity of all. So did McKeon. Ernest, can nothing be done about that canker in the rosebud?

Olaf has been probing Coble, ever so gently, as to the chances of getting a position here for Peter in Mathematics. This is a sort of last resort, but I think that as things are at present, a position in mathematics would be better than a return to Oppenheim. I don't think that the chances here are very great, but it looks as though there will be a vacancy....

Love from both
Eileen

Dear Edith and Ernest: Urbana is not quite (but almost) as bad as Eileen makes it out to be. On the whole, in spite of its drawbacks, we are quite happy here, though admittedly only because there is reasonable hope that we are not going to finish our days here. What do you think of our summer plans? We have heard so much of Vermont, it seem to be a wonderful place. Unfortunately, so we have been told by some people, you are eaten up by mosquitos before you get a chance to look at the place. Is that so? Be seeing you soonish, I hope! Olaf

For the following academic year, both Helmer and Hempel moved to New York. The Nagels invited the Helmers to visit them in Vermont. Ernest helped Peter Hempel to get a job at City College. The Hempels also lived in the Nagels' New York apartment during the summer of 1939 while they were in Vermont.

Chicago, June 13, 1939

Dear Ernest,

It was very good of you to write me even in the turmoil of your departure, and to talk to Bronstein about my prospects. Yesterday, I got a letter from Bronstein telling me that the Board of Higher Education has again delayed action "on the whole evening session because of certain technicalities", but that the recommendation of my appointment has been accepted by the administration of the evening session, and approved by the president, and that "for all practical purposes, favorable action of the Board can be regarded as a foregone conclusion".

You and Edith can, no doubt, imagine how delighted we were to get these news! But we realize that an unfavorable decision may still be made by the Board; therefore, I have not given up, but only delayed, the negotiations with Prof. Tyler. Of course, I would by far prefer the job at the NYCC, even though the total amount I would be paid for the nine months (6 hours per week) would be about 900 dollars only; but we could bridge the gap between this sum and our actual expenditures by our savings, and we are also quite confident that Eva could make some money in addition. Both of us would be very glad to live in New York with its many facilities and possibilities for practical and theoretical needs and interests. And I would be so happy finally to get a start and have an opportunity to teach, and that at good school like NYCC. I am so grateful to you, Ernest that you told Bronstein about my looking for a position, when I was last with you in New York.

As Bronstein wrote me, there is even a certain probability for me to get a position in the summer session, beginning July 5; in this case, we would only too gladly accept your offer to live in your apartment during your absence. We shall write you about this as soon as we know the final decision of the Board.

We hope that you had a fine trip to Vermont, and that you are now enjoying restful days; we are sure, both of you need them!

Many thanks again, and best wishes to both of you from Eva and myself.

Yours,

(signed) Peter [See Letter 22 in this volume's Chap. 4]

Daniel Bronstein of City College thanked Ernest for recommending Hempel. This letter shows that they were also interested in hiring Ernest himself.

June 27, 1939

Dear Ernest,

I am very glad we were able to help Hempel and also to obtain a first-rate logician for the summer and evening sessions, something we needed badly. I am only sorry the position is so inadequate. We did get $4 per hour for him which is high for City College. He will have 12 hours this summer and 9 hours next year in the evening.

We have also recommended White for the readership and [Lewis] Feuer has been appointed at $5000 per year for one semester. You see, it is the wrath of Gotschall (?), Mead et al. to get a "big shot" for the Philos. Dep't to replace Overstreet – in which case Feuer will be dropped at midyear. We'll have to be careful if we don't want to get somebody who doesn't fill the bill. I know we would all welcome you, and it would solve our problem if you could accept. Of course, I don't know what terms would be offered – but with the new administration, I feel an offer satisfactory to you would be made. Or is it out of the question? ...

Sincerely,

Danny [See Letter 25 in this volume's Chap. 4]

The Hempels enjoyed staying at the Nagels' apartment, but there were some drawbacks:

July 5, 1939

Dear Edith and Ernest,

Here I am, sittting in your living room, the typewriter on the telephone table, looking on the Hudson, and enjoying life even though writing letters! You have no idea how much we enjoy this apartment. There is only one drawback: living here will spoil us for ever for any furnished apartment. ...

By the way, we really admired the orderly and efficient way in which you put this apartment into its summer sleep! And the cleanness and neatness everywhere. It is quite a treat, after the Chicago apartment, to have lamps where one needs them, to have matching dishes, and as many as one wants, to change towels more often than once a week ...

(signed) Eva

Dear Edith and Ernest,

last night, I came home at midnight; so I did not begin writing to you, and today, we were downtown all day long; and since I want this letter to reach you before the end of the vacations, I shall have to make it short this time, and I hope you will forgive me.

First, I want to mention that Eva overlooked a second very serious drawback of your apartment, and that is your library, Ernest. As soon as I settle down to work, my glance falls on some interesting book-title, I get up again and begin reading a book which certainly is not absolutely necessary for preparing my classes in elementary logic. It is really a wonderful library, and surprisingly complete, and I am very glad to have an opportunity of using it. ...

We got a letter from the Helmers today, telling us how much they liked your country home; it must be a beautiful place! We hope you are having as good a time as we are having in your apartment! ...

(signed)

Peter [See Letter 26 in this volume's Chap. 4]

2.9 Susan Stebbing in 1935–1941

While life in the U.S. was relatively peaceful, the political situation in Europe was increasingly precarious. Ernest was well aware of events in Europe through his correspondence with people like Susan Stebbing and Otto Neurath as well as his relatives in Austria-Hungary and Italy. Susan's letters to Ernest and Ann Hook paint a picture of what life was like during the years before World War II and how the turmoil of those years affected the lives of academics of that era.

Although Ernest, Edith and Ann had only stayed a few months in London, Susan formed a lasting friendship with them (Fig. 2.26). In a letter she sent to Ernest and Edith on August 15, 1935, shortly after they had left England, she wrote (Nagel 2020, Box 5, Folder 12):

> I do not know at all how to tell you how much your friendship means to me & how happy your letter makes me – for I, too, have a strong affection for you both. I have always liked you, Ernest, and admired your work but now, as you'll perhaps understand, 'like' is not the proper word – not strong enough.
>
> It does seem all wrong that I saw so little of you while you were in Europe. I wish you had been over here in a less bad year for me, but both School and College pressed upon me.

She also wrote to Ann Hook. Since the Hooks and Nagels were friends and even neighbors in Vermont in the summers, they often shared the letters they received from Susan with each other. Because she was very busy during the school year, she

Fig. 2.26 Susan Stebbing at her home in Tremorrab Tintagel in Cornwall. (Private Collection of Yvonne Nagel)

usually wrote during the summer or winter breaks, often while vacationing in Cornwall or other places such as Ireland and, once, from Norway.

Her letters reveal what academic life was like as Britain and the rest of Europe descended into war. In addition to her heavy teaching and administrative duties (she served as Dean of the Faculty), Susan was heavily involved in helping refugee scholars (especially members of the Berlin and Vienna Circles) flee the Continent and to find jobs in Britain or elsewhere. She and her friends also bought a house in Cornwall (at Tremorrab Tintagel) and moved the Kingsly school there to protect the students from wartime bombing. Every year, she promised her friends she would come to America "next year" and every year, something happened to make this impossible—her illness or that of her friends and sister, the bombings. These letters tell this story movingly.

Susan went to the First Unity of Science Congress held in Paris in September 1935 after much prodding by Otto Neurath, who had started the Unity of Science Movement as a way of unifying, promoting and popularizing the ideas discussed among members of the former Vienna and Berlin circles. She had not wanted to go because she was busy and disliked Paris. However, she ended up enjoying it as she wrote in a letter to Ernest on October 18, 1935 (Nagel 2020, Box 5, Folder 12):

> I shall be interested, Ernest, to read your article about European impressions. I've not seen it yet. Whatever you write is good. I hope if you have reviewed Reichenbach yet. I met him in Paris & liked him. I did wish you had been there. It was pleasant. There was a friendly atmosphere at the conference at the Sorbonne. Bertrand Russell was there – as I told you in my last note from Paris. What I enjoyed most was going with Carnap & some others to a small village in the Belgian Ardennes – Justin-sur-Meuse – we went there for four days – Carnap, Dr. & Mrs. Hempel, Prof. Frank, Dr. Oppenheim, Dr. & Mrs. Grelling, Prof. & Mrs. Jørgensen, Dr. Naidering and myself. I did find it most interesting and pleasant. We had set discussions in the mornings & once in the evening, & went for walks in the afternoon. Carnap lent us his paper on 'Testability & Meaning' to read in typescript - & then we discussed it. All the time we talked in German & I did not always follow, though I improved a lot in [a] week, & I believe if I were with German speaking people for 3 or 4 weeks I could be fairly fluent. I must try to get clear about the various points raised, & write to you again.

During the years 1935–38, Susan was very involved in administrative affairs (she was Dean of the Faculty), helping refugees from the Continent relocate to England or other countries (she was a member of the Academic Assistance Council) as well as busy with teaching at Bedford College and running the Kingsley school. In addition, she suffered from Menière's disease, a disease of the inner ear which affects hearing and balance. It led to what she called "Labyrinth Attacks"—falls in which she did not realize she was falling until she hit the ground. Some of them were quite serious, resulting in broken ribs and other bruises (Nagel 2020, Box 5, Folders 12–13).

> December 9, 1935
>
> I am glad, Ernest, to think that Carnap and Reichenbach may go to Chicago; it would be a happy solution for them both and good for you. But I can't help being a bit sorry for myself – if all the good people go to America. Prof. Moore says there isn't any chance of a chair being created in Cambridge. I wish Carnap could come here. I have been trying to discuss Der Logische Syntax in my discussion club at College, but Mace doesn't read German and most of them can't, & so I read the two first papers and the discussion was not very good. I wish I could get clearer about the principle of tolerance. I want – very badly need – a few weeks

leisure to be able to think about these things. This has been such a very hectic time owing to my having been really ill at the beginning of term and so I got more and more behind with work. I only get 3 weeks at Christmas which no doubt seems a long time to you, but is short for an English University. I have had very heavy Committee work. Being Dean of the Faculty is no joke! I have never had so many Boards and Committees and other kinds of administrative work, & I have to take the chair at three Inaugural lectures, one next week and two next term, as we have 3 new heads of departments this term, & the Dean has to take the chair. The one next week is Reader in Sociology, Dr. Henry Mess(?); next term is the Prof. of Latin, & then the Prof. of English Literature – both of them men…

However, I had one very good weekend – at least it was only from Friday to middle Saturday. Moore asked me to spend the night at his house as Bertrand Russell was reading a paper at the Moral Science Club. The paper was on ‘The Limits of Empiricism’. The paper was not very good, but the discussion was. (It was in Braithwaite’s room at Kings, & 68 people crowded into it & the fire was huge – one nearly expired of heat) But what I liked best was at Moore’s house. Russell was also staying there. We did discuss the paper, we also talked about old times – the war and the conscientious objectors. I have not seen Russell so sincere and fine for a long time. Somehow it touched me really, in a way you might not understand. I did so much admire him during the war, & in the past I knew him pretty well. In the course of our after dinner conversation I remarked that I had heard him debate with Rabbi Wise in New York. (Do you remember?) He asked me what impression I had had. I said ‘Do you really want me to tell you?’ He said ‘Yes’. So I replied that I’d wished he had talked philosophy and not been so cheap – whereat he said very gently that he had to make a living & it was not possible for him to do it by teaching philosophy. So Moore said that Russell could easily have got one of the Cambridge Professorships (both Moore and Broad have been appointed since the war) - & Russell said ‘no – I couldn’t. How could I have dined night after night with the Fellows at Trinity knowing that they had wanted to turn me out’. Well, Russell is no doubt not quite sincere about all that - & I daresay you’ll think this anecdote pretty feeble. But I was really rather moved. I think Russell has been so fine and he means much to my generation. I learnt so much from his lectures in logic and I also learnt from his attitude toward the war…

I am glad to hear of your discussion group. I wish we could manage something like that in London. I must stop. It’s late, & I have still to write to Ann…

Tintagel, September 3, 1936

At the Joint Session in St. Andrews an appeal was made to keep Popper in this country. I promised to contribute if anything could be done. But it seems doubtful. There are so many refugees here and most of us already have committed ourselves to supporting more than it is easy to keep going…

Yes, I knew the details about the shooting of Schlick. I gather it was not at all political. It’s indeed sad; his death in that way and the end of the Wiener Kreis. Indeed it is a sad world. It looks now as though the Spanish government may be defeated by the rebels. I do so hope that may not be so. If there is a Fascist dictatorship in Spain then the whole of Europe may be in for a bad time. If there is a general European war, of course I won’t be able to go to America; and if it seems likely that there will be a war, I can’t very well leave home. That is one reason why I can’t be definite about coming next year. My friends are apprehensive & I couldn’t leave them if this country is likely to be plunging into war. I am unable to decide whether we are likely to go to war or not.

One of your old girls (from Kingsley school) was killed a few days ago in Spain. She was fighting (actually fighting in steel helmet & all) for the government forces, & she & 9 others made for a raid, & 4 were killed, including Felicia Browne (the Kingsley old girl). She was becoming well known as a sculptor. She left school some years ago, & went to Moscow & became, I believe, a communist. She had a touch of genius. She was a boarder at the school in the early days, when I taught a lot in the school & I knew, & liked, her very much. I don’t quite understand how English volunteers can be fighting in the Spanish army, but it seems

> they did – at least Felicia did. But she spoke Spanish and may very well have not revealed her identity. I can believe she was keen to lead a sortie into the rebel ranks. It is very sad – this futile waste of life & human values. I agree with you that it seems as though we were in for a bad time – the white races of the world – what fools we are that we can't control the few who seek to make war. Sometimes I am inclined to think that in a few years we in Europe will be either dead or destroyed in mind. Those of us who lived through the last war cannot – I should think – have illusions about the next.

In 1936 and 1937, Susan was very busy writing a book she had contracted to do for Metheun's on "Philosophy and the Physicists". She was also preoccupied with her sister Helen's long illness. Helen suffered from "an erosion of the bone and spine" and was receiving radium treatments at Lambeth Hospital where their brother George worked (he was a radium specialist). She reported that she visited Helen in the hospital every day during most of June and July. Susan took a vacation in Norway and Sweden for the month of August but hurried back when Helen took a turn for the worse. She and her sister Edith stayed with Helen until she died on September 14, 1937. Tragedy struck again when her brother Bertram suddenly died on January 31, 1938 from an ulcer in his pancreas.

Susan was very busy during 1938. She helped academic refugees through organizations like the A.A.C. (Academic Assistance Council) and the Baldwin Fund. She and her friends also bought a house in Cornwall (at Tremorrab Tintagel) in case they had to move the Kingsley school out of London. She was also applying for Moore's chair as she wrote to Ernest (Nagel 2020, Box 5, Folder 16):

> January 18, 1939
>
> There is no chance of my going to the Conference at Harvard. I am bitterly disappointed. The political situation here has upset all our plans. We (my friends & I) have spent a great deal on the refugees. We gave £100 to the Baldwin fund & have adopted three Jewish children & that means we keep them altogether – clothe them etc, & have them all the year not only in term time. & since Jan. 9th I've paid out £21 to help a really destitute man with two children - & we are already overdrawn at the bank. I have my salary at the end of each term & would normally have a good deal now – of the December cheque – but I haven't, & I've to pay £102 in income tax. So, you see, we have to be very economical - & I couldn't afford to go to U.S.A. That is decisive. Then there is the factor that I shouldn't go so far so long as war may break out at any moment. It is my job to be in this country when we are at war. I am quite without hope. Meanwhile, we do what we can; it is sad beyond words. I've been spending some time even speaking at meetings & doing what I can (so very little) to rouse people against Chamberlain. One hates to be English - & this country must suffer for our disregard of all principles.
>
> You may be amused to hear that I am a candidate for the Chair of Philosophy at Cambridge (which will be vacant by Moore's retirement at the end of this session). I've no chance of being appointed (& I don't really want it for most things – except for more leisure!) & had no intention of applying, but I was persuaded. Moore wants me to apply. Keep all this confidential please. The electors meet on Feb. 10 & applications have to be in by Jan. 28. I've not written mine yet but I have promised I will.
>
> I wish we could have someone good to succeed Moore, but there are no outstanding philosophers in England now. I wish you were having Moore's chair.

Unfortunately, she did not get the job (Nagel 2020, Box 5, Folder 16):

February 14, 1939

Wittgenstein has been elected to the Cambridge Chair. He is a distinguished successor to Moore, but I hate to have Moore go, & I don't think Wittgenstein is good for the students or lecturers! I am not disappointed that I didn't get it. I never expected to do so. I was surprised to hear (informally, from an elector) that the only candidates seriously considered were Wittgenstein & myself - & that if he had not had it; I should have. I realize so well that I am not fitted for it; that I cannot be sorry that Wittgenstein has the Chair, but I could wish there had been someone else. I am having a fantastically busy time - & doing a lot of extra work in connexion with refugees. Such heart rending cases, so little we can individually do.

During this period, students and lecturers at the University of London were also forced to move. Apparently, Susan's Bedford College classes were now held at Cambridge (Nagel 2020, Box 5, Folder 16).

July 1, 1939

I hope that we shall not be at war before you get this. I fear we shall not get through the summer without war. The news today that Eden & Churchill are to join the Cabinet (tho' not yet confirmed) suggests that Chamberlain takes a serious view at last. That there should be such useless slaughter destruction of buildings & everything, is sad: still, I suppose we must do it. I can't be a pacifist when I am so much surrounded by the results of the Jewish persecution. I know so much of the awful suppression in Germany itself…

I agree with you that the governments should shoulder the job of supporting the refugees. We private people can do so little. I am glad Hempel will have some job - tho so meager a job. Still it's something & also on the other side of the Atlantic. Please give him & his wife my greetings - & also Olaf Helmer & his wife if you see them.

At the end of the year, she wrote this letter from Tintagel (Nagel 2020, Box 5, Folder 16):

December 21, 1939

This war is indeed most terrible. I have been so utterly sad about it I hardly knew how to write. Well – I should have to write long letters, & then I felt that you would know all there was to say and would largely agree – at least I hope so. I do not think we could have avoided it; I never agreed that it was right to make the Munich Pact, & I felt very impatient when people, especially the other side of the Atlantic (not you) --- blamed this country for not defending Czechoslovakia & also for going to war. The fate of Poland is terrible. How the Finns manage to hold out rouses one's admiration. These are sad times for nearly everyone in Europe.

Our school came here in September. I was so certain there would be war I could not go to the Harvard Conference, as you know. It seems a lifetime since the end of August; when our holidays and much of our happiness ended together. We have difficult times here. The children flourish, but we are so desperately hard up I don't know how long we can last. We've had so much capital expenditure in moving the school & making this place possible, & our expenditure exceeds our receipts as so very many either can't pay or pay only the smallest nominal fees. But the children are very happy & their education progresses. My friends are very tired but feel it is worthwhile.

I do not like being in Cambridge at all. I think it is all the more difficult because it is familiar to me under such different circumstances. Also, if Moore had still been professor it would have been better. I find Wittgenstein's attitude most trying, but Wisdom is very friendly, so are Braithwaite & Ewing (?). I live at Newnham College. Our students are in many different places - we lecture in various University lecture rooms & also in Merton House so I have 1 1/4 miles to walk each way to my lecture room on days I go to Merton House - & I feel tired and depressed. People are very sociable. & tho they mean to be kind I find that very tiring also - I mean so much chatter & so little said…

I've done a little work since I came home but a good deal of time goes in dusting, making beds & doing odd jobs. The school staff went home for the holidays last Tuesday & only H.G. (Hilda Gavin); V.S.S. (Vivian Shepherd) two matrons & I are left. There are about 40 girls, many under ten, here for the holidays.

I wish you could see this village – & in peace time. We love it so, & count our great blessing in the friendly people here. Our friends, the Browns, are so good - such a fine type, so wise & quiet, and full of a serene dignity. He is a farmer, in quite a small way, but he has the bearing of - well, I don't know how to put it - a man whose ancestry is rooted in the soil. Another farmer, Will Baker, 6 miles away, is like Harvey Brown. We often visit them, & I find great strength from having a quiet evening with them - this country life suits us all - never any fuss but a great welcome that does not need to be spoken in words. These yeomen farmers possess something in their serene spirit which, I fear, may be lost in a collectivized state. They are not at all well off, but they are without fear & without avarice. Sometimes we can laugh together notwithstanding our sadness. I hope when I go back to College we have gained some hopefulness from being here - the hopefulness one needs in teaching the undergraduate young – It is well that they can believe, as I did in 1916, that it lies in their power to build a new world.

Susan and the other British philosophers struggled to keep the University and their various associations going even though their lives were disrupted. Some of the younger men were "called up" to serve in the armed forces (Nagel 2020, Box 5, Folder 17):

February 10, 1940

I have not seen much of Prof. Wittgenstein – only at Moral Science Club meetings, he has never attempted to speak to me. I have only been to one meeting this term, owing to my being ill – & not really supposed yet to be doing as much as I am. Concussion, even if not very severe, takes a time to get over & until last Tuesday (today is Saturday), I had to have a 'taxi' whenever I went out. I went to the Aristotelian meeting, held at Trinity, when Acton read a paper on the 'Philosophy of History' – not a very good paper. There had been an earlier one, when Joad read a paper, & I hear that at that meeting Wittgenstein was 'brilliant' & made a long speech! . . .

We are anxious to keep the Aristotelian alive, & to publish Proceedings, so we (the Committee, I mean) agreed to have papers read at any place where the speaker could go & an audience be found. In practice that means the Universities. We keep to the programme already fixed last July.

Duncan-Jones, Regle, & I also hope to keep Analysis going – but – I fear costs may so increase as to make it impossible. I've not seen the Cambridge philosophers - except at that one Moral Soc. meeting I attended – this term . . .

It is ironical that, seeing how often I have wanted to spend a sabbatical year (which I DON'T get!) in Cambridge, that I should now be here, & yet see so little & hear almost nothing from the Cambridge philosophers – would that Moore were still professor. He was lecturing once a week last term in Oxford, but now he has returned he may not lecture here. Ayer has toured the Welsh guards, & Rifke has, only he has not yet been called up, & thinks he will be before May. He is trying to speed it up. He is keen to join & go to the front, but as he is 39, it isn't easy for him to get accepted. . .

I love to hear from you, & to hear news of all that happens. The school goes on, but the financial difficulties are increasing, & whether we or the war comes to an end sooner, we don't know. Only one thing is certain, Kingsley will never again be in Hampstead. But I do not try to look ahead for more than a week or so. Two such wars in the lifetime of our effort with the school – the first when we were setting out so hopefully, & I was still in the twenties, & now when, for the first time, I feel more than middle-aged! But I hope to recover my energy . . .

Letters which arrived later in 1940 and during the war were opened and marked "censored". Susan described the wartime raids during which they had to go to underground shelters. She also talked about Neurath's escape from The Hague (Nagel 2020, Box 5, Foler 17).

August 8, 1940

Dear Ernest, It is good to have your letter. I wish I used a typewriter. You would find it easier to read, & I am sure the censor must prefer your letters to mine! . . .

You need not fear that we are really suffering any serious deprivations . . . I was amazed, as well as touched, by your offering to send us tea! If it is permissible, I should like some Orange Pekoe tea. . .One night at College I had just made the tea & poured out a cup when the sirens went; so I had to go to the shelter & leave the tea, & when, 4 hours later I returned, the tea was wasted! . . .

I forgot in my letter to Ann, just closed, to tell her about Neurath. I quite hope he will be released from internment. He is so clearly a friend of this country, and an anti-Nazi, & there seems enough proof of that, that I think it should not be long. Meanwhile he is well & calm, & he & M. Reidemeister have been allowed to meet. They will probably manage to leave together if they can go to U.S.A. or the Latin Americas. I have arranged with our solicitor to see to all the details. I sent clothes to him & M. Reidemeister & various other things they needed. I had to buy them as best I could – but I hear the clothes fit. Miss Reidemeister was easier than Neurath; he is so huge & I had difficulty in getting big enough shirts. They escaped from Holland in a lifeboat, & were then picked up by a British warship & they had nothing with them except the clothes they wore. I think I may have told you this before. They are now both of them well & comfortable, & Neurath says, 'calm in mind,' but both of them are longing to be at work. I hope that they will manage to go somewhere where they can get the books & all they need for their work.

I am glad to hear that Hempel has a job, & I hope he will like it, & that Olaf Helmer has been in N. York. It is a pity that his wife so much dislikes Urbana. Do you know whether Helmer's mother is with him, or has he no news of her? Do you know whether Dr. Oppenheim was in Belgium when war came? I have not been able to get news of Prof. & Mrs. Clay, or Heyting, or van Dantzig. They were all very kind to me when I was in Holland. I fear that some & perhaps all, will not have survived. Amsterdam, as well as Rotterdam, was very heavily bombed by the Germans.

I did not know that Max Black had left England. He has not written to me since last November, and I did not know for certain that he was able to accept the post in Illinois. He will be glad to be safe in U.S.A. with his wife & children. I am sorry you have lost so many people from your Department. . .

I hope Tarski will get a post. J. Wisdom very much wants to go to U.S.A. He has for sometime been dissatisfied with Cambridge. I was conscious there of a curious sterility; . . .Perhaps you will think that a foolish comment but I have been so much struck by the mental immaturity & lack of spontaneity of the Cambridge M.Sc. students this last year. . .

You must think of us as going on as well we can with our job. It is so clear that we must look after these children & try as far as possible to keep them from horror and suffering. They are very happy. Two of them are leaving tomorrow, & three have already left, to be ready to go to U.S.A. We are sorry to lose them & they all five have wept to go. But they can't judge what is good for them. It is hard to know what is best. These children's parents are all fairly well-off & can pay for them, & they are afraid what may happen here. But at present the children here are happy enough. I am not afraid that we shall be defeated; but I am sure that much suffering lies ahead of us, & it will be increasingly difficult to carry on well for the children. I do not think the end of the war is in sight; indeed, it will get worse, & we must expect reverses, but we shall defeat Hitler in the end. Sometimes, in fact most nights, I feel a curious sadness in thinking that before morning someone or other will be killed – perhaps a child, by a bomb. I know that the 'deaths from motor accidents' have been more numerous, & as a matter of fact I am one of those who used to be deeply shocked by these 'casualties'. But there is a difference in one's attitude; I feel horror at the air raid casualties because they mean 'successes' for the bombing side & they are deliberate. In one air raid I was in I saw some children crouching on the ground, evidently in an attitude they had been taught as most likely to be safe, & I felt horror that this should be so - in a battle each side is actively doing something; here there is passive waiting - odd that it is just [the kind of] situation in which Aristotle would say there is no place for courage since nothing can be done to avert the danger!

I hope Roosevelt will be re-elected - at least I'd like it. I don't expect it as I fancy the 3rd term plus New Deal will prevent it. But the world needs an experienced statesman in such a position as your president. I hope Edith has got on well with her experiment. I had a letter from Carnap this week asking me what I could do for Dr. Wittgenstein, who, he hears, is interned, so I shall see about that.

Forgive [my] scrawl. Much love to both from Susan.

It is interesting to compare this with Neurath's own account of his escape from The Hague. The following is excerpted from a letter by Otto Neurath sent to Peter Hempel on March 27, 1941 and which the latter then shared with the Nagels, Hooks and Schapiros (Nagel 2020, Box 4, Folder 44).

March 27, 1941

You know we had to escape without luggage etc. At the last moment we succeeded in jumping into a motor life boat which was leaving the shore. Three students managed the boat, no skipper on board, about 50 persons instead of 20 – nothing happened. Harry Hack – that is a real story name – was our commander in chief through mine fields etc. We must admit that we were a little interested in the trip in spite of danger and sorrow (red clouds over Rotterdam and exploding tanks etc. at The Hague). Waves sprayed over the backside of the boat where Mary and I were sprawling in big white rubber boots. I said we were now on the way to marry, that would not be so bad, and then to start lecturing in England would not be so bad either. Mary answered that such ideas were insolent, "they have been waiting for you" ... Afterwards we were interned, long weeks of separation, then monthly meetings, intervention by American and British friends, Stebbing and Kaempffert – especially, Julian Huxley, Albert Einstein etc. doing what they could. Finally the British Academy made a statement that we were scientists of sufficient reputation, and the tribunal found out (with courtesy, I have to maintain) that we are pro-British, then we were released and are now, as you know, staying at Oxford. We rented a nice room with a view on trees and flowers, no bombing, no noise, no traffic, only singing birds, nice parks and gardens, nice buildings at Oxford and nice people. We feel no discrimination at all – we are in a particularly good position because I am lecturing at the university and that gives here a certain reputation.

Susan wrote the following letter to Ann Hook and asked her to share it with the Nagels (Nagel 2020, Box 5, Folder 17).

Tremorrab Tintagel, October 3, 1940

I have been home for just over three months, but I have achieved very little....

I have let myself be disturbed by noises & bustle that in other days I should have ignored. My brothers & my sister Edith are so far safe; they have had bad times - are having them - in London. George & his wife lead a strenuous life; his hospital has been bombed twice, when Reggie wrote to me last week, he was still operating at 3:30 a.m. & a raid was then in progress. They have not had their clothes off since Sept 15th as they sleep in his sitting room in the hospital ready dressed for emergency. So with my sister Edith. We are very quiet here - having only occasional warnings. Hilda is ill just now – a recurrence of her old serious illness, but she is getting better - the new treatment is doing her a lot of good, we have a very good nurse from Plymouth. Vivian is extra busy, with Hilda out of school. I have been helping a little. We are fortunate to be quiet & able to carry on at all.

Last week some people we know at Hampstead were killed – two families, in each case a father (& two ?) children killed – the mother being only injured. That is common enough. It is sad. And this is happening in many parts of the world. . . .

College buildings have been bombed. The council had decided that we should return to London – they decided that the very day the raids on London began. They did not make the decision till Sept. 19th after 1066 windows had been broken - & various damage to roofs etc. – & now more damage has been done & it isn't habitable. So we return to Cambridge. The prospect does not please me – with Moore away & all sorts of difficulties. But – I never affirmed the decision in London because the students have many of them such long journeys to make, & because of the noise. Newnham can't take any of our stuff this time, & I shall be living at Springfield, Sidgwich Avenue - Cambridge.

In Fall, 1940. Ann Hook organized a fund-raising effort to raise money for the Kingsley School. Ernest helped with this effort by writing to several philosophers he knew. They were able to send £100 to Susan for the school. Here is Susan's response (Nagel 2020, Box 5, Folder 17):

Springfield. Sidgwick Avenue. Cambridge. Nov. 16, 1940

Ann, my dear, how wonderful you all are. I do not know how to thank you. So many calls upon you and then such generosity to our school and refugees. I shall in a moment, write a letter for you to show your committee - a formal letter like yours some – but I must write this as well. The money is a great help, and has come at a most convenient moment because we shall shortly have to shew our school balance sheet, for the school year ends September, to the bank - as they hold the mortgages on our houses. We have now pledged all our possible securities, and if we exceed £ 3,500 overdraft we can't go on. We hope to hold out; & we shall certainly be safe for another year. I shall say something about the school in the more formal letter. I shall also write to Ernest. . . Evelyn Clarke, & those whom I know – but I mean to catch this post now, for after all one never knows how long one will be alive. So if I get snuffed out before I write a mere letter, you must please tell them all how grateful [we are] & that I mean [to] write if I can.

It was a red[-]letter day for by the same post I had a letter from home saying that the doctor says Hilda is making a wonderful recovery, & he quite hopes she will be able to write letters a little - & so on. When I left home on Oct. 14, I did not think she would be alive by the time I am due home. I am so happy about that, so warmed & cheered by your letter & all this great kindness. Please tell everyone how grateful. I dare say Vivian will write [later] on. I'd love to be home when she gets the letter with the news! I banked the cheque dashing up between lectures to the Bank here. One does not like to keep cheques or paper money in large amounts in the house, lest incendiary bombs destroy them!

Last night we had a long raid - first a short one in the evening, then one beginning about 1 a.m. & lasting till 6:40 a.m. I do not think much damage was done in the town itself. I am sad to learn how badly Coventry has suffered. I feel sad at times at the destruction of some buildings, the Middle Temple Hall, & the Old City Halls, & Churches. It is a different sadness from one's sorrow at the loss of life & awful maimings. It is odd this feeling, that perhaps one will be dead before another day has passed. Of course that is always possible – but now it seems very near, & one has a sort of surprise when one wakes up in the morning to find one is alive. I have been deeply depressed by this awful destruction & slaughter, here & on the continent. I still am convinced that we have to fight; that Hitler and Mussolini will be defeated, but that does not make one unsad.

In the more formal letter, she wrote to the committee, she described some of the children they were helping:

My friends, Hilda Gavin and Vivian Shepherd, and myself are greatly cheered by your appreciation of what we are trying to do for refugees. I wish that you could see the children. I will send you some details and perhaps some snapshots, if that is possible, later. Perhaps the best think I can do at the moment is to tell you about one or two of the children. Heinz Rosenbaum has been with us just over a year. He was six when he came. His father, a Jew, had been in Dachau concentration camp; he managed to get out and come to this country, a complete nervous wreck. We do not know what had been happening to Heinz, but he was terrified. He would scarcely speak, never smiled, and shrank away in terror from all strangers. Now he has begun to talk to grown-ups, and sometimes to smile, and he looks likely to become a cheerful child in time. I notice the difference when I go home in the vacations. We have two children whom we have wholly adopted, one from Prague, and one from Vienna, who have been with us for two years. They came to us nearly two years ago; she is Jewish; her father was killed. She is in a very nervous state but she is happily settled at the school. I do not know the exact numbers, as I am away from home, but I believe we have at least 50 foreign children, most of whom have scarcely any money, and some have no parents.

Of course we have English children as well, and are indeed supposed to be an English School. Some of them pay no fees at all - since the war, because their parents have lost their means of livelihood in one way or another. They are of varied kinds, some of these parents are clergy with minute incomes, some business men, some serving in the forces. Some children have no fathers and their mothers have tiny incomes.

We also have three "mothers" now living in our village - and having their expenses from us - because they have had their homes in London bombed.

I hope this hastily written account has not been boring. I wanted to suggest to you how we use our money. This large gift will make it more possible for me to spare more of my salary for refugee scholars who are interned in this country.

Paul Oppenheim, Peter Hempel, Olaf Helmer and Meyer Schapiro were among the people who contributed to the fund. Later, she wrote to Ernest (Nagel 2020, Box 5, Folder 17):

Springfield. Sidgwick Avenue. Cambridge. Nov. 28. 1940

I do not know how to thank you both for your generous and kind gift to help our refugees. You are, indeed, too generous - and not only in money but in time. I realize how much time you & Ann must have spent in organizing this fund. You are very good to me. I wish you could have known how greatly we were cheered, as well as helped, by this gift. Maybe you can guess. Vivian & I have been much worried by the fear of having to close the school. We shall (not) need to do so; we can go on some time yet. We constantly get new children - & last week, two who will really need the fees! We get a good many whose parents have lost so much -- cannot be - & many are in debt to us for 2 years or more - and of course the number grows. But - lately we have had some fees - more than we'd counted - & so we were able to take more who were destitute. I do wish I could tell you how grateful we are to you...

At present we are having an 'alert'; I have heard the planes a lot - German planes - but so far our windows have not rattled - nor have I heard bombs falling - Sometimes they fall here, sometimes it is fairly quiet here during a raid. This awful destruction is indescribably sad - so much of Europe is being laid in ruins - here & on the continent...

But I do not really despair - I believe that it is possible to rouse people to believe that reason counts. It's a struggle. We need to struggle to keep alive (in many cases it has first to be kindled) a passionate awareness of the importance of intellectual freedom. Of course many distinguished professors in this county (as in others) do not really believe in it; they are ready to condemn men to be always irrational & to scorn international ideals. But - we can & must protest. I try to make clear to my students that nothing else matters so much as intellectual freedom. I am cheered sometimes when I reflect that even now a great deal that is profoundly critical of the government and of the utterances of certain public men is published here constantly. We need not be cynical -- we have already lost the struggle for freedom. Mr. Joseph Kennedy is wrong, his statement that democracy is dead in England is a travesty. I trust you to realize that & will not say more about it now....

I have not once been to the Moral Sciences Club....now that Moore is not here I don't mind so much not being able to go to the M. Sciences Club. His successor [Wittgenstein] is now so dominant I hear that barely anyone else speaks, & never any but a lecturer! It is a sad decline. Moore certainly often led the discussion - in the old days - because he was the clearest and best one there. His successor is, I fear, The Leader - & does not really like anyone to differ from him.

It is interesting to note Susan's strong belief in intellectual freedom, her love of her country, efforts to help refugees, and struggle to keep up her intellectual efforts despite the bombings and wartime deprivations. During this period, she managed to write four books: *Philosophy and the Physicists* (1937), *Thinking to Some Purpose* (1939), *Ideals and Illusions* (1941) and *A Modern Elementary Logic* (1943). In the following letter she described her writing method. She wrote both letters and books very quickly (Nagel 2020, Box 5, Folder 18).

January 18, 1941

I had a good time at home - for the Xmas vacation. It was so good to be home - & to have Hilda alive - & so I was very thankful. She is very much of an invalid, but as I had never expected her to recover at all, it is splendid she is up for a part of every day. I worked very hard during the vac, and finished my book. I think you know I had signed a contract to write a book called 'Ideals & Illusions' - for Watts & Co. -

It was supposed to be an 'expansion of a lecture into a book of 60000 words'. Forvan(?) Whyte (a director) heard a lecture I gave in Oct. 1939 at the Conway Hall - & he asked me to do it. I did try to do it in the summer, but could not get down to it & then Hilda's serious illness & the school worries absorbed my time & I wrote only 2 chapters. During the 3 weeks of my vacation I wrote the remaining 7 chapters - far too fast, & indeed I only finished it at 11 p.m. on Sunday last, & I was leaving home at 6:30 a.m. Monday. So, you see, it was very hurried & I did not read it through - & if ever it does get published it won't be worth reading. I was stupid, I think, to keep my word to send it in by Jan. 12 - for I was really too tired when I wrote the last two chapters. I had written 4 chapters in the one week. But it may not be published & that will be a gain. However, I quite liked writing it until I got too tired from writing all day. It did me good to have to think about these things...

We have had some raids since I came back, & a good deal of damage, but not to the Colleges, by incendiaries one night this week...

My brother's hospital was again bombed last week & 8 nurses, 3 sisters & 2 men staff were killed, but 150 patients were evacuated without any casualties. My brother did a good deal of --- work & giving morphine to those burned partly by the debris. Three of those killed were special friends of his & my sister-in-law, & they will be sad. But we must not expect to be spared. However, much one does realize that - we must not expect to be spared the loss of our OWN loved ones – one can't NOT be anxious about them. Miriam was on her way home the last time a large S.W. port was badly raided & I felt very anxious about her. She could not get back that night but she managed to stop en route at an hotel. I had persuaded her to go to friends for five days - just before I left home; she was due to return the day I left home, as we do not like to leave Hilda without one of us. It was 4 days before I had news of her - as the telephone down there was not available at first - & letters were slow. She was quite unharmed, and actually in the raid - some miles from it...

If peace comes to me, I'd love to go to you for a while see & hear you all talk. It would be heaven! It might happen.

I am not at all doubtful that in the end Hitler will be defeated, but I do not think it will be for a long time yet, & perhaps too late for me. But I am glad to feel that he will be defeated, &

there is work I can do now - I am not helpless at all - tho' sometimes I am angry with people who don't see how much more than wishes & vague kindliness is needed to rebuild the world - but I am sure there is reason to hope: we shall emerge - & by 'we' here, I mean the sane and humane people all over the world. In a way there is more reason to hope than in the last war. But I think you may think that foolish. Still, I have been in both, & I do see a very great difference - people (of course there are exceptions) do seem to feel that a change must, will, ought to be made, and that though heroism is fine there is something else that is needed...

2.10 The War in Europe (1936–1940)

While the situation in England was worse than that in the U.S. life on the European continent was even worse. After the *Anschluss*, or political union of Austria with Germany on March 12, 1938, life became desperate for the Jews there. The political union was really the annexation of Austria by Germany and so German racial laws were applied to the country. This affected not only people like the philosophers of the Vienna Circle, but many ordinary people like Ernest's relatives in Vienna, Vrbove and Nove Mesto. German hegemony spread rapidly. Germany annexed Czechoslovakia after the Munich Pact of 1938. After Germany and Russia signed a non-aggression pact, Germany bombed Warsaw and invaded Poland from the west on September 1, 1938. Russia invaded from the east a few weeks later and the two countries partitioned Poland. This marked the beginning of World War II in Europe as England and France declared war on Germany.

All these events brought fresh waves of refugees to America and England. Vrbové, where Ernest's grandfather and many other relatives lived, became a ghetto for the Jewish population of the Piestany province of Slovakia. The ghetto was liquidated by the Slovak Nazis known as the Hlinka and the German SS and most of the Jewish people were deported to the death camp at Auschwitz. Most of Vrbové's Jewish population was gassed in Auschwitz's gas chambers.

Several of the Nagels' relatives tried to emigrate to either the U.S. or Palestine. They wrote to Ernest and his mother for help with getting visas and, especially for the all-important affidavits which would show that they would have adequate means of support after coming to the U.S. Rudi Blum wrote to Ernest three times. Finally he proposed using the ruse of visiting the New York World's Fair in order to get a visa [translated from German]:

Vrbove, Dec. 2, 1938

Dear Ernö,

My pleas continue uninterruptedly and since I am convinced that you understand our situation I take it that you won't be upset about my plea today. I have just found out that one can easily get an entry visa based on a written invitation to visit the World Exhibit. However, you have to show the local U.S. Consul a return ticket. I can get this. I ask you therefore to send me the following letter in English language. This is all without any kind of obligation on your part and is a pure formality. It would be a small service on your part to me and I hope that you won't deny it in view of the current bad situation. Maybe the American Consul will ask you about the properness of this invitation and in this case I ask you to affirm it. I imagine that you would have to send me a letter like this:

Based on our last year's and continuing correspondence, I would be very pleased if it would be possible for you to visit me and my family for the New York World Exhibit. The Fair opening date is May 1, 1939. I think it would be a very happy reunion. I particularly regret that I couldn't see you during my last visit. I'm also convinced that you could combine the agreeable with the useful. If you bring Slovakian handcrafts, I believe that you could succeed in setting up new export possibilities. The Slovakian handcrafts are, as far as I know, very sought after and popular. I would love to see you visit us and please don't worry about food and housing. You will be my guest. How did it go in Egypt? I'm sure you have seen beautiful sights. The world famous Egyptian building must be wonderful. I look forward to seeing you again soon. I am sending my greetings to you and your loved ones, Yours,

[Ernest]

My uncle writes to me that a very well off cousin of ours lives there and that she can maybe send me a visa. I expect communications daily. Dear Ernö, please send me this letter in English language and I thank you already ahead of time for your effort and agreeableness. How is your dear family and what do you do? With us, there is much to be desired. We all try to get away from here as quickly as possible which you will well understand. I hope to receive news from you very soon. The best greetings and wishes from your

Rudi Blum

Isidore Weisz, the uncle who had been so hospitable to Ernest and his mother in 1934 when they were in Vienna, died of a heart attack just before his wife Anna (who was not Jewish) and their daughter Liesl escaped. They went first to Abbazia which is near Fiume, then to the United States with the help of Ernest and his sister Irene Finkel. Anna wrote to them (translated from German, Nagel 2020, Box 5, Folder 38):

Abbazia, 5/4/39

Dear Frida & children

You have already heard about our great misfortune from Ernö and now since yesterday I am with Leisl in Abbazia. I still cannot believe that we have lost the best father and companion. I cannot stop weeping. Wherever I go I see him in front of me – he who was so happy to be alive and who managed to overcome the worst things. He must have died a day before we left. If there is any comfort for me, it is that only people whom God loved die in this way – without pain and without any idea that they will die. The poor fellow had a heart attack the night before around 3:30 a.m. He woke me because of the excruciating pain and said I should give him Termopfen. Afterwards he was a little better but then after another hour the pain returned. I was with him until 7 a.m. and sat by his bed until he fell asleep again. At 9 a.m. He woke up and felt pretty good. He had an uncomfortable feeling but no pain. Frl. Fischer called her doctor for him. We brought him to the table and [the doctor] gave him an injection and told him to stay in bed for 2 to 3 days and assured us that it would all get better. When he got home he was happy to note how the injection had improved him. He wanted another one before they left. Towards evening, I told him "the doctor told you to go to bed and you have time for that." So he did lie down around 6 p.m. And in the evening Adi & Helene came to visit him and my sister Ida and also Gretl's father as well as Frl. Fischer with her sister and brother in law (?). He talked wonderfully with us. Mr. Fischer told him to get up and not let the doctor make him totally crazy since doctors don't understand anything anyway. He laughed about it and said that nothing bad could happen while you are resting. Around 9:30 we had our evening meal. I gave him sausage, butter and cheese in bed and he ate with great appetite. Suddenly, I was just about to take the cup from him when he gave a loud sigh. Frl. Fischer, who had recommended the doctor, had her back to him. When she turned around she said "Herr Weisz . . ." but everyone was looking at him and he was in the arms of my sister Ida who caught him at once and . . . . (?)

In 5 minutes a doctor was there. Unfortunately he was no more. The first question the doctor asked was "Was he Jewish?" and my sister said "yes" and the doctor said "that's how they all die." Death is not tragic...

The Almighty will care for us and the child whom he loved so much. She (Leisl) doesn't speak a word to me. Now that he is dead, I am not sure, for us two, what our citizenship (identity) is and how we can travel....

We are now at the seashore and she (Leisl) has not spoken for a long time. She doesn't speak for hours, sitting with me on a bench by the sea.

If I remember, dear Frida, you, too, lost your husband like a lightning bolt out of the sky. So I am asking you, dear Frida, write to me at once. I long to have letters and am hungry for news.

Your Anna

The Nagels were able to bring Anna and Liesl to New York. Anna worked in a hat shop on Madison Avenue owned by Greta (Margaret Rittner) and Gisi (Geraldine Gollen), the daughters of one of Anna's sisters. Liesl later married Howard Marr and had a daughter, Kathy. Liesl kept in touch with Ernest and his sister Irene.

He even received letters from complete strangers with the surname Nagel.

[in English, typed]

Vienna,
October 14th 1938
Ernest Nagel Esq.
610 W 142nd St.
New York City
Dear Mr. Nagel,

I owe your address to the information office of the various Aid committees here helping us as yet principally by advising to apply to our namesakes.

As per my signature below, my name is the same as yours, only I don't know whether I am so lucky as to find in your person a relative and rescuer. Let me hope that will be! I descend of a family whose parents and grandparents were originally of Slovakia (formerly Hungary), Pressburg and Neutra district. I am 46 years of age, married and have 3 children – below I beg to state the exact personal dates. - I am an expert in the shoe branch, both as manufacturer and as dealer, and have been active in Germany in leading positions. To-day I have become an emigrant that is a man who in spite of his knowledge, his capacity and willingness to work cannot help existing here.

I don't desire but to work, to produce, to do my best, demanding for it only my and my family's living, but here no means to do so only abroad, in U.S.A. Wherever in your country there is the possibility – provided that I might find a philanthropic rescuer.

Let you kindly induce to save me from perishing by the assurance that your affidavit is to serve only to the purpose to get me over there while you may feel sure that in no event whatsoever I shall be a burden to you.

When reading these lines you probably will say: "what about, I don't know this man at all!" So recollect, please, the great many moments when you render kind actions in daily life to people you do not know.

Be convinced that one day I shall be able to prove you that I am not unworthy of your kindness.

For us, German Jews, there is no rescue but emigration overseas, and this rescue is in inseparable connection with the word "Affidavit" a word which, in my case, would not engage you to any charge. Should you issue an affidavit to help me, for the fare and other expenses I should provide by myself. Thus, I implore you once more to lend me your helping hand.

I request you kindly to let me have your early news telling me whether I may count on your assistance. Anticipating my best thanks for your troubles and attention, I am, dear Mr. Nagel,

Very sincerely yourselves
Wilhelm Nagel
Vienna, II., Komödiengasse 1

Personal dates:
Wilhelm Nagel, born July 11th 1891 at Tyrnau, Czechoslovakia
Louise Nagel, "Nov. 2nd 1891 at Verlky Surany"
Edith Nagel, "Mar 24th 1920 in Vienna"
Götz Nagel, "Jan 12th 1924"
Josef Nagel, "Nov 4th 1926 in Berlin"

P.S. I want to point out that my wife is a perfect cook for any appertaining establishment; my daughter Edith is a skilled dress-maker.

Ernest spent much time helping refugees during those years. When they succeeded, the recipients were very grateful (Nagel 2020, Box 4, Folder 68).

From Dr. Ludwig Rittner
Wien, X. Gudrunstrasse 120
To: Frida Nagel
94, Hamilton Place
New York, NY
Wien, 26, April 1938

Very esteemed, dear gracious lady!
Dear Doctor!

It seems to be a sign of the times, that not only large but small events are rushed. I was just about to answer your letter of the 4th which arrived on the 22nd when your letter of the 15th arrived.

I had to read the letter several times in order to convince myself that it was true and not a dream. However, it was a truth as beautiful as a bold vision. I cannot describe what your lines mean to us spiritually and materially. Perhaps you will be able to appreciate it once we have met properly. In any case, you have given us new courage and hope to live, because I again see the opportunity to work and be a useful member of human society.

My love, I can only express my thanks to you in this way until we meet again when I can thank you personally and then pay off my debt to you til the end of my life. Only God can fully repay you for this humane act – the efforts of two people – a simple man cannot do it.

I notified Antschi and Turl immediately and like us they were both happy and thankful. Antschi will continue the business but she would like to continue to correspond with Ernest.

We are now waiting for the papers from the Consulate to be in order and when this happens we will soon begin dissolving our household. Until then, I want to learn a manual skill because, as time has taught us, spiritual work will never forsake us.

I can imagine, dear lady, that you now have a very active correspondence and I hope that this will soon end. I can assure you that your correspondence was not in vain but will be crowned by success. I shall do my best to prove that you have not wasted your efforts on an unworthy person.

I hope that you and your family are in good health and would be delighted if you would write back soon. You do not know how happy it makes us to have overseas friends.

I thank you heartily again for your goodness and efforts and remain with cordial greetings and hand kissing to you and your family.

Your thankful
Ludwig Rittner

Ina Carnap kept up a lively correspondence because she also was working to help refugee friends and often asked for Ernest's help. He was also part of the group of people who tried to find jobs for refugees like the physicist Philipp Frank (who succeeded Einstein as professor of physics at the German Charles-Ferdinand University of Prague), Felix Kaufmann (the privat-dozent whom Ernest had met in Vienna in 1934), Alfred Tarski and others (Nagel 2020, Box 1, Folder 35).

Chicago, Illinois April 15, 1938

Dear Edith, dear Ernest:

I have not written to you for ages. I just don't feel like writing letters. And I had to write so many in the last weeks, because appeals for help come in from Vienna faster than I can answer them. Good news on one point: Felix Kaufmann about whom Carnap had written to Coss (or more justly: to whom Ina had written and Carnap signed) is placed at the New School for Social Research. But there are so many more. Now I know it won't be much good if I ask you whether you know people who might be willing to do something, because these people you know will already have been approached from many sides. But as a good case worker I am following up every clue and one of them is called: Nagel. I do not mean philosophy jobs in the moment. I know, Ernest can't supply them. No I am meaning people who might be willing to sign an affidavit. Or I am trying hard to find an internship in a hospital for a promising internist or a job in an institute or a chance for a woman who is a goldsmith. By the way: this woman will arrive on a tourist visa in New York within the next weeks, which means arrival around the 28th. I gave her your address for the case that she might feel very lost and lonely. She will come to Chicago and I suppose stay with us. I do not know her personally but she has been warmly recommended to me by friends. I don't even apologize for giving your address, first because I know how good you are and second because now it is my business to approach people for help for immigrants. But it is a rotten business and not very encouraging. I suppose you have received letters of appeal also, especially since I remember Ernest to have relatives in Vienna. But things are not splendid here either. Erna Loewenberg has had a job for the last three weeks and I just began to breathe hopefully when she got notice and lost it again by today. Hempel and Helmer have not found a thing for the next year. So, on what basis does one help other people to come over? But on the other hand if one knows how they are persecuted on what basis can one deny to help them? . . .

ina

Carmel, November 26, 1938

Dear Edith and Ernest:

. . . We too are swamped with refugee letters and don't seem able to do anything for most of them. Zilsel has escaped to London by now and will wait there for his visa for the States. There are two different points about giving affidavits. In the beginning of the flood the Jewish Charities told me to encourage everybody to sign affidavits for prospective immigrants promising, they would take care for them, once they are in this country. They seemed eager to get in as many Jews as possible. If one could be sure about this status, i.e. be sure that the people coming would not be a financial burden upon you, there might be the way out to borrow money, depositing it in your name, and sign affidavits on the basis of this "property" for your relatives. But I am not sure how much one can depend upon these organizations. I do think, though, that it might be very wise on your part to make an appointment at the Jewish Charities and to talk over with them the cases of your relatives. The Chicago Committee told me it was absolutely wrong that I helped people to come over and then tried to work out plans for them as best as I could. They said, this was their duty and privilege and of course they can do it much more effectively than I ever could. Well, that's that. But I think, it might perhaps ease the financial burden on your shoulders if you send your relatives right away to these organizations and discuss their cases perhaps already before they come as well as the cases of your people whom you had to turn down. The Chicago Committee asked me to deposit with them all the pleas for affidavits which I cannot fill promising that they would try to find affiants. I have not heard yet that they have done something, but at least there is some hope. I know of another case in San Francisco, where the Jewish Charities have pledged themselves in advance that they will support a certain family for such and such an amount, if their relatives pledge to contribute so and so much (I don't know the percentage). So you certainly should try to find out before you assume the whole burden. Also: the Jewish Charities in Chicago pay Erna Löwenberg a certain amount each week since last Xmas (I believe $10 a week; if she earns something of which the Jews know, the amount earned will be deducted. It is not much but it is something she could live on); ... It is absolutely necessary that you do not even attempt to carry the full burden.

We don't know Dr. Wald. But Frank, who was here for a few days says that he is a very good man. But we know of many similar cases. Could we not perhaps found a College in order to be able to invite people in on a non-quota visa?

A very good friend of mine has arrived in New York and has found a position at the Mount Sinai Hospital. He is an excellent physician and whenever you should need such a man I should like you to go to him. His present address: Dr. Daniel Laszlo, 166 Second Avenue, c/o Jacobson. He has a wife and a little child. They do not find it easy to get acclimatised. I did not give them your address because too many people are taking your time already and because I am not so sure how you will like them. He is a very old friend of mine as a matter of fact, the first man who wanted to marry me. But I have hardly ever seen him since and I have a slight suspicion that he is only interested politically because he is a Jew. I am not sure about it though. He had to emigrate twice, first from Cologne, now from Vienna. ... Who writes the refugee letters in your family?

Whenever you should need help in the making out or notarizing of affidavits I know of a very nice man who will gladly help you: Rudolf Modley, 142 Lexington Avenue (Lexington 2-8149). He has done it for me several times. I do not know him personally but we have had a very intensive correspondence. (He and his friends have formed a group, called "AUSTRIAN AID" who specialize in help and legal advise for affidavits.)

A good friend of mine, Gusti Doderer, is to arrive on the Washington on December 8th; but this time I have found already somebody who sacrifices himself to fetch her from the boat. She will go to Chicago about two days later (and stay with us for the time being). May I give her your address just in case that she needs help?

Yours, ina

Several people who stayed in Europe too long had great difficulty leaving or (worse still) were unable to escape. One of these was Otto Neurath, who had been a founding member of the Vienna Circle. He had fled to the Hague during the Austrian civil war of 1934. Ever optimistic and enterprising, he kept himself busy with projects. Among these were the Unity of Science movement and the International Encyclopedia of Unified Science. He and Charles Morris of the University of Chicago were the principal organizers of this movement. He also promoted Isotype, as an international picture language. Although such graphs are now commonplace, it used pictures to graphically display statistics. For example, one could show the relative population of several countries by letting a row of figures of men represent the number of people in each country. The impressive letterhead he devised for the Unity of Science movement, shows that this movement included many members of the former Vienna Circle (Philipp Frank, Rudolf Carnap, Herbert Feigl and Richard von Mises) as well as Americans like Charles Morris, Ernest Nagel, William Malisoff and John Dewey. Other countries represented were England (Stebbing and Woodger), Poland (Łukasiewicz and Tarski), Italy (Federigo Enriques), The Netherlands (Gerrit Mannoury), Norway (Arne Naess) etc.

Neurath's wife Olga Hahn (the sister of mathematician Hans Hahn) died in June 1938 following an operation for kidney stones. Nevertheless, he visited the U.S. that fall to visit friends (including Ernest, Charles Morris and the Carnaps) and prepare for the Fifth Unity of Science Congress which was to be held in Cambridge, Massachusetts, September 5–10, 1939. The first Unity of Science Congress, which Susan Stebbing had gone to but Ernest was unable to attend, had been held in Paris in September, 1935.

Herbert Feigl described him thus:

> Carnap and Neurath also had a great deal in common in that they were somewhat utopian social reformers – Neurath quite actively, Carnap more "philosophically." Neurath, a man of great erudition, especially in history and social disciplines, was also a powerful propagandist of the Viennese positivist outlook. He was of powerful physical stature, extremely energetic, full of "enterprise," with great talents for organization. He was also a very witty man, using sarcastic dialectics most effectively in discussions and controversies...
>
> It was especially Neurath and Frank who envisioned and worked for a new era of enlightenment, propagating the Viennese form of positivism. In consonance with this idea, Neurath began the planning of the International Encyclopedia of Unified Science. (Feigl 1969/1981, 62)

As we saw in the last chapter, Otto Neurath was forced to leave The Netherlands when German troops began attacking The Hague on May 10, 1940. He felt quite welcome in England (Nagel 2020, Box 4, Folder 44):

> I made a luncheon speech and was invited to read papers etc. "As once in the nice month of May." We agreed to avoid contacts etc., but we have almost no day completely free. People who want to see us, lectures we have to attend, etc. It is a great pleasure to listen to Price and others when they speak about Logical Positivism, and all the things are mentioned we know so well. The New Statesman published a review of Russell's (not even pleasant) book by Joad in which Carnap, Hempel and Neurath are mentioned, and mentioned as people well known here.

Like many refugees from the continent, Neurath and Reidemeister were kept at the Isle of Man until the British authorities could decide whether or not they were enemy aliens. In this letter Neurath told Ernest about their quarantine on the Isle of Man. Once released, he and Marie Reidemeister rebuilt their institute and, later, married.

> March 28, 1941
>
> Dear Nagel,
>
> Many thanks for the very nice reference letter you wrote in my and Mary's behalf (I assume, you received in the meantime our wedding card). You know probably how many steps were necessary to secure our release, because we escaped in the last moment without particular permit or visa, very glad to be safed (sic.). We collected interesting experience in prisons and camps – the time reserved for these research work was rather long and sometimes dull and boring, but we used the opportunity to read many books, to prepare the next book and the monograph for the encyclopedia. The "intern the lot" had the result, that in our camp a whole academy was assembled, many people who were of the unity of science movement or at least interested in it. M. Strauss told us something about his logical analysis of quantum mechanics, Waismann read a paper on Mengenlehre, some Viennese discussed with us general scientific problems, there were economists too,

> physicists and biologists. The greatest disadvantage was that only two 24-lines letters weekle (sic) were permitted, plus a few additional special letters (mostly written to my solicitor, who was provided for us by Susan; she and Kaempffert were the leading powers who worked for our release, together with our good friends). After weeks of intellectual quarantine monthly meetings were permitted – a very strange institution; please imagine hundreds of married or engaged couples in a big dancing hall. I used my time to take care of my stature and after a period of increasing weight (I am assimilating carbohydrates very intensely) I reached a real normal British figure, slimm (sic) and tall. Now I am keeping my weight and feel very well. Mary taking care of her nurishment (sic) has red cheeks and used the camp period for recreation (no holidays during the last lustrum). She made transformation for the pictured encyclopedia, wrote a little play and succeeded in being an actress together with her house mates whom she ruled as their "captain". The whole thing similar to a boring college or cloister – you know certain mystery plays were allowed. After some efforts the British Academy found out that we were scientists of sufficient reputation (the famous White Paper of the government had created certain categories of people who could be released) and afterwards we were "tribunalized" too, (I should maintain with real courtesy) and a few weeks afterwards released. In the meantime Professors not only from Oxford, but from Cambridge too tried to get me free. I was invited by Prof. Radcliffe-Brown to deliver a course of lectures on LOGICAL EMPIRICISM AND THE SOCIAL SCIENCES. That was my activity during the last term, in the next term I shall have a Discussion Class. That will give me an opportunity to discuss some subject matters I am treating in my monograph.
>
> Now we are rebuilding our institute, and that needs some breathing space. We are very thankful that you and other friends were kind enough to help us to bridge the gap. I hope we will be in a position in the near future to repay the sum. Many thanks.
>
> We are very glad to stay here, because we like the whole atmosphere, we feel no discrimination at all and have much more contacts here that we had at the Hague. What a pleasure to listen to lectures in which the names of our people are quoted, and e.g. Price discusses the probability opinions of Mr. Nagel, Mises, etc. We are here well known, as I learned from a review (by Joad) of Russell's book, 1st March New Statesman, discussions in the Aristotelian Society etc. My Protocol Statements doctrine and proposals are often discussed in detail. I am disappointed that Russell seems to misunderstand my direction, but that he gives 10 pages to the protocol statement is significant. I hope Carnap and Hempel will tell me something about her (sic) opinions on Russell – what [do] you think about [it]?
>
> We are reading in all the first class libraries here where we can get the newest books on social sciences, anthropology, history, philosophy, biology etc.
>
> I delivered a lunch speech to some academic people and shall read papers in certain circles, I have to write etc, we have to prepare charts for the US. Life is going on very well, in spite of war and danger.

Neurath thoroughly enjoyed being in England. In addition to rebuilding his Isotype Institute in Oxford, he continued his activities with the Unity of Science movement, organizing Congresses with Charles Morris and others and soliciting articles for his Encyclopedia of the Unified Sciences. He was also very playful. He called Ernest "the Pineapple" and often signed his letters with a picture, which depicts him as an elephant spraying a pineapple (Ernest).

The invasion of Poland on September 1, 1939 was especially devastating. Many of the Poles Ernest had met in 1934 were forced to flee. Some were lucky enough to be out of the country at the time.

Alfred Tarski was one of the lucky ones. He left for New York on August 11, 1939, to attend the Unity of Science Congress which was to be held at Harvard University on September 5–10. By coincidence, the two Ulam brothers, Stanislaw

Fig. 2.27 Alexander Wundheiler and Alfred Tarski. (Private Collection of Yvonne Nagel)

("Stan") and Adam were on the same ship. Stan, who was a Polish mathematician from Lvov, had a lectureship at Harvard University and Adam was planning to enter Brown University that fall. Tarski, was only planning a short stay and so did not pack winter clothes. On the day after he arrived, Peter Hempel and Olaf Helmer, along with their wives, took him on a tour of New York and its environs (Feferman and Feferman 2004, 127). Tarski, the Ulams and the other Poles who had come to Cambridge, including Stefan Bergman and Alexander Wundheiler were dismayed to learn that Germany and the Soviet Union had signed a non-aggression pact. The Nazi invasion of Poland a week later, on September 1, left them stranded in the U.S. Ernest was one of the people who helped them get settled in the U.S., especially Tarski and Wundheiler (Fig. 2.27). As Ernest said in the 1983 Oral Interview:

> I joined the department in '31, and by that time Goedel's paper was out, and not only didn't anybody in the philosophy department know anything about that, but there was nobody in the math department. This was shown very clearly. Tarski had come over from Poland, and for a number of years he was footloose - he couldn't get a job - and so he went the rounds lecturing. And it was suggested that he ought to give a lecture or two to the department of mathematics. They didn't know who Tarski was, but he eventually was invited to give a talk, and what he had to say was of no interest to any of them. (Neuber and Tuboly 2022, 282)

Ernest did his best to help Tarski. Among other things, he invited him to give a guest lecture in his undergraduate philosophy class. Leon Henkin, who would later become a prominent logician at Berkeley, was a student in that class. He reported that:

> I had no idea who he was before Professor Nagel told us that a famous logician from Poland was coming to talk to us. He spoke on decision methods and on Gödel's incompleteness proofs; although he had a very heavy Polish accent, his English was understandable and I was terribly impressed. When he finished and it was time for questions, I raised my hand. I had a "real" question to ask but it was also clear to me that my purpose in asking it was not merely to gain information; I wanted to establish contact with him; I wanted him to notice me; I wanted to be more than just a passive member of the audience. (Feferman and Feferman 2004, 132)

For the next few years, Tarski split his time between apartments in Cambridge, Massachusetts and Manhattan. He had a visiting professorship at CCNY for a semester, a research appointment at Harvard and gave lectures at several universities including Yale, Brown, Dartmouth, Princeton, Chicago and Michigan.

Ernest helped several Poles including Alexander Wundheiler, Janina Hosiasson-Lindenbaum (whom he had met at Prague and Warsaw) as well as Alfred Tarski (Nagel 2020, Box 6, Folder 2):

> December 29, 1939
>
> Dear Nagel:
>
> May I ask you to do now what you have promised me so kindly, viz. to write out a testimonial saying that for the last twelve years until the outbreak of war I was teaching mathematics and mechanics at the University of Warsaw. It is also true that I taught for the last two years at a Teachers' Seminary in Warsaw; this is also considered here as a higher institution of learning.
>
> The second thing I should like to ask you to say is that I am really going to pursue the same vocation here and that I shall not take to any other job.
>
> I think that you will agree that such a certificate, if it has to be of any use at all, should be given as official a form as possible. I venture to suggest either a regular affidavit, or a testimonial on official stationery of your department with your signature testified by the department.
>
> Now, I realize perfectly how much you are now bothered and pestered by the refugee gang. But you will agree that people fall into two categories: those who don't help anybody, and those who help everybody. It is obvious at the first look that you belong to the second category. So you evidently have to submit and to take all the monkeying with other people business as an immanent part of your life. ...
>
> Yours most affectionately,
>
> A Wundheiler

Janina Hosiasson-Lindenbaum was not so fortunate. She had planned to leave Poland, to attend the Unity of Science Congress, a week after Tarski did, but by then the border was closed and the ship was not allowed to leave. She fled to Vilnius, Lithuania (Nagel 2020, Box 4, Folder 6):

> November 14, 1939
>
> Dear Mr. Nagel,
>
> Please forward the enclosed letter to Dr. Wundheiler. I describe to him our situation and ask him to see some people, amongst others you. I thank you and apologize in advance for the trouble it may make to you.
>
> We had fled from Warsaw as we stood; the way was full of danger and emotions. My husband is now under Soviet Occupation and will remain there; as to myself, I should like to

go to a western country as I wish to do freely my scientific work. U.S.A. would be the best place as there are people who are working on matters similar to those I am interested in, amongst others you and Hempel. I am very glad indeed and find it a chance for me that you got interested in probability.

I take this opportunity to thank you for your letter and your book on Probability. I did not thank at once as I wanted to read the book first. I am very fond of your work and should like to write a review of it. In a letter to Dr. Wundheiler before the Congress I asked him to ask you which journal will be willing, as you suppose, to print my review. The trouble now is that I have not the book and my notes with me, as I left all in Warsaw. The Encyclopedia volumes are not to be found here. I wonder whether I shall find them in Riga, where I intend to go next to my family, if I shall get the permit of the Latvian authorities. I hope I shall be able to live here or in Latvia until the spring, but it is not impossible that I shall have to return to the Soviet Occupation. I shall then wait there for indirect news from America, although I realize that to receive a permission to go out of there a strong invitation is needed. There remains of course the question of the permission to enter U.S.A.

With my best wishes
Yours sincerely
Janina Lindenbaumowa

Both Tarski and Wundheiler asked Ernest to write letters to secure a visa (and position) for Mrs. Lindenbaum. Later, Tarski wrote to Ernest about a committee chaired by Alvin Johnson of the New School for Social Research which was set up to by the Rockefeller Foundation to help refugee scholars (Nagel 2020, Box 5, Folder 27):

New York
September 2, 1940
Dear Ernest,

.... as you have perhaps already heard, there is now a possibility to bring many European scholars to this country ... there is a special committee, headed by A[lvin] Johnson, which is now in a position to bring about 100–150 scientists from Europe here and to secure them a salary 1500–2400 $ yearly (depending on their family situation) for two years; the money will be given by the Rockefeller foundation. The candidates should be good scholars, not too old, and should have had an academic position in Europe. On [Felix] Kaufmann's suggestion, this committee asked me to call the names of some Polish scholars who deserve to be brought here ... So far I have made formal proposals concerning three persons only : [Tadeusz] Kotarbiński, Mrs. [Maria] Kokoszyńska and Mrs. [Janina] Hossiasson (although in the latter case it is somewhat doubtful whether she may be considered a person with an academic position[12]) ...

In all three cases I named you and Carnap as those American scholars who could give authoritative references concerning my candidates (moreover, I named [Oscar] Lange in Kotarbiński's case, [Alonzo] Church in the case of Mrs. Kokoszyńska and [Bertrand] Russell in the case of Mrs. Hossiasson); ...

Always yours,
Alfred

Janina Hosiasson-Lindenbaum continued to work on philosophy while she was in Vilnius. She sent a manuscript to Ernest for publication in the *Journal of Symbolic Logic*. (Ernest and Alonzo Church were co-editors of this journal). Peter Hempel

[12]Janina Hosiasson did not have a University position. She worked as a high school teacher and translator.

reviewed it, and his wife Eva retyped it in July, 1940. Later Ernest asked about sending proofs to the author. Peter recommended sending them to Vilnius, but with a note saying that publication would not be delayed if they did not hear from the author.

Tarski wrote often to Ernest (and many others) about the desperate situation of his wife and two children whom he was trying to bring to the U.S. He did not succeed until 1946, after the war was over. Here are some examples (Nagel 2020, Box 5, Folders 27, 30, 31):

September 2, 1940

Two days ago I received several cards from Warsaw. It is clearly seen from my wife's cards that their situation got considerably worse in the last months; while in her previous letters she asked me not to send them anything, in the last cards she wrote that she would be happy if I could send them some food, especially fats and oils. She still hoped that she would be able to come here before winter (the cards were written in the middle of July). Unfortunately, I am scarcely able to share her hope. It is true that some people (for instance Mises' brother) succeeded in these last few weeks in coming here from Switzerland; but Poles are in a much worse situation.

November 22,1943

There is not much new as regards my personal problems. I wonder whether I mentioned to you that I got a few months ago a letter from my wife (mailed at the beginning of March). There was not much new in this letter; but I found out finally that Kokoszyńska is alive; she lives in Lwow with her mother and tries to write something. She was in Warsaw in February and stayed a few days with my wife. A few weeks ago I got again two letters from my wife - one from December 1942, the second dated March 30. These letters are not encouraging. My wife does not conceal any longer that their life is very hard; my help does not reach them. She says that the children are well. In her December letter she informs me that my father died in September 1942; and she seems to prepare me to a similar news about my mother. In her later letters she doesn't mention my mother at all.

January 27, 1944

I got a letter from my sister-in-law in Warsaw (she lives together with my wife). A letter without date, written presumably in July; the gloomiest letter I ever got from them - I don't understand why the Germans passed it. It begins as all other letters: we are well. But the next sentence refers directly to my wife: she looks worse than we all, grew terribly lean is completely exhausted, can hardly do any work. Then a long story about my little daughter. Of course at the beginning the assurance that the child feels very well, but the essence of the story is this: the child is often hungry and has to use all possible tricks to get from adults . . . a piece of bread (literally). then another bad news: their whole terribly modest budget seems to rest to a large extent on the salary of my second sister-in-law, who became ill a year ago, was to be operated and to loose her job. I know too well the author of the letter - she is very quiet, never exaggerates - the situation was certainly much worse than it looked in her description.

Kurt Grelling also remained in Europe. Like Peter Hempel, he had worked for Paul Oppenheim in Brussels in 1937. He was Jewish, but his wife was not. He received a doctorate in mathematics in 1910, advised by David Hilbert. He was a member of the Berlin Circle, working with Hans Reichenbach to plan meetings. He collaborated with Kurt Gödel. He was arrested on May 10, 1940, the first day the Germans invaded Belgium. He was interned in southern France under the Vichy regime. Although Oppenheim and Hempel secured a position for him at the New School for

Social Research, they were unable to get him out of Europe. He was sent to Auschwitz with his wife and perished there in September 1942.

Leon Chwistek, the Polish philosopher and polymath in Lwow who had so impressed Ernest in 1934, fled to Moscow after September, 1939. He was a Communist and thought he would be safest there. Oscar Lange was a Polish economist and diplomat who had immigrated to the U.S. in 1937 and become a professor at the University of Chicago in 1938. Lange served as a go-between between Joseph Stalin and Franklin Delano Roosevelt during their discussions about post-war Poland. Lange met Chwistek in Moscow in 1944 and wrote to Ernest about him (Nagel 2020, Box 3, Folder 67).

The University of Chicago
Department of Economics
August 8, 1944
Dear Nagel:

I enclose a number of reprints which Chwistek gave me, with the request to give copies to you, Mr. Church, and Miss Brodie. I should appreciate it if you would send the copies to them, since I don't have their addresses.

I found Chwistek in pretty good condition. He has aged very much and has become somewhat senile in his private life. But in matters of science, his mind works excellently. He is very well taken care of, lives in a villa some fifteen miles out of Moscow, coming to town from time to time. Professor Kolmagoroff has done everything to facilitate his work, and he reads occasional papers before the Mathematical Society in Moscow. He is writing a book on the theory of sets.

He literally embraced and kissed me when I told him that his book is translated and supposed to be published by the Columbia University Press. He wanted details about date of publication, etc., but I didn't know much. He asked me to convey to you his expression of deepest gratitude for your help in getting the book translated and published.

If there is anything concrete about the date of publication of the book, I would appreciate it if you would let me know.

With best regards to Mrs. Nagel and yourself.
Cordially yours,
Oscar Lange

August 15, 1944
Dear Nagel:
Thanks for your letter.

I forgot to write you that according to the available news, Mrs. Chwistek and her daughter seem to have been killed by the Germans. Professor Steinhaus has been shot, quite certainly.

Cordially yours,
Oscar Lange

2.11 American Academics during WWII

The Second World War also had a profound effect on American philosophers and academics in general. The U.S. entered the war after Pearl Harbor was bombed on December 7, 1941. Soon after, many young American men either enlisted or were drafted. Ernest corresponded with several young philosophers who were in the armed forces, both to keep their spirits up and to maintain their interest in philosophy. He kept several of their letters.

New Guinea

July 31, 1944

Dear Professor Nagel,

This must be in the nature of a fan letter, for you won't have the slightest conception of who I am either before you read what I have to say or after you will have accomplished the sordid task.... My most urgent reason for writing is that I am somewhat bored, so that a letter from a Columbia Professor of Philosophy might give some much needed stimulation. ... I sat at your feet during a 6 month course in Metaphysics at City College. I trust that you will not set me any problems involving time, space or reality, for the reason that you will thereby learn how poor a student I was or, perhaps, how poor a teacher you were. Now in retrospect and out of distance of your possible vengeance I can realize that there was only a little meeting of the minds. ...

My biography, if you must have it, is not too sensational. I went from college to law school and into practice. At the same time I studied music, singing and cantorship, and ended on a synagogue pulpit. With two honourable professors and a good uncle, I managed to survive so that I would be available to the Army. ...

The Army read and heard this entire account of a musical, scholarly and juriatric background, and immediately assigned me to a tank outfit. The good tank outfit heard the story anew and put me on K.P. [kitchen patrol] There were guard duty, digging details, and work on the garbage truck to minimize the monotony of housework.

In May we arrived in New Guinea and at long last I got a job, several jobs in fact. By the flourish of an adjutant's quill, I forthwith became a country editor, an impresario of musical extravaganzas, dispenser of literature and culture, and of all things, athletic director. Presently we are to open a recreation hall wherein I shall be, librarian, master of ceremonies, and janitor. Perhaps the bouncer as well. Lest leisure be permitted to lower my efficiency I serve a few hours as draft mule, stevedore, and human freight elevator. I might add that there are two religious services per week at which I practice vocalizations, and a patient audience accepts the torment as penance for their sins.

I doubt that I deserve an answer any more than you deserve this letter. But if you could afford a few lines, please let me know about Morris Cohen. I picked up a rumor somewhere that he had died and I should like you to verify it. Also, if you have some articles or printed essays of yours lying around I should appreciate your sending them on to me. Perhaps with the accumulated experiences of years, I may be able to understand something of what you're talking about.

Affectionately yours,

Irving Weinberger 32705334

Service Co. 44 TK. Br.

APO 322-1, San Francisco

Mason Gross (born in 1911 in Hartford, Connecticut, Fig. 2.28) was a young American Philosopher who had obtained his PhD at Yale in 1938 under Alfred North Whitehead. He then taught at Columbia University from 1938 to 1942 where he was a colleague of Ernest. The Nagels had invited him to visit them at their summer house in Vermont several times. Since his family lived nearby, he had taken advantage of their hospitality. In 1942 he served in the Army Intelligence Corps and was assigned to a bomber group based in Italy. Here is a sample of letters he wrote to them. The first was written before he joined the army (Nagel 2020, Box 2, Folder 34).

Fig. 2.28 Mason Gross. (Private Collection of Yvonne Nagel)

552 West 114 Street
New York City
August 7, 1940
Dear Ernest -

Many thanks for your letter, and for writing Hutten. Incidentally I am told that Somerville has refused the Vassar job and that Jo Myer of Harvard has been appointed. I have also heard that Gladys Corliss has resigned her job at Vassar. Your news about Miss Stebbing and Neurath was grand. I couldn't believe that the irrepressible Neurath would be defeated, and I'd like to see Hitler take on Miss Stebbing in personal combat!

I am still uncertain about my plans for the rest of the summer. It's very nice of you to repeat your invitation, and if I do go the rounds again this year, I will certainly save a place on my schedule for an effortless bath in your lightning-proof house.

I flew up to Cambridge last weekend to stay with the Goheens…

I had two good visits with Whitehead. He and Mrs. W. drove out to Lexington with Bill Frankena to pay a call on the Goheens Friday evening, and we had lots of fun. They were both bubbling with good spirits. Incidentally they are getting no money at all from England now. Saturday John and I repaid the call and had a good two hour chat with him, really pressing him hard on certain points. I found it very stimulating and good….

At present I welcome you gladly to any of the Whitehead lectures next autumn, but I know I shall lose courage when the time comes!

My very best to you both,
Yours, Mason

After joining the Army, he was sent to Houston to teach Nagel (2020, Box 2, Folder 35).

September 27, 1942

Dear Ernest -

I've owed you a letter for a long time, in answer to your most welcome and very newsy letter from Wardsboro. Julia has probably told you all my news - namely, that I am teaching very elementary physics to future bombardiers and navigators here at Ellington Field. The story behind that is briefly this:

Colonel Snyder, in charge of the Pre-Flight School here, got the bright idea last Spring that he could tap the resources of the colleges for teachers who would thus relieve flying officers from ground school duties. Physicists and meteorologists were needed most of all, but of course professional meteorologists were seriously needed in the Air Corps anyhow and all the physicists who could be spared from the colleges were doing some form of war work already. But the Colonel still believed that it was better to have a teacher teaching physics even if he wasn't a physicist. So when about twenty of us arrived here from Miami Beach we were at once asked if we would try teaching physics. Was I on the spot! But it sounded good, so I confessed that I had never had a course in physics but if they were willing, so was I! They were, so I crammed and was given some essential instruction from Wednesday to Monday, and on Tuesday I was teaching two two-hour classes. That has been going on ever since. It's a 30 hour course, given in four weeks, and I am now on the second lap, & feeling quite confident. We cover velocity & acceleration, hydrostatic pressure, gas laws, Bernoulli, etc. and endeavour to do two things, one, to give them a physical background for meteorology, and two, to make them familiar with the physical principles lying behind the airplane instruments which they use. The physics is, of course, childishly simple, but it is giving me a chance to read up on aerodynamics and aeronautical design, which are quite fascinating. I also hope to spend some time on mathematics - so all in all this should be a very healthy experience for a cosmologist! The best remark yet was made by a Colonel Miller who interviewed me at Randall Field. When he learned that I was a teacher of philosophy he said: "Well, philosophy is closely akin to mathematics. I think we'll ask you to teach mathematics." ...

Mason

Later he joined the Army Intelligence Corps and was sent overseas in February 1944.

As the war was ending, he wrote this letter. Clearly his experiences abroad and in the army had changed his perspective on life (Nagel 2020, Box 2, Folder 35).

10 January, 1945

Dear Ernest,

Greetings and salutations to you and Edith with my warmest affection, for the year 1945. Your letter was here to greet me when I got back from a holiday leave, and I am deeply grateful to you, not only for writing such a long and newsy letter but also for giving me just the sort of answer to my questions that I wanted. It is very seldom that one encounters such basic and considerate honesty, and I appreciate it very much indeed.

My leave, which had been postponed several times, took me on a trip to the Middle East, the main visit being to Tel Aviv in Palestine, from which we made excursions to Bethlehem, Jerusalem and the surrounding country. It was simply fascinating, not the least interesting place being Tel Aviv itself, which is a flourishing and delightfully clean and modern city, set down in the midst of a multitude of squalid Arab towns. The hundreds of refugees from central Europe have given it a culture which is most refreshing, and for me it was a peculiar pleasure to be able to browse in decent bookstores again. Palestine itself is a very interesting and in many respects a most beautiful country. The curious combinations of color around Jerusalem, the yellow of the stone buildings against the brilliantly blue sky and the olive green of the vegetation, and the rolling country with the mountains of Transjordan in the background, all are quite unlike anything I have seen before and most pleasing. The tour of the shrines of Christianity are rather upsetting to one who was born to a belief in the tradition behind them -- for me the Church of the Nativity in Bethlehem and the Church of the Holy

Sepulcher in Jerusalem were altogether too reminiscent of the Sibyl's cave at Cumae and the Druid's temple at Stonehenge. The buildings themselves give discouragingly little clue as to which of the different creeds is most worthy of belief -- in fact there is an appalling sameness of pattern about them all. I can imagine your wondering just why this struck me as odd - but it did.

This war is showing a disappointing unwillingness to wind itself up and quit. Nevertheless, thoughts of post-war matters are constantly forcing themselves on my mind. The very close association that I had while on leave with four other officers ten years or more younger than I am, as well as regular letters from Julia describing the progress in intelligence and education of my two children, and the awareness that said family has to be supported whether Columbia casts me adrift or not -- all these things make me realize that thoughts of what I am to do after the war are not purely matters for idle interest but really rather pressing affairs. Anyhow, from an economic point of view, this much is clear: unless something better turns up I will certainly return to Columbia for the year following my discharge from the army, willy-nilly; I believe that the Columbia policy towards its faculty members entails that whatever the department feeling may be. Jimmy, in a letter which I received the same day as I got yours, says that he hopes that whatever I do I won't wander far away from teaching because he thinks I am a good teacher. He should know me better. To the end of my life I will be a teacher, or possibly it would be better to say a preacher. I simply have to talk, and to persuade people to think as I do - no, I hope that isn't quite right; let us say, to see things in the pattern in which I see them, because that doesn't exclude their being able to see them in some other pattern as well. And as long as the teaching platform seems to be the one best suited (see paragraph 2 above) I shall undoubtedly go right on teaching somewhere and somehow. My questions to you about something like the Rockefeller Foundation weren't meant to exclude teaching in some form -- I couldn't do that really. It meant rather a desire to teach within a framework where something definite might be accomplished toward a worthwhile end. Somehow I can't feel that quarrelling with Jack Randall over the trivial nonsense of Mack's thesis is a worthwhile occupation.

Was it in my letter to you that I mentioned my early fascination by Stephen Leacock's Dean who sat in his garden and casually read Theocritus in Greek because it seemed to him the most pleasant way of spending an afternoon, as if it were the most natural thing in the world to do? You may not quite have grasped the point of the illustration. Education has always meant to me two things, one the discovery of values and, two, the acquisition of the means of absorbing those values so that the constant enjoyment of them is a spontaneous and automatic matter. It takes training to appreciate a Bach fugue to the full. Once discovered the fugue remains a constant possession; the job is the discovery. But that is only half of the story. While one is learning to appreciate such values, the world around one is going on. People have to eat and there are wars. The ultimate value so far as action in the present is concerned is the establishment of a world order in which people will eat and there will be no wars. If I believed that there was a radical distinction between enjoying a Bach fugue and stopping wars, I certainly never would teach again. Instead I firmly believe that there is as much an aesthetic quality about the kind of human relationships within which wars would not be possible as there is a moral quality about learning to discover the beauty in a Bach fugue. Now I am sure that this must sound like the most awful nonsense to you, or possibly worse than that, it may sound to you like Whitehead. I am by no means sure that it is Whitehead, and I don't really care. I am sure that I have felt something like this since long before I met Whitehead, but it may explain my preference for all the so-called Platonic philosophers as against the Aristotelians. I don't say that a thing is beautiful because it has a moral effect on one; rather I insist that the ultimate justification for pursuing beauty is because of its moral value. I know your devotion to the cause of truth, and your general belief that knowledge and more knowledge, accurate, scientific and clear, contains the clue to the solution of human problems. But then I revert to my account of education. Knowledge is important because it reveals values and facilitates their discovery, while on the other hand it enables us to acquire and absorb those values. The merely learned man is as incomplete as the pure aesthete. Knowledge of the roots of human problems will not solve those problems unless there is also the burning desire to eradicate the evils in favor of a solution which one clearly perceives. The difficulty then lies in learning how to recognize the solution when one sees it, and the answer to that lies in the direction of learning to perceive and appreciate values, and of developing a more refined and subtle sense of the aesthetic qualities of human interrelationships. Knowledge without the desire for value leads to Nazi brutality and to the destruction of knowledge by itself. The desire for value without knowledge leads to American sentimentalism and superficiality. And here you have my Platonism: beauty and the good are in a sense one and the same thing, with the good somehow something more than just appreciation. I believe firmly in Eros, the desire for the good manifest as a love of things beautiful, and I also believe that the pursuit of knowledge is the chief manifestation of Eros. Of course this does not lead me to any doctrine of innate ideas or of absolute standards. Possibly there are such things, but a search for them is no assistance to our present problem.

All this may be a long way of saying that I believe that if a man can really and truly enjoy Shakespeare and Mozart, he will be a useful citizen in the cause of preserving world peace, provided that he is one and the same man always, the man who enjoys Mozart and the man who can tip the balance in favor of peace or war. He must realize the social problem, and his love for Mozart must be so intelligent and so permanently a part of him that that sense of value dominates his political actions unconsciously. And because this is my belief, I shall go on trying to point out to people what there is in Mozart that they may see it for themselves, and what there is in human relationships which makes a solution of the problem of war and peace quite possible.

Enough nonsense for one letter. I'll write again soon. My love and best wishes to you & Edith - As ever - Mason.

After the war, Mason Gross taught at Rutgers University, becoming President of Rutgers University from 1959 to 1971. From 1949 to 1950 he was a panelist on the television quiz show, *Think Fast.* He was also a judge for the show, *Two for the Money* from 1952 to 1955. He retired from Rutgers in 1971 and became director of the Harry Frank Guggenheim Foundation where he served until his death in 1977.

Like Mason Gross, Milton Munitz, who had been one of Ernest's students at Columbia, was fortunate enough to be sent to OTS (Officers Training School) in Miami Beach, then assigned to teach. Munitz was in the Air Force (Nagel 2020, Box 4, Folder 32).

March 28, 1943

Dear Ernest, Edith,

The three weeks since I've left have slipped by so quickly it seems just the other day we walked along the Drive. And happily, so far, my most optimistic expectations about Army life have been justified. I've really been having a wonderful time. The hours have been busy, the work is hard but with all that I've welcomed every minute of it. First of all the locale and facilities here are ideal. Miami Beach is truly a bit of paradise and I still can't get over the sheer richness of natural beauty...

Our course here is in many ways a fascinating and truly educative one - we have for the six weeks something like 25–30 courses divided into "academic" and military (including physical training). Six days a week, eight hours a day of constant activity make the hours skip by like seconds....

In the academic wing we study Administration, Supply, Staff & Counsel, Mess Management, Military Hygiene, Military law, Transportation etc etc. In the military - which is always done in the field we have close order drill, extended order drill, small arms, military courtesy, camouflage, Identification of aircraft, interior guard etc etc. And then of course there are two hours every day of physical training - calisthenics & sports. Throughout we have exams - several a day sometimes, practically every day. The academic work hinges primarily upon note-taking and memory reproduction. But its darn interesting. And on the physical side while the first week or so put a tax on many of us - we have been put in pretty good shape - and I have never felt better in my life. The regular hours, adequate food and clock-like regularity of our existence really tend to make for a healthy psyche. The morale of the men is good - the men themselves are on the whole a swell bunch and previously of academic or professional backgrounds. We have some men here from City, Brooklyn and N.Y.U. - also from many other universities. So its quite home-like and congenial…

As ever, Milton

After Miami, he was sent first to Montgomery, Alabama where he was asked whether he wanted to teach navigation, meteorology or gunnery. He selected meteorology and started a two-month course in that subject. After 2 weeks he was asked to start teaching the course. However, on the first day, he was sent to Courtland Air Field in Alabama and told to teach "Aircraft Recognition" using the Renshaw flash system (the students had to identify airplanes or ships when their images were briefly shown on a screen). He found the base isolated but enjoyed hearing news about the Philosophy Department and Columbia. Ernest also gave him a subscription to the New York Times so that he could keep up with the news.

Ernest helped the war effort in another way. In 1944, he received the following letter from the publisher of the Introductory Logic book:

Harcourt, Brace and Company, Inc.
Publishers
383 Madison Avenue, New York
New York 17, N.Y.
January 18, 1944

Dear Professor Cohen and Professor Nagel:

We received today a request for a price on a special edition of 15,000 copies of AN INTRODUCTION TO LOGIC AND SCIENTIFIC METHOD, all to go to the armed forces. That's a wonderful testimonial for your book! Little did you think as you were writing it, I daresay, that it would ultimately fill a need in the training and instructional program of the American Army.

The request comes to us from the Editorial Staff of the United States Armed Forces Institute, an organization set up jointly by the American Council on Education and the War and Navy Departments. The books, if supplied, will be used immediately for "group" study among soldiers in non-combat areas and, later, by soldiers during the period between cessation of hostilities and demobilization. The first order, as I have said, is for 15,000 copies. There is a chance that an additional 15,000 copies will be ordered later.

But there is a hitch to this proposal. We are asked to bid a special printing on the basis of limiting profit to 6% overhead and cost to 20%, and royalties to 6%, with little if any allowance for plates. We are to print on a 40-lb. paper and bind with a paper cover; the over-all size is to be 5″ by 7 3/8″. The price to the Army will have to be some place around $.50.

We are ready to bid on the basis of 6% profit. Do you give us your permission to bid with the understanding that royalties on the 15,000 copies will be limited to 6% of what we get? My guess is that royalties on this first order will run to some place between $400 and $500.

Part of the understanding is that all of these books will be shipped overseas and, when the soldiers are through with the books, they will be destroyed. Hence the special edition will not get back into the market to compete with the regular book. We are getting a marginal sale in a new market with a smaller per-book royalty, yet an over-all royalty amounting to a substantial sum.

Will you wire or 'phone immediately, please, as to whether we have your approval to bid on AN INTRODUCTION TO LOGIC AND SCIENTIFIC METHOD and to manufacture the special edition with the understanding that royalties will be limited to 6%? We'd like to have your reply by Wednesday, surely by Thursday. We are supposed to get our bid in by Friday.

Sincerely,
Dudley Meek
DM:EK

P.S. Would you phone me, just as soon as possible, Mr. Cohen's present address?

In fact, the army intended to use these copies for a course in Logic. Ernest wrote about this to Cohen on July 6, 1944 (Nagel 2020, Box 1, Folder 54):[13]

> I have had a bit of chore in connection with the army adopting the Cohen and Nagel [logic book]. First I received a request to make a detailed outline of the course with the text as the basis, lesson by lesson, with hints to green-horns (so the instructions ran!) how to teach various parts. This took me a couple of days, for the plan calls for a course of 100 hours. I was barely finished with that when I received another request to prepare an end-of-course examination. The instructions called for multiple-choice questions, each with five possible answers. I had to construct 100 questions. The formal logic part was not too bad – but what can one do with multiple choice questions in the more philosophical parts of the course? One and only one of the five possible answers could be correct, and of course I had to supply a key to the correct ones. I sweated over this for the better part of a week. But then, that is about all the "contribution" I have made to the war-effort – or is it the post-war effort?

It is interesting to think of American G.I.s going off to battle well-fortified with not just guns and ammunition but a thorough grounding in the fundamentals of Logic.

Not everyone was as fortunate as these people. Ernest also corresponded with two former students, George Bowdery and his wife Barbara (Fig. 2.29). Toward the end of the war, George was put in charge of a region which included Luxembourg and part of northern France (Nagel 2020, Box 1, Folder 22):

> 29 Nov 44
>
> Dear Prof. Nagel,
>
> I'm in a Civil Affairs Detachment here in charge of a small chunk of Luxembourg. French, German, and Luxembourgeois are all spoken here. My wife (and son Allan G.) are now near Chicago. Best wishes to you and your wife.
>
> George J. Bowdery
>
> From: Mrs. George J. Bowdery,
> La Grange, Illinois
> To: Professor Ernest Nagel, Columbia University
>
> Yesterday I received a letter from George, from a prisoner of war camp in Germany, the first word of any kind I have had since he was reported missing in action as of 19 December 1944. George asked me to inform you of his present status, for, as you probably know, his correspondence is very limited (4 cards and 2 letters per month). Previous to his capture, George was a civil affairs officer in a city in Luxembourg Duchy, where he had been for several months. His civil affairs detachment had also administered 2 or 3 French cities, including a large Paris suburb. George had really enjoyed this work, and he felt that he was doing something constructive, helping the civilian population to restore ordinary public services and trying to get the local officials to proceed somewhat more democratically than

[13]Letter from University of Chicago Regenstein Library, Morris Cohen correspondence.

Fig. 2.29 Allan and Barbara Bowdery. (Private Collection of Yvonne Nagel)

they had under German domination. It is too bad that his army career had to be cut short, just before they got to Germany, for which he had especially been trained.

In his letter George wrote that he had been keeping occupied reading, working, and thinking. All these recent experiences, he said, had broadened his interests to include some more traditional philosophical problems. He said that he had congenial companions, and that he played chess frequently. But at the same time he is awaiting that happy day when he can return home, and take up his philosophy again as a full time occupation.

Our son, Allan George, is now 7 months old, and is a very great joy to both of us. George will be very proud of his little son when he can at last see him. He seems so bright, and acts as if he understood everything that goes on, even if he doesn't! He still is a full time job for me, but I hope it won't be long before I can continue my graduate work at Columbia, where I had been taking some work before Allan was born. So far as I have been able to discover, George's camp is in southern Germany - Hammelburg/Mainfranken. We hope only that the war soon will be over and that prisoners will be freed.

Very best wishes to you, and to Mrs. Nagel,
Sincerely,
Barbara Bowdery
From: Mrs. George J. Bowdery
La Grange, Illinois

Dear Professor Nagel,

I am writing to tell you that a war department telegram has notified me that George was killed in action on 5 April. Apparently he had been or was being moved from one camp to another when his death occurred. I have talked to the commanding officer of George's civil affairs detachment, who is now in an army hospital in Chicago, recovering from the effects of his imprisonment. He also was at Hammelburg, and he told me that George was in good health and spirits just before 27 March when the Americans made an unsuccessful attempt to liberate the camp. Immediately afterwards the prisoners were marched away to another camp.

Apparently the long delay in notifying me was due to the necessity of obtaining official information. I have been terribly worried for some time, for I knew that George would make every possible effort to notify me of his liberation as soon as possible.

We were an unusually happy couple; we enjoyed working together, and had so many plans to do things together when the strains and tensions of army life would be lifted from us.

I know that George would have made some worthwhile contributions to philosophical knowledge. Our 10 months old son, Allan George whom his Daddy never saw, especially will miss the fun and help which only fathers can provide for their sons. But I will do all possible for him.

My own plans are, naturally, as yet, indefinite. I plan definitely to come back to Columbia to finish the work necessary for my degree. This may need to be postponed, however, until Allan is a year older, or until I have worked a little first. At any rate, when the time comes I will surely contact you, and I hope that we may have some conversations on points relating both to philosophy and sociology.

My best wishes to you and to Mrs. Nagel,
Sincerely your,
Barbara Bowdery

Barbara Bowdery did return to New York to finish her degree. The Nagels kept up their friendship with her.

2.12 1941–1945

The war years were eventful ones for the Nagels quite apart from their activities on behalf of refugees. Edith received her PhD in physics from Columbia in 1941. She then worked at two jobs, one of which was teaching at City College. The following fall, she was hired to teach physics at Queens College in Flushing. In 1941 Ernest had helped Peter Hempel get a job at Queens and the Hempels, who were living in Flushing, welcomed Edith to Queens:

August 23, 1942

. . . Give my future colleague our very best, please, and tell her that all the class rooms are being washed, the floors polished, some places painted to get everything into proper shape for her arrival. Also, I looked into the physics lab the other day; they have some very nice things: apparatus for the Millikan experiment, for Laue X-ray diagrams of crystal structures, for measuring e/m, etc. [See letter 36 in this volume's Chap. 4]

Ernest had severe back pains which required hospitalization. He was also worried about his mother. She had long suffered from an enlarged goiter and an overactive thyroid gland, but she also had abdominal pains, so she had to submit to a number of tests to rule out a tumor or ulcer. Perhaps because of this, he was unable to visit Carnap who spent the 1940–41 academic year at Harvard. Bertrand Russell and Alfred Tarski were also there that year.

By 1943, Ernest's back had improved and he and Edith were expecting a baby. Eva Hempel was also pregnant and the two women shared notes and books

Fig. 2.30 Eva Hempel with the Nagels in New York. (Private Collection of Yvonne Nagel)

(Fig. 2.30). Unfortunately, Edith suffered a miscarriage in August 1943. The Hempels' son, Peter Andrew, was born in January 1944, but Eva died shortly afterwards.

Susan Stebbing, whose letters had kept them in touch with events in England and the continent wrote in January, 1942, that she had abdominal cancer and was seeking treatment in Cornwall. Despite her illness, she tried to keep up her busy schedule and even took on additional tasks. She promised to write an article on G. E. Moore for Paul Arthur Schilpp's *Library of Living Philosophers*. She had also signed a contract with Methuen's to write an elementary logic textbook and had plans to write two other books—a sequel to her *Ideals and Illusions* and a book on the *Scope of Conventions*. She did finish the logic text but not the other two. She described her life in this letter from Cornwall (Nagel 2020, Box 5, Folder 19):

August 25. 1942

My dear Ernest, It is a long time since I have heard from you & a long time since I have written. I want very much to hear how you are. I think much about you. I am afraid that spinal trouble must have meant agony for you, & I do so want to know how you are, & whether you left hospital in time to go back to work at College. I hope to hear soon. It is long also since I heard from Ann.

I have been at home since July 11th. I think I am a good deal better – whether permanently it is not possible to say – but I am able to read & write - & do a few small jobs. I don't seem able to walk much, for I tire easily & was told not to walk far. I get suddenly overwhelmingly tired – a sort of exhaustion which I have never had before, but is no doubt partly due to the intense or concentrated radium treatment. . . .

I had a very busy term last term, as I told you in my last letter, & I stayed up at Cambridge this year until all the Academic Examiners' Boards were over. Did I tell you that after I had called to Prof. Schilpp that I must withdraw - & not write my promised article on Moore, he persuaded me – . . . to 'write something however short'. So I did it in the few days between June 30th and July 8th. I got it typed hurriedly & sent by airmail on July 10 & he called he had received it on July 21st. I had no time to write what I had originally intended doing, so I wrote a hurried and utterly feeble article on 'Moore's Influence'. I told Professor Schilpp that if he

& Moore preferred it could be scrapped. It is not worth printing. I was sorry to fail so badly; there is nothing I so much wanted to do well.

Did I tell you that Methuen's persuaded me to sign a contract last December to write a short, & really elementary, really textbook, book on Logic! I signed the contract just before I knew about my illness. I don't think I could have if I had known; there are two other books I urgently want to write & now I doubt if I shall have time. One was a sort of background to 'Ideals & Illusions' & the other one on 'The Scope of Convention in the Physical Sciences' which I began just before the war & put aside after writing only a couple of chapters. . . .

Ever yours, Susan

Her operation and radiation treatments did not succeed. Her final letter to Ernest was written in August 1943 (Nagel 2020, Box 5, Folder 20):

Mount Vernon Hospital.

Aug. 14, 1943.

Northwood Middlesex

Dear Ernest, you will hear from Ann that I am seriously ill. I wanted to let you know myself but have not been able to write sooner. This is one of my 'good' days. It is five weeks today since I had the operation that was not a success. I am sad to know that we shall not meet. I wanted so much to see you all, and to see the end of the war, & to do a few jobs of work. But after all, at my age, I need not complain when so many young men and women are being killed in war. I have been fortunate in my life.

I had planned this summer vac. to do a revised and enlarged edition of 'Philosophy & the Physicists' - as this one is out of print & then to write my long thought of book on the scope of Convention in science - I have been thinking much about that topic and I felt I really could tackle it now! But perhaps in trying to do so I would find I couldn't!

I do not know how long this will take. The worst of cancer is that it does (in some cases) completely disable one & yet is so slow in killing. Today I have no pain, for the first time for weeks, and so I feel clearheaded & as if I could do much. But - I am very weak, & can't even hold a pen for long, & I can't of course be out of bed. I could wish it would be quick. It is sad for Vivian, just watching the slow process.

I send love to Edith and hope her work flourishes. I am glad we have had our friendship. I do so hope you are better. That your back is cured. With love from

Susan

Her friend Vivian Shepherd described her death and the memorial they planned for her to Ann Hook (Nagel 2020, Box 3, Folder 26).

January 7th 1944

Dear (Ann),

Doubtless you will have heard by now of the death of my greatest friend, Susan Stebbing on September 11th last year...

Please forgive a typed letter, but I feel I just cannot write the details of Susan's death over & over again.

She came home from examining in Wales in order to see Miss Gavin, who was very ill then, & she intended to proceed in a few days to Cambridge & then on to London for further examining. She arrived in Tintagel on June 25th but was herself taken ill there on the 30th & I took her to London on the 2nd to see a specialist.

The surgeon ordered an operation immediately but it was arranged for the 12th & I returned to Tintagel because of the school, though much against my will. However, she had a further attack on the 8th & had an emergency operation on the 10th. I returned to London on the 12th & remained with her until September 11th.

She was at Mount Vernon Hospital, Northwood for the entire illness, she liked her nurses very much indeed. She was herself, cheerful, courageous, dominant almost to the end, & always thoughtful for others.

I was with her every day & many nights & right up to the end: she suffered great pain with even greater fortitude. Her mind was clear nearly all the time. . . .

She died peacefully on Saturday, September 11th. . .

Of her loss it is not possible for me to speak. There have been many tributes to her in the newspapers, & periodicals, & I should indeed be glad to have any that may appear in American periodicals. . .

Susan's mind was clear almost all the time & she was delighted to see her 'Elementary Logic' appear. She was sad that she could not re-write 'Philosophy & the Physicists' for Methuen's as had been arranged, but she was much more grieved that she had not been given time to write her "magnum opus" as she & I liked to call the big work which she had hoped to write & which was already planned in her head & she had hoped to write last summer. However I tried to arrange for Methuen's to offer "Philosophy & the Physicists" to Penguin Books, Ltd. (as this firm had wanted to publish this work about a year or more ago) & I heard the night before Susan died that they were going to accept the offer provided she approved. . .

A panel in Girton College Chapel is to be carved in her memory & the Aristotelian Society hope to produce a volume of essays on various subjects in which she was interested, as a memorial. Bedford College are organizing a memorial in which Kingsley School is joining, & it is hoped that all her friends, both academic & personal from all parts of the world, will join.

The memorial is to take the form of "a substantial Scholarship for post graduate studies in Philosophy", leaving open for the present the questions how to be awarded & where to be held. . .

Vivian Shepherd

Ernest responded to these tragedies by raising money for the memorial Scholarship Fund for Susan and creating a fund to hire a housekeeper for Peter Hempel. He wrote to his friends, including many from the former Vienna Circle. Most of them contributed to both funds.

Although Peter Hempel found life difficult after his wife's tragic death, the Nagels' support and the excellent housekeeper they hired helped him immensely.

They hired a Mrs. Richmond, who helped Peter through this difficult time. After the spring term at Queens was over, his friends persuaded him to let Mrs. Richmond take Peter Andrew with her for the summer to Vermont, where her family lived. Peter Hempel was then free to travel to visit his friends and colleagues. He went first to Princeton where he worked with Paul Oppenheim. Afterwards, he went to Saranac Lake in the Finger Lakes region of New York with the Oppenheims for a vacation. He wrote to the Nagels from there:

From: C. Hempel
Hotel Ampersand
Saranac Lake 7/3/44
To: Prof. And Mrs. Ernest Nagel
South Wardsboro, VT.

We arrived in Saranac Lake two days ago and I am eager to send you my news at long last. The twelve days at Princeton proved enjoyable and intellectually very stimulating, especially thanks to three discussions which Oppenheim and I had with Goedel on the subject of how to define explicitly the degree of confirmation of a hypothesis with respect to a given description or protocol. I think I had mentioned to Ernest that Oppenheim was interested in the question, and that we had at least a nucleus of a definition (for lower functional calculus of one-place predicates only) which satisfied the probability postulates of the Hosiasson [Janina Hosiasson-Lindenbaum] article in J.S.L. Well, we discovered in Princeton that the def. involved serious difficulties, & we were ready to drop the matter; &

then Goedel of all people revived it by some very constructive suggestions, and now we are continuing our work on it. The whole matter smacks too much of an a priori approach to give me an easy conscience, but it involves interesting technical problems and some hope of a reasonable amount of positive results; so for the time being I feel better about helping O[ppenheim] with this than if we were engaged in the more general phases of his ideas. But it is too early as yet to tell what will come of it.

I found Goedel amazingly interested in the problem & very willing to give us his time for discussions, & I was very much impressed by his incisive and very constructive comments; often in following up a lead he outlined in 5 minutes, I discovered only after many hours of work the full implications of his suggestions. - He is somewhat shy & almost self-effacing, but seemed in good spirits and gave the appearance of enjoying these talks with us. - I also saw v. Neumann again, said hello to Church (whom I found on a sweltering day, lying on a couch & reviewing with his teen-age son such formulae as $a(b+c) = ab + ac$, which the son had to write on a little blackboard built into the wall of Church's office); I had a chat with Stace & two nice evenings with the Bergmanns (you may recall them: they stayed with us at the physical institute after Reichenbach's quantum lecture; he is a former assistant of Waismann; at their house I met the radium physicist Rosenblum (Rosenbloom?) who had been working at the Institut Curie at Paris on α-rays when the Germans took over; he told us the story of his escape with his little daughter & his pregnant wife, from Nazi-occupied France; it was pretty hair-raising; I would have given up twenty times along the way.

I also did some work at the library of Princeton Union & at Fine Hall; what facilities for scientific work; I turned yellow with envy. Also from Princeton I mailed my ms of the article on Confirmation to [G.E.] Moore; I had been wise in postponing my trip to Princeton for several days to finish all the planned changes & to take them to Queens College for retyping. All I had to do in Princeton was then the final "proofreading" [footnote: I forgot, in my letter to Moore to ask him specifically for reprints. Do you know whether one gets a certain number automatically & whether on request I might be able to get 75 of them?].

I have received no reply yet from the Am-Math-Monthly to whose editor I sent the ms on Geometry and Empirical Science on May 26; doesn't that seem rather unusual to you? Do you think I ought to ask back? A Prof. Schaaf of the Brooklyn College Ed. Dept wrote me spontaneously to ask whether he could have the ms of the lecture (which he had attended) for the Nat'l Math. Magazine, Teachers' Section. But he wanted a version which would point up the questions of teaching geometry at high school, about which I have no definite ideas. I told him about the Am. Math. Monthly & promised to let him know after I had received word from the latter, & I enclosed a copy of the ms & suggested that if the article was not to be published in the other magazine & he still wanted it, he might take care of the pedagogical aspects by publishing Hertog's article or a similar one in addition to mine.

Despite the weekend crowds, we had a rather smooth trip to Saranac Lake; we are staying in a comfortable cottage under hotel management. I have never been to the Adirondacks before and am very much surprised with the beauty of this part of the country. The hotel and our cottage are in a very attractive location, overlooking lower Saranac Lake which is surrounded by forests and beautifully set off against a background of blue mountain chains. The quiet of the place is very soothing; the contrast with Flushing strikes me over & over again.

The morning after our arrival Einstein turned up to say hello to the Oppenheims. He had come with his sister, who resembles him very much, in his sailboat from Knollwood, about 2 miles (as the boat sails) from [the hotel] Ampersand. We had a nice chat, & then the Oppenheims & I escorted him back in a canoe. - Today we rowed over again, & Einstein talked 1 ½ hour about some work he has just started in pursuit of an idea he has concerning the expanding universe; it was very fascinating. In the course of the conversation, he also endorsed the view that time's arrow can be determined only by means of the 2nd theorem of thermodynamics; when I asked about the hourglass counter-example that had come up in a talk with Wundheiler at your house recently, he pointed out that this process too involved the 2nd theorem because the kinetic energy of the grains of sand is transformed irreversibly into heat. I am not sure that that meets the point; perhaps we can take this up once more when we get together. - Einstein's new idea refers to the question of the initial stage of the expansion process & aims at eliminating a singularity which occurs in the solution of the differential equations for the expansion; this singularity seems to exist in a temporal starting point of the process, with all the (uranes?) of the universe forming a material point or the universe having the radius 0; but the time elapsed since this event turns out much too short (~ 10^9 years, ago of the earliest const) as computed in this fashion; E[instein] has now an idea of how to remove the singularity & make the whole process periodic; then there is no beginning & no end of the universe in time, and instead of the singularity there is a state of relatively great irregular speeds of the stars in relatively close distances rather than a state where the radius of the universe is 0. I know this is pretty vague, but it shows you at least roughly what the talk was about. E[instein] was really very nice, he also had had a haircut & looked much better. -

I have had good news about Peter Andrew from Mrs. Richmond, who, in her latest letter, sent along a small piece of chewing gum with a tiny crescent-shaped imprint on it – the bite of his first tooth! -

The Flushing apartment is sublet, & the people who have it now are delighted with it; also, they are keeping it in excellent condition, as I saw during a brief stopover in N.Y. on my way here. -

Tomorrow, I am to go over to the Brunswicks' house for the first time to tutor their daughter in German, & I am feeling a little jittery because I am not sure of my knowledge of the subject; I'll prepare some time today. The B's live some 6 miles from here, & one has to go there by taxi or in the B's car. Mrs. Brunswick is expected to arrive only toward the end of the month. -

Finally, a story that Einstein told yesterday, & that you might enjoy. He put it into the mouths of two Jews – one of them complains that he can't understand at all how a wireless telegraph works, & the other explains: that's quite easy. Imagine a dachshund that is so long its head is in Paris & its tail is in New York. Now when you pinch the dog's tail in N.Y., he screeches in Paris. That's how the ordinary telegraph works. Now just imagine exactly the same thing, but without the dachshund. That is wireless. -

I hope you are both well & enjoying your cottage. How is the garden coming, & in what condition did you find the house? Is the ms making progress? And are you getting tanned?

Best regards from the Oppenheims, & my love to both of you,

Peter [See letter 40 in this volume's Chap. 4]

This letter gives us some idea of the philosophical discussions in Ernest's circle of philosophical friends.

From time to time the Nagels visited Mrs. Richmond to check on Peter Andrew's progress. They would send reports to Peter. Peter Hempel continued to work with Paul Oppenheim at Saranac Lake. It seems to have been an idyllic time. He mentions that they saw Einstein (who was a good friend of Paul Oppenheim) and Wolfgang Pauli who were both vacationing nearby. Later, he took a long train ride to Santa Fe, New Mexico to visit the Carnaps who were in Santa Fe.

Rudolf Carnap had received a Rockefeller grant for the 1942–43 academic year (Fig. 2.31). He and Ina wanted to be in the mountains and so they went to Santa Fe, New Mexico. It turned out to be a difficult year for them because of Carnap's back troubles. They spent the winter in Texas, because they thought it might be too cold in Santa Fe. A doctor there advised them to visit the Mayo Clinic in Minnesota. In November, Ina wrote a long letter to Ernest describing Carnap's medical problems. Since Ernest also had serious back problems and had undergone an unsuccessful operation, they felt he would understand. Carnap ended up spending most of the next 6 months in bed. He was able to do a lot of work on his book (which Ina then typed), but they worried that he would not be able to return to teaching and tried to see if

Fig. 2.31 Rudolf and Ina Carnap with their dog Marnie in Santa Fe, New Mexico. (Private Collection of Yvonne Nagel)

they could extend the Fellowship to a second year. In May, Ina wrote a desperate letter. She writes in detail about Carnap's problems but then adds at the end that she herself has lost 36 pounds! (Nagel 2020, Box 1, Folder 38)

Santa Fe, N.M. May 18, 1943

Dear Edith and Ernest,

It's really a crying shame that I write you so rarely. I just see that I haven't even replied to Ernest's long and much appreciated letter of last November! As usual, I have excuses though you will have to be the judge as to their weightiness. Ernest's letter came when we were just moving from Santa Fe to El Paso where the only house I could find was a huge unfurnished place out in the country. Since we were already negotiating about buying the little house near Santa Fe where we are now, we had the courage to buy some furniture in El Paso with the idea of having them shipped here in the Spring. It all did work out just about that way, but it all meant a lot of chasing around since El Paso was a town with almost empty stores – due to the migration of soldiers' families from and to the town because of the big camps there. Well, when we were settled to some extent, we felt like consulting another doctor about C's back and we hit upon a man who thought he knew what was wrong and that it could be fixed at Mayo's. This negotiating and the decision kept us on tenterhooks for the next few weeks, and as you know we then actually went to Mayo's, unsuccessfully, and back to El Paso, and after a few weeks there we left for Santa Fe. And thus life rushes on, and there isn't much mood for letter writing: either a decision is pending or a trip is ahead or this and that, and each time one has the feeling that postponement of writing will be good because something may clear up – well, it never does, but that's the feeling anyway. Now in Santa Fe, I am again rushed for the time being, though it's rushing of a nice sort. We have bought a small house high up on a hill [footnote: absolute solitude, no passersby, haven't seen a soul up here in the whole month] with an enchanting view and a leaky roof and all that goes with the latter (we knew about the roof before we bought the house, but we liked it so well and it was so reasonable that we took it in spite of it). The house is "amateur built" [footnote: ??? has city water and electricity] (from rocks collected right here on the hills, stuck together with some sort of adobe) and it has stood empty for the past months – both mean quite a number of things to do right now. The roofing man will take care of putting on new roof layers and of getting cement over the adobe on the walls, but it is me who has to decide what should get priority among the things I myself ought to do: is it more important to prune the trees besides the road so that they won't scratch the paint of my car fenders or is there more hurry about putting fresh putty on the window next to Carnap's bed where the rain came in last time. Or is it more important to rebuild the stone steps in the garden (no there is no garden, it's in the midst of rocks and pine trees but anyway) which are now inconvenient for Carnap's short walks, or to relay flagstones on the porch where the rain is washing them out. You must admit that all these tasks have more urgency and more appeal about them than sitting down and typing letters (not to speak of tidying up the house which C. thinks the most important of all though I don't agree). From all this you see that I have been pretty much wrapped up in the Carnaps and their affairs during the past years. (I really meant to write "year", but there the plural stands and probably is quite correct). All the outward activity I am writing about such as the puttering about the house, is of course only an accompaniment of having C. in bed all the time. Reading your case history we had the definite impression that you actually were and are worse off than C. Though he is completely incapacitated since last July (and he is definitely another few degrees worse since the trip to Rochester), still the outstanding fact is that most of the time he has no pain which would interfere with his work, and that he is quite happy and cheerful because he can work. Now, an existence in bed isn't the ideal situation. Nevertheless, if (and that's a big if) our income didn't depend on his ability to teach, I think he would not be too unhappy about it all even if he knew that it would be a matter of years. Since C. always felt his writing more important than his teaching, he doesn't really miss anything in that respect by being laid up. What he misses are discussions with friends, but that he did not have in Chicago either. You should have seen him grinning from ear to ear when he was with Feigl for a few days after the stay in Mayo!

I suppose you have seen the report about Mayo's which I sent to Hempel. One does not quite know what to think of the whole matter. Our doctor in El Paso who had sent us to Mayo's was quite disgusted that they had sent C away without an operation because he felt strongly that no "spontaneous improvement" could be hoped for in a case which has gone on so long without being improved by rest in bed; he asked why C. had not implored the doctors at Mayo's to do the operation. Now, you are the one who will best appreciate why C. as a layman did not want to do that; actually it was your case which he had constantly before his eyes, the operation they did on you – being reasonably sure about their diagnosis before the operation, and then not finding the diagnosis confirmed during the operation. Now, it seems that in his case the doctors were not even too sure about the diagnosis, though they behaved as if they knew that there were two disks slipped. Still, we had the impression that they would be much happier about this diagnosis if C. showed any of the sciatic pain which normally seems to accompany slipped disks (at times I was inclined to envy you your clear symptoms – but what good did it do you?). There is [sic] a number of possibilities why they could have wished to postpone the operation: 1) possibly they really thought an improvement without an operation might happen; 2) they might have wished to wait for clearer symptoms confirming the slipped disk diagnosis, 3) they might have been just too busy at the moment and not have been willing to bother with a case who doesn't have sufficient discomfort; 4) they might have had not too good experiences with this kind of operation and therefore might try to limit the number of cases where at least it can't hurt (better: do not harm, because it does hurt body and pocketbook).

Since it seems impossible to get straight which of these reasons is the real one, the perplexity persists as to "what next". They dismissed C. with the vague notion that, if there is no spontaneous improvement, they might reconsider the operation in "say another six or twelve months". In general C. has the feeling that he would not like to urge an operation about whose advisability the doctors themselves are in doubt, and therefore, if there were no pressure to get back to teaching, he would give it time and wait a year. However, he is scheduled to teach in the fall (he was already scheduled to teach in the summer, but that has been changed), and September is the last month where the Rockefellers are paying his salary. As you will see from the enclosures we have tried to get a renewal from the Rockefellers but have received a pretty clear "NO" as the answer. Since it is quite obvious anyway, I will admit that I have pulled myself together to writing you today the letter which I meant to write for the past five months primarily for the reason that I wish you would give me your opinion on the matter with R. Do you think Hanson's letter has to be taken as the last word in the matter or do you think something might still be done? (Please don't be puzzled why I have picked just you as my consultant; it is not really because I think you have special information on the matter or anything the like. But I have to have someone who I think means us well without reservations, who does not think that he might endanger his own cause at the Rockefellers, who will think that C. will do worthwhile work during the year, who has a feeling for American customs and behavior, etc. etc. I do not wish to report in detail why I have ruled out other people [side note: I suppose your imagination tells you whom I have in mind primarily] as advisors, often it was just a matter of intuition which warned me against it; anyway, you are the poor chosen victim). If things could be talked over frankly with Hanson, the following could still be said: 1) he is wrong in saying that "undoubtedly at the end of the second year or even a third year, the situation would not be different from what it is at the end of the first year", because at the end of the second year C. would have carried out the project to the extent that there would be two manuscripts written by then. 2) If it were not so much a matter of "Foundation principle and practice" as a financial question, C. could live on a smaller grant (for this year R. is paying his normal salary to the university, and the university continues to pay C.), particularly if the grant were paid directly to him instead of to the university because in this case it were not liable to income tax (and that will be quite an amount the coming year). Now, I do not know what to think about mentioning that point. Would you think that the amount does matter to them? That is, would they be more willing to renew the grant if it would cost them less? And what about mentioning the tax business? It's true, they paid Feigl directly and thus he did not have to pay tax on it; but what about putting it to them in that form?? Here I am appealing to your American sensitivity in judging. There is also the point that if it were a question of money to them, C. would already be grateful for a part-renewal, because he thinks that in two more quarters he could at least finish one book and greatly advance the work on the other. If you think it might be tried, would you suggest that we name the lowest amount on which we could do it or what? It is, of course, rather stupid, that one can't make any use of the main reason why a continuation of the grant is to us so important right now, that is C's back, a condition which allows him perfectly well to do research work but prevents him from teaching. C. has the strong feeling that the R's should not know that he is ill because they might feel that that is the reason why he wishes support and not because he wants to do research. I see that this is going to be a long letter, but this is really the only point which needs an answer. Do you think you could send me a line on it, as soon as possible? If either the R. matter is not pursued any further (actually, Hanson's letter sounds rather forbidding but what does one risk by pressing the slim chance which might still lie in our additional reasons), or if it is definitely closed by an additional "NO", then comes the point of how to proceed with the university. But before I go into that, something more to the previous topic: it might or it might not be a good idea to have the R. matter urged by some big shot. But 1) we don't know many big shots, and 2) who is a big shot at the R.'s anyway, probably no philosopher (such as Russell or who?), more likely a mathematician (but there are no "big" names like Weyl and Birkhoff are against the kind of work C. is doing). So what??

Going back to the University business. There the trouble is of course, that there is no obligation on their side to continue to pay if C. does not teach in return. Probably they will continue to pay for a certain length of time, perhaps as long as his present contract goes (another two years) or as long as they expect him to resume teaching again or who knows. It isn't that the university does not have the money: they are better off now than before the war; it is not that they have to hire somebody else to do his teaching: if he can teach he has to teach math or physics because there are no graduate students in phil. for him. It's just that it is a matter of being at their mercy. C. feels if R. is definitely decided in the negative, he will have to urge the doctor at Mayo's to operate (perhaps in September when the half year is over which seemed the minimum postponement which they thought advisable) more or less in order to impress the university with the fact that he is doing all he can to get fit for teaching. He feels that we cannot go on living in Santa Fe and drawing his pay, simply awaiting the "spontaneous improvement". T.V. Smith happened to come to El Paso while we were there and he was out to our house and saw C. in bed and talked to him about Mayo and so on – but that's about all the department can lay their hands on. How does one prove in an almost symptomless illness that one is actually not able to teach?

Since you are still not out of the woods with similar trouble I am sending you two articles which may interest you. Love and Ghormley were the two main persons who saw C. at Mayo's. Dandy is in the army now and thus not available.

Thank you for the Partisan Review. We read your article with great interest. It's a bit like fighting a conflagration with a bucket of water, isn't it? Still, it makes one [feel] good to throw the bucket of water, I suppose. What is this new Conference on the Democratic Faith and the Scientific Spirit? Randall asked C. to become a sponsor. Not seeing your name on it, I advised against it, however. True there was Hook. But I have never gotten entirely clear about him. What do you think about his newest book? I only read its review in "TIME".

It seems, I could go on indefinitely talking. C. has a birthday today, and a touch of pleurisy since about a week (not high temperature, but it goes on and on, and breathing hurts). In November you mentioned that you two seriously consider having children; are you still thinking about it?

Have you seen vol. II - the Formalization business – and, do you like it better than vol. I ??? [handwritten note: just reread your card where you say you are halfway thru].

In case you should think of it: yes, I could earn money and the Carnaps would not starve. The trouble about it is that if C. does not improve we almost should have someone to take my place here at home because he cannot even go to the bath room without help, and thus most of what I might be able to earn would go out again to pay someone to replace me. Also, now it might be relatively easy for me to find a job, but now, while his salary is coming in, it's more important that I be at home than that I earn money; while on the other hand it might be hard for me to find a job in a year or two when there might be no more salary coming. There is a new hospital here near Santa Fe for war wounded. I suppose I could get a job there now. But what with the probability of a trip to Mayo's ahead in about three months, there would not be much sense in it. To forstall questions: I haven't done a stroke on my thesis. I have read some useful books, but I have never sat down to doing my own work. In addition to having fairly much practical work to do and having C. in bed I am rather below par myself since I had a minor operation a year ago; nothing striking, just lost 30 pounds and have trouble with sleeping etc. But surely that's enough about me, and enough about the Carnaps alltogether [sic].

Our warmest love to you both,
Yours,
ina

Ernest responded by helping them write a letter to the Rockefeller Foundation. He also wrote a letter strongly supporting Carnap's research after the Foundation asked him for his opinion of Carnap's research (Fig. 2.32).

South Wardsboro, Vt.
July 1, 1943
Mr. Frank Blair Hanson,
The Rockefeller Foundation,
49 West 49th Street,
New York, N.Y.

Fig. 2.32 The Vienna Circle circa 1953. (Private Collection of Yvonne Nagel)
l. to r. Herbert Feigl, Paul Oppenheim (?), Hilary Putnam, Rudolf Carnap, Herbert G. Bohnert, Wilfrid Sellars, Ernest Nagel, Peter Hempel. May have been taken at a conference On Physicalism at Princeton University in 1953.

Dear Mr. Hanson,

You letter about Professor Carnap just reached me here in Vermont, where I am spending the summer, and this will explain the delay in my replying to it.

I have read with care the first two volumes of Professor Carnap's *Studies in Semantics*, and indeed have reviewed them both (in the *Journal of Philosophy,* the second volume in the issue for June 10, 1943). I regard both of them as substantial contributions to the subject. The first volume is introductory in character, but at the same time effects a valuable systematization of a relatively virgin field of study and extends our knowledge of it in certain directions. The second volume is distinctly more original, and makes a noteworthy contribution to our knowledge of logical systems. Both volumes seem to me, in different ways, to be quite indispensable to serious students of the foundations of logic and mathematics; and certainly, at the present time, they are the only source of systematic information available in English concerning so-called "scientific semantics".

The significance of semantics for the study of certain phases of logic and mathematics seems to me beyond debate – the work of other students as well as Professor Carnap's second volume are sufficient evidence for this. I confess to some differences of opinion with Professor Carnap as to the scope of relevance of semantics, especially concerning the bearing of semantics upon the methodology of the empirical sciences. But this difference of opinion does not reduce my appreciation of the importance of Professor Carnap's researches; and in any event, the difference of opinion can only be resolved by pushing the type of analysis he is carrying on as far as it can be pushed, in order to determine concretely what contributions semantics can make to parts of logical theory other than formal logic and mathematics.

I find the remainder of Professor Carnap's "Project of Work" (devoted to the problem of modalities and the problem of probability and degree of confirmation) most interesting and significant. The logic of modalities is still a fairly undeveloped subject, and from Professor Carnap's indications it seems reasonable to conclude that he will do much toward completing it. I find even more suggestive his approach to the analysis of the notions of probability and degree of confirmation. The subject seems to me of great importance, and in spite of the studies that have been devoted to it, there is still a great [deal] of unclarity if not of confusion concerning the central concepts and distinctions belonging to it. If, as I have good reason to believe and as Professor Carnap's outline indicates, he can produce a systematic theory of these notions, he will have performed an extraordinarily valuable service. A book on this subject from him will, I am confident, have a relatively wide audience – not only among specialists in logic, but among methodologically conscious scientists and philosophers as well. Indeed, so important do I regard the matter, that I hope Professor Carnap may be persuaded to postpone his work on the logic of modalities until after he has completed his book on probability and degree of confirmation.

In my judgment, therefore, Professor Carnap's project clearly merits encouragement and support.

I hope my estimate of Professor Carnap's researches in semantics and the outline of my reasons for it will be of some assistance to you.

Sincerely yours,
Ernest Nagel
Professor of Philosophy

The Rockefeller Foundation
49 West 49th Street, New York
The Natural Sciences
Warren Weaver, Director
Frank Blair Hanson, Associate Director
Harry M. Miller, Jr. Assistant Director July 7, 1943
Dear Professor Nagel:

I appreciate very much having your letter of July 1st about Professor Carnap's work. We have had opportunity today for a group discussion of small grant-in-aid requests now before us, and the officers have agreed, on approval by the University of Chicago of a second year leave of absence, to give support over the year 1943–44 to Professor Carnap. I trust that the University will agree to the arrangement and that Professor Carnap may be able to continue full time work on his project in New Mexico, where he has been working for the past year.

Again, thanking you for your courtesy and cooperation, I am
Sincerely yours,
Frank Blair Hanson

Santa Fe, N.M. July 7, 1943
Dear Edith and Ernest,

Thank you ever so much for the long letter of May 28th and the post card of July 2nd. Do I have to say in detail how much we were touched about all the interest and trouble you have taken about our Rockefeller business? Even going so far as to draft my letter to R.! (by the way, it was a great help and I followed it pretty closely). Of course, I should have thanked you earlier – as usual – but again I thought that I could await a decision and present you with the result: Hanson replied very soon to our letter, saying that "It is difficult to get in touch with our advisors due to the fact that under war conditions everyone is so extremely busy and very often far from his usual place of occupation. We are now doing this actively, however, and I hope to be able to write you again, more definitely, in the not too distant future". Well, that was that, and it sounded pretty good as far as it went, that is to say, it gave us the feeling that at least now they were considering the merits of the project, but we were rather doubtful about what kind of person they would consult. Now it turns out that they asked you! They probably asked someone else besides, at least that's our interpretation why they asked you only now. C. always cautions me not to be too optimistic and he still estimates the chances only as 50:50. I always felt as you did that R. could be told that C. was ill but that his illness did not prevent him from doing research. Feigl and Carnap both, however, were strongly of the opinion that it might weaken his case and that they (R.) might think that he wanted the grant primarily because of his illness and not so much for the sake of the project. Well, it seems that his illness does matter to them....

There is one point in your letter where you are wrong. You write "how both of you can go on working under the circumstances..." well, I can't. I can do practical work on and around the house, but work on my thesis has completely broken off. I am far more keyed up about what's happening to Carnap than he is and I am actually exaggeratedly active doing practical things. How far that is an escape from being obliged to do theoretical work and how far it is due to physical trouble (I have lost 36 pounds during the past year) is hard to estimate. I just see that I have written about it already, just skip it. Good to hear that Edith has been in good health, and I do hope the baby will materialize.

I shall let you know right away when the final Rockefeller answer comes. I wish they would hurry up, not only for our peace of mind, but because such decisions as the renewal of our apartment lease in Chicago etc. depend on it (we have kept our apartment and sublet it furnished, but in spite of the housing shortage there we cannot even get for the furnished place what we pay for the unfurnished one).

Thank you again for everything. Isn't it very gratifying that with real friends one can take up where one left off quite some ways back? But I suppose that's what it means being friends. Have a good summer both of you!

In the end, Carnap was granted a second year's Fellowship by the Rockefeller Foundation. They encouraged the Nagels and their other friends to visit them in Santa Fe. Herbert Feigl, Peter Hempel and Willard Van Orman Quine did visit, but the Nagels did not. Here is Peter's account of his visit with the Carnaps.

Santa Fe, N.M.
August 22, 1944
c/o R. Carnap
P.O.B. 1214

Dear Edith and Ernest:

There are so many things of interest to tell you that I am afraid I'll just have to make a selection and leave the rest for oral report.

Thanks very much to both of you for your letter of July 30, which I received after my return from New York. The two days I spent there were extremely hot and sultry, but I got quite a few things done and thus was rather pleased with my trip. Among others, I saw the Goheens twice, met several friends of theirs, had a long and worthwhile discussion with Olaf, saw Eileen for an hour – she just returned that evening -, dropped in at our apartment and found again that it was being kept in fine shape. The main event, the awarding of citizenship in Brooklyn, was, I am sorry to say, an extremely dull & uninspiring procedure, which consisted of 3 ½ hours spent standing without interruption; 5 minutes of these were used for clerical details, about 20 by a speech of the Judge plus some general information; the rest consisted in just waiting in line. Well, I am very happy to be a citizen now!

The trip to Santa Fe, though rather long, was very fascinating; it showed me a really very impressive cross-section of this enormous country. The big industrial centers, gigantic and terribly ugly and slummy, the simply endless flat agricultural acres of the Great Plains, the Kansas oil fields, the ranch lands of New Mexico, and finally the high mountains. The college of the Carnaps has a superb location atop a mountain 2 ½ miles outside and well above Santa Fe; it commands a magnificent view in all directions, over rows & rows of mountain chains, over the whole picturesque and quiet city of Santa Fe, and 20 miles down toward the Rio Grande; the scene is quite unbelievable. The climate is magnificent; warm and sunny in the daytime, cool at night; & the air is almost always dry. The starry sky at night is simply overwhelming; the colors of sky, clouds, & mountains at any time of the day are simply wonderful, & I seem never to tire of admiring them. The vegetation is poor, according to the altitude (>7000 feet), with thousands of cacti; some needle trees & other evergreens, & plenty of sage brush. There are most fascinating animals around; wonderful instances of mimicking in insects; then there are horn toads, attractively ugly; they look exactly like a fat toad with a lizard's tail and legs and a cock's comb, really red, on their heads. It is very quiet up here, & on my regular walks with Carnap, I've never yet met a soul.

These walks, which I insist on extending gradually, take 1–1 ¼ hours in the morning, and another hour in the afternoon. C. always wears his brace when he is up, & on walks he uses a cane. He walks still a trifle hesitantly, but does remarkably well; two days ago, when we visited a ranch nearby, he followed me up to the top of a cliff which I had never expected him to make. He is in his usual well-balanced, confident & pleasant mood & is walking rather steadily (although my presence seem to have reduced the rate of his writing). Ina looked very much better than I had anticipated; but her health is by no means perfect, & running the house & attending Carnap is a very tough job, which certainly does not make things easy for her. The C's are planning on leaving here about Sept. 10 or 15 or so; they have found it impossible to be referred to even a single apartment by any of the ≥ 4 real estate agencies they contacted, & they are now seriously considering the purchase of a house; they plan to stay at a hotel first & then try to find something. The housing situation in Chicago really looks like a very tough proposition; the Chicago papers have lately been carrying ads offering a reward of $75 to $100 for being referred to an available apartment! -

I have had several talks with C about his work in probability and degree of confirmation, and I am also reading those parts of the ms which are already typed (about 500pp, & that is less than half of the total work, which C feels is his most important book to date.) His ideas on the subject seem to me to be very fruitful & important, & I hope to discuss them with you in some detail when we meet again. He had obtained many of the results that had come out of my work with Oppenheim this summer; and in many respects, he has of course gone far beyond them; but there are also some points which are, as it were, "on trial" with us (or with Goedel).

I have had good news from Mrs. Richmond about Peter Andrew; the only trouble is that he continues to have wheezing and coughing spells, especially in damp weather – It would be very nice if you could have a look at him again; but I do not know when Mrs. R. expects to take him back to Brattleboro; at present, she, her mother-in-law, & the baby are staying at Lake Spofford, N.H. (their address is the Brattleboro one). I expect to return to N.Y. On Sept 10 or so & suggested that Mrs. R. return with P.A. on Thursday, Sept. 14, after I have straightened out things in the apartment a little. When do you expect to come back?

Many thanks for Ernest's suggestion to Krik about sending me a copy of the Naturalism & the Human Spirit volume; I appreciate that very much and am looking forward to getting the book!

Finally, Ernest, I am honestly distressed & have been quite concerned about those expressions of your dissatisfaction with your own work, & when I came to Santa fe, I found the Carnaps feeling very much & very sincerely the same way about it; we all feel that this attitude is certainly without rational justification, & that in particular the book you are now working on is going to be important and valuable, & I feel that it may well exert a far greater influence than many of Carnap's writings. All this does not mean that I could not well appreciate the emotional depression about your work – which does not wait for a rational justification -; in fact, I think I can well realize what it is like, & I am very sorry you have to contend with these difficulties; but I do hope very much that you won't allow yourself to be discouraged by them!

I had expected to do various kinds of reading & perhaps writing up here, but I find that the time is simply flying, and I have not got half the things done I had counted on. But I think that the talks with C, walk, little trips, and the general physical & psychological atmosphere here will prove beneficial in their way, anyhow.

The Carnaps ask me to convey to you their very best, & also to say that there is absolutely no hurry about returning the ms on extension [n/a]; that may well wait until after you get back to N.Y.

I do hope you've both been well! Is Edith going to teach at Q.C,. this fall?

Love to both of you, and my best regards to the Hooks if they are around!

Cordially,

Peter [Letter 43 of this volume's Chap. 4]

Otto Neurath died suddenly and unexpectedly on December 22, 1945, in Oxford, England. His widow Marie continued his work on the Isotype project and Rudolf Carnap and Charles Morris continued work on the Encyclopedia for a while. The last Unity of Science Congress was held in Chicago in 1941. Otto had avoided capture in 1938 by choosing to go to The Hague instead of returning to Vienna during the Austrian Civil War. His son, Paul was not as fortunate. He tried to flee to Czechoslovakia but was arrested and turned over to the Gestapo. At the time he was a law student and a member of the socialist party. It is not clear why he was arrested. He was sent to Dachau in April 1938, one of the first Austrian prisoners to be sent to a German concentration camp. Later, he was transferred to Buchenwald. Lucie, his girlfriend at the time secured his release by promising that he would leave the country. He went to Sweden, which remained neutral throughout the war and from there moved to New York City. Ernest befriended him when he first arrived by inviting him to his apartment and introducing him to Meyer Schapiro and other friends of his father. Paul Neurath earned a PhD in Sociology at Columbia. His dissertation was a sociological analysis of his years at Dachau and Buchenwald. It was published in 2005, after his death, as *The Society of Terror, inside the Dachau and Buchenwald Concentration Camps* (Paradigm Publishers). After Ernest died in 1985, Paul Neurath wrote to Edith from Vienna.

6. October 1985

Dear Mrs. Nagel:

... I don't know whether you knew that, even though contact was only loose, I knew your husband from my first America year, 1941 or 1942. It was he who called on me when he learned that the son of Otto Neurath had arrived in New York and it was he who at that time introduced me to the then still functioning New York Philosophical Circle. And from that time on I loved Ernest Nagel. I don't mean to say that I just "liked" him, I really mean to say that I "loved" him. And that was a love that, despite of our meeting only on rare occasions, stayed with me for these altogether now 44 years. To have to acknowledge that he is no more cuts a deep hole in the web of my own life. ...

We met last at the Paul Lazarsfeld Lecture at Columbia in April and I wrote to Ernest shortly before I left New York, reminding him that he had agreed to write me some recollections of Paul Lazarsfeld. Like so many things we plan, this will now remain unwritten...

While in Vienna, I am also the Director of the Paul F. Lazarsfeld Archive and it was in this connection that I had asked Ernest for those recollections. Because he worked so closely with Lazarsfeld in their Seminar that they ran together, Ernest's name crops up again and again in my work on the Archive. And it always brings to mind this lovely softspoken, sharpminded but so utterly kindhearted man...

(signed) Paul Neurath

2.13 1945–1970

After the war, the Nagels' lives were more settled. They had two sons—Alexander Joseph, born in 1945 and Sidney Robert, born in 1948. Alexander was named after his maternal grandfather and Sidney after Sidney Hook and perhaps Ernest's brother Robert who had died in Hungary before the Nagels emigrated. The family called them Sandy and Bobby respectively. They moved into a Columbia apartment on Claremont Avenue where they lived for the rest of their lives. During the war, Edith worked for the Manhattan Project at Columbia University. She found the work satisfying and did not miss the long commute to Queens College.

Ernest continued teaching at Columbia. He continued to work on his book (started in 1940) on the structure of science, but there always seemed to be distractions. It is interesting to note that although it took him 20 years to finish his magnum opus, *The Structure of Science*, he continued to write many journal articles and reviews and the two books he wrote with co-authors (the *Introduction to Logic* with Morris R. Cohen and *Gödel's Proof* with J. R. Newman) were done fairly quickly.

James Roy Newman (born 1907) was a lawyer with strong interests in philosophy and mathematics. He took a class with Ernest in 1936 and thereafter kept in touch with him. His story illustrates Ernest's own wide interests and openness to talking with others outside his profession. Ernest first met him when he took his course in the Philosophy of Mathematics. At the time, he was a lawyer with an office on Fifth Avenue in New York. He had strong interests in Mathematics as we can see from this letter (Nagel 2020, Box 4, Folder 45).

> May 29, 1936
>
> My dear Dr. Nagel:
>
> I take this somewhat belated, but nevertheless, welcome opportunity to express my sincere appreciation of the lectures you gave in your course in Philosophy of Mathematics, which I was privileged to attend.
>
> I got a great deal of satisfaction out of the material and the manner in which it was presented, and I only hope that I shall be able to take the other half of your course next fall.
>
> I am not certain that I understood what you wanted those students who took the course for credit to do.
>
> I may want to make use of credits for this course towards an M.A. or PhD. degree in the field of the History of Science, and if there is any work that you would suggest me doing, so that you would feel justified in giving me a passing grade, I would appreciate your communicating with me.
>
> I hope that you will have a pleasant summer.
>
> Sincerely yours,
>
> James Newman

During World War II James Newman held several positions in the Federal government, including Chief Intelligence Officer in the U.S. Embassy in London. He later worked for the Foreign Economic Administration in Washington D.C. and later served on the U.S. Senate Committee on Atomic Energy. In 1940 he wrote (with Edward Kasner) *Mathematics and the Imagination*. The authors invented the terms *google* and *googleplex* in this book. He had a larger project in mind, though, and in 1944 he wrote to Ernest for advice (Nagel 2020, Box 4, Folder 45).

From: James Newman
Foreign Economic Administration
Washington 25, D.C.
Return address: 1321 Holly St. N.W.
Washington, D.C.
To: Prof. Ernest Nagel
Department of Philosophy
Columbia University
116 St. and Broadway
New York City
July 12, 1944
Dear Professor Nagel,

I have just returned from spending over a year in England on service for the FEA. It was an exciting and satisfying year but I am glad to be back with my family even though the work is less gratifying and the climate in Washington leaves something to be desired.

With the feeling of security which stems from being home again I have given thought to writing a book contracted for a few months before Pearl Harbor. It is to be a treasury of popular mathematical writings made up of chapters from well known works of popularization – Sylvanus Thompson, Whitehead, Dantzig, Hogben, Felix Klein etc; also, a portion composed of famous essays on the philosophy of mathematics and various other aspects; another, on historical and biographical material; another, on applications, paradoxes, games usw. The aim is a volume to stimulate and partly satisfy the appetites of the many who love mathematics without understanding or ever having glimpsed its multicolored tapestry. If I can also attract those who understand mathematics without loving it my reward will be so much increased.

If you are at all interested I will send you a rough outline of what I have in mind for your suggestions and criticisms. I can well understand if you have more important things to do than to go into this trifle. My only excuse is that your teaching is an important factor in my deep and continued interest in mathematics. I need scarcely add how much I learned from your lectures.

I have one specific request in regard to the subject of Probability. I intend to use for my chapter on probability the one in your excellent book with Cohen. I do not write, however, to trouble you with the details about permissions as these will be taken up in due course between S&S and Harcourt, Brace. I wondered whether you had suggestions as to additional material on Probability for a book such as I describe. My present intention is to use, in addition to your chapter, Peirce's Doctrine of Chances, a piece from Von Mieses' Probability, Truth and Statistics, and from the Fermat-Pascal correspondence. Pretty sketchy, as you can see, but more than an anthology would usually have. Last night I was rereading your volume in the Encyclopaedia of the Unified Sciences when I realized that that was better than almost anything else I had in mind but perhaps a little too technical. What would your view be? Is there some other essay that I may have read and forgotten or that I never did read, which ought to be in such a collection?

I trust I have not overburdened you with this request for advice. If you are in the midst of something just drop a penny postcard saying "Newman you are a nuisance, sincerely Ernest Nagel". I will understand. If, on the other hand, my little problem interests you as Mr. Holmes was wont to say, I should be deeply grateful for an extended reply or, any kind of a reply.

Sincerely,
James R. Newman

Ernest responded positively to his request, and Newman wrote three more letters outlining his ideas. His enthusiasm for mathematics and desire to assemble the best original sources is palpable. Here are some excerpts from these letters (Nagel 2020, Box 4, Folder 45).

July 20, 1944

Dear Dr. Nagel,

I am most grateful for the prompt and helpful reply to my letter. I shall certainly make it my business to come to New York after you have returned in the middle of September to spend a few hours with you . . .

1. The outline is now in something of a hodge-podge. I keep changing my mind about what I want. I will make a decent copy for you in a few days, setting forth the plan and a few of the tentative selections, and send it on. I am ashamed to send on what I have now especially since you have generously offered to look at it. . .

4. I like a few of Poincare's pieces on probability. Of course, I like almost everything he ever wrote whether I agree with it or not. He must have been a wonderful guy. There is the most marvelous photograph in Dantzig's Number book which makes me like P. all the better. Have you ever read his humorous section in the essay on Mathematics and Logic. The piece on Pasigraphy. I shall certainly include that. I also have a very funny piece by Stephen Leacock poking very scandalous fun at the mathematical physicists. Do you think Poincare's essay on Chance in the Foundations is a good choice? . . .

6. I have almost given up the idea of the Frank book. Knopf originally had the mss. And sent it to me – why me of all people I cannot imagine – for criticism. I hope, if you are a friend of Frank's this will not hurt you, but I would be less than candid if I did not say I thought it was terrible. Long, boring, confused, badly written, turgid, Germanic. I told Knopf it would have to be entirely rewritten if not scrapped but since Frank was a distinguished man this would have to be handled delicately etc etc. Knopf practically rammed the thing down my throat and said I had to do it, he couldn't get anyone else usw. I assured him, what would be evident to everyone but a publisher, that I had not the slightest competence in this field. I didn't know Einstein or modern physics or about Zionism. He just wouldn't take a no and insisted I meet Frank. This I did and we had a pleasant chat during which he was much too kind and I felt so ashamed of my sharp remarks about the mss. that I resolved to try my hand. I took the thing to England and in between raids I made a few attempts. But it was no go. I had neither the spirit nor the requisite knowledge. I had no books at my disposal. I had no background of personal experience. And above all I thought it was a pretty bad book, certainly in its present form. As you can well imagine there is no more thankless task than to rewrite someone else's bad writing. It is hard enough to wrestle with one's own. On one of my bomber trips across the Atlantic (I made 7 in one year) I met Knopf and told him he better find someone else but he pleaded with me and I again unwisely permitted myself to be flattered into it. All is vanities. Now I feel in a difficult position and don't quite know which way to turn.

What would one do a book like this for, I ask myself. Money? There is nothing in it. Renown? Certainly none of that. All I am is a glorious rewrite man. Well then for the sake of scholarship or culture, whatever that may mean. I admit these are by far more important considerations than the first two I mentioned. I should not want you to believe that one of your pupils had turned complete philistine and was deaf to the needs of learning. I can understand the real scholarship in your things and I can appreciate and envy your selfless devotion to them. But this is different. I have no creative contribution to make and I think the work is bad.

James R. Newman

Later, he sent an outline for his proposed series of books. It would consist of four parts. Part I would be didactic and cover Arithmetic, Algebra, Calculus, Geometry (both Euclidean and non-Euclidean), Trigonometry, Analytic Geometry, Transfinite mathematics, Probability and Topology. Part II would consist of Games, Curiosities, Paradoxes and Puzzles. Part III would have Miscellaneous pieces on Series, Function, Variable, Continuity, Number, Space, Time, Motion, Measurement, etc. Part IV would have longer essays on various topics. Below are excerpts from his last letter on this topic (Nagel 2020, Box 4, Folder 45).

September 5, 1944

Dear Professor Nagel,

I have been meaning to write for some time to thank you for your most helpful letter. I pray forgiveness for my rude neglect. The book is coming on so-so. That is not to say the enthusiasm has waned; merely that the more I do the more I find should go into the book and I have now got myself into a state where I am accumulating material fast enough for an encyclopaedia. What I will do when it is all gathered I will leave for Archimedes to decide. No living man could. . . .

Agreed that I should have something on group theory. I looked through Carmichaels book on your advice. It is good but too difficult. I am using a chapter from a book by Eddinton, New Pathways of Science, which I recall as very good. . . .

You are quite right, there is a great fund of material in De Morgan. I have him in the back of my head all the time but have settled on nothing yet. I must look at his Double Algebra and also at his Study of Mathematics . . . .

The Jourdain book (Nature of Math.) is an excellent suggestion. I know the book well having read it a number of years ago with great enjoyment. It surprises me it was never printed in this country. . . .

Back to my knitting. The Jourdain book on the Philosophy of B*rtr*nd R*ss*ll is very funny. I looked at it the other day just before your letter arrived. I intend to use pieces . . .Thank you many times for the lead to Speiser. His books are wonderful. Not only his chapter on group theory and ornamentation and design, a portion of which I shall certainly use, but also one of his excellent little books Klassische Stucke der Mathematik . . .

I will look at the Royce Essay in the Encyclopaedia of the Philosophical Sciences. I have a great deal on this subject, very good and mostly Nagel. Your piece before the Statistical Association was superb. Now I must have something from that, from your Encyclopaedia piece, something from your text. I'm glad you haven't written another text book on probability. Then I would really be confused. (I have you down, somewhat presumptuously I admit, for a book in the 25 cent series of people's books I mentioned before, on probability. I think you would enjoy doing it and I can scarcely wait to read it.) ((May I have your views on my presumption)). . . .

Best regards,

James R. Newman

The project he discussed in these letters was finally published in 1956 by Simon and Schuster in four volumes as *The World of Mathematics*.

Ernest's collaboration with Newman on *Gödel's Proof* began later. They wrote a popular exposition on the proof for the *Scientific American* in 1956. This article later appeared in Newman's four volume series *The World of Mathematics* which became a best seller and is still in print today. They decided to expand the article and have it published by NYU press. They wanted to include a translation of Gödel's 1931 paper on undecidable propositions together with notes for lectures he had given at Princeton in 1934. Their editor, Allan Angoff, approached Kurt Gödel with their request and Ernest even paid him a visit in Princeton. Gödel responded by setting three conditions which they would have to satisfy: 1) that he be allowed to write an introduction to the Appendix in order to mention advances made since publication of his (Gödel's) book, 2) that he be given a share of the royalties of the book and 3) that he be allowed to see the manuscript and make editorial changes.

Ernest, who was usually described as "mild mannered" by those who knew him, reacted angrily. He found the demand for a share of the royalties distasteful and greedy but was prepared to accept it. However, what really made him furious was Gödel's demand for editorial control of the book. Angoff tried to avoid direct conflict, but finally Ernest wrote to Gödel that he found the latter's demands unacceptable and the he and James Newman refused to accept them:

> ... I must say, quite frankly, that your ... stipulation was a shocking surprise to me, since you were ostensibly asking for the *right to censor* anything of which you disapproved in our essay. Neither Mr. Newman nor I felt we could concur in such a demand without a complete loss of self-respect. I made all this plain to Mr. Angoff when I wrote him last spring, though it seems he never stated our case to you. I regret now that I did not write you myself, for I believe you would have immediately recognized the justice of our demurrer ...

In the end, the book was published without the translation or any other contributions by Gödel (see Feferman 2009).

During the academic year 1959–1960 Ernest took a sabbatical leave from Columbia to visit the Institute for Behavioral Sciences at Stanford University. During that year the Nagels renewed their acquaintance with Stanford Physicist Richard Hofstadter who had taken classes with Ernest at Columbia and the Nagels' son Sandy met the Hofstadters' son Doug in the High School they both attended there. Doug was also eager to meet Ernest because he had read the book *Gödel's Proof* which Ernest had written with James Newman and Doug shared his interests in language, symbolism and logical puzzles. The Nagels invited Doug to visit them in Vermont during the summer of 1960. He later wrote about this visit in the preface to the 2001 edition of *Gödel's Proof.*

> For me, autumn in New York, 1972, was above all an intense time of communing with Ernest and Edith Nagel, and it solidified our friendship in an indelible fashion...
>
> I left New York in 1973, and that marked the beginning of my own intense personal involvement with Gödel's theorem and the whole rich sphere of ideas surrounding it. Over the next few years, I developed an idiosyncratic set of explorations on this nexus of ideas, and wound up calling it "*Gödel, Escher, Bach: an Eternal Golden Braid*". It is certainly fair to say that the parents of my sprawling volume were Nagel and Newman's book, on the one hand, and Howard DeLong's book, on the other.

Douglas Hofstadter became well-known for his work in Artificial Intelligence at Indiana University. Although he had been inspired by Nagel and Newman's book *Gödel's Proof* he differed with some of their ideas later because of his own work with computers and artificial intelligence. As he wrote:

> My book, despite owing an inexpressible debt to the ideas of Nagel and Newman, does not agree with all of their philosophical conclusions, and here I would just like to take a few lines to point out our principal differences of opinion. In their "Concluding Reflections", Nagel and Newman argue that from Gödel's discoveries it follows that computers – "calculating machines", as they call them - are in principle incapable of reasoning as flexibly as we humans reason, a result that supposedly ensues from the fact that computers follow "a fixed set of directives" (i.e., a program)...
>
> To be sure, we have not yet come close to producing a computer program that has anything remotely resembling the flexibility and power of the human mind, and it is in this sense that Ernest Nagel and James Newman were exactly on the mark in declaring, in their inimitable poetic fashion, that Gödel's theorem "is an occasion, not for dejection, but for a renewed appreciation of the powers of creative reason." No one ever said it better.

During his sabbatical leave at Stanford, Ernest visited Rudolf and Ina Carnap in Los Angeles. Carnap had been hired by UCLA after Hans Reichenbach's death in 1953.

They had lived in Los Angeles since 1955. Ernest and Rudolf Carnap had a disagreement over his contribution to the Schilpp volume on Carnap and had not talked much since then. However, they both urged him to visit (Nagel 2020, Box 1, Folder 44):

December 12, 1959

Very dear Ernest (and Edith, if she is with you, which I don't know):

Somewhat miraculously you are rather near us, have braved the unknown dangers of coming West of the Mississippi – and I hope you found it well worth it. We have come to like the climate, don't miss the city winter weather. Of course, now that you are so near, we want to see you very much...

I have heard all sorts of nonsense about people having given you the idea that Carnap is greatly put out (or whatever the word was) over your Schilpp-contribution. Well, you have seen his reply, but that sort of wipes the slate clean. What did annoy him was the fact that a few talks between you two, in the years when we were in Princeton, thus even nearer you than now, would have cleared up so much of the points of controversy, which will now be eternalized in the Schilpp volume. (Though I just read that due to chemicals in today's paper most books now printed will no longer be in existence in 50 years – a rather short eternity). I think, the feeling was that – knowing that he was a careful worker – you should have known that most of your points had been considered by him (and for one reason or another either been taken into account or dismissed, as the case may be, you know that I don't understand details). But since we both are very fond of you and have all our previous meetings in excellent memory, there does not seem to be any reason why this Schilpp business should in any way come between us in our personal relationship! Let's meet while there is still time and for once we are not separated by a whole continent! We have seen you several times on TV (and there are not many people for whom I would turn it on at 8 A.M. on a Sunday), and there was always this joyful recognition that there is our good friend Ernest! ...

Well, I hope you will say that you will come – it seems to me we have not had good long talks since that first summer in Cambridge when you stayed with us and I made your life miserable by making you clean up your plate in the good economical European fashion.

Love and our very best, Yours,

ina

[handwritten addition by Carnap]

Dear Ernest I hope very much that we will see you here before or after the S[anta].B [arbara]. Meeting; at the meeting there will hardly be time for relaxed, unworried talks. Cordially yours, Carnap

Ernest did not attend the meeting in Santa Barbara, but he visited them for a weekend in April 1960. It may have been the last time he saw Ina. She committed suicide on May 26, 1964. Apparently, she had suffered from severe depression for many years (Creath 1990, 39).

The Nagels continued to spend summers at their house in Vermont. It gradually expanded from a two-bedroom cabin to a rambling, but crooked 5-bedroom house with the addition of children and aging parents. In this country setting they were able to continue their friendships with old friends and colleagues, several of whom moved away from New York. Sidney and Ann, for example, moved to Stanford after he retired from NYU. Below are pictures of Ernest, Edith (Fig. 2.33) and Sidney (Fig. 2.34) taken by Sidney's son Ernest (Ben) Hook in Vermont in 1983.

Fig. 2.33 Edith and Ernest Nagel in 1983. (Private Collection of Yvonne Nagel)

Fig. 2.34 Ernest Nagel and Sidney Hook. (Private Collection of Yvonne Nagel)

2.14 Later Life

Ernest's professional career flourished. He was elected to the National Academy of Sciences and, after a brief move to Rockefeller University in 1966–67, became a University Professor at Columbia. In later life, he received many honorary degrees.

He gave the Carus Lectures in 1963 on "The Dimensions of Critical Philosophy". These are a prestigious series of three lectures funded by the Open Court Publishing Company. Each speaker is supposed to write up his or her lectures for publication. The first lecture he gave was on the philosophy of natural sciences, the second on the philosophy of law and the third on the philosophy of history.

He worked on writing them up during his retirement and succeeded in revising the lecture on philosophy of law, expanding it from the original lecture of 20 pages to about 50 typewritten pages. He talked about his plans in the 1983 Oral Interview:

> Now the book on the structure of science, I had a contract to do the manuscript in 1941 and the book wasn't published until 1961. So there was a long delay. In addition to that, the original plan was to have either a tremendously large one volume or a two volume book. Now in the second volume what I planned to do was to deal mainly, although not exclusively, with the problem connected with probability and induction. I felt really very regretful that I haven't been able to do that and now it's too late for me to do that because the literature in that area has grown tremendously and I couldn't control it any more. It would be a considerably younger man than I am who would have the energy to go through it and make some systematic presentation of the main ideas and of my own views on this. So it would have to be one of these truncated affairs, just a one volume rather that a two volume book.
>
> I suppose a substitute for writing the second volume is the Carus lectures and there again I let time slip by without paying attention to the things as soon as -- many people who are Carus lecturers pretty much had the book written before they gave the lectures. I mean this I think was probably true for Dewey who was the first Carus lecturer. It was also true of the late philosopher at Chicago, George Mead. It was certainly true for Professor C.I. Lewis. He gave just three lectures and they were on rather abstract things and they were already written, they are now chapters in this book called *Analysis of Knowledge and Valuation*. It's a tremendously written book. It's probably over-written, but he had no problem in publishing it pretty shortly after he gave the lectures.
>
> Whether there's enough time for me to do the edition, Lord knows and He won't tell. (Neuber and Tuboly 2022, 306–307)

Here is an excerpt from the biography of Ernest Nagel written by his former student Patrick Suppes, a distinguished philosopher at Stanford University:

> Nagel's many contributions to the philosophy of science are discussed below, but what is most important to emphasize about his more than forty years' association with Columbia University is the central role he played in the intellectual life of Columbia, and more generally, of New York City. To many generations of students he was the outstanding spokesman of what philosophy could offer in terms of analysis of the scientific method, as it is practiced in many different sciences, and in the relation between science and perennial problems of philosophy such as those of causality and determinism. What is important about this influence is that it was not simply students of philosophy, but students of many different disciplines whom he influenced in a way that many of them still remember. He saw his principal role as that of a philosophical critic of ill-conceived notions from whatever quarter they might come. It is this critical spirit of analysis and reflection that he especially communicated to others. He was properly skeptical of philosophical edifices built

> independent of detailed scientific considerations. But he was equally critical of the writings of scientists who too blithely thought they could straighten out their colleagues on fundamental philosophical questions without proper knowledge of the many issues involved. (Suppes 1994/2012, 470–471)

Suppes went on to describe Ernest's wide-ranging philosophical interests including: Causality, Explanation and Laws, the Foundations of Measurement, Geometry and Physics, the Theories of Induction, Scientific Explanation in Biology and Methodology of the Social Sciences. He also said:

> His closest colleague, personally and philosophically, was probably Sidney Hook, who also taught in New York City for many years, primarily at New York University. Like Dewey and Hook, Nagel also enjoyed the wider arena of intellectual and political life in New York. He wrote extensively for such publications as *Partisan Review* and *The Nation*, as well as for the standard scholarly journals. With these many different interests and engagements he occupied a position, especially in the intellectual life of New York City, that extended far beyond the boundaries of academic philosophy. Within the university Nagel interacted with colleagues in the sciences in a way that was unusual then, and is unusual now, for philosophers. For example, he gave for many years a famous seminar with Paul Lazarfeld on the methodology of the social sciences, which was widely attended by social scientists as well as philosophers at Columbia. His interest in current research in physics continued well into retirement. It is not common practice for philosophers to be elected to the National Academy of Sciences, and there is no special section to which they naturally belong. His election was a tribute to Ernest Nagel's wide-ranging interests and extensive substantive knowledge of many different branches of science. It is fair to say that the range of his scientific interests and knowledge exceeded that of any other philosopher of science of his generation in the United States. (Suppes 1994/2012, 471)

Ernest retired from Columbia in 1970 but continued to participate in seminars. He spent the Spring semester of 1976 at the Hebrew University in Jerusalem. Edith accompanied him. It was their first overseas trip since they visited Europe together in 1935.

Ernest died in New York on September 20, 1985, and Edith on March 18, 1988. They are buried in the cemetery at South Wardsboro, Vermont, near their summer vacation home.

Acknowledgments For over 20 years, we stored two boxes of letters, photos and other memorabilia from my husband's parents in our basement. These boxes moved with us, unopened, silent witnesses to another era. Finally in 2008 a letter from Susan Goulian, Sidney Hook's daughter, prompted me to open the boxes and read their contents.

The main focus of this book was to try to convey something of the intellectual environment that Ernest Nagel worked in—especially during the years 1934–1950. Ernest thrived on philosophical discourse. For many years, he and other scholars met to discuss philosophical ideas in the self-styled "New York Philosophical Circle" inspired by similar groups in Europe such as the Berlin and Vienna Circles and the Aristotelian Society in England. During the years 1934–40, Nazi invasions of other European countries sent a flood of refugees to England and the U.S. Many of them passed through New York. Ernest helped many of them, as well as his own relatives from Czechoslovakia, with the all-important affidavits and visas. He also helped them get settled in the U.S. and sometimes found jobs for them. The academic refugees would often join the discussion group.

I received much help and encouragement from my family, Alex, Sidney, Kathy and Rebecca Nagel, Sidney Hook's children, Susan Goulian and Ernest Hook, as well as from Dale Reed, manuscript cataloguer at the Hoover Institution of Stanford University where Sidney Hook's papers reside. In addition, I also drew heavily on the Oral Interview of Ernest Nagel conducted by Remmel Nunn of Columbia University in February and March of 1983. I would also like to thank Elaine and Alice Morton, the daughters of Miriam (Bidner) Morton.

Like many others, I found information through Google, Wikipedia and other on-line resources. However, my primary sources are the letters Ernest wrote and the letters he received from others. I have tried to quote from them, when possible, to give the reader some idea of what they were like and the times they lived through. This biography covers the period from 1900 to 1950 and especially the years 1919–1945 when Ernest grew from a bright but unformed high school student into a Columbia Philosophy professor.

I would also like to thank Adam Tamas Tuboly who read the manuscript carefully and provided many useful comments.

The help of the Archives of Scientific Philosophy shall be acknowledged too, where Ernest Nagel's letters are stored now: Ernest Nagel Papers, 1925–1985, ASP.2020.01, Archives of Scientific Philosophy, Archives & Special Collections, University of Pittsburgh Library System.

The photos used and reproduced here are from the the family's private collection. All rights reserved.

References

Cohen, Morris R. 1949. *A Dreamer's Journey: The Autobiography of Morris Raphael Cohen.* Glencoe: The Free Press.

Creath, Richard, ed. 1990. *Dear Carnap, Dear Van: The Quine-Carnap Correspondence and Related Work.* Berkeley: University of California Press.

Feferman, Solomon. 2009. Gödel, Nagel, Minds, and Machines. *Journal of Philosophy* 106 (4): 201–219.

Feferman, Anita, and Solomon Feferman. 2004. *Alfred Tarski: Life and Logic.* Cambridge: Cambridge University Press.

Feigl, Herbert. 1969/1981. The Wiener Kreis in America. In *Herbert Feigl: Inquiries and Provocations, Selected Writings 1929–1975*, ed. Robert S. Cohen, 57–94. Dordrecht: D. Reidel.

Hook, Sidney. 1987. *Out of Step: An Unquiet Life in the 20th Century.* New York: Harper & Row.

Nagel, Ernest. 2020. *Ernest Nagel Papers, 1925–1985.* ASP.2020.01, Archives of Scientific Philosophy, Archives & Special Collections, University of Pittsburgh Library System.

Neuber, Matthias, and Adam Tamas Tuboly. 2022. Interview with Ernest Nagel by Remmel Nunn. In *Ernest Nagel: Philosophy of Science and the Fight for Clarity*, ed. Matthias Neuber and Adam Tamas Tuboly, 257–314. Cham: Springer.

Quine, W. V. O. 1970. Homage to Rudolf Carnap. *PSA* 1970:xxii–xxv.

Suppes, Patrick. 1994/2012. Biographical Memoir: Ernest Nagel, November 16, 1901 – September 20, 1985. *The Journal of Philosophy* 109 (8/9): 470–478.

Verhaegh, Sander. 2022. Nagel's Philosophical Development. In *Ernest Nagel: Philosophy of Science and the Fight for Clarity*, ed. Matthias Neuber and Adam Tamas Tuboly, 43–65. Cham: Springer.

Verhaegh, Sander. 2025. Columbia Naturalism and the Analytic Turn: Eclipse or Synthesis? In *American Philosophy and the Intellectual Migration: Pragmatism, Logical Empiricism, Phenomenology, Critical Theory*, ed. Sander Verhaegh, 77–99. Berlin: De Gruyter.

Chapter 3
Another Life in Letters: Ernest Nagel and Philosophy of Science in the Nagel-Hempel Correspondence

Roberto Gronda and Adam Tamas Tuboly

Abstract This chapter serves as an introduction to the Nagel-Hempel correspondence. It contextualizes the relevance of philosophical correspondences, provides details about Hempel's life and background leading up to his emigration, and discusses the changing relationship between Nagel and Hempel with respect to the twentieth century American philosophy of science. Finally, it provides a narrative about the demise of Nagel in the 1960s.

Keywords Ernest Nagel · C. G. Hempel · Analytic philosophy · Logical empiricism · Pragmatism · Thomas Kuhn

3.1 Introduction

The community of historians and philosophers (of science) now has an easier and broader access to some of the most important correspondence relating to twentieth-century scientific philosophy. Many archives have already scanned (some of) their materials, and numerous scholars have put work into publishing the most frequently cited exchanges. Thus the Carnap-Neurath correspondence is now available in its entirety (Damböck et al. 2024; parts were published in Cat and Tuboly 2019), just like the letters between Moritz Schlick and Ludwig Wittgenstein (Iven 2015), and two volumes of Paul Feyerabend's correspondence have also appeared recently (Collodel and Oberheim 2020, 2024). One can also think of the Viennese initiative (under Christian Damböck's lead) of the "virtual archive of logical empiricism" (VALEP), which contains in open access form countless letters between major

R. Gronda
University of Pisa, Pisa, Italy
e-mail: roberto.gronda@unipi.it

A. T. Tuboly (✉)
Institute of Philosophy, ELTE Research Centre for Humanities, Budapest, Hungary

Institute of Behavioural Sciences, Medical School, University of Pécs, Pécs, Hungary

R. Gronda, A. T. Tuboly (eds.), *Ernest Nagel on Science and Philosophy*, Vienna Circle Institute Library 14, https://doi.org/10.1007/978-3-031-99458-6_3

figures of logical empiricism.[1] One aim of the present volume is to make the long and rich correspondence between Ernest Nagel and Carl G. Hempel available. In this editorial introduction to the letters, we have the following goals.

In Sect. 3.2, we will provide a brief background to Carl G. Hempel; Nagel's life has been dealt with extensively in Yvonne Nagel's biography in Chap. 2.[2] The section will provide information to allow a better understanding of what happens in the correspondence, and it provides some context to assess those events that Hempel and Nagel report about and react to in their letters.

Then, as an example, in Sect. 3.3, we show how Hempel was trying to acclimatize in the States and how he reached out to Nagel for help and assistance. Nagel's role in helping emigrant scientists and philosophers is still underrated (in contrast to Warren Weaver, James Conant, or Harlow Shapley—see Jones 1984). But because of his broad knowledge of other movements, traditions and schools (Verhaegh 2025b, 16), Nagel was in a suitable position to help others with policy recommendations and to assist them in their "job-hunt", as Hempel called it. Interestingly, however, as time went on, and European scholars slowly took over important positions in scholarly associations, journals, and departments, greenhorn emigrants became leading advisors to native philosophers (on that process, the best available and comprehensive account is provided by the chapters in Verhaegh 2025a).

The Nagel-Hempel correspondence is, of course, more than just a diary of who did what and when. One might look at all their reports of meetings with colleagues and friends, the reading-experiences of each other's papers, conference-reports, travelling, and similar data from a traditional point of view and declare them as philosophically non-relevant existential matters. But if one attributes any importance to the role of historical context in understanding how to assess certain arguments, or looking for conceived, but rejected alternatives, or even unconceived paths, then a correspondence such as the Nagel-Hempel, could be a goldmine. We do not state, of course, that such historical, or nuanced archival materials should or could replace systematic philosophical argumentations; but we do claim that to understand the success or failure of philosophical arguments, positions, and schools, it is inevitable to consult such contextual and historical materials that could shed new light on important facts of the matter. We release the Nagel-Hempel correspondence in the hope that scholars and colleagues will find within it something of interest and importance for their research—whether they are historians of analytic philosophy, of philosophy of science, of logic, or of American philosophy.

[1] See https://valep.vc.univie.ac.at/virtualarchive/public/Archive/a0. We would like to express here too our indebtedness to the staff of the Archives of Scientific Philosophy in Pittsburgh for their help and work on these materials. For the relevant correspondence of Hempel and Nagel see Nagel (2020) and Hempel (1999).

[2] For longer treatments of Hempel's life and work, see Fetzer (2022) and Hempel (2000).

3.2 Carl Gustav Hempel

Carl Gustav Hempel was born on January 8, 1905, in Eden bei Oranienburg, near Berlin. He studied in the newly established Realgymnasium in Berlin-Pankow, where he graduated on March 3, 1923. He started his university-education in Göttingen for two semesters between the Easter of 1923 and 1924; he enrolled at the University of Heidelberg for the summer semester of 1924 (to study mathematics and natural science), then he moved to the philosophy faculty of the University of Berlin in October 1924 till late 1928 (for eight semesters). There, he studied experimental chemistry, physics, mathematics, and philosophy. His teachers in Berlin included Max Planck, Max von Laue, John von Neumann, Richard von Mises, Wolfgang Köhler, and after the summer semester of 1927/28, Hans Reichenbach. In Göttingen he had studied with David Hilbert, Paul Bernays, Emmy Noether, Heinrich Behmann, and Richard Courant.

In early November 1929, Hempel "passed with distinction" his written and oral exams for teaching mathematics, physics and philosophy at secondary schools (his topics were the status of the decision problem in mathematical logic and causality and the problem of free will in contemporary German natural philosophy, including Reichenbach, Schlick, Driesch, and Planck). While Hempel enrolled at the philosophical faculty of the University of Vienna in the same month—and took courses on philosophy of nature with Schlick, introduction to philosophy with Carnap, and finished his studies in March 1930—he started to teach in different schools in Berlin between 1931 and 1933. A year later, in February 1934, he completed and defended his doctoral dissertation summa cum laude in Berlin.[3] Hempel was working on his dissertation (*Beiträge zur logischen Analyse des Wahrscheinlichkeitsbegriff—Contributions to the logical analysis of the concept of probability*) with Reichenbach, but after Hitler seized power in 1933, Reichenbach was dismissed (and went to Turkey) and Hempel was nominally taken on by two of his previous teachers, Wolfgang Köhler and Nicolai Hartmann (Jeffrey 2000, viii; the dissertation was published as Hempel 1934).

Without much chance for a job, Hempel went to Brussels where he was hired by Paul Oppenheim in April 1934. Oppenheim was a Jewish scientist and a wealthy industrialist who emigrated to Brussels from Germany after the Nazis seized power in 1933. He had a wide scientific and philosophical interest, though he always combined the two and worked on such topics as the comparison of the sciences, concept formation in various sciences, psychology, logic, and the philosophy of science. As Nicholas Rescher (1997, 339) recalled, "a person of broad intellectual and scientific culture, Oppenheim was a collector of people", who worked mainly in collaborations. It was Hans Reichenbach who established the acquaintance between Oppenheim and Hempel in 1933 (Oppenheim 1969, 1), and they developed such a friendly and collegial relationship that Hempel became "a privately supported

[3] Information about Hempel's education is taken from his diplomas and certificate that are stored in his archive, in Hempel (1999) from box 2, folder 1 to 20.

research associate to Oppenheim, and lived in his establishment as a houseguest" (Rescher 1997, 340).

In Brussels they worked together on various topics, the most important and fruitful being the "notion of type in the light of the new logic" (Hempel and Oppenheim 1936). In their joint book, they analyzed various non-quantitative concepts of orders, types, comparison, and categories—especially in the psychological and medical sciences—to show that with the new logic even the fluid and multidimensional fields of experiences could be grasped. When Nagel (1936, 611) reviewed the book, he concluded by saying that for students of psychology and social sciences, "this book should be required reading."

In 1937, Carnap negotiated the chance of having assistants at the University of Chicago, and after some troubles he secured 1000–1000 dollars to hire Olaf Helmer and Hempel in April, and another 1400 dollars from the Rockefeller Foundation in May 1937.[4] In September 1937, after the *Third International Congress for the Unity of Science* in Paris, Hempel boarded a ship and sailed to the United States where he became a research associate to Carnap at the University of Chicago until June 1938.[5] Although he was not officially able to have his own courses as an assistant, he did lead courses with Carnap on, for example, "Logistik", "Semantik", and "Problems in the Logical Foundation of Mathematics" at the University of Chicago.

During his stay, Hempel regularly gave lectures in Nagel's philosophical group in New York, and in Morris' in Chicago. Although he tried to establish himself as a young scholar to become visible on the national job-market as well, Hempel did not have much of a chance in the States in general, and especially not in Chicago. As Sander Verhaegh (2025b, 1) notes, between 1933 and 1945, more than 6000 scholars tried to apply officially for some kind of help and assistance in the States, but "only a fraction of them, less than six percent, were granted financial assistance." Although Hempel did find help and a position, it is understandable that as a young postdoc he was in a much worse situation than many of his senior colleagues (like Philipp Frank, who was first on a major lecture tour in the States, living from the royalties of his Einstein book—that were paid in advance before publication—until he secured a permanent position (Reisch and Tuboly 2025)).

In Chicago it was even more difficult due to the already longstanding conflict between Thomists and positivists that had begun with Carnap's appointment in 1936. In his diaries, Carnap noted that Richard McKeon, dean of the Humanities in Chicago, was concerned about the growing influence of positivism—a concern that was also shared by the Philosophy Department, where Hempel was considered as a representative of "dogmatic positivism." Carnap had the feeling that Charner Marquis Perry—a philosopher and educator in the department—and McKeon would do anything to prevent Hempel from gaining a foothold in Chicago.[6]

[4] See Carnap's diary entry on May 3, 1937.

[5] See Hempel's immigration documents in the archive Hempel (1999), box 1, Folder 3.

[6] See the entries of Carnap's diary on October 8, 1937, February 23 and March 2, 1938.

Due to the animosity of the Thomist leaders of the University, Carnap was unable to secure further funding for Hempel, who thus had to return to Brussels in June 1938. Hempel worked there again with Oppenheim until January 1939 and then travelled back to the States on SS Bremen. Without any concrete plans or offers of a position, only Nagel was waiting for him in the dockside of New York.

Between March and June 1939, Hempel was again working with Carnap and attended his classes on semantics as an assistant. After the job with him, Hempel went back to New York where he was finally able to settle as an Instructor at City College for 1939–1940, and in 1940 he became an Instructor at Queens College, Flushing, NY. After a few very productive years (mainly with Oppenheim, working on confirmation and explanation), Hempel became an Associate Professor at Yale University in 1948, and moved to Princeton in 1955 when he was appointed Stuart Professor of Philosophy. He retired there in 1973, though he did not stop teaching or working and became a professor again, this time at Pittsburgh, from 1977 to 1985, the year Ernest Nagel passed away.

During his long and fruitful career in the States,[7] Hempel not only published several papers, a few monographs, and a large collection of his essays, but also set the agenda for scientific and analytic philosophers for decades. In that respect, he was perhaps even more influential than Nagel, who was raised intellectually and socially in the States and had very good relations with various local philosophers and movements from the very first day of his professional career.

Numerous fields and topics have been associated with Hempel's name: the concept of explanation (especially the deductive-nomological model), the philosophy of history, the notion of confirmation (and its paradox), the critical revision of the positivist criterion of meaning, concept formation, science and values, and it was Hempel who made the story of Ignac Semmelweis famous among philosophers of science in his book on the philosophy of natural science. All these issues produced countless responses, discussions, criticisms and refinements, and many of them were used to define what *analytic* and *scientific philosophy* was in the second half of the twentieth century.[8]

As the editor of one collection about his philosophy stated,

> [Hempel's] influence was immense. At least twelve of his papers were instant classics, which not only exerted profound influence upon professional philosophers of science but spawned dozens and dozens of critical commentaries and learned replies. His influence upon teaching was similarly enormous. His *Philosophy of Natural Science*, for example, became the instant text of choice for introductory courses worldwide. Page for page, his achievements were stunning. And as a human being, he was exemplary. As a philosopher, as a teacher, and as a human being, "Peter" was a paradigm. (Fetzer 2000, vii)

---

[7]It must be noted, though, that Hempel had published numerous papers while he was still in Europe (in English, French, and German). He wrote about the concept of "truth" in the protocol-sentence debate within the Vienna Circle, probability theory, formal logic, and the philosophy of psychology.

[8]This is very well-documented in Hempel's *Festschrift* (Rescher 1969), regarding *who* wrote about *what* topics.

It is an important topic for historians of philosophy of science to understand how different characters exerted different types of influences in the formation of twentieth-century philosophy, and how—counterfactually speaking—could have been able to develop the profession in different directions if socio-political, cultural, and even perhaps scholarly life had taken an alternative road. Hempel was a much more focused philosopher who undertook exploratory work and thus gave impetus to the new profession of (rather technical) philosophy of science. Nagel, on the other hand, was a traditional intellectual with a broad range of knowledge about culture, natural, social and human sciences, who connected various dots from the history of human thought and thus whose work cannot be categorized as easily as "analytic philosophy". As Hempel (2000, 33) recalled later, "I thought of [Nagel] also as a modern kind of Aristotle. He was enormously widely read; I could not touch him in this respect." But because of that, Nagel's influence is very often restricted to those issues and topics that dominated the analytic tradition (like reduction, induction, explanation) not least due to the much more technical works of Hempel.

3.3 Changing Roles and Acclimatization

The letters between Hempel and Nagel show that their relationship had changed slightly over time.[9] Until the end of the Second World War, Nagel was not only a friend to Hempel, but also a local philosophical authority. Hempel had countless questions about philosophy, institutional policy, cultural habits, personal relations, and manners of the American way of life and most of his achievements and actions were shaped by Nagel's efforts. Hempel noted after Nagel's death that he was a "wise and sympathetic counselor", who put him "in touch with stimulating thinkers and lively professional groups" (letter 188). On a more personal level, it was Nagel who helped Hempel to arrange and pay for a nanny after Eva's death in 1944 to take care of the young Peter Andrew (letter 188). In his Application for Certificate of Identification, Hempel gave Carnap and Nagel as references to vouch for his loyalty.[10]

Soon after his arrival in October 1937, Hempel asked questions about submitting his paper to the conference of the Eastern Division of the American Philosophical Association (letter 5) and Nagel lobbied for Hempel and helped him to be accepted at the conference (letter 6). There was not much Hempel could offer in exchange, and he was shy enough to apologize to Nagel repeatedly and deeply in almost every letter. But he did not have many choices and as Nagel's "advice would be particularly valuable" (letter 5), he took the trouble and contacted him with countless

[9] Although we do not have the early letters of Nagel, we could infer their contents from Hempel's letters, and that would still support our claims in the main text.

[10] See the file in the Hempel Archive, Hempel (1999), Box 3, Folder 5.

questions. "Ernest would not accept a word of thanks, regarding it all as a matter of course", said Hempel later (letter 188).

Besides solving the trouble of attending the conference, Hempel asked Nagel to integrate him into his philosophical and scientific circles (letter 6) in order to socialize and make new acquaintances for his soon-to-be-commenced search for a permanent position.[11] As Fons Dewulf (2022, 70) noted, "Nagel immediately integrated Hempel in the flourishing scene of New York Philosophy," invited him to his lectures and courses at Columbia, and thus provided important perspectives, and more—for example the topic that made Hempel famous among analytic philosophers even today, the nature of explanation.

Hempel was active on his own too, however, as he established a circle, a "Scientific Method Group" in Chicago, that basically grew out of Carnap's seminar (letter 7) and provided Hempel a chance to meet some of the people in Chicago who despised Thomism and showed any interest in scientific philosophy.

Hempel realized that logical positivism was facing serious attacks in the States and as they were new in the country, they had to familiarize themselves with the local routines and customs to provide sufficient and meaningful replies. Thus, Hempel again asked Nagel about how to handle criticisms against the movement, whether it was worth attacking back, whether one should just simply reply in a neutral tone, and how to navigate in this new landscape of philosophers in general (letter 10). After all, Hempel remained a firm interpreter of positivism, who delivered numerous lectures to various audiences,[12] calmly explained the main insights of scientific philosophy, sought out peaceful connections by familiarizing himself with the literature and traditions (see below) and avoided heated public debates. In other words, Hempel wrote the following to Nagel after meeting Howard Dykema Roelofs (professor of philosophy at the University of Cincinnati) for a reason:

> Nevertheless, Roelofs told me, afterwards, that he had enjoyed my talk, and that I was the first positivist, after Schlick, with whom he found it possible to discuss; and he would be glad to see me come to Cincinnati. (Letter 19)

Schlick was twice a visiting professor in the States (at Stanford in 1929 and at Berkeley in 1931–32), he was held in high esteem, and several of his papers appeared in American journals; his book on the problems of ethics was translated by one of his students, while another wrote a dissertation that was connected to Schlick in various ways (see Tuboly 2021, 32).[13] Hempel thus followed Schlick's

[11] Hempel sent a similar request to Nagel for help on the East Coast when he was about to familiarize himself with the local philosophers and logicians (letter 7).

[12] Hempel also took every chance to lecture in Chicago, for example at the Social Science Research Group, or in Charles Morris' group (letter 9). Later he was one of the speakers at the Eastern Colleges Science Conference at Vassar College, in New York, in late February 1947 (a series that continues to this day). See "Scientists to Address Conference on Social Implications of Science", *Vassar Chronicle*, February 7, 1947, p. 4.

[13] On Schlick's visits and their impact for the future of logical positivism in the States, see Verhaegh (2020).

footprints and caused less turmoil in general than Carnap or Neurath did (on the latter see Burke and Tuboly 2025, Chap. 2).

By December 1938 Hempel had noted to Nagel that they were not "greenhorns" in New York anymore, perhaps referring to the infrastructure and way of city-life there, but one could also read the passage as indicating that the Hempels were familiarizing themselves relatively quickly with the *American* way of life (letter 15). But living in the States was one thing, working there permanently was another. After his second stay in the States, but still before World War II, Hempel started "job-hunting" and asked again for Nagel's help (letter 21). Seemingly Nagel was able to intercede for Hempel as he was offered a temporary teaching position at the City College of New York for the summer of 1939 (letters 21, 22, 23). To make the situation scholarly productive and economically bearable, the Hempels took the Nagels apartment for the summer, while Ernest and his wife Edith went to Vermont, their usual summer residence (see letters 23ff).

As soon as the CCNY job was over, Hempel needed another, more solid position in order to remain in the States, and yet again it was Nagel who wrote a "long and illuminating" letter about Hempel to Paul Klapper (letter 31), who was the first president of Queens College. Klapper hired Hempel as an instructor, then assistant professor at Queens College in New York, and Hempel remained there until 1948. During his first long-term and substantial position in the States, Hempel was still relying on Nagel's expertise and knowledge in designing his lectures and structuring his participation in the "contemporary civilisation" course (letter 30).[14]

Such a state of profound asymmetry between Nagel and Hempel did not last long. Indeed, Hempel was very successful in building upon the institutional connections that Nagel had laid out for him. He quickly rose to philosophical prominence, a status sealed by his appointment first as Associate Professor of Philosophy at Yale in 1948 and then as Stuart Professor of Philosophy at Princeton in 1955.

For almost a decade, Nagel and Hempel walked almost hand-in-hand to exert their influence on the field of philosophy of science. It was not unusual for them to receive offers from the same institutions or to be invited to the same conferences—so they often discussed their choices in their letters, exchanging information about departmental environments, salaries, costs of living, and so on (see, for instance, letters 64, 66, 77, and 78 on Princeton, or 108 on Stanford). They most often agreed on the judgments of their colleagues, and shared an agenda on what philosophy of science should look like.

The latter point is made particularly clear in their exchange about Tarski's request to Nagel—who was then (1963) President of the Philosophy of Science Association (PSA)—to "appoint some member [sic] of the PSA residing in the vicinity of Berkeley to help him administer the distribution of the grant" for which the US National Committee of the International Union of History and Philosophy of Science

[14]The topic was familiar to Nagel as he took a long "philosophy of civilization" course at City College from Morris R. Cohen (see Nagel and Nunn 2022, 310). That course was called Morris' "most influential course" (Randall and Herman 1949, 306).

(IUHPS) had applied (letter 140). The issue was strictly financial: it was about who should receive funds to cover travel expenses to the 3rd International Congress for Logic, Methodology and Philosophy of Science, organized by Yehoshua Bar-Hillel in Jerusalem in 1964. But Nagel's concerns were institutional: he feared that giving Tarski all the power to decide on the allocation of funds would lead to the selection of the positions and approaches to be represented at the Congress being too narrow. In Nagel's words, "the names that occur to me of qualified members of the PSA who are residents of the Bay Area are so completely dominated by Tarski's views on what is significant work in the philosophy of science (e.g. Pat Suppes), that designating one of them to advise Tarski would in effect leave the choice of suggested recipients of travel funds entirely up to Tarski himself" (letter 140).

Nagel's efforts to ensure that philosophy of science was not reduced to a single voice—which he considered too formalistic—were in keeping with his pluralistic temperament and training. In his search for help, Nagel turned to Hempel: he asked his friend—who was a fellow at the Center for Advanced Study in the Behavioral Sciences at Stanford for the year 1963–64—to step in and to act as an informal advisor to Tarski. In his reply, Hempel accepted, somehow reluctantly, to work with Tarski on the basis of a common understanding of the situation. "Concerning the matter of advising Tarski", Hempel wrote, "I fully agree with your sentiments" and concluded that "[i]f it should seem desirable to you, however, that you accede to his request, I would be willing to act as an informal adviser" (letter 141).

There is no evidence in the correspondence of disagreements or misunderstandings that could have jeopardized the relationship between Nagel and Hempel; their personal and intellectual friendship seems to have strengthened year by year. This does not mean, however, that it did not change in some of its most important aspects. Nagel remained a valuable source of information for Hempel on recent philosophical work, both within and outside their field. Hempel genuinely admired Nagel for the latter's capacity to "span so many fields" (letter 155) and to keep abreast of the latest trends (letter 138).

All this notwithstanding, it is apparent from the letters that Hempel became the driving force between the two: it was Hempel who started keeping Nagel updated of what was going on in the philosophy of science, as it was Hempel who came to occupy a more central position in the field—although Nagel retained a privileged role as gatekeeper as he served on many committees for the allocation of grants and funding (see, for instance, letters 109, 162, 173). This change in balance was the net effect of several factors. From a relatively young age, Nagel suffered back problems (letter 18); later in life, the deterioration of his back condition affected not only his quality of life but also his ability to work (see, for instance, letters 91, 128, 132, 135). From the early 60s onwards, Hempel became increasingly concerned about Nagel's mental and physical well-being. On several occasions, he warned Nagel about the negative effects of overworking, recommending that he take a break from his chores (letter 135). Nagel's exhaustion was a direct result of his workload and work habits (letter 134).

At the same time, Nagel's wife Edith was going through serious physical problems, which caused Nagel great emotional distress (letters 124, 125, and 127;

see also letter 157). Nagel felt obliged not to "leave Edith and the boys", as he did not want to put additional pressure on Edith's shoulders (letters 107 and 108). Because of that, Nagel missed many conferences and events, which partly contributed to his (relative) loss of centrality in the field of philosophy of science.

Nagel's personal and professional relationship to Columbia University also became an issue. Nagel was educated at Columbia and was, in a sense, one of the best examples of what Columbia naturalism amounted to. However, from the 1940s he became increasingly dissatisfied with the direction that the philosophy department was taking and, more generally, with the status of the university as a whole (letter 62). One main source of dissatisfaction was the criteria for recruiting new members of the faculty: Nagel wanted to hire people more versed in analytic philosophy and logical empiricism, while his colleagues opted for profiles more closely related to Columbia's traditional philosophical approach (Verhaegh 2025c, 91). The letters show that Nagel's dislike of Columbia peaked in the mid-1950s, when he repeatedly expressed his desire to leave his *alma mater* (letters 98 and 107). The lack of a competitive and stimulating philosophical environment—compared with, for example, the one that Hempel enjoyed at Princeton—was another major contributing factor to Nagel lagging behind his friend in academic recognition (Verhaegh 2025d, e).

Hempel tried to cheer Nagel up by commenting positively on recent hiring by Columbia (letter 136) or by inviting him to apply for fellowships that would enable him to take a break from his duties and to work with greater ease (letter 138). But things were not quite so simple. When Nagel had the chance to leave Columbia for Rockefeller University, where he taught for the academic year 1966–67, he plunged into a state of deep emotional distress (letter 161): leaving Columbia was a "traumatic experience" for Nagel (letter 164). With his usual understatement, Nagel commented that the best thing that came out of his time at Rockefeller University was that it made room for new appointments (letter 164).

Nagel's statement was deeply unfair to his philosophical value and status, and it would have been interesting to read how Hempel handled such a negative remark—unfortunately, we could not find the letter, if it existed, which Hempel wrote in reply to his friend. Nagel was well known among his friends and relatives for being too hard on himself (see Yvonne Nagel's Biography in Chap. 2, letter by Sidney Hook to Nagel, dated Nov. 27, 1928, and letter by Edith to Nagel, October 17, 1934). Especially in the early 1960s, this sense of intellectual inadequacy grew into a "lack of drive" and a state of near depression that severely hampered his work (Letter 137). At one point he was so desperate about his condition that he wrote to Hempel that "Maybe my 'time has come'" (letter 134).

In the next section we will reconstruct in more detail the events that led to the publication of Nagel's masterpiece, *The Structure of Science* (1961); we will then assess in what sense and for what reasons it did not secure its author the place he had hoped for and deserved, given the excellence of the work. For now, we will limit ourselves to saying a few words on the long-term consequences that writing *The Structure of Science* had on Nagel. Indeed, it seems that working on the book consumed all his energies: he planned to prepare a sequel to *The Structure of*

Science, which was never published (letter 144). It is unclear from the correspondence how much Nagel worked on the second volume, as he only mentioned it once.

Hempel was as supportive and comprehensive as ever: he wrote to Nagel that the latter had no right to complain about his contribution to the field after the publication of *The Structure of Science*, since this was "a great achievement" (letter 145, see also letter 173). And he often expressed to Nagel his genuine admiration for the quality of his work (as in letters 145 and 155). Nagel, in turn, turned to Hempel for an impartial judgement of his intellectual production, as in the Laudan *affair* (Letter 182). By then, the roles had reversed; Hempel was now what Nagel had been to him when he arrived in the USA—the person to turn to for advice, help, guidance and reassurance.

3.4 The Demise of Nagel?

The turnover in the Nagel-Hempel relation reflects, of course, various things about the history of twentieth-century philosophy of science, the institutionalization of the profession, and the interests of readers, students, and colleagues across departments. It could be surely explained by numerous factors, but we will focus here on one issue that came up several times in the correspondence; and that is Nagel's much-awaited monograph, *The Structure of Science*.

By the 1950s, Hempel was established; he published some very technical, but to-the-point papers that set the agenda for generations. His textbook about concept formation published by The University of Chicago Press in the *International Encyclopedia of Unified Science* was widely used by philosophers and scientists alike (Hempel 1952). Although it was in the making for a few years, people—as Nagel indicates in the correspondence, for example, Paul Lazarsfeld—used it for years already before publication (see letters 48ff). It is not surprising therefore, that Hempel had some definite success amongst philosophers (of science), and philosophically minded and inclined scientists.

Nagel was striving for that position too, but he was shyer and slower. Many letters are filled with his insecurity, self-criticism, and lack of self-confidence. Though Hempel always tried to reassure Nagel, the latter was much frustrated with both the progress and quality of his unpublished manuscripts and published output. That certainly did not help much with his writing practices and outlook. Already in 1944, Hempel was noting that there was nothing to be afraid of for Nagel: "in particular the book you are now working on is going to be important and valuable, & I feel that it may well exert a far greater influence than many of Carnap's writings" (letter 43; see also letter 46). As we do not have Nagel's letter, we do not know for sure which book Hempel mentioned here but given that Nagel's only single-authored monograph after this date was *The Structure of Science*, this may indicate that perhaps he was working on the book already in the early 1940s.

Nagel was not idle during these years, of course, and two volumes of his selected writings (mainly reviews and popular papers) would appear (Nagel 1954, 1956), although neither of them had the function or purpose of becoming a textbook (as Hempel's monograph did) or of presenting a systematic narrative.[15] But during all these years, Nagel was devoting much of his time and energy to the big project of *Structure*.

It is clear from the correspondence that Nagel was well into the process of writing the book by 1950. In a letter dated September 7, 1950, Nagel informed Hempel that he was writing a long chapter on explanation for about a month and that he was not satisfied with what he had written in the previous years (letter 60). A year later, in a letter dated September 1, 1951, he wrote to Hempel that he was working on what would become the final part of the book, the one devoted to social sciences and historical inquiry (letter 62).

From the same letter we learn that at that time Nagel thought that the book should consist of three parts. While the first part, which was to correspond more or less to *The Structure of Science*, was in a rather advanced stage of completion, the second (on concept formation) and the third (on probability and induction) parts were at the first draft stage (letter 62). Nagel was insecure about how to proceed from there: on the one hand, he felt that the material from Parts II and III should be included in the book, as they covered topics that were widely debated in those years; on the other hand, he was aware that the book would be too long with these additions, and he worried about the time and effort that would be required to complete the remaining two parts and reduce the word count to an acceptable length. For the first time, he expressed concern that he might not be able to finish the book (letter 62).

However, all these difficulties notwithstanding, Nagel kept working on *The Structure of Science.* Six years later, he was still complaining to Hempel about the poor quality of the manuscript and the "snail-like pace at which the work proceeds". Once again, he vented his frustration that the undertaking was too difficult for his means (letter 95). Hempel was a constant and steady source of encouragement and support, both on a personal and professional level (letter 96).

Nagel did not give up. In July 1958, he was once again rewriting and revising the final chapter, which would become chapters 14 and 15 of *The Structure of Science* (letter 107). And in April 1959 he announced to Hempel that he was about to complete the final chapter of "the first part of the book on the philosophy of science"; for this reason, because he wanted to give the completion of the book priority over all other possible initiatives and enterprises, he had to decline Hempel's offer to write "a monograph on a critical appraisal of the work in the philosophy of science of the last 30 years or so" (letter 118).

The book finally appeared in 1961 as *The Structure of Science: Problems in the Logic of Scientific Explanation*, published by Harcourt, Brace & World, Inc. in New York. Hempel welcomed it as "There is absolutely no doubt in my mind but that this book is going to be the standard work in the field" (letter 125). In 15 long

[15] Even before *Structure*, Nagel has published an influential book about Gödel's proof with James R. Newman, though the influence of that book remained somewhat restricted in the history of analytic and scientific philosophy to a well-defined circle of scholars (see Nagel and Newman 1958). The book is discussed in the correspondence, see letters 110ff and 121.

chapters Nagel discusses such topics as explanation, scientific laws and the cognitive status of theories, mechanical issues, geometry, determinism and causality, reduction, biology, social science, and history. On more than 600 pages, Nagel went into tremendous details and provided minute arguments for his special positions on all these questions.

Two things seem to be striking even at first glance. Firstly, the book's size: it was huge and heavier than a usual textbook, which it obviously aspired to be.[16] It is not a book to read from cover to cover. As J. J. C. Smart (1962, 217) noted, "The chapters of the book are long, often of almost monograph length." Another reviewer, Adolf Grünbaum (1962, 294), called the book an "intellectual cornucopia", referring to the wealth of the philosophical material, although possibly not entirely without some resentment about how that material overflowed. Beyond that, even from a stylistic point of view, it had certain shortcomings. "It is not an *easy* work to read", Michael Scriven (1964, 403) made clear, "the phrasing being rather complicated, and the topics at times repetitious and intertwined."[17] Whatever lucidity, vividity and catchiness Nagel had in his reviews and popular pieces, *Structure* was "a scholarly tour de force" (Scriven 1964, 403).

On the one hand, as Scriven (1964, 403) noted, great works are not meant to be easy texts. On the other hand, many topics that Nagel treated in the most detailed ways are typical issues in the *tradition of logical empiricism*. Explanation, geometry, space and time, reduction, logical analysis, realism and instrumentalism connect Nagel to some old-fashioned traditional debates that defined the 1940s and 1950s. Because of this, Nagel's book might have given the impression to some (progressive) readers that he was still taking up the fights that his colleagues had been engaged with during the two decades that he was writing the book. It is very plausible, one might argue, that Nagel was just simply *too late* with the publication of his magnum opus. By the time it appeared, philosophy of science was about to change radically.

In fact, Mary Hesse (1963, 429) made it clear to her readers that Nagel's book is "the last" of its kind that defends "the commonplaces of empirical philosophy". Hesse read Nagel's book as a major, systematic, and impressively detailed argumentation for old empiricist and positivist philosophy; while some of these features could be taken with a grain of salt as appraisals, the salt is more important here:

> I say perhaps the last, partly because it is not only philosophers of science who now challenge these commonplaces, but also because to exhibit the whole massive argument, as Nagel here does so brilliantly, is itself to prompt the reflection that it is perhaps time Ptolemy met his Copernicus. (Hesse, 1963, 430)

---

[16] Nagel's book was used, for example, by Isaac Levi as a textbook in graduate philosophy of science courses at Columbia even in 1979. In Chicago, a few years later, Howard Stein explicitly declared not to use Nagel's book. Email communication from Sahotra Sarkar, March 5, 2025. See further Sarkar (2024).

[17] Nagel's intention was, though, to present a book "for a wider audience than that of professional students of philosophy", and for that reason he even avoided "highly formalized presentations of analysis", "symbolic notation", despite their merits. See Nagel (1961, ix).

Just as one could tailor and adjust Ptolemian theory to meet experience and new scientific standards via *ad hoc* hypotheses and incidental adjustment for a while—although they would make the original theory almost unrecognizable—positivist philosophy of science did struggle for years to keep itself alive and viable in the face of actual scientific practice. As Hesse (1963, 429) says, "the theoretical character of modern science has been a considerable embarrassment to positivist philosophies of science" as science does not work the way that the positivist paints it. It was time to drop the positivist philosophy of science, which included Nagel.

Just a year after Nagel's *Structure* appeared, another, more famous "Structure" hit the light from Thomas Kuhn, *The Structure of Scientific Revolutions* (Kuhn 1962). A much thinner, more lucid book, packed with engaging and easily memorizable concepts such as "paradigm", "anomaly", "crisis", "puzzle solving", and even "revolution", Kuhn's book became an instant classic, even outside history and philosophy of science. As K. Brad Wray have noted recently in his edited volume on the sixtieth anniversary of Kuhn's *Structure*:

> The influence of the *Structure of Scientific Revolutions* (SSR) has been remarkably wide-ranging. Thomas Kuhn, the author of the book, was honored by the History of Science Society, the Philosophy of Science Association, and the Society for the Social Studies of Science [. . .], three very different academic societies. (Wray 2024b, 1)

We have no intention to assess or even indicate the real influence of Kuhn here; as Wray continues, "Nothing approaching a comprehensive literature review is possible for a book that has been cited more than 135,000 times."[18] Kuhn published his book after a decade (he was commissioned to write the book in 1952), but as soon as it appeared, many scholars recognized its potential to transform traditions, movements, and to set education and inquiry on radically new paths. And that is what Kuhn's *Structure* did.

In a certain sense it was easy for Kuhn and others to wash away Nagel's book because Nagel had published a lot about all these things in the previous decades. As May Brodbeck (1961, 997) noted, "for many of us the book contains few surprises." Nothing new appeared in the book, and thus it only provided another occasion for many to take up the fight with individual, smaller points. If one goes through the reception of Nagel's *Structure*, two things become visible. One the one hand, every reviewer was respectful with Nagel, honored his person, his achievements and general role in the community, and praised his scholarly virtues. On the other hand, every paper was a typical sandwich review: between two appraisals (at the beginning and at the end) there was significant and substantial criticism. The most important reviews are, on average, around ten pages long, a very uncommon phenomenon already, and most of the pages were reserved for painstaking and

[18] The interested reader might consult the still growing Kuhn-literature with benefits. See, recently, Reisch (2019), Wray (2021a, b, 2024a), Shan (2024), and Giri et al. (2023).

substantial criticisms.[19] As Nagel's book did not break ground but summarized previous points, it was much easier to overcome this *Structure*.

Another point in favor of Kuhn's program—showing the wrong timing of Nagel—concerns the *metaperspective* on science. After World War II, philosophy of science was seen by many as a collaborative, interdisciplinary enterprise, whose main aim was to critically investigate the sciences and build bridges between various endeavors. The war had mainly shown two things: firstly, that science and its technological dimension has an immense impact on society, and secondly that the new funding structure (like the National Science Foundation) requires a much better understanding of the scientific practice from the side of the public. It is not an accident, that Philipp Frank—a colleague, friend, and collaborator of Nagel—was able to convince figures such as James Conant (President of Harvard and leader of the General Education Program), Warren Weaver (Director of the Division of Natural Sciences at the Rockefeller Foundation), and finally Raymond J. Seeger (Assistant Director at the National Science Foundation) to fund his Institute for the Unity of Science in 1947. The rationale behind the Institute was that recent departmentalization and specialization "have become a serious threat to our culture and to the very goal of liberal education" and if philosophy of science does not provide bridges, and thus integrates science, philosophy (values) and society, then the next generations will be "eager to accept the integration from any sources" (Frank 1947, 160). Through emphasizing cross-connections, interdisciplinary fields, unifications, and pushing "logico-empirical and socio-psychological analysis" of science at the same time (1947, 166), philosophy of science earns a special place in this *socio-cultural project*.

Nagel was very much in agreement with this as a foundational member of the Institute.[20] When he spoke about the difficulties of integrating science, morals, values, and liberal education in an unpublished lecture, he spoke about the responsibility of scientists as well:

---

[19]Ayer (1961) 5 pages, Brodbeck (1961) 3 pages, Grünbaum (1962) 12 pages, Hesse (1963) 13 pages, Madden (1963) 7 pages, Scriven (1964) 22 pages, Smart (1962) 8 pages. The major exception is Feyerabend's (1966) review or discussion paper: although it is 13 pages long, it did not really place his criticism within appraisals. Similarly, in the correspondence, Nagel was highly critical of Feyerabend, denouncing some of his comments (about Hempel) "a new record of cheap philosophical writing" and called it even "quatsch" (see letter 153). Later Nagel published a review of Feyerabend's *Against Method*, calling basically some of its assumptions "nonsensical" and some of its points "incompatible with just about every action in which we must engage to survive" (Nagel 1977, 1134).

[20]Before the Institute was launched, Frank even asked Nagel to become the "Managing Director" and thus be the "real centre of the Institute", with the offices in New York. We do not have Nagel's reply, but when the offices were established in Boston (in the building of the American Academy of Arts and Sciences) in 1950, Nagel did not take the role, although he played an important role in helping Frank. See Philipp Frank to Ernest Nagel, June 25, 1946 (Ernest Nagel Papers, 1925–1985, ASP.2020.01, Archives & Special Collections, University of Pittsburgh Library System).

> The scientific community as a whole must take at least part of the blame for the persistence of the widespread conviction that while the study of scientific subjects may serve a narrowly utilitarian purpose, and is desirable for those planning to make a career in some branch of science, it contributes little to realizing the aims of liberal education – that is, to the development of critical standards for appreciating the diverse achievements, aspirations, and problems of the human spirit.[21]

Specialization and professionalization, however useful they might be in the regular practice of science, pose serious problems and threats to our general and public understanding of science's role in our life and socialization. Nagel goes on:

> For it is pertinent to note that many intelligent men continue to have serious misconceptions about the nature of science, despite considerable exposure to formal instructions in it; and surely this fact alone suffices to indict the way the sciences are commonly taught. In any event, as a number of distinguished scientists have recognized, far too little effort has been devoted in the teaching of scientific subjects to the explicit analysis of the structure of scientific ideas, to making clear just how highly abstract scientific theories are related to matters of familiar experience, or to exhibiting the logic of the intellectual method that is the ultimate basis for the credibility of scientific claims.

Scientific instruction as such does not reflect much about the *development, process, and social embeddedness* of science. The relationship between science and common sense (a major topic of Nagel's *Structure* and Frank's philosophy in the United States), is largely absent from the education of scientists, but even of the humanities. But besides not receiving the right education, scientists either do not develop in themselves an inclination towards philosophy (of science) to conceive subjects outside their technical profession or, when they do, it usually just makes things worse:

> In point of fact, comparatively few scientists show signs that they have reflected deeply on the logic of their disciplines, on the source and scope of its authority, or on its significance for the development of a liberal society and a rational morality. When they *do* express themselves on such broad but important questions, they frequently make pronouncements they have not thought through, and which are likely to be echoes of some old wives' tale they have accepted uncritically.

Scientists are indoctrinated by their immediate socialization, by their reading experiences, by society, by Church, or by the State. As Frank (1930/1949) said, the statement of not being influenced by any philosophy is also a form of (highly uncritical) philosophy. What Nagel calls "echoes of some old wives' tale", Frank had called "school philosophy" since the late 1920s. Both of them thought that one goal of *philosophy of science* as a discipline and form of reflection was to integrate science and philosophy (where philosophy involves cultural studies, values, social sensitivities, and historical determinants), to make scientists more reflective and humanist (that is, sensitive to the human agency in science), and make philosophers

[21] All three following quotes are taken from Ernest Nagel, "Science and Culture" (Ernest Nagel Papers, 1930–1988, Series III: Arranged Manuscripts, Box 7, Rare Book and Manuscript Library, Columbia University. All rights are reserved). The manuscript is forthcoming in Gronda and Tuboly (2026).

and the humanities more scientific (that is why Frank (1957) called his own textbook *Philosophy of Science: The Link between Science and Philosophy*). Nagel wrote in similar fashion,

> if the nature of the scientific enterprise and its place in contemporary society are to be properly understood, the types and the articulation of scientific statements, as well as the logic by which scientific conclusions are established, also require careful analysis. This is a task – a major if not exclusive task – that the philosophy of science undertakes to execute. (Nagel 1961, 14)[22]

Given their convergences, it might be important to note that the referee for Frank's manuscript in the 1950s was Nagel, who appreciated not only Frank's knowledge in philosophy and science, but also his broad *humanist perspective*, and thus recommended the book for publication (see Reisch and Tuboly n.d.).

When Kuhn entered the scene with his second book (the first appeared in 1957 about the Copernican revolution), he painted a very different picture of science and its relation to philosophy and the humanities. He praised *professionalism* and *specialization* as the driving forces of normal scientific inquiry (a point made succinctly and analyzed in Kuhn's historical context by Reisch 2019), and in that *Zeitgeist* Nagel's broad viewpoint was not just strange, but even old-fashioned. As Kuhn argued, normal science did not require philosophical sophistication and foundational analysis, but even worse, according to Kuhn the practice and rules of normal science even informally forbid such reflection. As he wrote in his own *Structure*, "scientists have not generally needed or wanted to be philosophers. Indeed, normal science usually holds creative philosophy at arm's length, and probably for good reasons" (Kuhn 1962, 88).

Nagel's monumental volume, however, was the time's most detailed treatment of all such scientific questions that had the slightest philosophical interest and relevance. Nagel thought that philosophy shall be integrated into science and not kept at arm's length. But it was not an easy and smooth thing to do. As Michael Scriven (1964, 403) has noted, "few philosophers of science, besides Whewell and Aristotle, have had Nagel's command of the mathematics, the history and the content of such a range of sciences." To *genuinely* integrate science and philosophy, a broad vision was needed, and that vision had to be rooted in a massive empirical material, provided by the philosophy, history, and sociology of science. But by dealing with such a vast number of references, details, programs, arguments and even diametrically opposed perspectives, Nagel's monumental book was bursting. "The great tension underlying this work", says Scriven (1964, 413) "is an aspect of that between the humanistic and the scientific approach which Nagel so willingly and so well synthesizes."

---

[22] Something very similar—a balance between logico-analytical and socio-historical analysis—was advocated by some members of the Vienna Circle, called "the left wing". This view—accepted in different forms and degrees and dubbed by Thomas Uebel (2022) as "the bipartite metatheory of science"—was defended by Rudolf Carnap, Otto Neurath, and Nagel's close friend, Philipp Frank. On this idea, see further Bentley (2023).

Kuhn did not have such a struggle in his views on science and its educational background. Regarding historical scrutiny and reflection, he wrote later the following:

> These [science] students need to acquire command of the current tools of the profession, both conceptual and instrumental; history is at best a slow and inefficient way to do that. Learning alternate possibilities [. . .] these would be irrelevant distractions. Conceivably, it might even undermine the student's grasp of and faith in the tools required for current practice. Though the sort of history I have practiced should, I feel, be taught to a variety of audiences, including scientists, it is not an appropriate route to professionalization. (Kuhn 1987/2022, 88)

What some colleagues considered Nagel's strength (his broad perspective and integration of numerous sciences) was considered his weakness by others (the tension of bridging the humanities and the sciences), and it was entirely rejected and overcome by Kuhn. Professionalism prevailed and Kuhn was able to explain and motivate its role, presence, and value. During normal scientific research, when a paradigm determines scholarly work, says Kuhn (1962), science tracks the truth, makes progress, solves puzzles, and thus the pursuing of any philosophical questions, criticism, or questioning the fundamentals (hallmarks of social scientific and humanities works) means departure from truth-seeking, from progress, and thus from concrete scientific work. The scientific community had good reasons, then, to focus and specialize and it was reassured by the new funding system of the National Science Foundation as well (see, for example, Vaesen and Katzav 2019). The decade that saw the publication of Nagel's and Kuhn's "Structures" marked the beginning of the transformation of a socially more engaged, liberal-critical, public discipline of scientific philosophy (done by people like James Conant, Philipp Frank, John Dewey and others) into a professionalized technical field of philosophy of science.[23]

Nagel's timing was also unfortunate for another reason. Many of his contemporaries were already dealing with topics such as induction and probability, something initiated by Carnap's masterpiece on *The Logical Foundations of Probability* (see Carnap 1950) and continued when he came up with new systems (e.g., in Carnap 1952, and Carnap and Jeffrey 1971). Nagel (1961, ix) promised that the questions of inductive and probabilistic inferences and their validity would be discussed in another volume, which was "in active preparation", but he was never able to complete that book. Unfortunately for him, as soon as he published the first volume, many members of his targeted community were either no longer considering seriously the problems he treated or knew Nagel's solutions and approaches very well, or were already pursuing something else, and as a result, they were much more eager to learn from the second, never finished volume.[24]

---

[23] Despite some initial reservations, Nagel also had a very interesting and engaging public and socially sensitive philosophy of science—see Schliesser (2022).

[24] See Grünbaum (1962, 305); Ayer (1961, 203); Brodbeck (1961, 997); Smart (1962, 217); Scriven (1964, 403).

All the above considerations point to a somewhat negative picture of Nagel, but his situation was not that dramatic. He had a good relationship with Kuhn, and although Kuhn's name did not come up often or in great detail in the correspondence, Nagel noted in 1963 that "though I disagree with a good deal in his recent book on revolutions in science, I find him stimulating" (letter 140). This is also confirmed by the personal correspondence of Nagel and Kuhn (see Reisch 2019 for some insights), where Nagel expressed a truly positive opinion about Kuhn's book that he had read in manuscript form (for which he is thanked by Kuhn in his preface, 1962, xii):

> I have read your manuscript with unbounded admiration and personal profit, and have no doubt that its publication will be a major contribution to the literature of the philosophy as well as the history of science. The material you have included for analysis and the issues you have discussed seem to me of greatest importance and I am confident that the book will have a wide audience of readers, not restricted to the professional historians or philosophers. (Quoted by Wray 2021a, 69)

Nagel obviously knew about Kuhn's importance and tried to join the discussion on several occasions later when positivist and post-positivists narratives were on the table.[25] But, as noted above, their ways parted, and the atmosphere did not favor them both equally. Nagel had started to focus on more specific issues in his publications (like teleology and reduction) and offered specific insights into the logical analysis of science. That was emphasized in Brodbeck's (1961, 997) review as well: "Some lessons, some insights are peculiarly slippery and, it seems, must always be learned anew by each generation. Many of those lessons and insights will be found in this volume." Nagel thus became the teacher of generations, bringing to the table not new insights and revolutionary narratives, but such fundamental ideas that are often forgotten and require renewed attention.[26]

Perhaps it is time now to reevaluate Nagel's contribution, to analyze and understand his original contexts, relations, working moral, mission, and perception of the field to provide a balanced judgement about his roles and influences in the history of philosophy of science.

3.5 Locations of Letters and Editorial Issues

None of the letters presented here have been published before. They are to be found in various archives, and we have used copies from the following ones:

[25] Most obviously, for example, in his unpublished talk "Current Critiques of Science as a Cumulative Enterprise", read before the American Philosophical Society in 1964. The manuscript is contained in the Nagel Archive at Columbia and will be published in Gronda and Tuboly (2026).

[26] Another reviewer, J. C. C. Smart (1962, 223) compared Nagel to Socrates given their similar educational agendas for the public and the profession, although for Smart Nagel was offering more: "But unlike Socrates, Nagel does not only make one aware of one's ignorance; he proceeds, so far as possible, to remedy it."

- Carl Gustav Hempel Papers, 1903–1997, ASP.1999.01, Archives of Scientific Philosophy, Archives & Special Collections, University of Pittsburgh Library System, Box 28, Folders 1–2.
- Ernest Nagel Papers, 1925–1985, ASP.2020.01, Archives of Scientific Philosophy, Archives & Special Collections, University of Pittsburgh Library System, Box 2, Folders 52–57, and Box 3, Folders 1–4.

Most of these letters are also contained in Nagel's archive at Columbia University:

- Rare Book & Manuscript Library, Ernest Nagel papers, 1930–1988, Series II: Arranged Correspondence, Box 1, "Hempel, C. G. ('Pete')".

For some reason, all of Nagel's letters to Hempel seem to be lost until 1950. While that seriously restricts our abilities and chances to provide a thoroughgoing and comprehensive interpretation of Nagel's early relationship to Hempel and to logical empiricist émigrés in general, one can still infer many issues from Hempel's well-documented responses.

As we did not prepare a critical edition here, only minor changes and editing have been introduced into the correspondence. Our editorial additions, like a few names to make references in a letter unambiguous are added within [[double brackets]]. We indicate at the beginning of each letter whether they were handwritten or typed; as the former occasionally caused trouble when we were deciphering the handwriting, we used {curly brackets} to indicate handwritten additions to a given typed letter, either within a given sentence in the main text or as marginalia, or as a late addition at the end of a given letter.

Permissions and Acknowledgments We gratefully acknowledge our sincere indebtedness to the families of Ernest Nagel (Alexander and Sidney Nagel) and Carl Gustav Hempel (Miranda and Peter Andrew Hempel) for giving us permission to reproduce the correspondence here. The whole project would not have been feasible and manageable, however, without the full support and knowledge of Yvonne Nagel who provided us many letters, much information, and continuous encouragement.

We would also like to express our sincere indebtness to the staff of the Archives of Scientific Philosophy at the University of Pittsburgh Library System for their help and permission.

During the editorial process Adam Tamas Tuboly was supported by the MTA Lendület Values and Science Research Group.

References

Ayer, A. J. 1961. Review of Nagel's The Structure of Science. *Scientific American* 204 (6): 197–203.

Bentley, Joseph. 2023. *Logical Empiricism and Natural: Neurath and Carnap's Metatheory of Science*. Cham: Springer.

Brodbeck, May. 1961. Structure of Science. Review of Nagel (1961). *Science* 134 (3484): 997–999.

Burke, Christopher, and Adam Tamas Tuboly. 2025. *Otto Neurath in Britain*. Cambridge: Cambridge University Press.

Carnap, Rudolf. 1950. *Logical Foundations of Probability*. Chicago: The University of Chicago Press.
Carnap, Rudolf. 1952. *The Continuum of Inductive Methods*. Chicago: The University of Chicago Press.
Carnap, Rudolf, and Richard C. Jeffrey, eds. 1971. *Studies in Inductive Logic and Probability. Volume 1*. Berkeley: University of California Press.
Cat, Jordi, and Adam Tamas Tuboly, eds. 2019. *Neurath Reconsidered: New Sources and Perspectives*. Cham: Springer.
Collodel, Matteo, and Eric Oberheim, eds. 2020. *Feyerabend's Formative Years. Volume 1. Feyerabend and Popper*. Cham: Springer.
Collodel, Matteo, and Eric Oberheim, eds. 2024. *Feyerabend's Formative Years. Volume 2. Feyerabend on Logical Empiricism, Bohm & Kuhn: Correspondence and Unpublished Papers*. Cham: Springer.
Damböck, Christian, Johannes Friedl, and Ulf Höfer, eds. 2024. *Rudolf Carnap / Otto Neurath: Briefwechsel*. Hambug: Felix Meiner.
Dewulf, Fons. 2022. The Emergence of Scientific Explanation as a Problem for Philosophy of Science: Aristotle, Nagel, and Hempel. In *Ernest Nagel: Philosophy of Science and the Fight for Clarity*, ed. Matthias Neuber and Adam Tamas Tuboly, 67–87. Cham: Springer.
Fetzer, James H. 2000. Editor's Preface. In *Science, Explanation, and Rationality: Aspects of the Philosophy of Carl G. Hempel*, ed. James H. Fetzer, vii. Oxford: Oxford University Press.
Fetzer, James. 2022. Carl Hempel. In *The Stanford Encyclopedia of Philosophy*, ed. Edward N. Zalta and Uri Nodelman. Stanford: Stanford University. https://plato.stanford.edu/archives/fall2022/entries/hempel/.
Feyerabend, Paul. 1966. Review of Nagel's The Structure of Science. *The British Journal for the Philosophy of Science* 17 (3): 237–249.
Frank, Philipp. 1930/1949. Physical Theories of the Twentieth Century and School Philosophy. In *Modern Science and Its Philosophy*, 90–121. Cambridge, MA: Harvard University Press.
Frank, Philipp. 1947. The Institute for the Unity of Science: Its Background and its Purpose, *Synthese* 6: 160–167.
Frank, Philipp. 1957. *Philosophy of Science: The Link between Science and Philosophy*. New York: Prentice Hall.
Giri, Leandro, Pablo Melogno, and Hernán Miguel, eds. 2023. *Perspectives on Kuhn: Contemporary Approaches to the Philosophy of Thomas Kuhn*. Cham: Springer.
Gronda, Roberto, and Adam Tamas Tuboly, eds. 2026. *Ernest Nagel on Science and Philosophy. Volume 2. Selection of (Un)published Writings*. Cham: Springer, forthcoming.
Grünbaum, Adolf. 1962. Discussion: The Structure of Science. *Philosophy of Science* 29 (3): 294–305.
Hempel, Carl G. 1934. *Beiträge zur logischen Analyse des Wahrscheinlichkeitsbegriffs*. Jena: Universitäts-Buchdruckerei G. Neuenhahn, G.m.b.H.
Hempel, Carl G. 1952. *Fundamentals of Concept Formation in Empirical Science*. Chicago: The University of Chicago Press.
Hempel, Carl G. 1999. *Carl Gustav Hempel Papers, 1903–1997, ASP.1999.01*. Pittsburgh: Archives of Scientific Philosophy, Archives & Special Collections, University of Pittsburgh Library System.
Hempel, Carl G. 2000. Prologue: An Intellectual Autobiography: Carl G. Hempel. In *Science, Explanation, and Rationality: Aspects of the Philosophy of Carl G. Hempel*, ed. James H. Fetzer, 3–35. Oxford: Oxford University Press.
Hempel, Carl G., and Paul Oppenheim. 1936. *Der Typusbegriff im Lichte der neuen Logik. Wissenschaftstheoretische Untersuchungen zur Konstitutionsforschung und Psychologie*. Leiden: A. W. Sijthoff's Uitgeversmaatschappij.
Hesse, Mary. 1963. Review of Nagel's The Structure of Science. *Mind* 72 (287): 429–441.
Iven, Mathias. 2015. Er 'ist eine Künstlernatur von hinreissender Genialität'. *Wittgenstein-Studien* 6 (1): 83–174.

Jeffrey, Richard. 2000. Preface. In *Carl G. Hempel: Selected Philosophical Essays,* ed. Richard Jeffrey, vii–x. Cambridge: Cambridge University Press.

Jones, Bessie Zaban. 1984. To the Rescue of the Learned: The Asylum Fellowship Plan at Harvard, 1938–1940. *Harvard Library Bulletin* 33 (3): 205–238.

Kuhn, Thomas. 1962. *The Structure of Scientific Revolutions*. Chicago: The University of Chicago Press.

Kuhn, Thomas. 1987/2022. The Presence of Past Science. In *The Last Writings of Thomas S. Kuhn*, ed. Bojana Mladenović, 23–89. Chicago: The University of Chicago Press.

Madden, Edward H. 1963. Discussion: Ernest Nagel's The Structure of Science. *Philosophy of Science* 30 (1): 64–70.

Nagel, Ernest. 1936. Review of Hempel and Oppenheim (1936). *The Journal of Philosophy* 33 (22): 611.

Nagel, Ernest. 1954. *Sovereign Reason and Other Studies in the Philosophy of Science*. New York: The Free Press.

Nagel, Ernest. 1956. *Logic Without Metaphysics and Other Studies in the Philosophy of Science*. New York: The Free Press.

Nagel, Ernest. 1961. *The Structure of Science: Problems in the Logic of Scientific Explanation*. New York: Harcourt, Brace, & World.

Nagel, Ernest. 1977. Review of Feyerabend's Against Method. *The American Political Science Review* 71 (3): 1132–1134.

Nagel, Ernest. 2020. *Ernest Nagel Papers, 1925–1985, ASP.2020.01*. Pittsburgh: Archives of Scientific Philosophy, Archives & Special Collections, University of Pittsburgh Library System.

Nagel, Ernest, and James R. Newman. 1958. *Gödel's Proof.* New York: New York University Press.

Nagel, Ernest, and Remmel Nunn. 2022. Interview with Ernest Nagel by Remmel Nunn. In *Ernest Nagel: Philosophy of Science and the Fight for Clarity*, ed. Matthias Neuber and Adam Tamas Tuboly, 258–314. Cham: Springer.

Oppenheim, Paul. 1969. Reminiscences of Peter. In *Essays in Honor of Carl G. Hempel: A Tribute on the Occasion of his Sixty-Fifth Birthday*, ed. Nicholas Rescher, 1–4. Dordrecht: D. Reidel.

Randall, Jr, and John Herman. 1949. Review: Morris R. Cohen's the Meaning of Human History. *Journal of the History of Ideas* 10 (2): 305–312.

Reisch, George. 2019. *The Politics of Paradigms: Thomas S. Kuhn, James B. Conant, and the Cold War "Struggle for Men's Minds"*. Albany: SUNY Press.

Reisch, George, and Adam Tamas Tuboly. 2025. Reviving the Unity of Science Movement: Philipp Frank's Journey to Harvard. In *American Philosophy and the Intellectual Migration: Pragmatism, Logical Empiricism, Phenomenology, Critical Theory*, ed. Sander Verhaegh, 169–190. Berlin: De Gruyter.

Reisch, George, and Adam Tamas Tuboly. n.d. *Humanistic Philosophy of Science: Philipp Frank's Life Between Science and the Humanities.* Unpublished manuscript.

Rescher, Nicholas, ed. 1969. *Essays in Honor of Carl G. Hempel: A Tribute on the Occasion of his Sixty-Fifth Birthday*. Dordrecht: D. Reidel.

Rescher, Nicholas. 1997. H2O: Hempel-Helmer-Oppenheim, an Episode in the History of Scientific Philosophy in the 20th Century. *Philosophy of Science* 64 (2): 334–360.

Sarkar, Sahotra. 2024. *Vienna Circle: The Story of Logical Empiricism*. New York/London: Routledge.

Schliesser, Eric. 2022. Philosophy of Science as First Philosophy: The Liberal Polemics of Ernest Nagel. In *Ernest Nagel: Philosophy of Science and the Fight for Clarity*, ed. Matthias Neuber and Adam Tamas Tuboly, 43–65. Cham: Springer.

Scriven, Michael. 1964. The Structure of Science. *The Review of Metaphysics* 17 (3): 403–424.

Shan, Yafeng, ed. 2024. *Rethinking Thomas Kuhn's Legacy*. Cham: Springer.

Smart, J. J. C. 1962. Review of Nagel's The Structure of Science. *The Journal of Philosophy* 59 (8): 216–223.

Tuboly, Adam Tamas. 2021. To the Icy Slopes in the Melting Pot: Forging Logical Empiricisms in the Context of American Pragmatisms. *HOPOS: The Journal of the International Society for the History of Philosophy of Science* 11: 27–71.

Uebel, Thomas. 2022. The Bipartite Metatheory Conception of Philosophy. In *The Routledge Handbook of Logical Empiricism*, ed. Thomas Uebel and Christoph Limbeck-Lilienau, 361–370. New York/London: Routledge.

Vaesen, Krist, and Joel Katzav. 2019. The National Science Foundation and Philosophy of Science's Withdrawal from Social Concerns. *Studies in History and Philosophy of Science Part A* 78: 73–82.

Verhaegh, S. 2020. The American Reception of Logical Positivism: First Encounters 1929–1932. *HOPOS: The Journal of the International Society for the History of Philosophy of Science* 10 (1): 106–142.

Verhaegh, Sander, ed. 2025a. *American Philosophy and the Intellectual Migration: Pragmatism, Logical Empiricism, Phenomenology, Critical Theory*. Berlin: De Gruyter.

Verhaegh, Sander. 2025b. Introduction: American Philosophy and the Intellectual Migration. In *American Philosophy and the Intellectual Migration: Pragmatism, Logical Empiricism, Phenomenology, Critical Theory*, ed. Sander Verhaegh, 1–28. Berlin: De Gruyter.

Verhaegh, Sander. 2025c. Columbia Naturalism and the Analytic Turn: Eclipse or Synthesis? In *American Philosophy and the Intellectual Migration: Pragmatism, Logical Empiricism, Phenomenology, Critical Theory*, ed. Sander Verhaegh, 77–99. Berlin: De Gruyter.

Verhaegh, Sander. 2025d. The Analytic Turn in American Philosophy: An Institutional Perspective – Part I: Scientific Versus Humanistic Philosophy. *HOPOS: The Journal of the International Society for the History of Philosophy of Science* 15 (1): 61–98.

Verhaegh, Sander. 2025e. The Analytic Turn in American Philosophy: An Institutional Perspective – Part II: Analytic vs Continental Philosophy. *HOPOS: The Journal of the International Society for the History of Philosophy of Science* 15 (2), forthcoming.

Wray, K. Brad. 2021a. *Kuhn's Intellectual Path: Charting the Structure of Scientific Revolutions*. Cambridge: Cambridge University Press.

Wray, K. Brad, ed. 2021b. *Interpreting Kuhn: Critical Essays*. Cambridge: Cambridge University Press.

Wray, K. Brad, ed. 2024a. *Kuhn's The Structure of Scientific Revolutions at 60*. Cambridge: Cambridge University Press.

Wray, K. Brad. 2024b. Introduction: The Impact of The Structure of Scientific Revolutions. In *Kuhn's The Structure of Scientific Revolutions at 60*, ed. K. Brad Wray, 1–18. Cambridge: Cambridge University Press.

Chapter 4
The Nagel-Hempel Correspondence (1935–1985)

Ernest Nagel and Carl G. Hempel

Contents

Authors “Ernest Nagel and Carl G. Hempel” are deceased at the time of publication.

R. Gronda, A. T. Tuboly (eds.), *Ernest Nagel on Science and Philosophy*, Vienna Circle Institute Library 14, https://doi.org/10.1007/978-3-031-99458-6_4

Abstract This chapter contains the Nagel-Hempel correspondence from 1935 to 1985.

Keywords Ernest Nagel · C. G. Hempel · History of philosophy of science · Correspondence · Scientific philosophy · History of analytic philosophy

4.1 The Correspondence

[After the initial data of every letter, we also record their archive place as: Hempel (1999) and Nagel (2020) refer to the main places in the Archive of Scientific Philosophy at Pittsburgh. For details see the reference list at the end of the chapter.]

4.1.1 Hempel to Nagel, June 23, 1935

[Typewritten; Nagel 2020, Box 2, Folder 52]

Dr. CARL G. HEMPEL
IXELLES-BRUXELLES, le. 23/VI/1935.
Rue Saint-Georges, 26

Lieber Dr. Nagel,

Herzlichen Dank für Ihren Gruß aus London, der mich sehr erfreut hat, und, mit schrecklicher Verspätung, für Ihren freundlichen Brief aus Prag von Anfang Dezember.[1] Ich habe die Antwort immer verzögert, weil ich hoffte, Sie im Frühsommer hier zu sehen. Es ist wirklich sehr schade, daß sich dieser Plan nicht hat realisieren lassen, aber ich hoffe aufrichtig, daß sich in nicht zu langer Zeit unsere Wege wieder einmal kreuzen werden! Bis dahin bitte ich Sie, mich doch ständig über Ihre Adresse auf dem Laufenden zu halten. Ich sende Ihnen mit gleicher Post noch nach London zwei Separata; Abdrucke von späteren Veröffentlichungen, die Sie interessieren könnten, schicke ich Ihnen dann nach Amerika, darunter hoffentlich auch ein Separatum meiner Erwiderung auf Schlick,[2] die noch im Juni in der „Analysis“ erscheinen soll.

[1] Between the falls of 1934 and 1935, Nagel was traveling across Europe with a Guggenheim Fellowship. He visited prominent representatives of logical empiricism and scientific philosophy in Prague, Vienna, Warsaw, Lwow, and stayed for shorter periods in Italy, France, Holland, and England. He wrote a long and substantial report about his experiences that was published in two parts, Nagel (1936a, b).

[2] Moritz Schlick (1882–1936) was a German philosopher and physicist, leader of the so-called Vienna Circle. Hempel's paper about Schlick is “Some Remarks on ‘Facts’ and Propositions” (Hempel 1935a); the other article could have been “On the Logical Positivists' Theory of Truth” (Hempel 1935b).

Auf der hiesigen Bibliothek habe ich mit großem Interesse Ihre Besprechungen von Carnaps „Syntax“[3] und anderen Büchern gelesen. – Ihr Bericht über den Prager Kongress[4] ist bei weitem der beste, den ich kenne; er hat mir so gefallen, daß ich ein zweites Exemplar, das mir Carnap geschenkt hatte, einem Freund in Münster gelassen habe, der es nun dort zirkulieren läßt.

Nun noch eine sachliche Frage: Ist Ihnen vielleicht bekannt, ob es Versuche (vielleicht von den amerikanischen oder polnischen Spezialisten für mehrwertige Logik) gibt, eine quasi-mehrwertige Logik aufzubauen die nicht metrisch, sondern rein topologisch ist, die also nicht jedem Satz einen bestimmten von mehreren verschiedenen, genau festgelegten Wahrheitswerten zuordnet, sondern nur eine topologische Ordnung von Sätzen nach steigendem Wahrheitswert festlegt?[5]

Wir beide denken oft und gern an die schönen und interessanten Tage in London[6] zurück, und wir sind sicher, daß auch Sie sich drüben sehr wohl fühlen werden. Falls Sie Frau Professor Stebbing[7] sehen, richten Sie ihr doch bitte höfliche Grüße von uns aus!

Vor allem aber senden wir Ihnen und Ihrer Gefährtin herzliche Grüße und die besten Wünsche für gute Heimfahrt und alles Weitere!

Ihr
Carl G. Hempel

[English translation]
Hempel to Nagel, June 23, 1935.

Dr. CARL G. HEMPEL
IXELLES-BRUXELLES, le. 23/VI/1935.
Rue Saint-Georges, 26

Dear Dr. Nagel,

Thank you very much for your greeting from London, which gave me great pleasure – and, with terrible delay – for your kind letter from Prague at the beginning of December. I have always delayed replying because I hoped to see you here in early summer. It is truly a

[3] See Carnap (1934); for English translation see Carnap (1937b). Nagel has reviewed the original German edition (Nagel 1935a), and its English translation (Nagel 1937).

[4] Nagel has written a longer report about the congress (Nagel 1934). The Congress was organized between September 2–7, 1934, where Nagel gave the talk “Reduction and Autonomy in the Sciences”, published as (Nagel 1936c). Before the Congress, “The Preliminary Conference of the International Congresses for the Unity of Science” was organized between 31 August and September 2, which Nagel attended in the audience as well. About the conference see, Stadler (2015, 166–171).

[5] A year later, Hempel would publish his article on non-Aristotelian logic as Hempel (1936). A longer, English version was published next year, see Hempel (1937a).

[6] Nagel spent five weeks in England during his Fellowship, roughly from late May to mid July, thus presumably Hempel’s London-visit was just before Nagel’s. Hempel, on the other hand, participated (on the recommendation of Otto Neurath) at a small informal workshop on logical positivism in London, organized by Susan Stebbing, 5–6 January 1935. He presented his soon-to-be-published paper, “On the Logical Positivists’ Theory of Truth” (Hempel 1935b). About the workshop, see further Burke and Tuboly (2025, Chap. 5).

[7] Susan Stebbing (1885–1943) was a British philosopher, co-founder of the journal *Analysis*, and the first woman professor in England. She wrote many important logic textbooks during the 1930s, attended the Unity of Science Congresses, and helped Otto Neurath during his English years. While she organized lectures for Carnap and Hempel in London, and helped Neurath, Schlick and others with disseminating their views, she was quite critical of logical empiricism.

great pity that this plan could not be realized, but I sincerely hope that our paths will cross again in the not-too-distant future! Until then, please keep me informed about your address. I will send you two offprints by the same post to London; I will then send you reprints of later publications that might be of interest to you in America, hopefully including an offprint of my reply to Schlick, which is due to appear in *Analysis* in June.

I have read your reviews of Carnap's 'Syntax' and other books in the local library with great interest. Your report on the Prague Congress is by far the best I know; I liked it so much that I left a second copy, which Carnap had given me, to a friend in Münster, who is now circulating it there.

Now another factual question: Are you perhaps aware of any attempts (perhaps by the American or Polish specialists in many-valued logic) to construct a quasi-many-valued logic which is not metrical but purely topological, i.e. which does not assign to each proposition one particular of several different, precisely defined truth values, but only establishes a topological order of propositions according to increasing truth value?

We both often and fondly think back to the beautiful and interesting days in London, and we are sure that you will also feel very comfortable over there. If you see Professor Stebbing, please send her our kind regards!

Above all, however, we send you and your companion our warmest greetings and best wishes for a safe journey home and everything else!

Yours
Carl G. Hempel

4.1.2 Hempel to Nagel, July 7, 1936

[Typewritten, Nagel 2020, Box 2, Folder 52]

Brüssel den 7. Juli 1936.

Lieber Dr. Nagel,

mit gleicher Post sende ich Ihnen ein Buch zu, das ich hier in Brüssel zusammen mit Dr. Oppenheim geschrieben habe.[8] Da Sie selbst ja verschiedene Untersuchungen durchgeführt haben, die methodisch ganz ähnlich orientiert sind, so hoffe ich, dass unsere Überlegungen Sie interessieren werden. Sollten Sie sogar geneigt sein, die Schrift kritisch zu besprechen, so würde mich das natürlich ganz besonders freuen. Auf jeden Fall wird der Verleger dem Journal of Philosophy ein Rezensionsexemplar übersenden, und ich wäre Ihnen sehr dankbar, wenn Sie veranlassen wollten, dass das Buch in Ihrer Zeitschrift besprochen wird – auch wenn Sie aus irgendwelchen Gründen es nicht persönlich besprechen wollen.[9]

Meine Frau und ich sind gestern von unserer Reise nach Kopenhagen zurückgekehrt, wo wir den „2. internationalen Kongress für Einheit der Wissenschaft" besucht haben.[10] Die Reise hat sich gelohnt, obgleich natürlich nicht in allen Punkten eine Klärung erreicht worden ist. Das gilt besonders für das Zentralproblem dieses Kongresses, die Frage nach

---

[8] Paul Oppenheim (1885–1977) was an independent scholar, industrialist, who often worked with others, both in Europe (in Brussels) and after 1939 in the United States (mainly at Princeton). Oppenheim wrote together with Hempel *Die Typusbegriff* (Hempel and Oppenheim 1936).

[9] In fact, after all, Nagel himself reviewed the book in the journal, see Nagel (1936d).

[10] "The Second International Congress for the Unity of Science" was organized in Copenhagen, June 21–26, 1936, with the subtitle, "The Problem of Causality—With Special Consideration of Physics and Biology." See Stadler (2015, 178–182).

der Bedeutung der Quantenphysik für Psychologie und Biologie. Da haben nämlich einerseits die Physiker (an ihrer Spitze Niels Bohr[11]) mancherlei interessante Dinge über die Quantenphysik und über ihre logische Struktur erzählt, und andererseits haben die Psychologen (u.a. Rubin[12]) und Biologen (besonders Haldane[13] und Rashevsky[14] – der blendend vortrug) über die Methoden ihrer Gebiete und besonders über die Gesetzlichkeiten biologischer Erscheinungen gesprochen; aber von den Zusammenhängen zwischen diesen beiden Fragenkreisen war nur wenig die Rede. Diese Frage wurde eigentlich nur in einem Vortrag von Frank[15] und dann in dem Manuskript des auf so schreckliche Weise ums Leben gekommenen Schlick[16] kurz berührt; sie wurde aber in der Diskussion nicht weiter behandelt. Interessant war das Auftauchen der Vertreter einer weiteren Schule analytischer Philosophie, der schwedischen Schule von Uppsala,[17] deren Begründer der vor einigen Jahren verstobene Prof. Phalén[18] ist, und die jetzt besonders von Prof. Tegen[19] vertreten wurde. Da müssten Sie eigentlich noch ein Nachwort zu Ihrem Januar-Aufsatz schreiben![20] Ich habe Ihren Artikel inzwischen mit großer Freude nochmals gelesen und ihn auch schon weiter verliehen und empfohlen; er ist wirklich sehr interessant und aufschlussreich!

---

[11] Niels Bohr (1885–1962) was a Danish physicist known for his contributions to atomic and quantum theory. He won the Nobel Prize in Physics in 1922. Bohr is also known for his philosophical reflections on philosophy of science, and played a key role in the Copenhagen Congress, where he spoke about causality and complementarity.

[12] Edgar Rubin (1886–1951) was a Danish psychologist who is mainly known for his work on optical illusions; he presented a paper at the Congress about our knowledge of other people.

[13] J. B. S. Haldane (1892–1964) was a British biologist and mathematician, working mainly on genetics. At the congress he spoke about "Some Principles of Causal Analysis in Genetics".

[14] Nicolas Rashevsky (1899–1972) was an Ukrainian-born, theoretical physicist, who later emigrated to the United States and became known for his work on mathematical biology and biophysics; he closed the Copenhagen congress' panel on "Biology".

[15] Philipp Frank (1884–1966) was an Austrian physicist and philosopher, a key figure in logical empiricism. He held Einstein's chair in Prague for 25 years, and after his emigration to the United States (1938), he taught at Harvard for a decade and led the Institute for the Unity of Science that was established in 1947. At the congress he spoke about the philosophical interpretations and misinterpretations of quantum mechanics.

[16] A former student shot Schlick on June 22, 1936, thus the news of his death reached the audience of the Congress and the event's second day. Schlick was supposed to talk about quantum theory and the knowability of nature and his talk was read from his manuscript. On the death of Schlick, and the events afterwards, see Stadler (2015, Chap. 13).

[17] The so-called Uppsala School was led by Axel Hägerström (1868–1939) and Adolf Phalén (1884–1931). Members of the school tried to break ties with traditional idealistic Swedish philosophy and their program (developed before the official appearance of the Vienna Circle) resembles the ideas of logical empiricism in many ways.

[18] Adolf Phalén (1884–1931) was an Uppsala based Swedish philosopher who wrote a doctoral dissertation on Hegelian philosophy in 1912. He taught in Uppsala, and worked mainly on the philosophy of physics, geometry, and conceptual analysis

[19] Einar Tegen (1884–1965) was a Swedish philosopher and sociologist, with a doctorate in philosophy from Uppsala University (1918). He worked as a professor of practical philosophy first at Lund University (1937–1951) and later at Stockholm University. During the war, he was conducting social psychological studies in the United States.

[20] Hempel presumably refers to the second part of Nagel's "Impressions and Appraisals of Analytic Philosophy in Europe," which was published in January 1936 (see Nagel 1936a, b); Nagel indeed did not discuss any logical positivists or scientific philosophers from the Nordic countries in his report.

Mit herzlichen Grüßen, auch von meiner Frau, bin ich
Ihr
Carl G. Hempel

[English translation]
Hempel to Nagel, July 7, 1936.

Brüssel, July 7, 1936.

Dear Dr Nagel,

I am sending you by the same post a book that I wrote here in Brussels together with Dr Oppenheim. As you yourself have carried out various studies with a very similar methodological orientation, I hope that our considerations will be of interest to you. If you are even inclined to critically review the paper, I would, of course, be particularly pleased. In any case, the publisher will send a review copy to the *Journal of Philosophy*, and I would be very grateful if you would arrange for the book to be reviewed in your journal – even if for some reason you do not wish to review it personally.

My wife and I returned yesterday from our trip to Copenhagen, where we attended the "2nd International Congress for the Unity of Science". The trip was worthwhile, although of course not all discussion points were resolved. This applies in particular to the central problem of this congress, the question of the significance of quantum physics for psychology and biology. On the one hand, the physicists (led by Niels Bohr) said a number of interesting things about quantum physics and its logical structure, and on the other hand, the psychologists (including Rubin) and biologists (especially Haldane and Rashevsky – who gave a dazzling presentation) spoke about the methods of their fields and especially about the laws of biological phenomena; but there was little mention of the connections between these two areas of enquiry. This question was actually only briefly touched upon in a lecture by Frank and then in the manuscript by Schlick, who died in such a terrible way; but it was not dealt with further in the discussion. It was interesting to see the appearance of representatives of another school of analytical philosophy, the Swedish school of Uppsala, whose founder was Professor Phalén, who died a few years ago, and which was now represented in particular by Prof. Tegen. You should actually write an epilogue to your January essay! I have since re-read your article with great pleasure and have already lent and recommended it to others; it is really very interesting and informative!

With best regards, also from my wife, I am
Yours,
Carl G. Hempel

4.1.3 Hempel to Nagel, September 2, 1936

[Handwritten, Nagel 2020, Box 2, Folder 52]

9-2-36.

Dear Dr. Nagel, I thank you very much indeed for having sent me your article on Analytic Philosophy in Europe;[21] I like your paper just as much as your book-reviews

[21] See Nagel (1936a, b).

which I read rather regularly, and I find your account extremely interesting and instructive, also for European readers.

Since a few weeks, I have got another address which I haven't indicated you so long.

My wife and I hope that you are going well and send you our best greetings!

Yours very sincerely,

C.G. Hempel.

4.1.4 Hempel to Nagel, November 14, 1936

[Typewritten, Nagel 2020, Box 2, Folder 52]

Brüssel, den 14.XI.1936
84, rue Timmermans.

Lieber Dr. Nagel,

nehmen Sie von meiner Frau und mir allerherzlichsten Dank für Ihren so reizenden Brief vom 4.August; wir waren beide ganz gerührt über Ihre grosse Anteilnahme an meinem Missgeschick und haben uns wirklich besonders über Ihre freundlichen Worte gefreut![22]

Seien Sie bitte nicht böse, dass ich Ihnen erst so spät antworte; zu der Zeit, als Ihr Brief eintraf, musste ich noch ganz ruhig liegen und durfte nicht schreiben; und als ich – ungefähr Mitte September – wieder schreiben konnte, hat mich zunächst die mannigfache liegengebliebene oder während meiner Krankheit aufgelaufene Arbeit nicht zu einer Antwort kommen lassen; Sie werden sicher Verständnis dafür haben!

Zu meiner Freude kann ich Ihnen berichten, dass ich jetzt völlig wiederhergestellt bin, und dass keinerlei Spuren von dem ganzen Unfall zurückgeblieben scheinen, nicht einmal Kopfweh oder gar Narben.

Es hat Dr. Oppenheim und mich sehr erfreut zu hören, dass Sie von unserm kleinen Buch eine günstige Meinung haben, und wir beide danken Ihnen ferner vielmals dafür, dass Sie so freundlich waren, es zu besprechen;[23] wir freuen uns schon auf die Lektüre Ihrer Rezension.

Ich hoffe sehr, Sie werden nicht vergessen, mir die in Aussicht gestellten Abdrucke Ihrer Arbeit über Wahrscheinlichkeit und Ihrer Reichenbach-Besprechung für „Mind" zu schicken;[24] Sie würden mir damit eine grosse Freude machen, und der Gegenstand interessiert mich nach wie vor sehr!

---

[22] Just after learning how to ride a bike, Hempel had a serious accident and was taken into the nearest hospital with a fractured skull. At first, the doctors claimed that nothing can be done and Hempel will not make it to the next day. Paul Oppenheim arranged some special conditions and medical care for him, and by the next morning Hempel's state has been stabilized. Information taken from Fons Dewulf's blog, see https://www.fonsdewulf.be/post/at-death-s-door-hempel-s-years-in-brussels (retrieved on May 29, 2025).

[23] See Nagel (1936d).

[24] Hans Reichenbach (1891–1953) was a German philosopher who earned his PhD from the University of Erlangen with a thesis on theories of probability in 1915. Reichenbach was the leading member of the so-called Berling Group of logical positivists, and after the rise of Nazism, he emigrated to Istanbul and later in 1938 to the United States, where he taught at UCLA until his 1953 death. Reichenbach's work in question is his *Wahrscheinlichkeitslehre* (Reichenbach 1935a). It was translated into English as Reichenbach (1949). Nagel reviewed the original German edition; see Nagel (1936e). Hempel referred perhaps to Nagel's book manuscript on probability that was published in 1939 as *Principles of the Theory of Probability* (Nagel 1939).

Ich habe zwar seit längerer Zeit mich nicht mit Wahrscheinlichkeitslehre beschäftigt, habe aber im Zusammenhang mit eigenen Überlegungen über nichtaristotelische Logik[25] die Reichenbachsche Wahrscheinlichkeitslogik unter rein formalen Gesichtspunkten nochmals genauer durchgearbeitet und dabei eine Reihe von Eigentümlichkeiten festgestellt, die Sie möglicherweise auch interessieren werden. Auf jeden Fall schicke ich Ihnen einen Durchschlag meiner Notizen, den ich nicht zurückzuhaben brauche.[26] Reichenbach, dem ich ebenfalls ein Exemplar geschickt habe, hat dem Inhalt dieser Überlegungen zugestimmt; für den Aufbau seiner Wahrscheinlichkeitslogik entscheidet er sich für die auf Blatt IV oben angegebene Festsetzung und nimmt in Kauf, dass die Wahrscheinlichkeitswerte der Aussagenverbindungen seiner n-wertigen Logik nicht sämtlich zu β_n gehören, oder: dass β_n nicht abgeschlossen ist mit Bezug auf die arithmetischen Operationen, mittels deren er in seiner Werttafel die W-Werte der verschiedenen elementarlogischen Verknüpfungen definiert.

Kürzlich ist mir noch folgendes Weitere aufgefallen: Reichenbachs „Werttafeln der Wahrscheinlichkeitslogik“ (S.381)[27] legen nicht, {wie die üblichen} Werttafeln n-wertiger Systeme, den W-Wert jedes durch elementarlogische Verknüpfungen aus Elementarausdrücken gebildeten Komplexausdrucks fest, z.B. nicht W(φ > (φ v ψ)). In der Tat braucht man hier noch außer p, q, u den Wert von W(φ » (φ v ψ)), der durch die erstgenannten drei Zeilen nicht festgelegt ist. Entsprechendes gilt für *sämtliche* Komplexe aus Elementarausdrücken φ, ψ, …, die nicht mit einem der 5 in den Werttafeln festgelegten Ausdrücken identisch sind oder die Form ~ ~ ~ … ~ φ haben. – Es sei allgemein ein komplexer Ausdruck A (formal entsprechend einer „Wahrheitsfunktion“) gegeben, der mittels der Zeichen „~“, „v“, „.“, „>“, „≡“, aus Elementarausdrücken φ, ψ … aufgebaut sei. s sei die Gesamtzahl der in A auftretenden Verknüpfungszeichen, die nicht Negationszeichen sind. Dann sind zur Berechnung von W(A) außer den W-Werten der verschiedenen Elementarausdrücke (und unabhängig von deren Anzahl) genau s Angaben nach Art der in den Tabellen vorgesehenen Größe zu machen. Die Ausführungen S.379u/380oben[28] bedürfen also einer Verallgemeinerung in diesem Sinne.

Ich weiss allerdings nicht, wie weit Sie gerade dieser formale Aspekt der Wahrscheinlichkeitslogik interessiert: ich selbst habe mich eine Zeitlang genauer mit mehrwertigen Systemen beschäftigt und eine topologische Verallgemeinerung entwickelt, die, wie ich hoffe, im Journal of Symbolic Logic erscheinen wird.[29] – Allerdings muss ich Ihnen gestehen, dass ich bezüglich der Fruchtbarkeit der mehrwertigen Systeme nicht allzu optimistisch bin; sie scheinen mir aufklärend und anregend für die allgemeine Theorie der deduktiven Systeme zu sein, und vielleicht auch nützlich zur Auflockerung eines Apriorismus, der bestimmte Gesetze der Logik für denknotwendig halten möchte oder für unentbehrlich beim formalen Aufbau irgendeiner Logik; aber „Anwendungen“, auch in den abstraktesten Gebieten, scheint es nicht zu geben; und der Reichenbachsche Fall ist insofern interessant, als seine Logik, die eine gewisse anschauliche Interpretation zulässt, von den n-wertigen Logiken, wie sich nun zeigt, in mehrfacher Hinsicht wesentlich abweicht.

---

[25] See Hempel (1936). Previously, Hempel defended his doctoral dissertation that concerned probability; see Hempel (1934).

[26] One extra page was attached to Hempel's letter, see (Fig. 4.1) at the end of this letter.

[27] See Reichenbach (1935a).

[28] See Reichenbach (1935a).

[29] See Hempel (1936, 1937a).

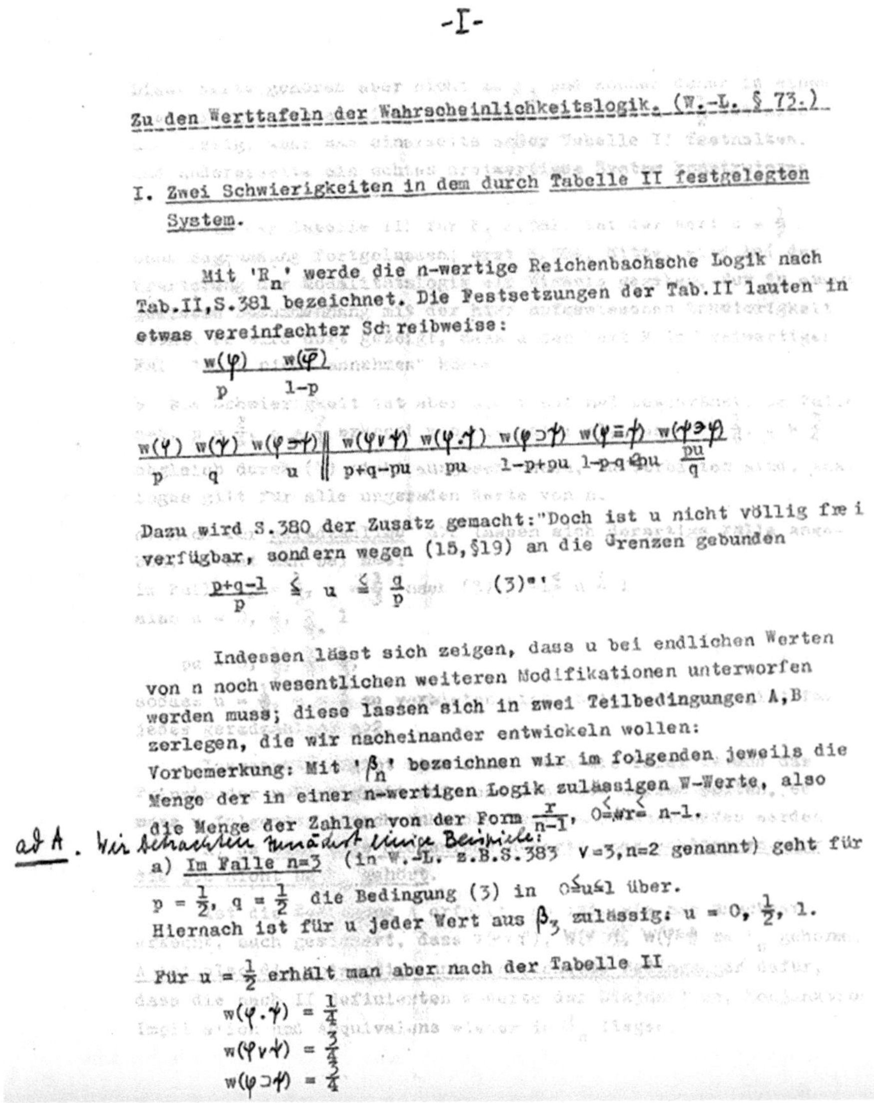

-I-

Zu den Werttafeln der Wahrscheinlichkeitslogik. (W.-L. § 73.)

I. Zwei Schwierigkeiten in dem durch Tabelle II festgelegten System.

Mit 'R_n' werde die n-wertige Reichenbachsche Logik nach Tab.II, S.381 bezeichnet. Die Festsetzungen der Tab.II lauten in etwas vereinfachter Schreibweise:

$w(\varphi)$	$w(\bar{\varphi})$
p	$1-p$

$w(\varphi)$	$w(\psi)$	$w(\varphi \ni \psi)$	$w(\varphi \vee \psi)$	$w(\varphi . \psi)$	$w(\varphi \supset \psi)$	$w(\varphi \equiv \psi)$	$w(\psi \ni \varphi)$
p	q	u	$p+q-pu$	pu	$1-p+pu$	$1-p-q+2pu$	$\frac{pu}{q}$

Dazu wird S.380 der Zusatz gemacht: "Doch ist u nicht völlig frei verfügbar, sondern wegen (15, §19) an die Grenzen gebunden

$$\frac{p+q-1}{p} \leq u \leq \frac{q}{p} \qquad (3)\text{"}$$

Indessen lässt sich zeigen, dass u bei endlichen Werten von n noch wesentlichen weiteren Modifikationen unterworfen werden muss; diese lassen sich in zwei Teilbedingungen A, B zerlegen, die wir nacheinander entwickeln wollen:

Vorbemerkung: Mit 'β_n' bezeichnen wir im folgenden jeweils die Menge der in einer n-wertigen Logik zulässigen W-Werte, also die Menge der Zahlen von der Form $\frac{r}{n-1}$, $0 \leq r \leq n-1$.

ad A. Wir betrachten zunächst einige Beispiele:

a) Im Falle n=3 (in W.-L. z.B. S.383 v=3, n=2 genannt) geht für $p = \frac{1}{2}$, $q = \frac{1}{2}$ die Bedingung (3) in $0 \leq u \leq 1$ über. Hiernach ist für u jeder Wert aus β_3 zulässig: $u = 0, \frac{1}{2}, 1$.

Für $u = \frac{1}{2}$ erhält man aber nach der Tabelle II

$$w(\varphi . \psi) = \frac{1}{4}$$
$$w(\varphi \vee \psi) = \frac{3}{4}$$
$$w(\varphi \supset \psi) = \frac{3}{4}$$

Fig. 4.1 Hempel to Nagel, Nov. 14, 1936. (Reproduced with the permission of Ernest Nagel Papers, 1925–1985, ASP.2020.01, Archives & Special Collections, University of Pittsburgh Library System.)

Ich habe kürzlich Poppers „Logik der Forschung“ nochmals gelesen, um eine Besprechung darüber zu schreiben.[30] Mir scheint, die Arbeit ist ein sehr wertvoller Beitrag zur Wissenschaftslogik, und Reichenbach ist in seinem polemischen Artikel in der Erkenntnis nicht gerecht und nicht fair mit Popper verfahren;[31] und ich finde es unerhört, dass er z.B. in seinem Artikel im Journal of Philosophy Popper (in einer Fußnote) mit dem Hinweis erwähnt, der ungefähr besagt, dass er P.s Buch nicht wissenschaftlich ernst nehmen könne.[32]

In einiger Zeit wird ja nun Helmer[33] nach Amerika kommen. Er schrieb mir ganz kürzlich sehr erfreut, dass Sie sich bereits so freundlich um ihn bemüht hätten; das ist wirklich sehr nett! Aber ich denke, Helmer, mit dem meine Frau und ich schon von Berlin her befreundet sind, verdient es auch, dass man sich seiner annimmt, und wir hoffen sehr, dass es ihm gelingen wird, in Ihrem zukunftsfroheren Lande Fuß zu fassen! (Und natürlich wären wir froh, eines Tages hinterherkommen zu können.)

Nochmals vielen Dank für all Ihre Freundlichkeit, und die herzlichsten Grüße und Wünsche von meiner Frau und mir!

Ihr Carl G. Hempel

[English translation]
Hempel to Nagel, November 14, 1936.

Brussels, 14.XI.1936
84, rue Timmermans.

Dear Dr Nagel,

Please accept my and my wife’s most heartfelt thanks for your lovely letter of 4th August; we were both very touched by your great sympathy for my misfortune and were particularly pleased by your kind words!

Please don’t be angry that I am replying to you so late; at the time your letter arrived, I had to lie very still and was not allowed to write; and when I was able to write again – around the middle of September – I was initially unable to reply due to the many things I had left undone or had accumulated during my illness; I am sure you will understand!

I am pleased to report that I am now fully recovered and that no traces of the accident seem to remain, not even headaches or scars.

Dr Oppenheim and I were very pleased to hear that you have a favorable opinion of our little book, and we both thank you very much for being so kind as to review it; we are already looking forward to reading your review.

I very much hope that you will not forget to send me the promised reprints of your work on probability and your Reichenbach review for ‘Mind’; you would make me very happy, and I am still very interested in the subject!

Although I have not dealt with probability theory for a long time, I have worked through Reichenbach’s probability logic again in more detail from a purely formal point of view in connection with my own considerations on non-Aristotelian logic and have noticed a

[30] Karl Popper (1902–1994) was an Austrian-British philosopher, mainly known for his work on the philosophy of science and political philosophy. After receiving a doctoral degree in psychology, Popper got closer to the Vienna Circle, though he was never an official member and never attended their regular meetings. He published in 1934 his famous *Logik der Forschung*, which was translated into English only in 1959 as *The Logic of Scientific Discovery*. Hempel reviewed the book, see Hempel (1937b).

[31] See Reichenbach (1935b). For an English translation see Reichenbach (1935/1978).

[32] See Reichenbach (1936, 153, n. 16).

[33] Olaf Helmer (1910–2011) was a German-American logician and mathematician, known mainly for his work at the RAND Corporation and for being the co-founder of the Institute for the Future.

number of peculiarities that may also interest you. In any case, I am sending you a copy of my notes, which I do not need to have back. Reichenbach, to whom I have also sent a copy, has agreed with the content of these considerations; for the structure of his probability logic he decides in favor of the determination given on sheet IV above and accepts that the probability values of the propositional connections of his n-valued logic do not all belong to β_n, or: that β_n is not complete with respect to the arithmetic operations by means of which he defines the W-values of the various elementary logical connections in his table of values.

Recently, I noticed the following: Reichenbach's "Value Tables of Probability Logic" (p.381) do *not* specify value tables of n-valued systems {like the usual}, the W-value of each complex expression formed by elementary logical operations from elementary expressions, e.g. not W(φ > (φ v ψ)). In fact, in addition to p, q, u, the value of W(φ » (φ v ψ)), which is not determined by the first three lines mentioned, is required here. The same applies to all the complexes of elementary expressions φ, ψ, ..., which are not identical to one of the 5 expressions defined in the value tables or have the form ~ ~ ~ ... ~ φ. – Let there generally be a complex expression A (formally corresponding to a 'truth function'), which is constructed by means of the characters '~', "v", '.', '>', '≡', from elementary expressions φ, ψ .... Let s be the total number of linking characters occurring in A that are not negation characters. Then, for the calculation of W(A), in addition to the W values of the various elementary expressions (and irrespective of their number), exactly s specifications of the type provided in the tables must be made. The explanations on p.379 down/380 up therefore require a generalization in this sense.

However, I do not know how much you are interested in this formal aspect of probability logic: I have been studying many-valued systems for some time and have developed a topological generalization which I hope will appear in the *Journal of Symbolic Logic*. However, I must confess that I am not too optimistic about the fruitfulness of many-valued systems; they seem to me to be enlightening and stimulating for the general theory of deductive systems, and perhaps also useful for loosening up an apriorism that would like to consider certain laws of logic as necessary for logic or as indispensable in the formal construction of any logic; but 'applications', even in the most abstract areas, do not seem to exist; and Reichenbach's case is interesting insofar as his logic, which allows a certain descriptive interpretation, differs substantially from the n-valued logics in several respects, as is now apparent.

I recently re-read Popper's "Logik der Forschung" in order to write a review of it. It seems to me that the work is a very valuable contribution to the logic of science, and Reichenbach did not treat Popper fairly and justly in his polemical article in *Erkenntnis*; and I find it outrageous that in his article in the *Journal of Philosophy*, for example, he mentions Popper (in a footnote) with a reference that roughly says that he cannot take P.'s book scientifically seriously.

Helmer will be coming to America in a while. He wrote to me very recently that he is very pleased that you had already made such a friendly effort to help him; that is really very nice! But I think Helmer, with whom my wife and I have been friends since Berlin, also deserves to be taken care of, and we very much hope that he will succeed in gaining a foothold in your more forward-looking country! (And of course we would be happy to be able to follow him one day.)

Thank you again for all your kindness, and warmest regards and wishes from my wife and me!

Yours, Carl G. Hempel

4.1.5 Hempel to Nagel, October 10, 1937

[Typewritten, Nagel 2020, Box 2, Folder 53]

Chicago, Ill., October 10th., 1937
c/o Faculty Exchange, The Univ. of Chicago

Dear Mr. Nagel,

since last Sunday, we are settled: we found a nice furnished flat in an apartment-hotel right near the university; and though, as far as we can so far see, we should probably like better living in New York than in Chicago, we got more and more accustomed to our new life and enjoy it progressively. Of course, this is considerably due to the fact that the Carnaps and the Helmers are living here; we meet each other very frequently.[34] The Helmers succeeded in finding a particularly nice unfurnished apartment which they furnished within three days and nevertheless in a very attractive kind and at a relatively small expense. The Carnaps are following the Helmers' example: Ina just found a conveniently situated very nice flat, and now they start buying the furniture.

As to Helmer's and my activity at the university, I cannot yet tell you very much. At any case, it seems, that in the department there is a decided opposition against our giving lectures or seminars of our own which would have to be announced in the time-schedule. It is to be supposed that this opposition is caused by the apprehension that positivistic tendencies might increase too much, and on the other hand, Helmer thinks the department will perhaps express by this attitude that it does not wish to be obliged to help us finding another job when this year is over (whilst practically it feels obliged to do so with its regular instructors or lecturers when their activity at this university comes to an end). Anyhow, in this quarter, we shall confine ourselves to leading a discussion group in connexion with Carnap's research-seminar; there, we are dealing with special chapters of modern logic and reading publications which are supposed to be known to the participants of Carnap's seminar. Besides that, we shall probably have a sort of office-hours in which students of philosophy may ask us for explanations or informations.

At the first session of the research-seminar, there were about 20 participants, amongst others also several members of the staff: Morris,[35] Benjamin,[36] Eckart[37] and a mathematician whose name I have forgotten.

The other day, I learnt from Prof. Hartshorne,[38] that the next meeting of the Eastern Division of the Amer. Philos. Soc. will be held in the end of December, in Princeton. Unfortunately, there is at the same time (29th dec.) a meeting of the Association for Symbolic Logic in Indianapolis which is not just in the neighborhood of Princeton. As to myself, I suppose it will be wiser (also for purely practical reasons: at the end, I must try to find a job for the time after this year) to attend the philosophical meeting and to read a paper at that occasion. In this context, I have several questions; will you be so kind to answer them as far as possible and to give me your advice with respect to the normative ones amongst them? That would be awfully kind of you.

[34] In 1937, the Department of Philosophy at the University of Chicago planned a research seminar under Carnap's supervision, and he had a chance to hire two assistants for a year. After all, he was able to get Hempel and Helmer and thus Hempel arrived in Chicago.

[35] Charles Morris (1901–1979) was an American philosopher, mainly known for his work on semiotics. He was a key figure connecting American pragmatism and European logical empiricism and became one of the leaders and organizers of the Unity of Science Movement.

[36] A. Cornelius Benjamin (1897–1968) was an American philosopher who wrote his dissertation on Bertrand Russell's logical atomism. He became an assistant professor at the University of Chicago in 1932 and stayed there until 1945. He wrote many books on science and philosophy of science.

[37] Though Hempel wrote originally "Eckardt", he presumably had in mind Carl Eckart (1902–1973), a well-known American physicist, who worked at the University of Chicago since 1928.

[38] Charles Hartshorne (1897–2000) was an American philosopher, working mainly on religion and metaphysics. He worked at the University of Chicago between 1923 and 1955.

1. Can you give me the exact date of the meeting?
2. I was told that those who want to read a paper at the meeting have to send an outline of 300 words to Prof. C. Krusé[39] by Nov.1th. Is that the latest date? Would a text which I should eventually submit, be accepted without further formalities, or would it be preferable to join an introducing letter say, of Carnap? – Are the papers presented at such a meeting intended to be published, and if so, has one to furnish the complete text for a later date?
3. How long is such a speech supposed to be? 10 to 20 minutes? Is it preferable to present something relatively new? Would it be wise to give the paper a more general character avoiding the use of any logistic expedient, or would just a condensed and very specialized paper be preferred? – Do you think I might perhaps discuss some problems connected with the concept "probability of a given theory" – even though I am not able to propose satisfactory solutions for each of these problems but should have to confine myself, with respect to certain points, to merely making the difficulties explicit? or to giving provisional suggestions?
4. Are you also likely to attend that meeting? I should be too glad to meet you on that occasion and to talk to you more in detail than it was possible during our short stay in New York.

I apologize for asking you so many questions; but just your advice would be particularly valuable to me, and so I hope you won't be too angry at my giving you this trouble.

With the best greetings and wished from both of us to yourself and Mrs. Nagel,
I am, yours very sincerely,
Carl G. Hempel.

4.1.6 *Hempel to Nagel, October 15, 1937*

[Typewritten, Nagel 2020, Box 2, Folder 53]

Chicago, 15th. of October, 1937.
c/o Faculty Exchange,
The University of Chicago.

Dear Ernest,

you see, I'm very glad to follow the suggestion you gave me in your P.S.;[40] if you don't mind, I should like to address you with your forename as Olaf [[Helmer]] does. As to myself, I do not like very much my forename 'Carl Gustav', and almost nobody is using it. Eva[41] gave me the name 'Peter' instead, which, amongst others, also the Helmers are using; so you may now decide yourself whether you like better calling me 'Peter' or 'Hempel'; please use the name you prefer.

It was very kind of you to answer so quickly, and it turned out to be very useful for me: I received your letter just before the beginning of Carnap's seminar, and so I could immediately afterwards talk to Carnap and Morris about your suggestions. Morris very kindly promised me to write to Krusé that very night; but since he is not a member of the Eastern Division, I should be very glad if you too would write a few words to Krusé. According to Morris' advice, I also wrote myself asking Krusé whether, though not being a member, I might be allowed to read a paper on the probability of empirical hypotheses and theories; and I added that this problem has recently been very much discussed by Reichenbach, Carnap and Popper. – Unfortunately, the connexions between the considerations which I should be

[39] Cornelius Krusé (1893–1978) was an American philosopher who taught at Wesleyan University; he acted as the president of the APA in 1947 and was the Secretary-Treasurer of the Eastern Division from 1937 to 1939.

[40] Nagel's letter seems to be lost.

[41] Eva Hempel (1908–1944, née Ahrends) was Carl G. Hempel's first wife.

in a position to submit, and the problem of causality are rather loose; so I did not dare to mention that point in my letter. But I shall keep your suggestion in mind, and I think I shall be able to make at least a few remarks concerning the question you mention in your letter.

May I still ask you to give me some more information in connexion with the problem of my eventual paper? I believe that I read in one of your articles or reviews a remark to the effect that the idea of a statistical interpretation of probability as given in von Mises'[42] and Reichenbach's theories would not be so new to American students as it seemed to be to European ones – since this type of interpretation had already been put forward by certain American authors.[43] Do I remember this right? And if so, would you perhaps let me know who the authors in question are (Peirce?),[44] and where they published their ideas about probability? And secondly, I should be very glad to know whether there are American authors who have seriously dealt with the question how to define the probability of empirical theories.

Yesterday, there was the first meeting of our discussion-group.[45] There were ten attendants, amongst them the chemist Prof. Senior[46] who seems to be very seriously interested in the logical analysis of empirical science. The discussion was very lively and, as Olaf and I think, also quite fertile. And for ourselves, I think, this is an excellent exercise.

Will you please remember us to the Mindells? I hope to write to Mindell[47] on one of the next days.

Once more many thanks for your help and love 2:2!

Yours cordially

Peter Hempel

4.1.7 Hempel to Nagel, November 14, 1937

[Typewritten, Nagel 2020, Box 2, Folder 53]

Chicago, Nov.14, 1937.

Dear Ernest,

thank you very much for all the trouble you have taken for me!

Krusé wrote to me very kindly indeed, and I sent him the full text and an abstract of my paper in time. So far, I do not yet know the decision of the committee; but I think it will be

[42]Richard von Mises (1883–1953) was an Austrian mathematician who worked on statistics, probability and the application of mathematics. He was a professor in Berlin until he went (similarly to Reichenbach) to Turkey in 1933, and finally emigrated to the United States, where he worked at Harvard. Von Mises is also known for his work on aerodynamics and was a recognized expert on Rainer Maria Rilke.

[43]Presumably Nagel (1933); see further Verhaegh (2020) on the differences between German and Anglophone literature of probabilities.

[44]Charles Sanders Peirce (1839–1914) was an American philosopher, logician, and mathematician, known mainly as one of the founders of American pragmatism.

[45]According to (the yet unpublished parts of) Carnap's diaries, Hempel had a discussion group at his home with Olaf Helmer that was seemingly connected to Carnap's research seminar on semantics. The group usually had 5–10 members, such as Carnap, Eva Hempel, Charles Morris, A. Cornelius Benjamin, and others.

[46]James. K. Senior (n/a) was an American chemist who was educated at the University of Chicago in 1917, and worked later in the Kent Laboratory of the university.

[47]Possibly Joseph Mindell (1910–1996), who was a well-known American broker in New York. He authored a textbook on stock markets but initially had a BA in philosophy from New York University in 1930 and later studied for two years at the Columbia Law School and worked at the School of Business, also at Columbia.

advisable for me to attend that Princeton meeting even in case they do not accept my paper. Probably, Eva will accompany me, and we are looking forward very much to seeing the two of you and to talking with you in detail about our impressions and experiences.

We get more and more accustomed to Chicago and like living here much better than in the first week. We got acquainted with several very nice people, and everybody we met so far has been very friendly with us.

Olaf and I enjoy our activity at the University very much. Carnap's lecture course on logistic which I regularly attend seems to me a masterpiece from every point of view, and I am learning a good deal from it. The seminar is, of course, of a particularly high interest; it would be so fine to have you here for it.

Besides, Helmer and I am glad to lead the discussion group in connexion with the seminar. A few weeks ago, another group which has been given the title "Scientific Method Group" has been founded on the suggestion of Mindell's friend Passin[48] and a second student, Krader;[49] the attendants are students of different fields who are interested in scientific philosophy and in the rejection of metaphysics (particularly in the Adlerian form).[50] These people are very nice and interested; unfortunately, some of them overestimate their degree of logical insight and training and are a little too eager to make a criticism of metaphysics which at the bottom is still very naive. Helmer and I try to do our best in order to inform and train the people a little and to prevent the discussion from taking a merely negative and critical aspect. In the first meeting, Carnap came himself and gave a talk on empirical science and metaphysics;[51] and now we are discussing some questions of the logic and methodology of physics. The next point will be a short analysis of mathematics and logic, and then we shall try to apply the points of view thus obtained to different other branches of empirical science, beginning perhaps with psychology. In the last meeting, a student of Adler's was introduced; he seems to be a nice and reasonable man, and we hope that his argument will be stimulating in the discussions.

I think it would interest Mindell if you occasionally told him these few details about the "S.M.G.".

Now I have to trouble you again with a question. Carnap and Olaf suggested me to visit, after the Princeton meeting, some more eastern universities in order to get personally acquainted with logicians and philosophers; they think this might also be helpful for finding a job. So I shall try to meet Prof. Hull[52] in Yale and perhaps Prof. Köhler[53] whom I know

[48]Herbert Passin (1916–2003) was an American scholar of contemporary Japan. After serving in the Army during World War II, he was appointed as Chair of the Sociology Department at Columbia University and its East Asia Institute.

[49]Lawrence Krader (1919–1998) was an American anthropologist and ethnologist who studied with Carnap at the University of Chicago in 1937–38.

[50]Mortimer Adler (1902–2001) was an American philosopher, known mainly for his work in the Aristotelian and Thomistic traditions. During his work at the University of Chicago, Adler attacked the positivists severely.

[51]According to Carnap's unpublished diary (see at https://valep.vc.univie.ac.at/valepPlus/), the first meeting of the "Gruppe für Methodologie" was held on November 1, and Carnap talked about metaphysics. He also noted in his diaries that the group consisted of "Trotzkists".

[52]Clark L. Hull (1884–1952) was an American psychologist who worked on learning and human behavior. He was associated with Yale University and developed a special institute and seminar series there. Hull had a balanced relationship with logical empiricism: he invited Otto Neurath to his research seminar and even contributed a paper to Neurath's *Einheitswissenschaft* series. For the English translation, see Hull (1937/1987).

[53]Wolfgang Köhler (1887–1967) was a German psychologist, contributing to Gestalt psychology. After working in Berlin, he left Germany in 1935 and got a professorship at Swarthmore College in Pennsylvania. Köhler was to launch and edit a special journal for scientific philosophy with Carnap and Reichenbach already in 1923 and he maintained a close relation to logical positivists.

from Berlin and who is now at Swarthmore College. Do you think there are still other people whom I ought to see? If so, it would be very kind of you to let me know; I then should write them beforehand and ask whether I may come and see them in the first days of January. But I think, most of the people I should have to see will come to Princeton, and if you do not know of a particularly important exception, then please, *do not trouble with writing*.

Finally I want to thank you very much for your informations about English and American supporters of the statistical theory of probability; these indications are very valuable for me!

Love from the two of us to you and Edith!

Very cordially,

Peter

P.S. I am joining Krusé's letter; many thanks!

October 20, 1937
Professor Ernest Nagel
Philosophy Hall
Columbia University
New York, New York

My dear Professor Nagel:

Thank you very much for your letter about Dr. Carl G. Hempel. I have had a letter from Professor Morris about him, and have written both that any paper Dr. Hempel would wish to submit to the program committee would receive serious consideration. Unfortunately, at this late date there are relatively few places still open. In the event that we should be able to find a place for him, I am very happy to know that you would be willing to sponsor him. I had asked Professor Morris whether he would be willing to do this, but in view of the fact that he belongs to the Western Division, it would be perhaps better for you to appear on the program as his sponsor. I am sure you will understand that it is impossible to assure a place on the program for Dr. Hempel. Nevertheless, the committee will explore all possibilities to see what can be done. We are always appreciative also of having the names of interesting and able people brought to our attention. I am wondering whether Dr. Hempel is to be permanently in this country or is just here on a visit this year. Thank you very much for your interest in this matter.

Sincerely yours,

Cornelius Krusé

P.S. Could his paper, I wonder, be worked in the session on "The Nature of the Causal Relation in the Light of Recent Physics"? Morris Cohen[54] & and Northrop[55] reading papers in this session.[56]

C.K.

---

[54]Morris R. Cohen (1880–1947) was an American philosopher and lawyer; he was professor of philosophy at the City College of New York from 1912 to 1938. He was an important teacher for Nagel, who often helped him with scholarships and with money when Nagel was in difficult existential situations. They wrote together *An Introduction to Logic and Scientific Method*, New York: Harcourt Brace (Cohen and Nagel 1934), though presumably Nagel did most of the writing.

[55]F.S.C. Northrop (1893–1992) was an American philosopher of science, working also in areas of philosophy of cultural studies.

[56]Morris Cohen gave a talk on "Causality in the Light of Recent Physics", while Northrop talked about "Causality in Field Physics in its Bearing on Biological Causation". See the abstracts in N/A (1936–37).

4.1.8 *Hempel to Nagel, December 3, 1937*

[Handwritten, Nagel 2020, Box 2, Folder 53]

3-12-37

Dear Ernest, this is to tell you that my Princeton paper has been accepted by the Program-Committee; it was so kind to mediate in this case; thank you very much![57]

Eva and I are probably arriving in Princeton on the 27th in the evening. From Dec. 30 to about Jan. 2 we are likely to stay in New York. During this time we hope to see the Mindell's too. I was very glad to hear from Lillian[58] and from Mindell about the new discussion-group plans and about [n/a] having become a member of your special group.[59]

I wonder whether I shall have to pay a participation fee for Princeton, and whether we can perhaps be lodged in the university dormitory. I hope appropriate informations will be sent to the attendants in time, and I asked Krusé to send me a copy as information are issued.

In a hurry, love 2:2!

Yours,

Peter

4.1.9 *Hempel to Nagel, December 16, 1937*

[Handwritten, Nagel 2020, Box 2, Folder 53]

Dec 16, 1937
5514 Blackstone Ave

Dear Ernest,

I am very glad indeed to have all your interesting and helpful informations. Eva and I had intended to attend the presidential dinner if possible, and we wanted to thank you very heartily for your invitation. We gratefully accept your offer to introduce us to that dinner; but we really think you ought not to make economical sacrifices in this context; you know that this year, I am a Rockefeller Croesus; so won't you agree that we pay the charge ourselves? We are really touched by your kindness; but I think, we ought to be sensible people!

Of course, I should be very glad to attend a meeting of your discussion-group; and if you like me to read a paper, I shall be glad to do so. As to the subject, I think that perhaps several members of your group might be in Princeton and hear my paper on probabilities of theories there, so that it is perhaps advisable to choose a different topic. The other day, I read a paper

[57] For Hempel's abstract ("On the Probability of Empirical Theories"), see N/A (1936–37).

[58] Presumably Lillian Broadwin, wife of Samuel Broadwin who was an American philosophy student, disciple of Émile Meyerson, who spent some time in Vienna in 1929. Seemingly, Broadwin never obtained a PhD, but lived later in New York and became a member of various philosophical associations.

[59] Hempel presumably refers to a philosophy circle, which was "an informal group of faculty members of various universities and colleges in New York who met at the house of one or another person and read papers." Hempel mentions as its members logicians like J. J. C. McKinsey, and philosophers like Philip P. Wiener and Nagel. See Hempel (2000, 19).

in the Society for Social Research[60] on "The typological method in Social Science"; the title had been coined by the secretary of the society; the content of the speech was an outline of the main ideas of the book on Typusbegriff, but extended also to Weber's "Idealtypen", which require a slightly different treatment.

On the other hand, I might also try and give you a sketch of the general character and of the methods of semantics. In case you want me to speak on this topic, I should be grateful if you just sent me a line to let me know how far the symbolic method in logic and syntax is familiar to the members of the group. I could arrange to get along with a rather modest technical apparatus and add – if it is desired – some remarks in the discussion, concerning the formalization of the semantical investigation.

As to the date, one might think of Dec 30, if that is possible. Before going to Princeton, we shall probably not stop in New York or just for a short time – we are doing the trip without interruption in day-coach-class, leaving here on Dec 26 in the afternoon and probably arriving in N.Y. in the late morning Dec 27. So, on the 27th, I shall probably be a little worn out and not fit for discussions; on the other hand, I should like to be fresh on the 28th.

But I shall not attend the New York part of the meeting, on Dec 30 (unless you give me a different advice); so certainly on Dec 30 in the late afternoon or in the evening, I should be free to attend a meeting of your group. If this date is not convenient, we might perhaps choose one of the first days of January. There's only a little technical difficulty; I offered Prof. Hull at Yale to visit him in the first days of January and even asked him whether he could perhaps arrange a paid lecture for me; now, so far, I have not yet received an answer from him, and that makes it difficult for me to tell you exactly how long I shall be in New York. But I am rather sure that I could arrange to be in N.Y. still in the evening of the 3rd; otherwise, we might perhaps make an arrangement for a date after my return from New Haven to N. York.

The Carnaps & and the Helmers send you & Edith their best regards & many good wishes for nice holidays. The Carnaps will stay here during the vacancies, & the Helmers are going to the mathematico-logical meeting in Indianapolis.

Unfortunately, I have not received, so far, any technical informations from Krusé; and I should so much like to know some details about the sleeping-possibilities & book a room for us. Could you perhaps send us your copy of the informations just for a day? I would return it immediately.

As soon as our time-table is definitely settled, I shall let you know the time of our arrival in New York; perhaps you & Edith are also going by train to Princeton; then we might perhaps go together (if you go already on the 27th).

I hope that you will enjoy nice & restful days before the meeting; you certainly need them, I suppose!

We are so much looking forward to seeing you both! Love from the two of us:

Yours, Peter

4.1.10 Hempel to Nagel, February 18, 1938

[Typewritten, Nagel 2020, Box 2, Folder 54]

[60]Hempel refers presumably to the Department of Sociology's "Society for Social Research" that was founded in the early 1920s for graduate students and faculty members. Carnap noted in his diaries (December 13) that Hempel gave a talk in the "Social Research Gruppe", although the special scientists there did not get much from the talk.

Chicago, Ill., Feb.18, 1938.
5514 Blackstone Avenue.

Dear Ernest,

I just read your paper on the philosophy of logic in Phil. of Sci.[61] Your remarks are of great interest for me. First of all, you can imagine, I was very glad that the attempt at defining a degree of confirmation for empirical hypotheses, about which I reported at Princeton, goes so well together with the general approach you outline on pp. 50–51, though I would not think that there is any difference in principle between speaking of the truth of a singular statement and of the truth of a theory; for there is certainly no essential difference between the concept of truth as applied to universal sentences and the concept of truth applied to singular sentences; and a theory may be considered as a class of sentences some or all of which are universal; hence it may be called true, if each of its members is a true sentence. And also with respect to empirical confirmation, there seems to me to be only a gradual difference between singular sentences and universal sentences or even entire theories. But perhaps you do not mean to deny this in your paper. – A second point of great interest for me is your discussion of the conventional character of logic. This is really a very important question, and it is astonishing to see how many authors still claim some sort of ontological content for the »laws of logic«; I am thinking, for instance, of Nelson's[62] Princeton paper, of the symposium on Truth by Convention in Analysis vol 4,[63] and of some remarks in McGill's paper on Logical Positivism and the Unity of Science in Science and Society, vol 1., No.4.[64] When I was in New York, McGill gave me a copy of that paper and asked me to let him know my objections to his theses. I am joining a copy of the remarks I wrote to him; perhaps you like to read them; I do not want that copy back, I got it made for you.[65] (Incidentally, do you think I ought to publish an answer to McGill's paper, following the lines indicated in my letter to him? Are public replies to such criticisms of Logical Positivism desirable, or will they not do any good, in your opinion?)

Some days ago, I sent the outlines of a paper "Is Logic descriptive of the World?" to Prof. Castell[66] and asked him whether there would be an opportunity for me to read that paper at the April Meeting of the Amer. Philos. Ass. in Urbana. I am afraid it is very unlikely for the paper to be put on the program, since only one afternoon is reserved for short papers, and all the rest of the time for panels, and there are no parallel sessions at all. – So much the better for me that I had, grace to your intercession, that fine opportunity at Princeton!

---

[61] See Nagel (1938a).

[62] Everett J. Nelson (1900–1988) was an American philosopher who obtained his PhD under the supervision of C.I. Lewis. He worked on the philosophy of logic and metaphysics at the University of Washington. At Princeton he talked about "the foundations of logic", arguing in his abstract for the "necessity of a metaphysical foundation for logic, to suggest such a basis, and to indicate why linguistic syntax, dispensing with metaphysics, is untenable as a theory of logic". See N/A (1936–37, 673).

[63] See Ayer et al. (1936).

[64] Vivian J. McGill (1897–1977) was an American philosopher, who obtained his PhD from Harvard in 1925. He studied in Europe as well, with philosophers such as Edmund Husserl and G. E. Moore. He taught philosophy and psychology at Hunter College between 1929 and 1954. After writing on logical positivism, he published books on Schopenhauer and happiness as well. Hempel refers to McGill (1937); see further McGill (1936).

[65] Hempel's note does not seem to survive; he did not publish his comments as a reply, though he published a year later a short review of McGill's "Concerning the Laws of Contradiction and Excluded Middle" (see McGill 1939).

[66] Alburey Castell (1904–1987) was a Canadian-American philosopher, interested in logic, philosophy of mind and education. Between 1931 and 1941, he taught at the University of Minnesota.

We think very often and with much pleasure of those days in the East which we really enjoyed in every respect. The drive back was very pleasant for us, in spite of the changing weather conditions; Lillian is really an amazing driver. It was very fine for us to have Lillian in Chicago for some weeks, and the Carnaps, the Helmers and we enjoyed her company very much. Almost three weeks have passed since she left us; I think you will have seen her in the meanwhile, and she will have told you lots of Chicago-gossip. And now we are very anxious to learn your decision as to the visit of the meeting here in Chicago, which Carnap has in mind. We would be so glad to have you here!

In this quarter's Seminar, Carnap is treating the foundations of logic and mathematics, and we have been discussing some rather difficult stuff; just now, we are dealing with Gödel's discoveries in the form in which he represents them in the Princeton lectures;[67] you may know the notes of these lectures which were taken by Rosser[68] and Kleene.[69] In the parallel discussion group, we were concerned with the paper by Carnap and Bachmann[70] on Extremalaxiome;[71] before that, we dealt with the basic concepts of set theory and particularly of the theory of ordered sets and thus prepared the discussion of the more difficult questions in the seminar. The theory of ordinal numbers will soon become a prerequisite for everybody who deals with formalized languages, it seems; for it seems to be convenient for many purposes – also for those of semantics – to abandon the strict system of types and introduce a more general and more liberal system containing "nomadizing variables" of level ω and of higher transfinite levels; those of level ω running through all finite levels.

We were very glad when we learned from Carnap that Edith is now through with the strenuous examination; please, give her from Eva and myself the most hearty congratulations and very sincere wishes for the further phases of her study!

From Turner and from Miss Higgins I got word that there is no probability of an opening at their colleges; to Miss Higgins I had sent an application form which McGill had procured, and she wrote to me that she had put it on file. – Carnap has written to several people on my and Olaf's behalf; but so far, nothing popped out. I also got registered with the American College Bureau in Chicago, and now we wonder, whether we shall be lucky enough to find something before the end of the academic year. For us, the situation is of course not so bad as for the Helmers, since we might go back to Brussels; however, that return would have to have a provisory character, and we would always have to try and find our way back to this country. – In any case, I want to thank you once more for all the trouble you took for me in this context during our stay in New York!

With love 2:2
yours,
Peter

---

[67] Kurt Gödel (1906–1978) was an Austrian logician and mathematician, mainly known for his incompleteness theorems regarding mathematical logic. He emigrated to the United States in 1939 (after several visits) and worked at the Institute for Advanced Studies at Princeton. Hempel refers presumably to Gödel (1934).

[68] John B. Rosser (1907–1989) was an American logician, known for his work on the lambda calculus and number theory.

[69] Stephen Cole Kleene (1909–1994) was an American mathematician and logician, student of Alonzo Church. He worked on recursion theory and computable functions.

[70] Friedrich Bachmann (1909–1982) was a German mathematician, working on geometry and group theory. After working in Marburg, he moved to Königsberg and Berlin in 1943. He is also known for his work on the rehabilitation and rebuilding of mathematics in Germany after WWII.

[71] Carnap and Bachmann (1936); for an English translation, see Carnap and Bachmann (1981).

4.1.11 Hempel to Nagel, April 22, 1938

[Typewritten, Nagel 2020, Box 2, Folder 54]

Chicago, April 22, 1938.

Dear Ernest:

I was very glad to have your interesting letter of February 26, and the offprints of two of your articles. (May I ask you quite generally, on this occasion, to send me offprints of your forthcoming papers too, as far as possible? I should be really grateful!)[72] One of them, from the J. of the Amer.Statist.Soc., I had not yet read before, and it was very welcome to me when I started preparing a series of lectures on probability which I am giving at Carnap's Seminar. Last time, even Morris Cohen attended the meeting.

Cohen's public lectures on logic are very successful; even in the big social science assembly room which contains, I guess, 350–400 seats, some people have to stand. Eva, Carnap, Helmer and I went to hear the first two of the lectures, and I shall at least attend one more: that devoted to probability.[73] Our impression is, that Cohen is a very good speaker, he presents this rather dry subject matter in a very interesting and amusing way. Last time, he discussed what he calls the linguistic interpretation of logic;[74] aiming at the position held by Carnap (whose name, however, he did not mention) and others. I think that he did not really do justice to the formal and syntactical conception of logic, and sometimes confused the issues; he attacked views which I think nobody holds. (It was, by the way, very surprising for us to hear him say that he did not think the paradox of the liar was a real and serious logical problem; he dismissed it with quite insufficient arguments. But you probably know that.)

Cohen also leads a seminar on logic which is frequented by surprisingly many students, as I hear. – On the whole, it seems that his activity is very stimulating and that his position is not so far from ours as that of many other philosophers.

This feeling was particularly strong with me at the recent meeting of the Western Philos. Division at Urbana. Olaf and I were taken to Urbana by Benjamin, and Cohen went in the same car. We had some discussion during the trip, and he was quite nice; and in the discussions {at the meeting} he made several really clear and helpful remarks when the issues became vague and the ideas confused.

The papers read at the meeting were not very suggestive for us from the theoretical point of view; but it was extremely interesting and reassuring to see the very strong antifascist tendencies that found theoretical expressions in a discussion about liberalism and in the presidential address.[75]

---

[72] One of the papers is Nagel (1936f), while the other paper is not known to us.

[73] Carnap noted in his diaries (11 April) that in the first lecture Cohen spoke about "Logic and Psychology" and clearly rejected psychologism.

[74] Cohen spoke about "Logic and Linguistics" on April 18. Like Hempel, Carnap also noted in the diaries that Cohen was misrepresenting their views.

[75] The presidential address was given by Joseph Alexander Leighton on April 15, 1938; see Leighton (1939). Two papers were devoted to liberalism: "Liberalism and the Quantitative Conception of Happiness" by M. S. Everett and "Liberalism and the Sovereign State" by D. S. Robinson. Daniel Sommer Robinson (1888–1977) was an American philosopher; he studied at Yale and got his PhD from Harvard University, where he was influenced by Josiah Royce and William Ernest Hocking. He wrote on many different philosophical topics, ranging from logic (*The Principle of Reasoning: An Introduction to Logic and Scientific Method*, New York: D. Appleton & Company, 1924) to ethics (*The Principles of Conduct: An Introduction to Theoretical and Applied Ethics*, New York: Appleton-Century-Crofts, 1948) to philosophy of religion (*The God of the Liberal Christian*, New York: D. Appleton and Company, 1926).

Except that, the meeting was very uninteresting; but it was perhaps worthwhile going there because some addresses of smaller Universities and Colleges with a vacancy in philosophy turned up; so Carnap wrote some more letters. But I think, the probability (whatever that may be) of finding a job for next year is nevertheless very low. Olaf has also got some new tips and is now anxiously waiting for a decision.

Many thanks for your writing to the Amer.Coll.Bureau; I had given you as a reference, even without asking your permission.

What you write about McGill interests me very much. Of course, I shall follow your advice and shall not try to publish a reply. By the way, McG. has not answered me so far; that is, of course, alright with me; I hope he could not feel offended by my criticism?

As to Reichenbach's reply,[76] I didn't find it very satisfactory either. Incidentally, Carnap told me today that Rb. is rather vexed by the fact that, apparently by an error of Malisoff,[77] some polite passages of this paper were left out, so that it sounds more aggressive in tone Rb. would have liked.

I read most of Rb.'s book[78] in the meanwhile and found it highly stimulating, though I certainly have many objections to it; in particular it does not seem to me to be very clear in the fundamental questions, though it pretends in a sense to be a much deeper analysis of the problem than any "positivist" ever made before. The book will partly be discussed in Carnap's seminar. Are you going to review it?[79]

We were very sorry to hear of your bad grippe. We do hope that you are now perfectly alright again, and that Edith is well, too.

The best wishes from Eva and myself for you both!

Yours,

Peter.

4.1.12 Hempel to Nagel, May 28, 1938

[Handwritten, Nagel 2020, Box 2, Folder 54]

Chicago, May 28, 1938.

Dear Ernest,

thank you very much for your letter. Already last week I intended to inform you about our travelling plans; here they are: We shall arrive in New York either on Thursday, June 9, about 7 pm or Friday morning about ten. Our boat, the Westernland, leaves on Saturday, at noon. We do hope that it is possible for you and Edith to make some hours of your time free for us on Friday; perhaps we could have lunch together? We shall give you a ring as soon as we arrive; then we shall know more exactly just what we have to do in New York and have to

[76] See Reichenbach (1938a).

[77] William Malisoff (1895–1947) was a Russian-American scientist and philosopher. From 1934 to 1944 he was the editor of *Philosophy of Science*. After 1940, Malisoff fell under FBI's radar as he turned out to be a spy for the Soviet Union.

[78] See Reichenbach (1935a).

[79] Nagel did review Reichenbach's book, see Nagel (1936e). In fact, Nagel wrote two reviews of the book; one was published in *Mind* (referred to above in letter 4), and the other published in 1935 in *The Journal of Philosophy*, see Nagel (1935b).

arrange our time schedule for Friday. (Maybe I shall have to see Weaver[80] of the Rockefeller Foundation; then, we would like to see the Broadwins, and, if possible, the Mindells.)

Unfortunately, we cannot come earlier to N.Y., because we (Eva, too!) are attending Carnap's seminar which has its last session on Wednesday; so we can leave, at earliest, Wednesday night.

The Helmers are to Urbana, today; for – we are very happy to tell you – Olaf has been appointed instructor in mathematics there.[81] Today, they are looking for an apartment. They are going in their own car (which they purchased on the very day of Olaf's appointment). The Helmers, too, are leaving New York on the 11th of June, on the Nieuw Amsterdam; Olaf will come to New York either with us by train or alone, in his car. Eileen[82] has different things to arrange in N.Y. and will come some days earlier; she will probably stay with the Broadwins.

As to myself, I have not got any job, so far. But just two days ago, a very small chance turned up that I might perhaps get a half time job at the New Bauhaus[83] here, as a kind of assistant to Prof. Eckart (Physicists of the U. of Chic) and Gerard[84] (Physiologist of same); I would then have to teach elementary mathematics and maybe a little physics. Unfortunately, the New Bauhaus has already got so many foreign teachers that the students seem to revolt against this situation, and it seems that just that point makes it very doubtful whether Moholy-Nagy[85] and his board of trustees will be willing to accept me; Eckart apparently is going to recommend me to him. Now we are, of course, very anxious to learn the result of his negotiations; before next Wednesday, nothing will be decided, and even then, at the very best, Moholy might assign, by an informal personal statement, just a certain weight to the prediction that I shall got that job. [Footnote: The final decision will take much more time, since it depends upon the agreement of the board of trustees.] So, keep this please confidentially for yourself and Edith, for it is not yet ripe for any propagation.

This afternoon Eva and I are going to Iowa City to see the Feigls,[86] and to discuss with Lewin.[87] I am interested in the logical and methodological characteristics of Lewin's topological psychology and I hope, that talking with Lewin, Feigl may be helpful to "make my ideas clear".

I am very much looking forward to discussing with you, too; in particular, we shall have to talk about Reichenbach's book,[88] which – as I wrote you – I find in many points pretty

[80] Warren Weaver (1894–1978) was an American scientist, mainly known as the Director of the Division of Natural Sciences at the Rockefeller Foundation (1932–1955).

[81] Olaf Helmer taught mathematics at the University of Illinois at Urbana between 1938 and 1941.

[82] Eileen Helmer was Olaf Helmer's first wife.

[83] The so-called "New Bauhaus" was a special continuation of the German-based Bauhaus movement. It was established in Chicago in 1937. Charles Morris and Rudolf Carnap also had some ties to the Chicago-based Bauhaus.

[84] Ralph W. Gerard (1900–1974) was an American neurophysiologist and behavioral scientist. He worked in the Department of Physiology at the University of Chicago between 1928 and 1952.

[85] László Moholy-Nagy (1895–1946) was a Hungarian painter and photographer, known mainly for his work in the Bauhaus.

[86] Herbert Feigl (1902–1988) was an Austrian philosopher who studied with Moritz Schlick in Vienna, and emigrated to the United States in 1930. He established the Minnesota Center for Philosophy of Science in 1953.

[87] Kurt Lewin (1890–1947) was a German-American psychologist, known mainly for his work on social psychology. In Berlin, he was involved in the so-called Berlin Group of logical empiricists around Hans Reichenbach; he emigrated to the United States in 1933, where he worked at Iowa and MIT.

[88] See Reichenbach (1935a).

objectionable, though I think it may be excellent as a basis for discussion in a philosophical seminar. I think you are right with your objection against R's probability-theory of meaning, if you really take 'weight' in the sense in which Rb. defines it;[89] but if you take it in a somewhat broader sense, then one can perhaps save Rb's criterion; Carnap to whom I talked about the question yesterday, admitted that when he wrote to you he had this broader meaning in mind; but he thinks, that in the first parts of the book this is also true with Rb; so that the objection then would [n/a] the following form: Either 'weight' is defined in the restricted sense of a probability in a statistical series – and then your counterexample is valid – or 'weight' is understood to be something more comprehensive – then it is hard to see how Rbch would define it within his theory.

I hope we can talk about this point and many others on Friday, June 10!

The best wishes from the two of us to you and Edith!

Yours Peter.

4.1.13 Hempel to Nagel, September 2, 1938

[Typewritten, Nagel 2020, Box 2, Folder 54]

Brussels, September 2, 1938.
26, rue de Monastère.

Dear Ernest:

When I sent you my last letter, on August 16, I forgot to enclose the snapshots I took of you and Edith when we met last time. I am sending them now and hope you will like them, although they certainly might have been much better. For us, they are a welcome souvenir of the nice day we enjoyed in your company!

My main reason for writing you to-day is to ask you for an information. – One of the teachers who were trusted with my pedagogical formation at Berlin high schools, a very sympathetic and able teacher of mathematics and physics who, as a Jew, has lost his position at that Berlin gymnasium, is now teaching at a purely Jewish school which, by a special agreement with the University of Cambridge (England), cannot only confer the German Abiturium, but also the so-called "School Certificate" which entitles the bearer to begin studying at the University of Cambridge. Some of his pupils who are going to pass their final examinations, want to go directly to the United States and go through their academic studies over there, probably in New York. Now, that teacher wrote to ask me the following questions which I in turn must ask you to answer me:

1. Will those pupils who possess the Cambridge School Certificate (also called "Matric") be unconditionally admitted at an American University, particularly in New York? – If not, how long would the necessary preparatory courses last?
2. Same questions for those pupils who only have the German Abitur, i.e. a certificate that they passed the final examination of a German high school.
3. Is there any advantage in having passed both examinations, that for the School Certificate and that for the Abitur?

As it seems to me, either of the two examinations would probably entitle the applicant to start his studies at any Amer. College, since the last two years of a German high school correspond roughly to the first two years of American College. But what the boys seemingly

[89]Hempel presumably refers to Nagel's review of Reichenbach; see Nagel (1936e).

hope is, that their certificates entitle them to enter a real University. If the New York universities require only two years of previous College studies, this might not be impossible.

I realize that it may not at all be easy for you to answer these questions. But you are really the most competent man in this matter I can think of. And you will certainly understand the importance of the questions for the poor boys in whose behalf they are asked, and therefore excuse me for troubling you. The sooner you are able to give me some information, the better for the boys who, as far as I know, have to pass their examinations in about 4 weeks. If you could ask the secretary of, say, Columbia University, to send a general prospectus indicating the prerequisites and the standard of the courses, this would be very helpful.

The best thing to do would be to make it mail directly to the teacher in questions:

Studienrat i.R. A.Heckscher, Münchener Str. 18, Berlin W.30.

In order to save time, you might also send him the information I asked you to give me. But if this is not convenient for you, I shall of course be very glad to transmit the content of your answer.

When these lines reach you, you and Edith will perhaps already be back from your country place. We hope that you are both nicely tanned and well rested, and that the new academic year will be satisfactory and successful for either of you.

With love 2:2

Yours Peter.

4.1.14 Hempel to Nagel, September 19, 1938

[Handwritten, Nagel 2020, Box 2, Folder 54]

Brussels, Sept. 19, 1938.

Dear Ernest, we just had, for a few days, the visit of Tarski[90] from Warsaw. When talking about probability, he mentioned the work of Mazurkiewicz;[91] he thinks that it is by far the most important and formally correct achievement in the field of an axiomatization of the theory of probability which has so far been obtained; the paper is in German and has been published in "Monatshefte für Math. u. Physik", most probably in 1934. (It is the periodical in which Gödel's papers appeared). I hasten to inform you about it because it might be good to mention it in your contribution to the Encyclopedia.[92] –

Another contribution to the problems of prob is contained in Neyman, J.[93] Lectures and Conferences on math. statistics. Mimeographed: Washington, Grad. School of the US. Dept. of Agriculture 163 pp; $1.25. 1938. Just today, I read a very favorable review of it in "Zentralblatt f. Math", vol 18, 265/6. It seems to be relevant for the »Anwendungs-problem«. (I know you mention Neyman already in your MS.) – Have you read Scholz'[94] most

[90] Alfred Tarski (1901–1983) was a Polish logician who made revolutionary contributions to logic and semantics. He emigrated to the United States in 1939 where he taught at the University of California, Berkeley until his 1983 death.

[91] Stefan Mazurkiewicz (1888–1945) was a Polish mathematician. The paper Hempel mentioned above is Mazurkiewicz (1934).

[92] See Nagel (1939).

[93] Jerzy Neyman (1894–1981) was an American mathematician and statistician, directing the Statistical Laboratory of the University of California, Berkeley. Previously he taught statistics at the University College of London and at the University of Warsaw.

[94] Heinrich Scholz (1884–1956) was a German philosopher and logician. Besides establishing in Münster an institute devoted to mathematical logic, Scholz was also known for his historical work on logic. See Scholz (1938).

extraordinary paper "Die math. Logik und die Metaphysik"? [Philos. Jahrbuch d. Görres-Ges., 51, 257–291].

We hope you and Edith are well! Love 2:2

In haste: Peter.

4.1.15 Hempel to Nagel, December 14, 1938

[Typewritten, Nagel 2020, Box 2, Folder 54]

December 14, 1938.

Dear Ernest,

thank you very much for the detailed information you gave me about the conditions for admission at Columbia and City College, and for the lot of trouble you took to obtain it yourself. I forwarded all your material without delay to Mr. Heckscher, and it must have reached him in time to inform those young people who were to pass their final examinations in October. However, I have not got any answer from him, which is rather unusual and therefore somewhat disquieting; I shall try to contact him by phone when I am in Berlin at Christmas.

Eva is going to stay in Brussels; but I have to see my mother and to talk many questions over with Eva's parents. Her father was kept in a concentration camp for 3 1/2 weeks; he was released a few days ago. Since he can no longer earn his living in Germany, he is forced to emigrate; he has applied for a visa for the US, but probably he would have to wait 2 to 4 years before his case could be decided; so he is considering emigration to Australia. For the time being we are trying to obtain for him a permit for a temporary stay in England where he might wait until the Australian visa is granted, which may last half a year or so; but it is very difficult and it lasts long to get it {i.e. the English permit}. His family would have to wait in Germany which will [be] possible as his wife is "Aryan". At least during his intermediary stay in England, we shall have to support him, and after that, we shall have to contribute for a while to the means of living of his family; this obligation adds to my strong personal inclination a material urge towards finding a position in America which offer certain chances of stability.

Besides, Grelling has now definitely emigrated from Germany, whereas his wife, who is "Aryan" is staying with the children in Germany for economical reasons.[95] Now and then, she comes to see him. But both of them, who are very likeable people, are now, so to speak, waiting anxiously for our definite departure, because then G's financial status with Oppenheim will improve, and his wife might leave Germany for good too.

Yesterday, we received the good news that my mother has been allowed by the Devisenstelle to present us with the steamship tickets; this is why we shall cross on a German boat, the Bremen, which is due in New York on February 2. Please do not trouble to

[95] Kurt Grelling (1886–1942) was a German mathematician and philosopher, working on logic and epistemology. He was a member of Hans Reichenbach's Berlin Circle of logical empiricism. Grelling went to Brussels twice, first in 1937, then in 1938. When the Germans invaded Belgium, Grelling was imprisoned in May 1940, and he was deported within days to France. He was interned in Vichy France for 2 years. In 1942, he was deported to Auschwitz where he died in September 1942. Grelling's (second) wife was called Margarethe Alma Berger, often called Greta; they had two children, Karin and Klaus-Peter. Grelling's children survived the Nazis because they were sent in 1939 to a private boarding school in Switzerland.

come to the pier; we are no longer greenhorns in New York, you know; and we shall ring you up when we are in.

We intend to stay in New York for some time; Eva will even try to find some job for herself and settle down when I am travelling around. I should be very glad to meet the members of your discussion group and whomever you think it might be good for me to meet for scientific or for practical reasons. I shall also try to contact Kaufmann[96] whom I know from Vienna and whom we met last in New York a day before sailing back to Europe. – My further program has not yet been fixed; it will depend for a good deal, on the suggestions and eventual introductions I might receive in New York.

As I do not yet have any definite address, I asked some of my friends and acquaintances to send my mail to you during the first weeks; I do hope it will not trouble you too much!?

Apart from the terribly depressing political events of the last weeks and months, we are feeling quite well here. The presence of Grelling whom Oppenheim invited, upon my suggestion, to become his collaborator after our departure in 1937,[97] is very stimulating for me, and also personally he is very sympathetic, and we are having a really enjoyable time with him. Whereas Grelling and Oppenheim are continuing their studies on the concept of Gestalt, I am still busy with an analysis of confirmation and disconfirmation I began with Oppenheim already in 1937. We have been getting on quite nicely, but since I tried to write it down, several new difficulties turned up, and I do not yet know in what state the MS will be when I arrive in New York; at any rate, it will not yet be finished.[98]

In the meanwhile, I have read Carnap's MS for the Encycl. on the Foundations of Logic and Mathematics.[99] It is of C's usual clarity and precision and contains a lot of new material; further it seems to me to be an important contribution to a number of questions which have been much discussed in recent years, and which concern the conventional character of logic, the relation of syntax and semantics, etc. However, the representation is very condensed and partly highly technical. Thus, the book resembles in my opinion rather more a textbook than a general survey for a wider group of readers.

I also read Woodger's monograph[100] which I liked very much. But it seems to me to be partly too technical too; and then the specific importance of the method for *biology* is not made explicit. But the representation as a whole is much less esoteric than I had expected; the book contains many fine remarks of a more general kind on formalization, axiomatics, logical analysis etc., and it gives also good concrete examples for some of the general considerations developed in Carnap's and Morris' monographs.[101] I wonder what you will think of it. Perhaps you have also seen the MS?

---

[96] Felix Kaufmann (1895–1949) was an Austrian philosopher, specialized in the philosophy of law, though he made his living as a manager of the Anglo-Iranian Oil Company in Vienna. He was known for his work to find bridges between logical empiricism and phenomenology; after his emigration to the United States, he worked on the relationship between pragmatism and positivism.

[97] Grelling and Oppenheim wrote various papers together, among others Grelling and Oppenheim (1937/38).

[98] Hempel and Oppenheim will publish jointly one paper about confirmation, "A Definition of 'Degree of Confirmation'" in 1945 as Hempel and Oppenheim (1945).

[99] Carnap's manuscript was published as *Foundations of Logic and Mathematics* (Carnap 1939).

[100] Joseph H. Woodger (1894–1981) was an influential British biologist and a pioneer in philosophy of biology. Hempel possibly refers to his book on *The Axiomatic Method in Biology* (Woodger 1937). A year later, Woodger also published a volume within Neurath's *International Encyclopedia of Unified Science* on *The Technique of Theory Construction* (Woodger 1939).

[101] Hempel refers to Carnap's previously mentioned *Foundations of Logic and Mathematics* volume and to Morris' *Foundations of the Theory of Signs* (Morris 1938) published also in Neurath's *International Encyclopedia of Unified Science*.

We hope that you and Edith are very well and that your work is progressing satisfactorily. We are very glad to think that we shall see you soon! In the meanwhile, we are sending you our very best wishes for good luck in 1939!

With love from both of us,
yours
Peter

4.1.16 Hempel to Nagel, January 19, 1939

[Handwritten, Nagel 2020, Box 2, Folder 55]

Brussels, Jan. 19, 1939.

Dear Ernest:

Carnap will have written you because of my MS "Vagueness and Logic". I should of course be very glad if it could be published in the Journal.[102] If so, please do not let the MS go to the Press before I arrive in NY; I should like to carry out some modifications of the text which have been suggested to me by the comments of Carnap, Morris and others; the modifications will not affect the length of the paper, and I could give you the modified version on or 1–2 days after our arrival in N.Y. At present, I have so many things to do in connexion with our imminent departure that I cannot find the time to work on the MS here.

Eva and I are well. Eva's father has been released; I saw him at Xmas. He is waiting eagerly for a permit to stay temporarily in England, and then for an Australian immigration visa.

We are looking forward to seeing you and Edith again!

Many good wishes 2:2
Yours Peter.

4.1.17 Hempel to Nagel, March 12, 1939

[Handwritten, Nagel 2020, Box 2, Folder 55]

March 12, 1939.

Dear Ernest,

thank you very much for forwarding my mail! Please keep the mail arriving after Monday, March 13, because I expect to be back to N.Y. late on Wednesday (though the Horkheimer[103] discussion will not take place before April, but there are some other things I should like to look after in N.Y. I shall tell you about it.) – I shall not leave N.Y. before Sunday; so if you think the Discussion Group would be interested in it, I should be glad to give my talk on Confirmation and Probability of Empirical Hypotheses to your Group on Friday or Saturday night. Will you talk it once with the man who is in charge of the program

[102] See Hempel (1939). By "Journal", Hempel presumably had in mind *The Journal of Philosophy*, nonetheless, the paper appeared in *Philosophy of Science.*

[103] Max Horkheimer (1895–1973) was a German philosopher and sociologist, known mainly as a member of the Frankfurt School of social research. After he emigrated to the United States in 1934, he moved the Institute to New York. It is uncertain what discussion Hempel had in mind.

(I guess: Gruen)?[104] But *please*, don't feel urged; I really would like to speak only in case it is perfectly convenient to the group. – Give my best love to Edith please! Auf gutes Wiedersehen!

Peter

I hope your back trouble has completely disappeared by now!

4.1.18 Hempel to Nagel, March 26, 1939

[Typewritten, Nagel 2020, Box 2, Folder 55]

Chicago, March 26, 1939.
5466 Kenwood Avenue
{Mailing address remains c/o Carnap!}

Dear Edith and Ernest:

This is the first day we can peacefully enjoy in our new home, and I seize this opportunity to send you our news.

On my trip to Cincinnati, I made a stopover at Pittsburgh and visited the "Cathedral of Learning";[105] quite impressive, but, in my opinion, a strange mixture of romanticism and "neue Sachlichkeit".[106] – In Cincinnati, where I arrived on Monday evening, I stayed at Roelofs'[107] house, about 12 miles out of town, in a very nice little village called Glendale; it reminded me somewhat of New Rochelle. Mr. and Mrs. Roelofs were very kind to me, and I really enjoyed the stay at their home. Both of them turned out to have strong interests in various branches of farming, and as I had spent a good deal of my childhood on farms, we had a nice time discussing such topics as raising fowl, different methods of growing and preserving fruits and vegetables, and many other things. The Roelofs have got a country residence in the State of New York, and as they have 6 children, 4 of whom are at home, and about 30 chicken, all of whom are taken along to that place, they use to move in two cars: an ordinary Ford person-car, and a small Ford truck. The latter is also used during their stay in Cincinnati, and in fact, R. took me to the University on it; it was quite exciting, the first time I rode on a truck. At the Univ., I met the members of the Department of Philosophy, and even the Dean of the College (a geologist, Prof. Barbour)[108] who seems to be an influential man,

[104] Hempel presumably refers to William Gruen, who was at that time a professor of philosophy at New York University. Carnap also notes in his diaries that Gruen visited him (with Hempel and Helmer) and reported about his group with mathematicians, where they studied extensively Carnap's *Syntax*-book. Carnap named "Mac Kinsey", though probably he had in mind the mathematician and logician J.C.C. McKinsey (see letter 32).

[105] The Cathedral of Learning is a 42-story skyscraper, built in 1926. It serves as a main building to the University of Pittsburgh, and also hosts the Department of Philosophy, and the Center for Philosophy of Science.

[106] The "neue Sachlichkeit" was a modernist movement that originated in Europe. Some of the logical empiricists, like Carnap and Neurath, had personal relations to its members.

[107] Hempel presumably had in mind Howard D. Roelofs (1893–1974), an American philosopher who earned his PhD at Harvard in 1925. In 1932 he became Professor of Ethics and head of the philosophy department at the University of Cincinnati, holding these positions until his 1960 retirement.

[108] George Brown Barbour (1890–1977) was a Scottish geologist and educator who became Associate Professor in Cincinnati in 1937.

received me in his office. It turned out that we had common acquaintances in Brussels, and I had a nice time with him. Then one of the philosophers, Prof. Ames,[109] took me for a sightseeing ride all round Cincinnati; it seems to me to be a particularly attractive place: the town is situated on several hills on the Ohio river; the scenery somewhat reminded me of Heidelberg, apart from the fact that the Ohio river has a dirty brown colour because all the suage[110] conducted into the river. This had most undesirable consequences during the big flood some years ago; the city's drinking water supply being spoiled for a fortnight; at that time, drinking water had to be brought to Cincinnati on trucks, and Roelofs actively took part in this service with his own truck. One of his outstanding features seems to be that of a practical Christian, I mean of one who does not "theoretically" profess a certain belief.

On Tuesday evening, most of the philosophy people and some other members of the Univ. staff had dinner at the Roelofs' house; after dinner, I gave a brief informal talk, and then there was some discussion with quite a lot of opposition, which made me feel rather bad because it was, at least partly, based on misunderstandings; this may of course have been my fault. Nevertheless, Roelofs told me, afterwards, that he had enjoyed my talk, and that I was the first positivist, after Schlick, with whom he found it possible to discuss; and he would be glad to see me come to Cincinnati; however, there would be no opening, if at all, before fall 1940; and he was not even sure whether they would then have a vacancy; for they have not got too much money, and the number of students of philosophy has decreased during the last years. – Thus, my chances seem to be almost negligible, in spite of their apparently friendly attitude.

On Wednesday, I went on to Chicago. I found Eva and the Carnaps very well. The Carnaps are pretty much concerned with a practical problem: As Carnap has to teach during the hot summer term, Ina has suggested that they buy a trailer and put it into the Dunes State Park, on the lake shore, about 30 to 40 miles from Chicago, where they may spend extended weekends; later on, they could go East with the trailer and go on using it when they go South in the free fall term. The other day, we inspected with them quite a series of different makes and also two trailer camps; the former were very nice and ingeniously designed; the latter, however, were not very encouraging (one of them even most distressing); but this may be due to the fact that they are near Chicago; there may be much nicer places in the East. So far, the Carnaps have not yet decided whether they are going to purchase a home on wheels. –

In the last days, Eva and I were lucky enough to find a nice little furnished apartment; it is quite light inside and has even got a tree in front of it; this means a lot in Chicago, where most apartments that are available at a moderate rent, are simply distressing. We plan to stay here for several months. I expect to attend Carnaps classes and to work on a manuscript; Eva has not yet succeeded in finding a job; but we are pretty sure that it will not last too long before she finds something; in the past weeks, she did not look around too eagerly, because she stayed with the Carnaps and helped them in their house. Ina will be very busy during the spring term; she is beginning her field work as a social worker, and is taking some classes in addition. – Carnap is working at a second edition of his "Abriss der Logistik";[111] he is

[109] Van Meter Ames (1898–1985) was an American philosopher who joined the faculty at Cincinnati in 1925 and became its chairman later in 1959. He was a founding member of the American Society for Aesthetics.

[110] Hempel possibly meant "sewage" here.

[111] See Carnap (1929). In 1938, Kurt Edward Rosinger (a philosopher from Harvard, later Princeton) translated Carnap's *Abriss* into English. Later Carnap wrote to Quine (October 25, 1938; in Creath 1990, 255) that the University of Chicago Press would like to publish the English translation. Carnap wrote also to John Cooley (September 22, 1938; Creath 1990, 256) that he aims at buying his mimeographed "Outline of Symbolic Logic" because he is "writing a new and entirely changed edition of [his] 'Abriss der symbolischen Logik.' Both the rewritten German version (*Einführung in die symbolische Logik*, 1954) and the new translation (*Introduction to Symbolic Logic and its Applications*, 1958) appeared much later. The book was translated, after all, by William H. Meyer and John Wilkinson.

feeling fine; however, he has to be cautious to avoid recurrence of his back trouble which has been diagnosed as a special form of arthritis; the doctors admitting, however, that this diagnosis means mainly introducing a name; they cannot suggest any kind of treatment apart from wearing a kind of girdle to support the vertebral column.

How is your back by now, Ernest? We do hope that it is perfectly allright again, and that you and Edith are feeling fine.

Thank you very much for forwarding our mail! It must be a terrible bore for you, especially the little package we got today, where you had to pay the duty and new postage and to go to the post-office etc. I am returning your expenses in the form of stamps; and I do hope that the amount of mail you have to bother with will decrease in the near future since all our relatives and friends got our Chicago address by now.

And many thanks again for all your kindness in New York!

With best love 2:2,

Yours, Peter.

Continued by Eva on March 27.

Dear Nagels,

I really wanted to write to you much sooner. But you know how it goes: the less one has to do, the less one accomplishes. That's why I feel like writing today, where I have already posted 5 letters to the different families!

The "nice little apartment" turned out to have bed-bugs in the studio couch!! There had to be some "Pferdefuss", it would have been too wonderful without that. Fortunately or unfortunately, it was my bed, and either I react to such things more violently than other people (I have had no experience with them before) or this particular beast had eaten a bit of poison before it had dinner with me: anyhow, I am all swollen, my forehead looks as if Peter has maltreated me, and I have the (fat) hands of a bishop (as Ina put it)! If it had at least chosen spots of my body, where it shows less. – Well, such is life, and at times, it is difficult.

Chicago has been much nicer than I expected it to be; it was a little bit like coming home, everything was quite unchanged, even the sales people in the shops were still the same. And I got along with the Carnaps wonderfully; we really did not go on each others nerves, which means something if you have to live together in a rather small apartment for 2 1/2 weeks.

But apart from the Carnaps, Chicago still does not have many lasting charms, and I would much rather live in New York. Peter told me of your very kind offer to give us your apartment in case he has to teach in Hunter in summer. This would simply be wonderful; but I don't think it will {materialize}.

The "Eileen"-job has not yet been decided on; the Helmers will come over for Easter, and then we will meet the Wolffs and talk about it. But I hope that I have found some part time job before that; it would be so much nicer if I could stay here with Peter (in case the exterminator got really rid off my little bed-mates), and do this household too.

How are your experiences progressing, Edith? Does the apparatus work satisfactorily by now? And how are the boys? Give my regards to Konstantin occasionally, please.

But I really should not formulate questions. For we don't expect an answer to this, since we know how busy you both are, and how you dislike writing letters. Thus, enjoy this letter, file it away immediately, and forget all about it!

Best love,

Eva.

4.1.19 Hempel to Nagel, April 3, 1939

[Typewritten, Nagel 2020, Box 2, Folder 55]

Chicago, Ill., April 3, 1939.

Dear Ernest,

it was very nice of you to inform me so quickly about that galley-reading matter. Of course, I would have preferred to do it myself, but I think it will be allright as it is. I just wrote to Malisoff and ordered reprints; I am sending that letter by airmail and do hope he won't tell me I placed the order too late.

Carnap was perfectly amazed by the goodness of your informations when he heard that you know already about that exchange professorship he is going to have in 1940/41 at Harvard;[112] he thought it was a perfect secret yet as it became definite only a few days ago. The Carnaps are both very much looking forward to seeing you and Edith; in fact, they plan to go to New York after the Congress.[113] Besides, Ina yesterday pictured to herself how grand it would be to go up to your country place and put the trailer on a meadow right near your house. But I'm afraid that this is not likely to materialize since the Carnaps have to stay here during the summer term, and after that, you will be back to New York, and Ina will no longer be so much interested in putting the trailer on your meadow. (Incidentally, the trailer has not been purchased yet, and the Carnaps have even not yet come to a decision, but Eva und myself think they are going to buy one, after all.)

Thank you very much for your information about Nelson. Both Carnap and Morris wrote immediately to Savery,[114] and Carnap also wrote to Nelson in my behalf. – After having mailed my last letter to you, I got a letter from Roelofs asking me what courses I would be prepared to give, especially of the kind usually given at Colleges. Thus it seems that they have not turned my application down completely. Roelofs is very much opposed to Positivism (do you happen to have seen his criticism of the Encyclopedia in Kenyon Review?),[115] but before I left Cincinnati, he told me that I was the first positivist after Schlick with whom he found it possible to discuss; so he seemed to like me in some sense, and it is now a perfect puzzle for me to figure just what my prospects are in Cincinnati. At any event, R. told me – and that was confirmed by another, confidential, information – they are not going to appoint anybody for an earlier date than fall 1940. In my answer to R., I mentioned that there is a small probability that the Emergency Committee for Displaced German Scholars[116] would pay a one year's salary for me, in case a Univ. or a College offered me an official teaching position and declared it would take into consideration (without obligation) to keep me after that year. I wonder what will come out of this suggestion.

I have been asked to become a contributing editor of the Philosophic Abstracts, to be edited by Montague,[117] McGill etc. Do you have any more precise information as to the

---

[112] About Carnap's time at Harvard, see Frost-Arnold (2013).

[113] Hempel referred to the *Fifth International Congress for the Unity of Science* at Harvard, September 3–9, 1939, after which Carnap indeed went to New York with Ina.

[114] Hempel presumably had in mind William B. Savery (1875–1945). Savery received his PhD from Harvard University and was then appointed Professor of Philosophy at the University of Washington from 1902 until 1945.

[115] Hempel refers to Vivas et al. (1939).

[116] The Emergency committee in Aid of Displaced German Scholars was established in 1933 to assist and help scholars who tried to emigrate from the Nazis. Soon it was expanded to include other scholars as well, and thus it was renamed as "Emergency Committee in Aid of Displaced Foreign Scholars". The committee functioned until 1945.

[117] Hempel presumably refers to William Pepperell Montague (1873–1953), known for his critical realism, who was a philosopher at Columbia University between 1903 and 1947. The "Philosophic Abstracts" was meant to be a "quarterly review of philosophical books and periodicals in form of brief excerpts and synopses" (as announced in the July issue of *The Philosophical Review* (vol. 48, no. 4, p. 449)). It covered books from North American and Europe, and periodicals from Australia, China, and North and South America. The general editor was Dagobert Runes (see next footnote), and Hempel's name seems to appear in the second volume in 1940.

background of this enterprise? The editor Runes[118] wrote to me on the suggestion of Sidney Hook;[119] thus I think it will be allright if I accept. In fact, I am very much interested in a periodical of that kind and think it might be found very useful, in case it is run well. I know how valuable analogous publications in mathematics are.

In case we are in the East in summer we shall be very glad to accept your kind offer for which Eva and I thank you very much. Just what we are going to do in summer, is of course essentially determined by the decisions the next months may bring about my possibilities for next year. If nothing turns up, Eva might not go East, and I may just come for a relatively short stay. The Carnaps will be away from Chicago from September to the end of December, and they offered us to use their apartment during their absence.

The Carnaps have not forgotten you; they are just rather busy, and they are not too good correspondents anyhow, you know! They asked me to give you their very best love; they are thinking and talking of you very often – Eva sends her love, too. She is going over to the Carnaps place to prepare lunch since Ina is busy with social work. – This afternoon, Eva is to go to one of those emergency committees for another vocational interview; she got registered with them about 4 weeks ago and was now invited to come; let's hope they have really got something for her.

All of us do hope that you will not have too boring Easter days after all. Is Edith going with you to stay at that camp?

The best wishes for both of you,

yours Peter.

Many thanks for forwarding a second parcel; we do not dare to refund the postage and feel very badly about it!

4.1.20 Hempel to Nagel, May 10, 1939

[Typewritten, Nagel 2020, Box 2, Folder 55]

Chicago, May 10, 1939.
5466 Kenwood Avenue.

Dear Ernest:

Carnap told me this afternoon the great news that you were appointed associate professor. I am sincerely delighted to hear that, and so is Eva, and we hasten to send you our warmest congratulations! You must be feeling a very justified satisfaction at this advancement which, we are quite confident, will be followed by further steps up the ladder of academic possibilities. Good luck to you!

Carnap also gave me the reprint of your article of Reichenbach's book.[120] Many thanks, I am glad to have a copy of my own; for I am sure I shall frequently use your paper. I read it for

---

[118] Dagobert D. Runes (1902–1982) was an American philosopher who originally came from the Austro-Hungarian Monarchy, and later obtained his doctoral degree under Schlick in Vienna in 1924. He is known as the editor of the 1942 *Dictionary of Philosophy*, and he was the founder of the successful *Philosophical Library*.

[119] Sidney Hook (1902–1989) was an American philosopher, known for his commitment to pragmatism. A student of Morris R. Cohen and John Dewey, Hook received his PhD from Columbia University in 1927 and taught at New York University from 1927 until his retirement in 1969. Hook was initially a socialist, but after the Second World War he was converted to conservatism and became an ardent anti-communist.

[120] Hempel probably refers to Reichenbach's latest book, *Experience and Prediction* (Reichenbach 1938b) that Nagel reviewed shortly after its appearance (Nagel 1938b).

a first time immediately after it had appeared, and I liked it very much. I agree with almost every point of your criticism, except one or two minor questions. I suppose, though, that Reichenbach will be rather upset when he reads your paper; there are a few admonitions in it (such as in the very last line of the whole), which, though being formulated with extreme kindness, will probably make him rather furious.[121]

As to our situation, the most important news is that Eva has a job; today she started the fourth week of it. She is taking care of the children of Prof. Adams of the Theological Seminary connected with the U. of C.;[122] all three of them are girls, 10, 8 and 1 1/2 years old respectively. Eva has mainly to look after the youngest child, and she does it quite successfully. In addition, she sometimes helps her boss to translate French or German philosophical articles into English. On the whole, Eva likes her job very well: the people are very nice with her, and the house is of course very cultivated; the drawbacks are the relatively small salary ($ 10 a week) and the very long hours of work: she leaves at 7.50 every morning (at 8.50 on Sundays), and comes back at 6.45 p.m., except on two afternoons where she is free already from 12.30. Sometimes she is pretty worn out when she comes home, especially on those days when the older girls do not go to school and hang around at home. And the "free" afternoons she has to use to clean our apartment, do laundering and mending or entertaining people, which we restrict under these circumstances to the utmost minimum. As to myself, there is only the Cincinnati prospect which has at least not yet vanished completely. I had an opportunity to talk to Roelofs at the Midwestern philos. meeting in Columbia, Mo., and I told him that the Emergency Committee might perhaps be willing to pay a year's salary for me if the U.of Cinc. declared that it would consider continuing me after that year. Roelofs, who is extremely conscientious, was very hesitant about any engagement on the part of the university. Nevertheless, he promised me to write to Tillich[123] who is in some way connected with the committee and ask him about the details of the possible arrangement. As I heard from Tillich, he really did so, but I have not yet heard from Roelofs whether he and Dean Barbour (Geologist) who seems to be the decisive man, are going to follow up the possibilities which Tillich might have shown them. Do you happen to know anybody at the Emergency Committee?

Both Carnap and Morris had also written in my behalf to Savoury and asked whether I could substitute Nelson during his year of absence; he got a Guggenheim fellowship. Carnap recently got a negative answer; they are paying to Nelson part of his salary, and are not taking anybody to substitute him.

We hope you and Edith are well. How did you enjoy the Easter vacation in the camp? How is the discussion group getting along? Eva and I are sending our very best wishes to both of you!

With love, Peter.

{I haven't heard anything from McGill about Hunter.}

---

[121] Hempel might refer to this closing line of Nagel's (1938b, 272) review: "His book gives evidence that although pragmatists have something to learn from him, he too has something to learn from them".

[122] Possibly James Luther Adams (1901–1994), who graduated from Harvard and joined the Theological School in Chicago in 1937. He was one of the most influential Unitarian theologians of the mid-century.

[123] Hempel refers to Paul Tillich (1886–1965) who was a German-American Christian philosopher, a Lutheran Protestant theologian. From 1933 until 1955 he taught at Union Theological Seminary in New York.

4.1.21 *Hempel to Nagel, June 2, 1939*

[Typewritten, Nagel 2020, Box 2, Folder 55]

Chicago, June 2, 1939.
5466 Kenwood Avenue

Dear Ernest,

may I trouble you with a question in connexion with my job-hunting?

The present situation is that the Cincinnati project which had looked quite promising for a while, has been turned down; if I read correctly between the lines of Roelofs' long and somewhat involved letter, it seems that some of the members of the Dept. had objected to having me there. He promised me to try to make an arrangement for 1940/41 – provided that the Emergency Committee is willing to pay my salary; and even then, it seems, he would restrict the arrangement to one year. I think that this is a very poor prospect indeed, though a sign of Roelofs' good will.

Also McGill wrote to me quite recently that the plans for Hunter College had been turned down.

On the other hand, I have meanwhile been introduced to Prof. Tyler[124] here, who is the Chief Examiner at the U. of C., and is in charge of three large research projects in connexion with the Progressive Education Association. There is a slim chance that I might get a half time job there, provided they find that I am able to develop new tests for mathematical and logical thinking abilities (not: knowledge). I am now trying to find something to suggest to them. I would, of course, be glad to get the job, though I do not have the impression that this is the kind of work I could do best or would like best.

I spoke to Carnap about the situation, and under these circumstances, he thought it would be justified if he told me about what you had written to him about a very slim chance that I might get some evening teaching at NYCC. He also told me that he had got a request from NYCC to send in information about me. You certainly imagine how much more I would like to teach than to develop methods of testing high school students;[125] but you may really be sure that I do not overestimate the probabilities involved here, and that I do not in the least count on that possibility before I have got an official offer from NYCC.

Now, I am writing to ask whether you could tell me anything about the probable date of a decision in this matter, and, of course, I would be very much interested to learn what you think my prospects are in this case. The reason why I am asking is that if a decision from NYCC could be expected within a short period of time, I would try to postpone making definite arrangements with Tyler. You will, no doubt, realize that I would feel rather uneasy about urgently applying for a job and then spending only some weeks in his offices and leaving just after I got accustomed to that kind of work. (All this presupposes, however, that they are at all willing to take me; I have seen the decisive people twice so far, and it will take at least one more interview before they come to a decision.) So, if you could give me any dope on this and also your opinion as to what I should do before NYCC has made its decision, I would be very grateful to you. But, please do restrict yourself to a very few lines;

[124] Ralph W. Tyler (1902–1994) was an American educator who earned his doctoral degree at the University of Chicago where he conducted several research programs about education. In 1953, he became the first Director of the Center for Advanced Study in the Behavioral Sciences at Stanford.

[125] Tyler was leading the so-called "Eight-Year Study", a national experimental research about secondary schools. They were interested in how high school students would prepare for college when they are released from the usual restrictions of college admissions. Hempel refers presumably to this job.

you are about going to leave New York by now, and I am anyhow feeling rather badly about troubling you at this moment.

The day before yesterday, I gave my first talk over the radio;[126] it was a 15-minutes talk about modern philosophy of science, in a series of lectures on modern philosophy, sponsored by the U. of C. The preparation took me a lot of time, but then, everything went very well, and it was a very interesting experience for me.

Eva is very busy with her job; but both of us are well and quite confident as to our future. – Eva's father has definitely got his immigration permit from Australia, which is good news, in spite of all the open problems behind it.

Eva and I send you and Edith our love and our best wishes for a restful, and enjoyable time up in the forests!

Yours Peter.

4.1.22 Hempel to Nagel, June 13, 1939

[Typewritten, Nagel 2020, Box 2, Folder 55]

Chicago, June 13, 1939.

Dear Ernest,

It was very good of you to write me even in the turmoil of your departure, and to talk to Bronstein[127] about my prospects. Yesterday, I got a letter from Bronstein telling me that the Board of Higher Education has again delayed action "on the whole evening session because of certain technicalities", but that the recommendation of my appointment has been accepted by the administration of the evening session, and approved by the president, and that "for all practical purposes, favorable action of the Board can be regarded as a foregone conclusion".

You and Edith can, no doubt, imagine how delighted we were to get these news! But we realize that an unfavorable decision may still be made by the Board; therefore, I have not given up, but only delayed, the negotiations with Prof. Tyler. Of course, I would by far prefer the job at the NYCC, even though the total amount I would be paid for the nine months (6 hours per week) would be about 900 dollars only; but we would bridge the gap between this sum and our actual expenditures by our savings, and we are also quite confident that Eva could make some money in addition. Both of us would be very glad to live in New York with its many facilities and possibilities for practical and theoretical needs and interests. And I would be so happy finally to get a start and have an opportunity to teach, and that at good school like NYCC. I am so grateful to you, Ernest that you told Bronstein about my looking for a position, when I was last with you in New York.

As Bronstein wrote me, there is even a certain probability for me to get a position in the summer session, beginning July 5; in this case, we would only too gladly accept your offer to live in your apartment during your absence. We shall write you about this as soon as we know the final decision of the Board.

We hope that you had a fine trip to Vermont, and that you are now enjoying restful days; we are sure, both of you need them!

---

[126] Carl G. Hempel, "Scientific Philosophy, Radio lecture" [Radio-lecture, Chicago, 1939. May 31. 8:30–8:45 p.m.] The lecture will be published in Gronda and Tuboly (2026).

[127] Daniel J. Bronstein (1908–2003) was an American philosopher, associated mainly with the City College of New York. He taught mathematical logic, physics, and philosophy for 40 years. He is known for an edited volume on philosophical problems that was widely used during the 1950s and 1960s (see Bronstein et al. 1948).

Many thanks again, and best wishes to both of you from Eva and myself.
Yours,
Peter.

4.1.23 Hempel to Nagel, June 18, 1939

[Typewritten, Nagel 2020, Box 2, Folder 55]

Chicago, June 18, 1939.
5466 Kenwood Avenue.

Dear Edith and Ernest:

I have got further good news from Bronstein; he writes: "As a result of a new development, I can say definitely that you will have a summer position. Board meets on 20th for action on Eve. Session budget. But this is a mere formality. I will let you know."

We were awfully glad to get this news: The fact that I will get those summer courses means a big financial help for us; besides, I am so much looking forward to start a teaching job; and finally, Eva is rather relieved to give up her job as a governess; for however kind the people are, it was a very tiresome sort of work and took so much of her time, and also she felt somewhat troubled by the idea that all her time and energy were used to take care of just one single baby; she felt that she might just as well take care of three such children or do some other work which would give her an opportunity to make a more adequate use of her abilities and her training. (Of course, this consideration has not prevented her from doing her job very well; the Adams are very sorry to see her go, though they are glad for us because of our new prospects; the baby, which is an extremely charming child, loves Eva affectionately and even jealously; and I am afraid it will display great chagrin when Eva leaves.)

We expect to leave for New York on June 30 or July 1 and possible to make a stopover at Buffalo to see the Roelofs from Cincinnati, who have a farm up there and who have repeatedly invited us to visit them.

We shall probably arrive in New York in the evening of Sunday, July 2; and if it really suits you, we should be only too glad to live in your apartment. In this case, there are a few questions we would like to ask you:

1. How would we get hold of the keys?
2. Would we have to inform the janitor or anybody else before moving in?
3. Should we give to our relatives in Europe the address "Hempel, 610 W 142..." or "Hempel c/o Nagel, 610 W..." or have you left instructions with the post office to forward any mail with that address to Vermont? (I could of course have our mail sent to City Coll.; but according to what Bronstein wrote to me previously, I shall have classes on 3 days of the week only, and would, therefore, probably not go over to the Coll. every day. Incidentally, according to my plan of New York, it seems to be only a 20 minutes' walk from your house to the College; is that right?)
4. We have got quite a number of suitcases and 2 larger trunks (the bigger one measures about 1x2x4 feet); is there any storage space available in the basement of your house? (If not, we could leave the two trunks with a storage house.) If so, should we ask the elevator man or the janitor to help us storing the trunks?
5. Were gas, electricity and telephone disconnected before you left? If so, would you be good enough to give instructions to have them reconnected, and to have the bills for the period during which we occupy the apartment sent to us?
6. Which is the name and address of your laundry?
7. Any additional information you think to be useful will be very welcome.

We are really sorry to trouble you with all these questions!

The spring quarter at the UoC ended a few days ago, and many of our acquaintances have already left Chicago. The Carnaps have rented a log house north of Michigan City from Prof. Gideonse,[128] former UoC economist who became president of Brooklyn Coll. a week or so ago. They plan to spend their 4-day-weekends out there. They have been out there since Wednesday and are due back tomorrow from their first stay; we wonder how they liked it.

The Brunswiks were in Chicago for a few days on their trip to Cleveland where B. is giving summer classes at the Western Reserve Univ.

We saw the Cohens occasionally before they left, a week ago. They were very kind with us. We found Mrs. C. to be a very nice and sympathetic woman, completely unbigwiggish; apparently she has been liked here very much, even by Ina, which means a lot, you know.

We hope you are having a very pleasant time now. We wonder whether the Helmers have turned up with you; they planned to go to Vermont. In case you should see them, tell them to give us their address, will you?

Best love 2:2

yours,

Peter.

4.1.24 Hempel to Nagel, June 27, 1939

[Typewritten, Nagel 2020, Box 2, Folder 55]

Dear Edith and Ernest:

This is just a hurried line to tell you how touched we were by your detailed indications and advices, and how grateful we are for everything. You are really sweet with us! (As to the silver, we have been using our own all the time; so that does not constitute a problem at all.) – We are now already busy preparing everything for the departure; we expect to arrive in N.Y. on Saturday or Sunday morning. – Bronstein sent word that my schedule for 39/40 has been extended from 6 to 9 hours per week: which means a very welcome improvement of the economical aspect of the job. – The Helmers are now located in West Bridgewater, Vermont (Ina asked me to tell it to you, so you may forward her letter to them). – Many thanks again, and love 2:2

{June 27, 1939. Yours Peter + Eva}

4.1.25 D. J. Bronstein to Nagel, June 27, 1939

[Handwritten, Nagel 2020, Box 2, Folder 55]

1630 Vanderhoff Ave
N.Y.
June 27, 1939.

Dear Ernest,

I am very glad we were able to help Hempel and also to obtain a first-rate logician for the summer and evening sessions, – something we needed badly. I am only sorry the position is

[128] Harry D. Gideonse (1901–1985) was a Dutch-American economist, and the second President of Brooklyn College from 1939 to 1966.

so inadequate. We did get $4 per hour for him which is high for City College. He will have 12 hours this summer and 9 hours next year in the evening.

We have also recommended White[129] for the readership – Feuer[130] has been appointed at $500 per year for one semester. You see, it is the work of Gottschalk,[131] Mead et al. to get a "big shot" for the Philos. dept to replace Overstreet[132] – in which case Feuer will be dropped at midyear. We'll have to be careful if we don't want to get somebody who doesn't fill the bill. I know we would all welcome you, and it would solve our problem if you could accept. Of course, I don't know what terms would be offered – but with the new administration, I feel an offer satisfactory to you would be made. Or is it out of the question?

We're leaving town Friday. According to an announcement in Erkenntnis the congress takes place Sept 3–9.[133] How's your Esperanto?

I hope you have an enjoyable as well as productive summer.

With best wishes to Edith and yourself.

Sincerely,

Danny

4.1.26 Hempel to Nagel, July 5, 1939

[Typewritten, Nagel 2020, Box 2, Folder 55]

July 5, 1939.

Dear Edith and Ernest:

Here I am, sitting in your living room, the typewriter on the telephone table, looking on the Hudson, and enjoying life even though writing letters! You have no idea how much we enjoy this apartment. There is only one drawback: living here will spoil us for ever for any furnished apartment.

So far, everything has been perfectly allright. Mrs. Nagel came over to meet us here (we were so sorry that she had to wait for us here for at least half an hour: we had called her up before we got our checked baggage, and we had not counted on the holiday rush at the

[129] Presumably Morton White (1917–2016), who was an American philosopher and historian of ideas. White studied with Nagel at Columbia, although his official PhD advisor was Herbert Schneider (see White 1999, 30). As White wrote, "Nagel's courses were the only ones from which I learned anything of value while I was at Columbia preparing to write my thesis" (1999, 30; one of our referees has pointed this out to us, and we are truly grateful for this). He taught at Harvard before being appointed professor in the School of Historical Studies at the Institute for Advanced Study in Princeton. A proponent of a holistic form of pragmatism, he wrote extensively on the history of American philosophy.

[130] Lewis Samuel Feuer (1912–2002) was an American sociologist and philosopher. Feuer was appointed as instructor in the Department of Philosophy at the City College in 1939.

[131] Louis R. Gottschalk (1899–1975) was an American historian, working mainly on eighteenth century France. He obtained his PhD from Cornell University in 1921, and joined the University of Chicago in 1927, where he became a professor in 1935, and led the Department of History between 1937 and 1942.

[132] Harry Allen Overstreet (1875–1970) was an American psychologist and sociologist. Overstreet retired from City College as the head of the department in 1939.

[133] Bronstein refers to the *Fifth International Congress for the Unity of Science*, Harvard University, Cambridge, Mass. (USA), September 3–9, 1939.

counter there) and very kindly showed us around, switched the icebox on, etc. We then settled down very quickly. We only unpacked the things which we are going to need during these 8 weeks; I put the content of three of the drawers of your chest in one of our empty bags, so that we have plenty of drawer space now. I piled our five bags one on the other in the corner between your writing desk and the record library; it does not look too bad, and we can get at them if we want to. We only opened one blanket parcel: one blanket a person will be quite sufficient at this time of the year! We also unrolled and put into its place (we hope we did it right) the carpet in Ernest's room, since the moth powder smelled too strongly to leave it there. What a pity that you did not know for sure that we would come before you left, so that you had to do this rather difficult job for so short a time. By the way, we really admired the orderly and efficient way in which you put this apartment into this summer sleep! And the cleanness and neatness everywhere. It is quite a treat, after the Chicago apartment, to have lamps where one needs them, to have matching dishes, and as many as one wants, to change towels more often than once a week. And then, I enjoy being home all day long lying on the couch and reading, taking sunbaths on the roof, making raspberry icecream, …

Until now, we have been leading a very retired life. Peter had a new style when we got here, he had to prepare his lectures, and we both wanted a rest. We tried to get hold of Wiener,[134] but we could not reach him. So we explored City College all by ourselves on Monday and finally found the right building and even the rooms in which Peter was supposed to teach. It is a quaint feeling to enter these buildings without medieval clothes. This morning, Peter went there half an hour early, found an open office, got the cards of his students (22 in the morning class, 18 in the evening), went into his classroom and lectured. During the intermission, they asked him what his name was.

The class was very lively and eager to discuss; Peter enjoyed it very much. I wonder how the evening class will be; he is giving it while I am writing this. – This afternoon, Wiener suddenly dropped in: he had only just returned to the city; and being on the welcoming committee, he seemed a little bit concerned that the welcome came so late. But it really did not matter.

Well, so much for today. I'm too tired now, and Peter will probably want to add some things of his own anyhow. He is not yet home though it is already 11.30; Wiener has probably got hold of him again.

Many thanks again for everything, and best love from
Eva

July 6, in the evening.

Dear Edith and Ernest,

last night, I came home at midnight; so I did not begin writing to you, and today, we were downtown all day long; and since I want this letter to reach you before the end of the vacations, I shall have to make it short this time, and I hope you will forgive me.

First, I want to mention that Eva overlooked a second very serious drawback of your apartment, and that is your library, Ernest. As soon as I settle down to work, my glance falls on some interesting book-title, I get up again and being reading a book which certainly is not absolutely necessary for preparing my classes in elementary logic. It is really a wonderful library, and surprisingly complete, and I am very glad to have an opportunity of using it. (Incidentally: Do you remember whether the most recent number of JSymbolLogic, which

[134] Philip P. Wiener (1905–1992) was an American philosopher. He was appointed as a professor at City College in 1933 and remained there until 1968. He worked on pragmatism, Leibniz, and philosophy of science. He published an influential selection of philosophy of science papers (Wiener 1953) in the same year that Herbert Feigl and May Brodbeck published their own, more famous collection (Feigl and Brodbeck 1953). Wiener co-founded the *Journal of the History of Ideas*.

came out about two weeks ago, has been sent to me c/o your address? I have not received it yet, and unfortunately, I do not remember what address I gave to Baylis;[135] but I think it was yours.)

Thanks to your kindness, it has been very easy for us to get accustomed again to living in New York; it has been a very pleasant experience so far. Wiener is taking care of me in a very considerate manner. This morning, just when we were going to leave, he picked us up in his new car, a nice 1933 Ford Coupé, and took us to the library of the college, and introduced me to the various librarians; he also showed us the College Store, and I am feeling more and more familiar with the institutions and regulations at CCNY. So far, I have taught once in the day-session, and once in the evening school. I liked it very much. The students were very eager and interested; particularly in the morning class, there are some very bright chaps, and after the first three or four minutes, they began already asking questions and making comments, and we had a very pleasant discussion. (I got one very amusing answer. In order to illustrate the falsehood of a sentence on logical grounds, I used Neurath's example of the report on two newly born twins, Tom and Harry, who have the strange property that Tom is heavier than Harry as well as Harry is heavier than Tom.

The students all agreed that that was quite impossible; fortunately, I was cautious enough to ask why they thought so, and one of them said: Why, if they are real twins, they will have the same weight.)

Today we bought a very light summer suit for me; now, I feel able to face the hot summer weather which may come. Even in this weather, I get pretty hot when I teach, but I do not mind it.

We got a letter from the Helmers today, telling us how much they liked your country home; it must be a beautiful place! We hope you are having as good a time as we are having in your apartment!

Good night for this time, and many thanks!

Peter.

4.1.27 Hempel to Nagel, August 6, 1939

[Typewritten, Nagel 2020, Box 2, Folder 55]

Aug.6, 1939.

Dear Ernest:

The other day I got a letter from the Revue Internationale de Philosophie, which is published in Brussels, inviting me to write an article for an issue which is devoted to logical empiricism, and which is scheduled to appear after next January.[136] The editor wrote to me

[135] Charles A. Baylis (1902–1975) was an American philosopher and the managing editor of *Journal of Symbolic Logic* after its foundation.

[136] *The Revue Internationale de Philosophie* was not published between 1940 and 1948, thus no special issue appeared in that period about logical empiricism. The first issue of 1950, however, was devoted to logical empiricism and it contained some lasting and influential works. Besides papers by Bertrand Russell ("Logical Positivism"), Marcel Barzin ("L'empirisme logique"), Herbert Feigl ("Selected Bibliography of Logical Empiricism" and "The Mind-Body Problem in the Development of Logical Empiricism") it featured Carnap's "Empiricism, Semantics, and Ontology" and Hempel's "Problems and Changes in the Empiricist Criterion of Meaning".

that they had asked you, C.I. Lewis,[137] Popper and Reichenbach, to collaborate and that Carnap, whom they had asked too, had declined because of too much work. Upon the request of the editor, I asked Carnap whether he was not willing to change his mind, considering that after all, it should be appreciated that these people are going to devote an entire issue to Log. Emp.; but Carnap answered he had no time to write an article though he perfectly agreed that it would be an important and worthwhile opportunity to do so.

The reason why I am writing to you is that I am at a loss as to what to do myself: I would prefer not to write an article for them now, simply for lack of time. I have to work out a common manuscript for Oppenheim and myself, and as you know, accepted to write a pamphlet for the Encyclop.[138] The first of these two things ought to have been almost ready by now, but for various reasons, and mainly because of this unfortunate furunculosis I have not even been able to finish the first chapter, so far.

On the other hand I am sure that Neurath will soon exert a strong pressure on me to obtain the MS for the Enc., and since in general I am a very slow worker, I would really be glad if I need not assume an additional obligation.

The only reason why I don't send them a negative answer right away is the following: On the one hand, the good will of the editors of the review concerning logical emp. should be appreciated by us, and also should we, as it seems to me, avoid the impression of being highbrow; and on the other hand, I am afraid that Lewis cannot be considered as a representative of log.emp.,[139] and that Reichenb. will take the opportunity to write another article against positivists, and that finally Popper may not send anything in time. So I want to make sure, before I decline, that you have accepted their invitation, for in that case the cause of empiricism has found an excellent advocate, even if you have to add certain critical reservations.

Will you please drop me a postcard about this? I have waited, hesitating for a while, and I cannot postpone my answer to Brussels much longer.

We hope that Edith is quite well again; I have been in bed too, with a bad cold, but I am going to be allright by tomorrow.

Best wishes 2:2

Peter.

{Many thanks for your postcard; we are enjoying the shower very much! – The radio program of WQXR[140] for August has arrived, too, and we enjoyed several very fine concerts already.}

4.1.28 Hempel to Nagel, August 13, 1939

[Typewritten, Nagel 2020, Box 2, Folder 55]

New York, August 13, 1939.

---

[137] C.I. Lewis (1883–1964) was an American philosopher, associated with pragmatism and modal logic. He taught at Harvard for nearly his entire career.

[138] At first, in 1937 Feigl was tasked to write a monograph for Neurath's *International Encyclopedia of Unified Science* on probability. But a year later Hempel was asked to do the same thing for a while; at the end, Nagel wrote on probability, but Neurath always tried to keep Hempel in the line for something. Finally, his monograph on the *Fundamentals of Concept Formation in Empirical Science* was published in 1952 (see Hempel 1952).

[139] Hempel presumably had in mind Lewis' "Experience and Meaning" article (Lewis 1934) which was quite critical with logical empiricism. To that article both Carnap (1936, 1937a) and Schlick (1936) replied.

[140] WQXR-FM is an American classical radio station that started to broadcast in 1936.

Dear Edith and Ernest:

Many thanks for Ernest's letter! We were soooo relieved, and the negative answer to Lameere[141] has already gone off! I wish you could, as a little mouse, watch the birth of one of Peter's article; and then we could discuss who "agonizes over it"!!

We hope that Mrs. Nagel had an agreeable trip and is now enjoying her stay. She went off just in time: that Wednesday was by far the worst day we have had so far, hot and humid all day long.

Thank you very, very much also for inviting us to stay with you after the congress.[142] We would love to come; but there are several points which prevent our making a decision right now:

1. Carnap spoke in Chicago of arranging a second "Lustin" after the congress, that is to say to ask a small group of nice people to stay together after the congress, somewhere not too far from Cambridge, perhaps at some sea resort, and discuss personal and scientific problems there for 4–6 days. ("Lustin" was the place where we arranged something of this sort after the Paris congress.)[143] He hoped that you, Feigls, Neurath, etc. might join in.
2. We have not yet found an apartment. We have now decided to take a furnished one for the first year, and people just don't want to let their furnished apartments already now for September 15, (which we can understand very well). Thus, if we don't find something in the last week of August, at least one of us will have to come back after the congress and look for living quarters.
3. and most important: There is some chance that the doctors advise Peter to have this tonsils removed. So far, the injections have not had any bad consequences (the temperature he had came from a cold), but not too good ones either; the styes have become smaller and less irritating, but unfortunately not less frequent. And though the doctor told Peter from the beginning that the treatment could not be expected to have its effects very soon, he himself asked him to see a throat specialist too. If that man thinks that the tonsils are infected, we are going to rush back after the congress and have them removed so that Peter has time to recover fully before his courses start on Sept.25.

But if all those points are settled so as to leave us free, we shall be delighted to come! And we hope that it will not be inconvenient for you to know our decision only in Cambridge? If, e.g., you would invite somebody else in case we can't come, please do so by all means, since the chances for our coming are comparatively small.

Could you give us the addresses of Wohlstetter[144] (New York or Cambridge, inclusive 'or')?

---

[141] Jean Lameere (1901–1964) was a Belgian philosopher, and editor of *Revue internationale de philosophie* (which was established in 1938). The negative answer refers to the invitation of the journal to submit a paper for a special issue on logical empiricism that was mentioned in the previous letter (27).

[142] The *Fifth International Congress for the Unity of Science*, September 3–9, 1939 at Harvard.

[143] Lustin is a small Belgian village where—among others—Rudolf Carnap, Susan Stebbing, Philipp Frank, Carl and Eva Hempel, Jörgen Jörgensen, Heinrich Neider, Kurt Grelling, and Paul Oppenheim met for 4 days to further discuss the philosophical questions of the First International Congress for the Unity of Science that had previously been held in Paris. For a photo, see Reisch and Tuboly (2021, 17).

[144] Albert Wohlstetter (1913–1997) was an American political scientist. In 1931 he studied at the City College of New York where he was a member of the league for a Revolutionary Workers Party. In 1934, Wohlstetter went to Columbia where he studied first law, then mathematical logic and philosophy of science. Though he did not finish his PhD studies, he published some review of logic-related works (e.g. of Woodger's *The Theory of Theory Construction* [Wohlstetter 1940]). During the Second World War, Wohlstetter started to work in government-related programs, and during the 1950 he worked at the RAND Corporation where he published several studies (e.g., Wohlstetter 1959), which became quite influential in the field of political foreign policy.

Now some questions for Edith: What should be done with the carpet in Ernest's room a) if we go away about Aug.26 and do not come back before you, that is to say before Sept.15; b) if we go away Aug.26, but return Sept.9, c) if we go away Sept.2 and return Sept.9, d) if we go away Sept.2 and return after you? We are very willing to clean it, put moth-powder on it and roll it up the way we found it; but since I think that it is more trouble to de-moth it again after it, I wanted to ask you about it before I do it.

How about the shower curtains? We have been very careful with them; but you probably want to have them laundered anyhow. But since they have to be prepared somehow specially, I wanted to know if you give them to the Champion Laundry too, or where they go.

Is there any special prescription for washing the mop? (Ina's, for ex., had to be washed in salt water), or do you dip it in any solution afterward?

Well, my dears, I hope that is all; and there is no hurry about answering my questions; a postcard some time before Aug.23 will do. And I am very sorry to force you into writing again.

Best love to both of you, and kindest regards to Mrs. Nagel

Eva and Peter {most gratefully!}

4.1.29 Hempel to Nagel, August 24, 1939

[Handwritten, Nagel 2020, Box 2, Folder 55]

8/24/39

Dear Edith and Ernest:

We were very impressed by Ernest's story: what a terrible thing![145] It must have been an awful shock to you both, and we are only glad that nothing worse happened. But we hope that you are going to [n/a] a lightening rod so that it can't happen again!

We are only living for the latest news from Europe at the moment. It is so depressing, we are feeling quite worn out.

Peter is almost through at City College; there is just the exam left to-morrow. But we are going to stay in New York until Sept. 2. There is another (and I hope the last) question: what shall we do about the keys? Shall we give or mail them to Mrs. Moran, deposit them with the super, or give them to Ernest in Cambridge? In case we don't hear from you, we bring them along to Cambridge.

Tarski arrived last Monday, Olaf (who was here for 3 days) and I met him at the [n/a], and we and Wiener, who is really very helpful and kind, took care of him for 3 days. He left to-day for Cambridge. But now, Jörgensen's[146] arrival is imminent. I like to see all these people, but I dream of lying in bed for 48 hours and rest.

Much love! Eva and Peter.

[145] As Nagel's letters from this period have not been preserved, we do not know what story Hempel might have in mind.

[146] Jørgen Jørgensen (1894–1969), sometimes written as Jörgen Jörgensen, or Joergen Joergensen, was a Danish philosopher, coeditor of *Einheitswissenschaft* and member of the "Committee of Organization" for the *International Encyclopedia of Unified Science*. In the English-speaking world he is mainly known for his *A Treatise of Formal Logic* (1931), and for his monograph *The Development of Logical Empiricism* (1951).

4.1.30 Hempel to Nagel, June 19, 1940

[Typewritten, Nagel 2020, Box 2, Folder 56]

New York City, June 19, 1940.
317 West 100th Street.

Dear Ernest and Edith:

many thanks for Ernest's long letter, or rather letters; we were very glad to hear from you, and we appreciate your detailed news so much the more as this should be a time for you to rest instead of going through a long correspondence!

Of course, I am only too glad to accept the invitation to join the staff of the consulting editors of the JSL,[147] and I am enclosing a formal letter of acceptance. Many thanks to you, Ernest, for arranging this for me; I am sure I shall like this sort of work, and it will also be useful to be thus obliged to study some of the relevant literature in the field more closely. If, as you suggest, assignments could be held off for some time – perhaps during the time of the summer vacations, I would appreciate it very much; during my illness, a number of things have accumulated which I have to take care of, and on the other hand, the physicians at the hospital advised me rather urgently to take things easy during the summer, and I mean to do it, particularly as I shall most probably have my tonsils taken out during the summer; they are in no good condition, and one of the doctors at Willard Parker believed even, that without them I should not have got into all that trouble.

But that is over now. Eva took me home on the 9th in Danny's car, and Feigl with the son of our superintendent carried me upstairs on their hands crossed in a way that is supposed to be a firemen's trick. At any rate, I stood the whole enterprise very well, and by the following Saturday, June 15, I was well enough for my first sortie which was to lead me to Klapper.[148] The Wohlstetters took me there, and Eva came along (but not up to Klapper's). The trip in itself incidentally, was very enjoyable; the view of the skyline in the morning sun from Triborough Bridge was extraordinarily beautiful; and on our way back, we made a luncheon-stopover at La Guardia airport, and during our meal, we saw at least ten big planes leaving and almost as many arriving; in addition to that, the new stratoliner was on exhibit, and I had a grand time, and the other three were quite patient with me in my indulging in my hobby of looking at planes.

Klapper was very nice to me. Quite generally, I had the impression that he is the ideal type of the warmhearted educator who is genuinely enthusiastic and optimistic about the possibilities of his work. He told me that he wanted the person who would get the job to devote some time to the students also outside of the classroom, to talk over their specific and frequently non-scientific problems. I feel that I could do that quite nicely, and also that I could satisfy his request that the viewpoint of the possible applications and of the bearing on empirical research be stressed in the introductory course in logic and sci. method. The only thing about which I am feeling somewhat uneasy is the course in contemporary civilisation,

[147] Consulting editors at the *Journal of Symbolic Logic* often served as referees, their names appearing on the cover of the journal as a sign of prestige and honor. As Church invited many logicians from across the world, consulting editors were also helping the journal by bringing to the attention of English-speaking scholars the latest results of non-English speaking countries. Nagel appeared as "consulting editor" since the second volume of the journal (1938), and he became editor along with Church in 1940 (Vol. 5. No. 1). He served as editor until 1946, when he was replaced by Max Black.

[148] Paul Klapper (1885–1952) was a Romanian-born Jewish-American educator, and the first president of Queens College between 1937 and 1948. Hempel was appointed as instructor, then assistant professor at Queens College in New York, and he remained there until 1948.

to which Klapper seems to attach a rather great importance. And he somehow has the idea that this course should not simply consist in presenting historical, economical, political material and treating it in the way which is characteristic of the corresponding special sciences, but that the more general – and probably philosophical – problems and aspects should be exhibited and discussed, and for this purpose, he seems to think, philosophers are more suited to give the course than the representatives of the sciences concerned. – But as I had learned from Goheen[149] that willingness to give one or the other of the four parts of this course would be a necessary condition for getting the job, I tried to learn something from my experiences with Northrop, and I did not object – in fact, he did not ask me about my opinion in this point, und thus, I did not have to express any particular cheerfulness about it, either, which was a relief. Now, Ernest, in case I should get that job, I shall ask you for advice; I warn you already now! – Klapper also said that in case I got the job he would want me not to continue my evening teaching, summer teaching or similar money-earning activities, but he would offer me a salary sufficient to get along without all that; it would probably begin at something around 2600, – which we think would be very nice. Kl. told me that he was going to see one other man, and he did not commit himself to anything; my impression was however, that he had quite friendly feelings for me; and Goheen, who saw him after our interview, confirmed this. Thus it seems as if there were a fair chance for me of getting the job, and Klapper told me I should know by the 25. already. In case I get it, he suggested strongly we should line in Flushing, which, of course, is very reasonable, and we mean to do it, though we do not like the idea of being so much farther away from where you live! But after all – what if we had to go to quite a small place in the middle West! – Now let us hope for the best!

Our plans for the summer have been tentatively fixed in the following way: we shall give up our apartment and shall leave New York probably around the 4th of July for Arlington, Mass.; we have accepted an invitation from the Wohlstetters who will have the Morgan's house all for themselves at that time. After 3 weeks, when they start out for their trip to Mexico, we shall continue, following conservatively the same line as last year, going to Rockport where Goodman[150] and the Leonards[151] will be, where the Oppenheims will probably come, and where we also hope to see the Quines[152] and the Feigls on weekend-visits from Cambridge. We would perhaps have preferred to go to the mountains, but as you say we would then need a car; in addition it might prove useful to have a big library in the neighborhood and a possibility of reaching it in Goodman's or somebody else's car. I am

[149] John Goheen (1907–1994) was an American philosopher (specialized in Greek philosophy) and educationist. He earned his PhD at Harvard in 1935. He taught there for a few years, and in 1950 he became the executive head of the Stanford Philosophy Department, where he taught until his retirement. He also co-edited, with John L. Mothershead Jr., the *Collected Papers of C. I. Lewis.*

[150] Nelson Goodman (1906–1998) was an American philosopher who was mostly associated with the University of Harvard, where he received his PhD in 1941 and where he was later appointed as a professor of philosophy (1968–1977). An extremely influential figure in American philosophy, he wrote extensively on the theory of knowledge, the problem of induction, and the philosophy of art.

[151] Henry S. Leonard (1905–1967) was an American philosopher who received his degrees from Harvard, and taught at the University of Rochester from 1930–1937, and later became the chairman of the philosophy department at Duke University.

[152] Willard Van Orman Quine (1908–2000) was an American philosopher. He earned his PhD at Harvard in 1932 under the supervision of Alfred North Whitehead and taught there for most of his career. He encountered logical empiricists in the early 1930s, while visiting Europe thanks to a Sheldon Fellowship. A towering figure in the history of analytic philosophy, he established himself as a leading American philosopher with his influential article "Two Dogmas of Empiricism", and with his books *From a Logical Point of View*, *Word and Object*, and *Ontological Relativity and Other Essays.*

afraid it is a little too far from your place to ask you to come over for a weekend; it would be such a fun to have you both there!

So far, I have read only very little of Carnaps ms;[153] but I expect to continue tomorrow. I do not know anything about where it is going to be published; I understand that C. is trying to make arrangements with the U.o.C. Press to take over that Neurath-series and also the Journal of Unified Science.

I have just finished the two articles by Bergmann in the last no. of Phil.of Sci.[154] I found them quite stimulating but they seem to me not to be sufficiently well "durchgedacht." Several of his points will not stand any closer examination, I am afraid. But somehow I think it is refreshing to see him come to closer grips with the concrete material than most papers in this field.

Runes sent me Harold Jeffreys'[155] Theory of Probability for review, a very impressive volume. I have not read J.s earlier book you mention in your probability pamphlet.[156] Do you consider it as interesting and valuable?

Eva and I are rather sorry to think that both of you should spend most of your vacation working rather than resting and doing non-routine things. What is your "more serious business for the summer" this time?

Today we are expecting the visit of a Swiss friend of ours with his wife; he was at the University of Brussels, and the two of them left four days after the German invasion of Belgium had started. He told Eva over the phone that Grelling had been arrested by the Belgian authorities, and he thinks that they may have taken him to France.[157] We hope to get some more clues tonight, also about Mrs. Grelling who had joined her husband in Brussels only a few months ago. – No word from Neurath yet, nor from anybody else we know in the Low Countries or France. Oppenheim is very much concerned about his elder son who was fighting with the Belgian army. These are terribly sad times.

We shall let you know when we get news from Klapper. In the meanwhile, best wishes and much love 2:2,

Peter.

4.1.31 Klapper to Nagel, July 6, 1940

[Handwritten, Nagel 2020, Box 2, Folder 56]

Saturday 7/6/40

Dear Ernest Nagel:

---

[153]The manuscript of Carnap's *Formalization of Logic* was finished already in the autumn of 1938 (Carnap 1943, xv), and thus here Hempel refers to the manuscript of *Introduction to Semantics* (confirmed also in his next letter 32). In its preface, Hempel's remarks and help were acknowledged (Carnap 1942, x). Instead of the University of Chicago Press, where Carnap's previous book (*Foundations of Logic and Mathematics*) was published in Neurath's *Encyclopedia*-series in 1939, this one appeared with Harvard University Press, in Carnap's series on "Studies in Semantics".

[154]See Bergmann (1940a, b).

[155]Harold Jeffreys (1891–1989) was an English mathematician, statistician, and astronomer. Jeffreys played a major role with his 1939 book *Theory of Probability* in reviving the Bayesian view of probability (Jeffreys 1939).

[156]Nagel referred to Jeffreys (1931) in his *Principles of The Theory of Probability*.

[157]On Grelling see letter 15.

I am grateful beyond words for your long and illuminating letter. Of the many candidates we interviewed, Hempel is by far the most promising from every aspect, – scholarship, general maturity, personality, clarity of exposition. I haven't met a superior person who can be had at the rank and salary available, an instructorship at $2616 a year. Your letter suits my mind and an invitation will go to Hempel.

Please forgive my intrusion on your vacation. You are too generous and too understanding not to forgive me.

Sincerely yours,
Paul Klapper

4.1.32 Hempel to Nagel, July 9, 1940

[Typewritten, Nagel 2020, Box 2, Folder 56]

Arlington, Mass.
49 Norfolk Road
July 9, 1940

Dear Ernest:

Here it {comes separately} is at last, the retyped version of the Lindenbaum ms.[158] The typing was a desperately tough job, partly because of her unwieldy notation, and partly because of the inconsistent manner in which she applied it in the ms. One of the most annoying features was the system of underlining certain letters and figures, and after I had carefully copied her original, I induced poor Peter to spend the three first evenings here in Arlington proofreading it. We then decided that, in order to make the thing as consistent as possible, we had to underline many more signs, in fact practically every symbol in any formula. We felt rather angry when we realized that the underlinings might have been omitted just as well, saving me several thousand strokes and backings of the carriage. Peter also decided to insert quotes, and he asks me to tell you that he feels rather uneasy about the formulae on page 8/9 containing sentential functions such as 'X is A'. If that 'X' is supposed to be a variable, then by putting the expression into quotes this characteristic is of course destroyed. But Peter calmed his conscience by the consideration that the 'X' may as well be interpreted as a constant with undetermined meaning (parameter) and in fact may have to be interpreted that way, because Lindenbaum's functor '[n/a]' has sentences, not sentential functions as arguments. Nevertheless, Peter thinks that it is an interesting paper, and he would be glad if it were accepted by the Journal.

Cont. Peter's dictation:

This morning, we finally got Klapper's letter, after a long period of ups and downs in our hopes. We were very glad to hear that Klapper had asked you to give him your impression about me, and we thank you very much for giving him so detailed and favorable information. Klapper's letter is to the effect that he is prepared to recommend my appointment to the Board and asks for my acceptance, and I hope and rather believe that the Board will not make any difficulties. (My salary for the first year is to be 2616. -) We are really very happy that I got this much more satisfactory job, although frankly we are unable to feel jubilant about it

[158] Hempel refers to the manuscript of Janina Hosiasson-Lindenbaum (1899–1942), who was a Polish logician and philosopher. She published numerous papers in logic and translated the books of Bertrand Russell into Polish. Although she went to Vilnius, she was killed by German forces when the Nazis invaded Russia. The manuscript in question is her most famous paper, "On confirmation"; see Hosiasson-Lindenbaum (1940).

any more because the repeated changes in our prospects during the waiting period have somewhat exhausted our emotional capacities. I am very much looking forward to the work in Queens, and even more so since Goheen (who is at Harvard, too, for the summer) told me that all the students have to take a course in math and that there will be ample opportunity to build up a special course in symbolic logic and perhaps another one in foundations of mathematics, and that already in the second term of the coming year I will probably have to give a special course in scientific method. (I wished your book were available by then!) I feel concerned only about having to take a section of the CC-course. When I know which section it is to be, I will probably write you again for your advice!

We were very interested in your clipping from the Times and we feel quite sure that it refers to Neurath. We are very glad about the news. – The Carnaps are staying in Chicago until the 15th of this month and then plan to go east, probably to Maine, at a slow pace. There is a slim possibility of their coming to Boston on their way, because Ina might have an interview here in connection with her application for a job in the social service.

During the last weeks in New York we were very busy; we had guests almost every night (we did not see McKinsey,[159] though and have not yet made the acquaintance of his wife). Then we had to pack {our things} and to arrange for the storage of most of them, but we got over it all right, and we are now enjoying a very pleasant time indeed out here in Arlington. It is so striking to live among trees and flowers again, and in this peaceful place where the average density of the traffic is about 1/4 of a vehicle per hour in {front of "our" house.}

Olaf enjoys his teaching at CC; he and Eileen seemed to be quite happy.

Before leaving New York I read about 70 per cent of Carnap's ms on Semantics. In spite of the title "Introduction to ..." it is no easy reading,[160] and to enjoy it, one should know symbolic logic and if possible some of the "Logical Syntax". But then it is very interesting to see how many of the concepts C defined in Syntax appear in a new light in Semantics and that some of the concepts which C had previously tried to define {in general syntax} are now seen not to be amenable to a general and adequate definition {even} in Sem. Thus on certain logical questions of very general importance, C's attitude seems to have become much more cautious, whereas in the discussion of many questions of a more special character, I felt that great progress had been made since I heard C discuss these things in his seminar last year.

So much for today; it is too hot to write letters anyway. We came away from New York just at the right moment; here we do not feel it so much, sitting on the porch and with only the minimum amount of clothing on.

Love 2:2

Peter

{The Wohlstetters are sending their best regards.}

4.1.33 *Hempel to Nagel, August 16, 1940*

[Handwritten, Nagel 2020, Box 2, Folder 56]

Rockport, Mass. August 16, 1940.
10a Granite Street

Dear Ernest and Edith,

it was good to hear from you; many thanks for your letter. We are very good to hear that you are both well and enjoying your vacations at your "country estate". As to the work that

[159] J.C.C. McKinsey (1908–1953) was an American mathematician and logician. He had a PhD from the University of California, Berkeley. After the war he worked at RAND and became a professor at Stanford. He is mainly known for his work on game theory and modal logic.

[160] See Carnap (1942), and letter 31 above.

remained undone so far, we have similar feelings as you, but certainly with much greater justification! I divided my working time between Contemporary Civilization and Logical Analysis of Confirmation. As to the letter, I had begun writing an outline for the MS before my sickness, and I am now continuing it;[161] unfortunately, my main definition chains now look very enigmatic to me, and it takes a lot of worrying to find out why I framed them that way. The main purpose of writing the outline is to find out whether the definitions I have in mind can stand any criticism; should I get through with it, I meant to show it to a few people whose impressions would be highly important for me; you, Ernest, are one of them; but I shall not give you anything before having asked you whether you have time to read it. And it is by far not ready yet, anyway.

For the CC-course, I got an outline of the program of my section which is an introduction to what might be called culturology, followed by a rather detailed introduction into economics (special attention being given to contemporary economic problems of the U.S.). For every one of the periods, the subject to be discussed, and the readings, both required and elective, are given in the outline, and though I would have to be scientifically chaperoned in my logic class or similar courses, I find it quite convenient in this field where I do not feel I could build up a course on the basis of my own previous comparative evaluation of the various theories advanced in these fields. So far, I have been reading Goldenweiser,[162] Ruth Benedict,[163] O. Klineberg,[164] Anne Anastasi[165] (I found her book on Differential Psychology quite interesting and a useful source of material on both methods and empirical evidence in those parts of psychology which are accessible to statistical methods), Ogburn,[166] a little MacIver[167] and some Kimball Young.[168] The second part, which covers 36 periods, frightens me considerably; I have a bad conscience about teaching these things. On the other hand, I enjoy preparing the course, and I think it was not bad that I just had to take it; this obliges me to get acquainted with a lot of material I would not have gone into otherwise. The only troubles are the bad conscience which I shall have anyway, and the great amount of time I shall certainly have to give the course, also during the term. Have you ever taught specifically economic material Ernest? I hope to have a chance to speak to you about the course not too long after the beginning of the coming term! Of course, we are both very

[161] Presumably what became eventually Hempel's major study on the "logic of confirmation", see Hempel (1945a, b).

[162] Possibly Alexander Goldenweiser (1880–1940), a Russian-born American anthropologist. Goldenweiser earned his PhD from Columbia in 1910 and taught later in numerous American colleges and universities.

[163] Ruth Benedict (1887–1948) was an American anthropologist, President of the American Anthropological Association, and one of the most famous and leading anthropologists of her time. She studied at Columbia University, her book, *Patterns of Culture*, was published in 1934.

[164] Otto Klineberg (1899–1992) was a Canadian born social psychologist at Columbia University, working on intelligence.

[165] Anne Anastasi (1908–2001) was an American psychologist, president of the American Psychological Association in 1972, known for her work on psychometrics. She obtained her PhD from Columbia in 1929 and worked at Queens College between 1939 and 1947, when Hempel was there too. For her mentioned book, see Anastasi (1937).

[166] William F. Ogburn (1886–1959) was an American sociologist first at Columbia, then at Chicago. He was the president of the American Sociological Society and the American Statistical Association.

[167] Robert M. MacIver (1882–1970) was a Scottish sociologist who became the Lieber Professor of Political Science and Sociology at Columbia in 1929.

[168] Kimball Young (1893–1972) was an American sociologist, president of the American Sociological Association in 1945.

curious about your new winter clothes, and we would also like to see those blankets. Whether we are going to have a furnished apartment, or to get our own furniture, we do not yet know. For one more year, we would rather be in favor of a furnished place; the times are so uncertain, and buying furniture is quite an investment. Eva will probably go to N.Y. for some days with the Goheens who also want to look for a place; the date has not yet been fixed; we expect to see the Goheens out here some day next week.

Last week, the Quines were here for three days, and before that, the Stevensons,[169] it was very nice and [n/a][170] Maria and the boy a few days ago.

During our stay in Arlington, I also saw Stevens,[171] the psychologist, and C. Pratt[172] from Rutgers who lectured in the summer school. One evening, we had a very nice discussion at Stevens' place, with Pratt, Feigl, Quine and Albert Wohlstetter participating.

The Carnaps, however, turned up neither in Cambridge nor here. Today we got a letter from them urging us to come and visit them in New Hampshire; but that will hardly be possible because various people are expected to come out here in the time to come, among them possibly Oppenheim (who has heard from his elder son at last; he is well but a German prisoner). Besides, Eva thinks the whole thing is a typically Carnapian idea: why did they not come here, after all? It's they who have the car. So she decided not to come along even if I wanted to go.

You will have heard through the Hooks that Oppenheim got word from Grelling. We are so glad he is alive and in good health, though his situation is certainly very precarious. Should he go back to Brussels to join his wife when he is discharged from the Camp? That seems the only thing for him to do, and yet, I think, he will hesitate to put himself again under Nazi control. – We got a letter from Eva's father yesterday; he is interned in the Isle of Man[173] and thus separated from his family.

It would be wonderful if the plans to bring Neurath + Reidemeister[174] over here worked out!

I am awaiting the proofs for the J.S.L. Personally, I like to have proofs sent to me and would be glad 1) to pay the postage 2) to promise [n/a] return, if I could also get them in future. But if this is not possible, then, I shall, of course, be able to do without.

Well, this is about all our news for today. If Ernest's mother should come up to see you, please give her our very kindest regards!

Many good wishes to both of you form Eva and myself:

Peter.

---

[169] C. L. Stevenson (1908–1979) was an American philosopher, known mainly for his emotivism in ethics.

[170] The letter is thorn here and cannot be read.

[171] S.S. Stevens (1906–1973) was an American psychologist who founded Harvard's Psycho-Acoustic Laboratory. Stevens was working at Harvard where he was involved in certain discussion groups initiated by Carnap and Frank.

[172] Carroll C. Pratt (1894–1979) was an American psychologist, working on music and emotions. In 1937 he became the head of the psychology department at Rutgers University.

[173] The Isle of Man was a self-governing British Crown Dependency in the Irish Sea, a typical tourist target between the wars. During the two World Wars, however, the island was used as the site of numerous internment camps for so-called "enemy aliens", that is Germans, Italians, and Austrians living in Britain. Otto Neurath and Friedrich Waismann were interned there too (see Burke and Tuboly 2025, Chap. 3).

[174] "Reidemeister" refers to Marie Reidemeister (1898–1986), a German designer and social scientist. Together with Otto Neurath, she developed the Isotype project. They became married in Oxford after their release from internment in February 1941. After Neurath's death, she was instrumental in promoting the translation of his work into English.

4.1.34 Hempel to Nagel, August 30, 1940

[Typewritten, Nagel 2020, Box 2, Folder 56]

August 30, 1940.
Rockport, Mass.

Dear Ernest:

your letter arrived yesterday just after Feigl had turned up for a one-day stay before going to Peacham, Vt. where his wife and his boy are vacationing. So I had to delay my answer until today. Poor Ernest, you have a pretty tough job on your hands with this Hosiasson-MS; Eva and I certainly don't envy you! As to your questions, both of us think that a German, not a Gothic c was meant, and that Vilna, Lithuania would be appropriate as indication of Miss H's residence. Eva suggests as another alternative: Formerly Warsaw; present residence unknown; but that would perhaps be misleading, since it is very probable that she still is in Vilna. At any event, a note might be added stating that and why it was not possible to have the author read proofs. – I shall be very glad to read the proofs as you suggest. It might not be a bad idea to have an extra set of the galleys made and to mail them out to Miss H., informing her, however, that for obvious reasons, the publication of the paper would not be postponed if her proof sheets did not come back in time.

Thank you for writing to Church[175] about the question of proofs on reviews. The conditions stated by Church are certainly agreeable to me; I was sorry to have to make a slight alteration in one of my latest reviews; I tried to cause as few changes as possible by taking out two words in the line following that one in which I had inserted two words, so that the arrangement of the remaining part would not be upset.

Eva came back yesterday from a trip to Flushing where she found a furnished apartment for us; she went along with the Goheens who went to rent a house. Our new address will be: 168–10, Cocheron Ave; Flushing, L.I. – We shall probably return to New York on the 12th of September only; the apartment is not available before that time, and also Oppenheim is coming here only tomorrow and wants to have some opportunity to discuss with me.

Eva called up Eliseo Vivas[176] who told her that he and his daughter are leaving for Wisconsin these days; Sarah is so poorly off that she had to be taken to a nursing home; she is most of the time under morphine, and Eliseo told Eva that it would be better to write than to go and visit her. She is being looked after by her family. We are terribly sorry about all this.

We enjoyed Feigl's brief visit here very much; we enjoyed the nice weather, the very attractive scenery and Feigl's report about the Carnaps whom he saw recently in Bethlehem, and whom he expects to see repeatedly in the week to come. Nelson Goodman, Feigl, and I also discussed physicalism, reduction sentences and similar items of common interest; Goodman is a very competent man who has many pertinent and frequently constructive criticisms to make on these topics.

Many good wishes from both of us to you and Edith!

Peter.

P.S. Sarah Vivas does *not* realize the seriousness of her condition; hence, no mention should be made of that in writing her.

---

[175] Alonzo Church (1903–1995) was an American mathematician and logician. He taught at Princeton and was the editor of the *Journal of Symbolic Logic*. Church is best known as the inventor of the λ-calculus and for the so-called Church-Turing thesis in computability theory.

[176] Eliseo Vivas (1901–1991) was an American philosopher, born in Colombia from Venezuelan parents. He received his PhD in philosophy from the University of Wisconsin 1935 and taught at Northwestern University for most of his life. In addition to his work in philosophy, he also published some writings in the field of literary criticism.

4.1.35 Hempel + Neurath to Nagel, April 18, 1941

[Typewritten, Nagel 2020, Box 2, Folder 57]

April 18, 1941

Dear Ernest:

I am enclosing a copy of a nice letter of Neurath's which we received a few days ago. I want Schapiro[177] and Hook to see it too; please forward it to them in the order which you judge safer – e.g. which gives the higher probability of the 3rd receiving it soon.

How are you both? And when are we ever going to see you again. It is so wonderful out here now, bushes blooming, new leaves everywhere. But I just dare not invite you busy and night-lifish people at a time of day when you would enjoy these things. Anyway, if and when you ever feel like coming out, please let us know! And if not, we might perhaps get together for a lunch or dinner in town sometime soon?

Grelling has been released from Camp de Gurs,[178] is now near Marseille – it is not quite clear whether still in another camp of altogether free – and applying for a visa.

Love 2:2

Eva

COPY

64 Park Town

Oxford, England

March 27th, 1941

Dear Hempels,

Many thanks for letter of March 3rd. Please deliver our thanks to all the kind helpers. I hope we shall be relatively soon in a position to pay the money back. We have to bridge the gap, because we need a breathing space. It would not be wise to seek an intermediary job instead of concentrating our forces on the continuation of our work. The reconstruction needs some time. I am now in contact with various influential persons who will help us in getting rooms, collaborators etc. but: there is a war on. In spite of all difficulties we start here with more personal contacts then 1934 in Holland. It was a great pleasure for me, as you may imagine, that I was invited by Prof. Radcliffe-Brown[179] to deliver a course of lectures on THE LOGICAL EMPIRICISM AND THE SOCIAL SCIENCES. Next term a discussion class of mine is already announced. Until we got married we were invited to stay at Cole's[180] Institute of Social History (he is not only an economist but also a writer of detective stories) – a nice start. Each of us got a fine room and we were in a position to enjoy the books of the library. I found certain interesting quotations of which I had the main points in my brain. You know we had to escape without luggage etc. At the last moment we succeeded in jumping into a motor life boat which was leaving the shore. Three students managed the

[177] Meyer Schapiro (1904–1996) was an American art historian and critic. He received his PhD from Columbia University in art history in 1929 and taught there for most of his career. In the 1930s Schapiro supported the Dewey commission, which investigated the charges against Leon Trotsky. He was also involved in the activities of the *Encyclopedia of Unified Science*: as early as in 1942 he was commissioned to write a monograph entitled *Interpretation and Judgment of Art*, which he never completed.

[178] On Grelling see letter 15.

[179] Alfred Reginald Radcliffe-Brown (1881–1955) was an influential English social anthropologist.

[180] George Douglas Howard Cole (1889–1959) was an English political theorist, economist, and historian, as well as a detective story writer and journalist. He was Professor of Social and Political Theory at All Souls College, Oxford. Closely connected to the Guild Socialist Movement, Cole was one of the founders of labor history and a historian of the Labour movement.

boat,[181] no shipper on board, about 50 persons instead of 20 – nothing happened. Harry Hack – that is a real story name – was our commander in chief through mine fields etc. We must admit that we were a little interested in the trip in spite of danger and sorrow (red clouds over Rotterdam and exploding tanks etc. at The Hague). Waves sprayed over the backside of the boat where Mary[182] and I were sprawling in big white rubber boots. I said we were now on the way to marry, that would not be so bad, and then to start lecturing in England would not be so bad either. Mary answered that such ideas were insolent, "they have just been waiting for you"... Afterwards we were interned, long weeks of separation, then monthly meetings, intervention by American and British friends, Stebbing and Kaempffert[183] especially, Julian Huxley,[184] Albert Einstein etc. doing what they could. Finally the British Academy made a statement that we were scientists of sufficient reputation, and the tribunal found out (with courtesy, I have to maintain) that we are pro-British, then we were released and are now, as you know, staying at Oxford. We rented a nice room with a view on trees and flowers, no bombing, no noise, no traffic, only singing birds, nice parks and gardens, nice buildings at Oxford, and nice people. We feel no discrimination at all – we are in a particular good position because I am lecturing at the University, and that gives here a certain reputation. I made a luncheon speech and was invited to read papers etc. "As once in the nice month of May." We agreed to avoid contacts etc., but we have almost no day completely free. People who want to see us, lectures we have to attend, etc. It is a great pleasure to listen to Price[185] and others when they speak about Logical Positivism, and all the things are mentioned we know so well. The New Statesman published a review of Russell's (not even pleasant) book by Joad in which Carnap, Hempel, Neurath are mentioned, and mentioned as people well known here.[186] Waismann,[187] Strauss[188] and other

---

[181] Neurath and his wife Marie sailed on a boat (the Zeemanshoop) from the port of Scheveningen in The Hague to England on May 14, 1940. The Zeemanshoop was the last boat to leave before the Nazi occupation, and the Neuraths were likely the last persons to join it. After a British boat found them at sea and collected them, the Neuraths were classified as "enemy aliens" and then interned in two separate camps on the Isle of Man. The Neuraths were finally released from their internment in February 1941, thanks to the intervention of Susan Stebbing, who sent a lawyer to get them out of prison.

[182] "Mary" refers to Marie Neurath (née Reidemeister), see letter 34.

[183] Waldemar Kaempffert (1877–1956) was an American journalist and translator. He was editor of *Popular Science Monthly* and then editor of Science and Engineering for the *New York Times*. Kaempffert was Neurath's cousin and helped to spread Neurath's views on scientific planning through his articles in the *New York Times*.

[184] Julian Huxley (1887–1957) was a British biologist. He was one of the major proponents of the modern evolutionary synthesis, which he presented in his book *Evolution: The Modern Synthesis* (1942).

[185] Henry Habberley Price (1899–1984) was a British philosopher. He studied at Oxford, where he was later appointed as Wykeham Professor of Logic. He is most known for his work on the philosophy of perception: *Perception* (1932) is probably his most influential book.

[186] Cyril Edwin Mitchinson Joad (1891–1953) was an English philosopher who achieved considerable success in popularizing philosophy. Joad reviewed Russell's *An Inquiry into Meaning and Truth*. See Joad (1941).

[187] Friedrich Waismann (1896–1959) was an Austrian philosopher and mathematician. He studied with Moritz Schlick in Vienna before emigrating to the UK, where he taught at Cambridge (1937–39) and then at Oxford. A member of the Vienna Circle, Waismann was the closest of the group to Ludwig Wittgenstein. Their conversations on the philosophy of mathematics and the philosophy of language were edited by Brian MacGuinness in the volume *Ludwig Wittgenstein and the Vienna Circle: Conversations Recorded by Friedrich Waismann* (1979).

[188] Martin Strauss (1907–1978) was a German physicist and philosopher. Strongly influenced by Hans Reichenbach and Philipp Frank, with whom he studied at the University of Berlin and the German University of Prague, his work dealt with the theoretical basis of quantum mechanics. Some of his articles were published in the volume *Modern Physics and its Philosophy. Selected Papers in the Logic, History, and Philosophy of Science* (1972).

were with me in the same camp. Waismann, according to my opinion, not free from metaphysics. He read a good paper on Mengenlehre. Physicists, biologists, economists etc. were there too. We attended in Oxford already a lecture by Waismann on the Geometries in a fine college room; the college is famous through the fact that it has the biggest pieces of coal in its fire place. We could prove it. By the way, we are very pleased to create a fire every day. A fire place is not an apparatus to an end, but it is an end in itself. Isn't it so?

We are glad that you are now settled. I hope to get the outlines of the Encyclopedia manuscript very soon. I wrote Morris that he may give Zilsel's[189] ms to the Press. I hope he tried to fit his paper into the whole encyclopedic environment. It is not meant as an isolated interesting study but as an introductory chapter for an encyclopedia of a movement of which he will be, I hope, an active member. Feigl will write an additional chapter which bridges the gap created by Santillana's[190] attitude. I think Morris told you about that.

Very glad to hear about Freudenthal.[191] I do not know anything about Mannoury,[192] Clay,[193] van Dantzig.[194]

We heard the Bohrs[195] are all right. Now I hope the Journal and the Library will be continued here.[196] I combat the usual defaitism which thinks Europe and Great Britain are gone, at least for some years. Not at all, not at all. WE ARE GOING TO SHOW IT. We like to stay here, and we would have remained at The Hague in spite of that little bombing and fighting in the streets if not the Dutch army had been forced to surrender. Force is force, bravery may be good, but martyrdom in a concentration camp or even living under Nazi government counts for nothing. Go on and be active – that is what the day wants. My book

[189] See de Santillana and Zilsel (1941).

[190] Giorgio de Santillana (1902–1974) was an Italian-American historian of science. Born in Rome, he graduated in physics in 1925 at the University of Rome. He then began to work in the field of the history of science, in collaboration with Federigo Enriques. In 1936 he emigrated to the USA, where he was visiting professor at Harvard before being appointed as professor at MIT. Together with Edgar Zilsel, de Santillana contributed to the Encyclopedia of Unified Science with the monograph *The Development of Rationalism and Empiricism* (1941). His most influential works are *The Crime of Galileo* (1955), *Reflections on Men and Ideas* (1968), and *Hamlet's Mill. Essay on Myth and the Structure of Time* (1969).

[191] Hans Freudenthal (1905–1990) was a German-Dutch mathematician who made relevant contributions to the field of algebraic topology.

[192] Gerrit Mannoury (1867–1956) was a Dutch philosopher and mathematician. He was a proponent of a form of philosophical relativism and was one of the organizers of the Significs Group.

[193] Jacob Clay (1882–1955) was a Dutch physics professor, working at the University of Amsterdam. Strongly interested in philosophy, he was initially influenced by Hegelianism, but later moved closer to logical empiricism.

[194] David van Dantzig (1900–1959) was a Dutch mathematician. A specialist in topology, he was an active member of the Significs Group.

[195] On Niels Bohr, see letter 2. His brother, Harald Bohr (1887–1851) was a Danish mathematician.

[196] Neurath refers to the *Journal of Unified Science* (that was a continuation of *Erkenntnis*) published between 1939 and 1940. "Library" refers to the "Library of Unified Science," a continuation of Neurath's *Einheitswissenschaft* series. It consisted in a "monograph series" (including Heinrich Gomperz's *Interpretation. Logical Analysis of a Method of Historical Research* (1939)), and a "book series," including Richard von Mises' *Kleines Lehrbuch des Positivismus. Einführung in die empiristische Wissenschaftsauffassung* (1939) and Hans Kelsen's *Vergeltung und Kausalität* (1941). Von Mises's book was translated into English in 1951.

makes good progress, my monograph too.[197] Mary and I read more books than in any period of our lives – we are very pleased, she has red cheeks and I am tall and slim.

From both of us love to both of you,

Neurath and Mary

4.1.36 Hempel to Nagel, August 23, 1942

[Typewritten, Nagel 2020, Box 3, Folder 1]

Flushing, August 23, 1942.
140-10 Franklin Avenue.

Dear Ernest:

Yesterday, Felix Oppenheim sent me your letter with the comments on his analysis of legal language.[198] I was very much interested to learn your opinion, and you really have written him a most valuable analysis; it is like a private book review! I fully agree with your general impressions about the incongruity of technical apparatus and objective difficulty of the logical questions treated with it; I have told Felix that repeatedly, and I am glad you stressed this point, too. Also with several of the other criticisms I find myself in agreement; there are some points about different schools in the theory of law about which I know nothing at all, and about which I have therefore been unable ever to make any helpful suggestions to Felix; and finally, there are one or two points – such as the imperative rather than descriptive character of "If y is guilty of larceny, x is punishable" – where I think, Felix can defend his position if only he states it more clearly and explicitly. Felix wrote he regretted not to have asked you explicitly whether you thought the essay was worth publishing; but he adds that Prof. Corwin[199] of Princeton has sent it – apparently with his recommendation – to the Harvard Law Review and Felix is now anxiously waiting for word from the editor. I should be glad for him if it were accepted, although I have no idea just how the general caliber of the paper compares with the average standards of the material published in a good law review. – Well, Ernest, it was very good of you to have taken the trouble of reading the essay and commenting on it; Felix was very glad to have your comments, and I want to thank you, too. –

I also read your review of Carnap's book[200] with great interest, and I feel it is a good thing that you stated your reservations so explicitly; but I also think that some of your objections can be satisfactorily met. I would like to discuss the matter in detail with you when you are back; and when I read your review, I also thought that it would probably be very stimulating to have a discussion of some of the fundamental issues you raise at one of the meetings of our discussion group. I am sure, that many members would be greatly interested in that; the problem is, after all, one of the very central ones in logical and in general philosophical research at present. What do you think of that? Of course, the program has probably by now been laid down rather rigorously; but quite possibly some member

[197] See Neurath (1944).

[198] Felix E. Oppenheim (1913–2011) was a German-born American political philosopher. He studied in Berlin and Heidelberg and obtained a PhD from Princeton in political science in 1942. The manuscript in question was possibly what became Oppenheim (1944).

[199] Edward Samuel Corwin (1878–1963) was an American legal scholar, McCormick Professor of Jurisprudence at Princeton University.

[200] See Nagel (1942).

scheduled to give a paper pretty soon will be happy to get an additional 4 weeks for preparation if a discussion of semantics were to be put in.

I have just finished grading my final examinations for the summer term, and I had a pretty good yield in boners. Here are some of them: Many words in common English are ambitious, but logical symbolism is precise. – It is logically premisable to say – A definition (That must have been a difficult topic for that student) – Acquiring knowledge consists of 1 part reasoning and 1 part appeal to brute empiricism. – If a theory involves a contradiction, you cannot make heads or details of it. – In explaining the method of tenacity, one student said: We have held certain beliefs so long that we feel they are altruisms. – A propos this sort of fun with language: Yesterday, I read in Gunther's *Inside Asia*[201] a report on the Indian province of Sind, in which he says that the first English general who conquered that province, reported his success by wiring home "Peccavi" – I have Sind.

I still have not heard from the draft board, although I wrote them a letter asking them to act on my case as soon as possible, because I had to find another apartment, and a cheap one had been offered me, but the landlord would not sign the lease unless he had it in writing that I would be deferred. In the meantime, that apartment has been rented. But today, we found a fairly nice one in the directly adjacent house, just around the corner; it is a little larger, and the house is not quite as old as our present one and seems better kept. We have made a deposit, and the super did not ask any questions about my draft status; so we hope, it will work out allright. We think that even if I am put in 1A, it may be several months before I am actually drafted, and we have to have a place till then. And If I am drafted, the lease will just be – come void; I have been told by several people that that is the general practice.

In the fall, I am scheduled to give one section of Contemporary Civilization in a greatly changed form; the first half of the course is now arranged around Lynd's *Middletown* and *Middletown in Transition*,[202] which I shall now have to read to prepare myself – I have also begun reading Margaret Schlauch's new book[203] and am very much fascinated by it. – I did manage to use a good deal of my spare time this summer for pondering over a suitable definition and theory of confirmation, and it almost looks as if I had arrived at a fairly adequate solution, which presupposes, however, a rather simple structure for the language of empirical science: my method of defining confirmation is, at least at present, applicable only to languages which have the logical structure of the lower functional calculus without the identity sign. It was a very hard job to get the definition into shape and to prove certain theorems for the concept of confirmation which it determines; but I have now written down a skeleton of the logical part of the theory, and Eva is going to type it soon. I would be particularly glad to get your reaction to the whole approach, and I wonder whether you think you could manage to read the typescript of the outline; including some introductory material it might be about 60 pages, and not all of it would require close reading. It may be a week before it is typed and corrected, and possibly even a little more. If you think that you might have the time and inclination to look at the material, please let me know; but *please* be sure to tell me if you would rather not do it now; I certainly do not want to out a big burden on you after you have been compelled by your sickness to put off many things you may have planned to do.

[201] See Gunther (1939).

[202] See Lynd and Lynd (1929, 1937). Robert S. Lynd (1892–1970) and Helen M. Lynd (1896–1982) were husband-and-wife sociologists, both associated with Columbia University, whose studies of the transformation of American society were highly influential in the 1930s.

[203] Margaret Schlauch (1898–1986) was an American linguist and medievalist. She received her PhD from Columbia University in 1927 and was appointed full professor at New York University in 1940. She left the USA for Poland in 1951 after being accused by Bella V. Dodd of being involved in communist activities. The book to which Hempel refers is *The Gift of Tongues*, see Schlauch (1942).

We had a few days of most unpleasant heavy and hot weather, and it was almost impossible to get some sleep at night. Fortunately, this afternoon, the temperature went down some 8 or 10 degrees in about an hour, and it is raining again. The cool breeze that comes into the room is a most welcome relief; we feel like different persons altogether.

We hope you are enjoying your stay in Vermont and are having a pleasant time with the Hooks, the Finkels[204] and perhaps other friends or relatives. Give my future colleague our very best, please, and tell her that all the classrooms are being washed, the floors polished, some places painted to get everything into proper shape for her arrival. Also, I looked into the physics lab the other day; they have some very nice things: apparatus for the Millikan experiment, for Laue X-ray diagrams of crystalstructures, for measuring e/m, etc.

Best wishes to all of you from both of us:

Peter

4.1.37 Hempel to Nagel, September 2, 1942

[Typewritten, Nagel 2020, Box 3, Folder 1]

Flushing, September 2, 1942.
140-11 Ash Avenue.

Dear Ernest:

many thanks for your nice letter which came this morning with the first mail to reach us in the new apartment; in return, you are now getting the first letter I write from here.

The moving was not so bad. I may have mentioned that the new house is directly adjacent to our old one, and that Eva and I were able to carry our clothes and some other things directly over the roof without the need of packing and the necessity of wrinkling them. We still have not finished unpacking; but the place begins to look like a human habitation; the living room has a very nice view and is extremely pleasant; besides, we have some genuine cross-ventilation, which is very agreeable. The bedroom faces a large court, which is also faced by the apartments of many a family with radios and babies; and we have to get used to the noise they produce. We still have our old apartment, for which we have a lease until Sept.30, and since in the present apartment the refrigerator does not yet work, I climb over the wall separating the roofs every morning to get butter and other breakfast accessories from the ice box in our old place. Besides, John Goheen may use our old apartment; he is going to come back in a week or so from California and has then to look for a house; he leaves his wife and child in the West for another month. I do not envy him the prospect of having to teach, take care of all the things that turn up at the College in the beginning of a new academic year, and to find a house and move – all within two weeks!

We saw the Oppenheims a few days ago; they were just back from Rockport, where they had seen a good deal of Tarski. It seems that the action taken by the Univ. of Cal. was successful; he was deferred for occupational reasons. – When you are back, I would like to ask you where you think Felix might publish his paper if the Harvard Law Review does not take it.[205] When we talked with the Oppenheims the other day, they asked for my advice and also wondered whether I could get your opinion. I shall tell you about the matter when we see each other.

---

[204] Irene and William Finkel (1904?–1970). William Finkel graduated in philosophy at Columbia University and then joined the faculty of the Speech Department of the City University of New York. A scholar of Walt Whitman, he was also interested in argumentation and rhetoric.

[205] See letter 36.

I am very glad that you are willing to look at my confirmation material;[206] in fact so glad that I shall not allow the sun to set before I have mailed it to you. The material I am sending you consists of two parts: One carbon copy of this summer's work; that is the more important part by far. It is written in a slightly condensed manner and contains, so to speak, the skeleton of the logical theory as I see it now. The rest of the material was written two years ago, when I believed to have a satisfactory solution. Before I reached the point of presenting the solution, I recognized that it did not work. The pages of the older ms which I am sending you are enclosed because they provide a more general background for the considerations I worked out this summer. I think you might find it desirable to be able to look at them, and I also think you will find them easy reading; by the way, the part entitled "Introduction", or even only §§1,2,3,5,6 would suffice to describe a little more in detail the problem which the more technical part attempts to solve. This older ms has many shortcomings and I would not think of publishing it in its present form; but I think that most of the points made in it will have to be presented in the final version of the essay – if that is ever reached. But *please don't feel hurried* – I am having some qualms of conscience anyway about sending it to you while you are having a too short period of rest anyway!

Just one more word about the technical part: The fact that it is slightly – but only slightly – condensed is not to be an excuse for any sloppiness in the presentation or in the proofs. There are some proofs which I had a hard time to find and which make use of some considerations which I wish I could state more explicitly; but I do not feel sure that I could do it. A typical and simple example is the proof of 4.4. on page 22. Also the proof of 8.7 – by far the hardest for me – contains some such arguments. Perhaps we can talk about that some time. – And if you find time to read the stuff within the next six weeks, I shall be very glad, and if you do not get around to doing it until Xmas, no harm is done; I shall have little if any time now for some time to devote to this work.

Oh, also on my program for when I see you next time is the question of whether you think I ought to try to prepare a paper for the New Haven Congress. If so, I think I could only take the material I presented in our circle on the principle of empiricism, with a few improvements. On a whole, the program does not tempt me very much, and I do not know whether my paper would have much chance of being accepted; but I would like to get your reaction to that, at any rate.

Please give Edith our love; we are both absolutely sure that she will enjoy teaching at Queens very much! We are very glad to hear she is feeling well. – We, too, became interested in pressure cookers when it was practically too late to get one; the reports of what they can do sound almost too good to be true! But we would have liked to buy one if we had known about them earlier.

Yes, we saw the lunar eclipse; here in New York it offered the added attraction of an unannounced practice blackout of 40 minutes! In one respect, the eclipse disappointed me: at its climax, the moon looked a dirty red rather than coal black. I resented that and wonder what the explanation is.

Now I have to run off. We are going to see the Wohlstetters tonight and to admire their new apartment, a sort of penthouse on the 18th floor at Amsterdam and 108th. After lots of excitement and trouble they have gone through to get and then to decorate it, they may now have to go to Washington shortly! But I do not know the details yet, nor the probabilities involved.

With very good wishes 2:2

Peter.

[206] See Hempel (1945a, b).

Both of us are quite sure Carnap will *not* feel hurt by your review;[207] he is the last person who would. At some places, he will say to himself: Oh, really – well maybe I am wrong! And at others: But I am sure this is a misunderstanding; if I could explain it to him, Ernest would revoke his criticism. That is the way he reacts.

4.1.38 Hempel to Nagel, July 7, 1943

[Handwritten, Nagel 2020, Box 3, Folder 1]

140-11 Ash Ave.
Flushing, NY.
July 7, 1943.

Dear Ernest:
[...][208]
a bit and felt good. We left at 12, Ross[209] and some others stayed till 2^{30} am (from 4^{30} on). Ross has left Camp Upton by now but we haven't heard yet where he was sent.

We were surprised to hear that you have difficulties getting fresh fruit and vegetables. I thought that you would get those "in the country" much better than here. How about oranges and apples, you get good ones here now; would you want us to send you a parcel of those? (Peter says you'll get them in S.Warbr. too) {If not, let us know so we may send you some!) Other fruits would be [n/a] to send if you collect your mail only once a week, except perhaps unripe peaches which would arrive just right. Please let us know about this; I'd be very glad to do that. I have a lot of time on my hand since I decided not to work during the summer. This is not only nice for me but also for Peter who would otherwise have spent half of his free time doing the shopping and attending to household chores. This way we hope that he'll be able to get his confirmation article into shape before the end of the summer. He has a light schedule, only 6 hrs per week, on three days. He had a heavy dose of Oppenheim last weekend (3 afternoons in 4 days) but now the Os have left for Northern Michigan where their son is stationed, and they won't be back till Sept.

Dr. Brunswick has formed a substitute for me, quite a nice girl, somewhat sentimental; I think she will work out. I spent 2 days then last week breaking her in, and one day this week doing some bookkeeping. Thus I make a little extra cash and also keep a reservoir of goodwill.

Poor Moore,[210] Peter wrote him a card. (He tried to write a letter but then didn't seem enough material to tell.)

---

[207] Nagel (1942).

[208] This is a handwritten letter and its first page is mainly missing. What can be seen from the remaining parts is that it gives a general description of the weather and the garden.

[209] Presumably Ralph Gilbert Ross (1911–2000), who completed his BA studies in Arizona, and studied philosophy at Columbia under F. J. E. Woodbridge, and later at City College under Morris R. Cohen. He earned his PhD in 1940. He taught between 1940 and 1945 at Queen's College of the City of New York.

[210] Presumably George Edward Moore (1873–1958), a British philosopher, professor at Cambridge. Together with Bertrand Russell, he was one of the founders and most influential proponents of analytic philosophy. He retired in 1939, and in 1940 he became the Neilson Visiting Professor at Smith College in Massachusetts. He stayed with his wife in the United States until 1944. In the spring of 1941, he lectured at Princeton and, in the summer, he became a lecturer at the University of California in Berkeley. From late 1941 to 1944, Moore lived in New York and taught at Columbia and lectured also at Swarthmore College in Pennsylvania. He had a few illnesses in the States, so presumably that is what the letter refers to.

It's a pity that Thorndyke[211] is leaving. They'll have a hard time replacing him by a good man these days. Do they not intend to replace him? The army is really coming, 2.60 for basic training, and 1.20 Foreign [n/a] study. They are just supposed to arrive on July 18 but courses start only on Aug. 9. We were afraid at first that Peter might be asked to teach math, and thus would lose his Sept. vacation, but since army teaching is paid extra (but not high at all) it seems to be wanted by the mathematicians – about which we are very glad.

I am feeling wonderful now, and everybody tells me that I look very well. I am rather lazy most of the time, and spend many afternoons at the Goheens lying in a deck chair or playing croquet. John Cooley[212] (Lt. in Navy) was transferred to Fort Schuyler (?) here in New York on July 1st. He has taken a room out here in Flushing (his Fort is just across Whitestone Bridge, and since he has a B-card he can drive over there very conveniently) and plays croquet with me on weekends.

So, now I'll leave some room for Peter.

Love to you both,

Eva.

I am just back from a trip to the oculist (who found that my eyes have spontaneously improved; an unusual phenomenon), and the dentist, with an hour between appointments, which I used for a trip – my first – to the Museum of Natural History. I still have chills running up and down my spine about those super colossal mastodons and whatnots; I also inspected with great fascination some of their Indian exhibits, including totem poles etc. Thorndike recently asked me for a statement of opinion about Albert Wohlstetter, and I wrote him a very favorable letter. I should be very glad if he got the job. – Thanks a lot for the reprint of the Journal containing your reviews, Ernest![213] I am very glad to have them; I think they are excellent jobs, both of them! I am grinding through the Carnap volume rather slowly; once I interrupted my reading for several days, giving all available time to the confirmation work; and then I found I had lost contact and had to start all over again; a procedure which, however, also improved my understanding of the material. The amount of formal detail (f. ex. on extensible rules of inference) one has to go through is a bit exasperating; but then, I find it rewarding to assimilate much more thoroughly, than even before these two conceptions of propositional logic as calculus and as interpreted system, and to realize the complications of such a distinction. – Your review on Carnap's book, Ernest, has changed John Goheen's attitude (which was sort of "negative a priori"); he now wants to read the book carefully and even ponder its suitability, as reading material in connection with an introductory course. From what I have read of the book, it seems to me that it is hardly suited as an introductory text; but I have it on my list of items for that course in advanced logic and philosophy of language (which *may* not be given for the [n/a] though).

The other day, I started reading in the book Edith brought Eva when she was sick; it is really a remarkably well done selection; I was quite excited to find some of my favorite T.S. Eliot poems, and a piece by Gertrude Stein. (I wonder why James Joyce is missing); then also a good detective story by Dashiell Hammett who ranks perhaps highest with me among the writers of this type.

Eva is calling me to dinner for the second time!

---

[211] Presumably Edward Lee Thorndike (1874–1949) was an American psychologist. After studying with William James, he spent most of his career at Columbia University. His main fields of research were educational psychology and the theory of learning.

[212] John C. Cooley was an American logician who taught at Columbia and retired in 1961. His most important work seems to be *A Primer of Formal Logic*, a textbook designed for classes, published in 1942 by Macmillan.

[213] Possibly volume 40, number 12 of *The Journal of Philosophy* that had two reviews from Nagel: one about Susanne K. Langer's *Philosophy in a New Key* (Nagel 1943a) and one about Carnap's *Formalization of Logic* (Nagel 1943b).

Give our best to the Hooks, and have a good and restful time!
Love to both of you,
Peter.

4.1.39 *Hempel to Nagel, December 21, 1943*

[Handwritten, Nagel 2020, Box 3, Folder 1]

12/21/43

Dear Ernest:

Just a line to tell you that you will receive a small parcel from a bookstore called *Orientalia*; please do not open it until Christmas. It contains a book which, I am fairly confident, you do not yet know (which is quite something to be able to say of any book), and which I think is a little masterpiece of its kind; I hope you'll like it. I've been trying for weeks to get a copy of it for you, but the bookstores were all out of it, and when Orentalia finally informed me they had got a new shipment, I had to have it sent to you directly, without a nice Christmas wrapping, if it was to arrive on time.

I am up from a slight case of flu, but everything seems to be allright now.

Love and best wishes for a good Christmas and for a happy New Year, from both of us to both of you:

Peter.

4.1.40 *Hempel to Nagel, July 3, 1944*

[Handwritten]

Saranac Lake, 7/3/44
Hotel Ampersand.

Dear Edith and Ernest:

We arrived in Saranac Lake two days ago and I am eager to send you my news at long last.

The twelve days at Princeton proved enjoyable and intellectually very stimulating, especially thanks to three discussions which Oppenheim and I had with Gödel on the subject of how to define explicitly the degree of confirmation of a hypothesis with respect to a given description or protocol. I think I had mentioned to Ernest that Oppenheim was interested in this question, and that we had at least a nucleus of a definition (for lower functional calculus of one-place predicates only) which satisfied the probability-postulates of the Hosiasson article in J.S.L.[214] Well, we discovered in Princeton that the def. involved serious difficulties, & we were ready to drop the matter; & then Gödel of all people revived it by some very constructive suggestions, and now we are continuing our work on it. The whole matter smacks too much of an a priori approach to give me an easy conscience, but it involves interesting technical problems and some hope of a reasonable amount of positive results; so for the time being I feel better about helping O[ppenheim] with this than if we were engaged in the more general phases of his ideas. But it is too early as yet to tell what will come of it.

[214] See Hosiasson-Lindenbaum (1940).

I found Gödel amazingly interested in the problem & very willing to give us his time for discussions, & I was very much impressed by his incisive and very constructive comments; often in following up a lead he outlined in 5 minutes, I discovered, only after many hours of work the full implications of his suggestions. – He is somewhat shy & almost self-effacing, but seemed in good spirits and gave the appearance of enjoying these talks with us. – I also saw v. Neumann[215] again, said hello to Church (whom I found on a sweltering day, lying on a couch & reviewing with his teen-aged son such formulae as a(b + c) = ab + ac, which the son had to write on a little blackboard built into the wall of Church's office); I had a chat with Stace[216] & two nice evenings with the Bergmanns[217] (you may recall them: they stayed with us at the physical institute after Reichenbach's quantum lecture; he is a former assistant of Einstein); at their house I met the radium physicist Rosenblum (Rosenbloom?)[218] who had been working at the Institut Curie at Paris on α-rays when the Germans took over; he told us the story of his escape with his little daughter & his pregnant wife, from Nazi-occupied France; it was pretty hair-raising; I would have given up twenty times along the way.

I also did some work at the library of Princeton Union & at Fine Hall; what facilities for scientific work; I turned yellow with envy. Also from Princeton I mailed my ms of the article on Confirmation to Moore;[219] I had been wise in postponing my trip to Princeton for several days to finish all the planned changes & to take them to Queens College for retyping. All I had to do in Princeton was then the final "proofreading" [footnote: I forgot, in my letter to Moore, to ask him specifically for reprints. Do you know whether one gets a certain number automatically & whether on request I might be able to get 75 of them?]

I have received no reply yet from the Am-Math-Monthly to whose editor I sent the ms of my lecture on Geom. and Empirical Science on May 26;[220] doesn't that seem rather unusual to you? Do you think I ought to ask back? A Prof. Schaaf[221] of the Brooklyn College Ed. Dept wrote me spontaneously to ask whether he could have the ms of the lecture (which he had attended) for the Nat'l Math. Magazine, Teachers' Section. But he wanted a version which would paint up the questions of teaching geom. at high school, about which I have no definite ideas. I told him about the Am. Math. Monthly & promised to let him know after I had received word from the latter, & I enclosed a copy of the ms & suggested that if the article was not to be published in the other magazine & he still wanted it, he might take care

[215] John von Neumann (1903–1957) was a Hungarian and American mathematician, logician, physicist, computer scientist, and engineer. One of the founders of game theory, he made major contributions to many fields of pure and applied science.

[216] Walter Terence Stace (1886–1967) was a British philosopher. He was an influential scholar in the philosophy of mysticism.

[217] Peter Bergmann (1915–2002) was a German-American physicist. He is mainly known for his work with Albert Einstein at the Institute for Advanced Study of Princeton; before that he worked with Philipp Frank in Prague.

[218] Salomon Rosenblum (1896–1959) was a French nuclear physicist. Born in Poland, then under Russia's control, he lived in France before fleeing to the USA in 1941 to escape from Nazism. In the USA he worked at Princeton University, among other places.

[219] Moore was the editor of *Mind* from 1921 to 1944. The manuscript to which Hempel refers is probably "Studies in the Logic of Confirmation", see Hempel (1945a, b).

[220] The lecture was given at the Third Annual Meeting of the Metropolitan New York Section of the Mathematical Association of America, on April 22, 1944. The manuscript appeared later in January 1945; see Hempel (1945c).

[221] William L. Schaaf (1898–1992) was an American mathematician and educator. He taught at Brooklyn College from 1932–63 and wrote numerous books about math instruction from elementary to college levels.

of the pedagogical aspects by publishing Hertzig's[222] article or a similar one in addition to mine. –

Despite the weekend crowds, we had a rather smooth trip to Saranac Lake; we are staying in a comfortable cottage under hotel management. I have never been to the Adirondacks before and am very much impressed with the beauty of this part of the country. The hotel, and our cottage, are in a very attractive location, overlooking lower Saranac Lake which is surrounded by forests and beautifully set off against a background of blue mountain chains. The quiet of the place is very soothing; the contrast with Flushing strikes me over & over again.

The morning after our arrival Einstein turned up to say hello to the Oppenheims. He had come with his sister,[223] who resembles him very much, in his sailboat from Knollwood, about 2 miles (as the boat sails) from Ampersand. We had a nice chat, & then the Oppenheims & I escorted him back in a canoe. – Today we rowed over again, & Einstein talked 1 ½ hours about some work he has just started in pursuit of an idea he has concerning the expanding universe; it was very fascinating. In the course of the conversation, he also endorsed the view that time's arrow can be determined only by means of the 2nd theorem of thermodynamics; when I asked about the hourglass counter-example that had come up in a talk with Wundheiler[224] at your house recently, he pointed out that this process too involves the 2nd theorem because the kinetic energy of the grains of sand is transformed irreversibly into heat. I am not sure that that meets the point; perhaps we can take this up once more when we get together. – Einstein's new idea refers to the question of the initial stage of the expansion process & aims at eliminating a singularity which occurs in the solution of the differential equations for the expansion;[225] this singularity seems to exist in a temporal starting point of the process, with all the [n/a] of the universe forming a material point, or the universe having the radius 0; but the time elapsed since this event turns out much too short (~10^9 years, age of the earth's crust) as computed in this fashion; E. has now an idea of how to remove the singularity & make the whole process periodic; then there is no beginning & no end of the universe in time, and instead of the singularity there is a state of relatively great irregular speeds of the stars in relatively close distances rather than a state where the radius of the universe is 0. I know this is pretty vague, but it shows you at least roughly what the talk was about. E. was really very nice, he also had had a haircut & looked much better. –

I have had good news about Peter Andrew[226] from Mrs. Richmond, who, in her latest letter, sent along a small piece of chewing gum with a tiny crescent-shaped imprint on it – the bite of his first tooth! –

The Flushing apartment is sublet, & the people who have it now are delighted with it; also, they are keeping it in excellent condition, as I saw during a brief stopover in N.Y. on my way here. –

Tomorrow, I am to go over to the Brunswicks' house for the first time to tutor the daughter in German, & I am feeling a little jittery because I am not sure of my knowledge of the subject; I'll prepare some time today. The B's live some 6 miles from here, & one has to go there by taxi or in the B's car. Mrs. Brunswick is expected to arrive only toward the end of the month. –

---

[222] Morris Hertzig (1908–2002) had a master's degree in Mathematics and Education from the City College of New York. He was a teacher of mathematics in various schools in New York and became chairman of the Department of Mathematics at the newly opened Forest Hills High School in 1941.

[223] Maja Einstein (1881–1951) was Albert Einstein's sister and a German Romanist.

[224] Alexander Wundheiler (1902–1957) was a Polish physicist/philosopher, student of Tadeusz Kotarbiński, and belonged to the Lvov-Warsaw school. He emigrated to the United States in 1939.

[225] On the issue, see Goenner (2014).

[226] Peter Andrew Hempel (1944–) son of Carl Hempel.

Finally, a story that Einstein told yesterday, & that you might enjoy. He put it into the mouths of two Jews – one of them complains that he can't understand at all how a wireless telegraph works, & the other explains: that's quite easy. Imaging a dachshund that is so long its head is in Paris & its tail is in New York. Now when you pinch the dog's tail in N.Y., he screeches in Paris. That's how the ordinary telegraph works. Now just imagine exactly the same thing, but without the dachshund. That is wireless. –

I hope you are both well & enjoying your cottage. How is the garden coming, & in what condition did you find the house? Is the ms making progress? And are you getting tanned?

Best regards from the Oppenheims, & my love to both of you,

Peter

4.1.41 Hempel to Nagel, July 18, 1944

[Handwritten letter]

Saranac Lake, 7/18/44
Hotel Ampersand

Dear Edith and Ernest:

It was very good to receive Ernest's long and interesting letter, and I was of course especially pleased by your report on your visit with Mrs. Richmond and Peter Andrew. It was nice of you, Ernest, to mention that Mrs. Richmond seemed to wait for a reaction to the news of Peter's first tooth; in fact, I had written to her about one week before your visit in Brattleboro & was wondering why she did not answer. I wrote her another letter promptly, & it turned out that my letter had been misdirected, but it was finally forwarded to her. Since then, Mrs. R. has sent me a set of charming photographs of P.A. which she had taken on various dates up in Brattleboro, and which confirmed fully your report about his excellent condition and happy disposition. Recently P.A. has had a recurrence of those wheezing spells which he sometimes got in Flushing, and which I viewed with suspicion & concern from the very beginning, because the tendency to the asthma – hayfever – allergy complex has occurred several times in the family of my mother; two of her sisters, in fact, had asthma pretty badly. Despite these suspicions, it was somewhat of a shock to me to hear from Mrs. R. in her latest letter that she finally took P.A. to the hospital, & that Dr. Otis there diagnosed the case as asthma. It is a pitiful thought that at his young age, little P.A. should be troubled with anything like that, & besides, of course, I can't help wondering whether he will take after those two great-aunts of his who had such bad asthma. I myself have inherited that disposition in a very mild form, & if Peter A. does not get it worse than that, there is nothing to worry about. The sad thing is that at least in this respect he has not inherited Eva's strong & sound constitution which I wish him so very much. – The physician advised that skin tests be taken by an allergist, but suggested that this be postponed until after the baby's return to New York, because he may not have those symptoms there. He prescribed something to relieve Peter's cough at night; & I certainly hope the whole trouble will turn out not to be serious at all.

Mrs. R. & her mother-in-law expected to move to a summer house on a lake some 12 or 13 miles out of Brattleboro these days; I wonder whether you had a chance to drop in on them before their departure. Mrs. Gabriel is supposed to come to town twice a week to collect the mail; their address for me remains the Brattleboro one. –

I was very much interested in Ernest's report on your various activities & certainly wish you full success in every respect; & also I hope to be able to verify that success, from canned berries to various manuscripts [footnote: also please a copy of the army Logic-test!] and rested, sun-tanned Nagels! I am eager to get reprints of your Kaufmann-discussion & your

Russell contribution, Ernest;[227] I have not seen the Russell volume yet, but I rather expect that I shall have a chance to look it over at the Carnaps. Through the help of the father of a former student of mine, I've recently got tickets + reservations to Santa Fe! I'm scheduled to leave here on 8/7 in the morning, due at Utica at 2^{40}; then I have a reservation on the N.Y. Central crack all-coach Pacemaker to Chicago, which leaves Utica only at 9^{07} pm. [In the afternoon, I have proposed a meeting to Stuart C. Dodd[228]! He wrote to say that he is in the States for a month on a Gov't mission & wants to discuss with me. He is living at Hartford, Conn. & travelling back & forth between that place, Huntington, L.I. where he lectures at a service center, & Washington. The Utica meeting was the best combination I could propose; I wonder whether he will accept.] I am here in Chicago on Tuesday morning, 8/8, & the train for Santa Fe leaves at 11^{30} p.m. & it is due at [n/a] around 11 $^{\text{a.m.}}$ 8/10. For this train, on which I am to spend two nights & a day, I've got a Pullman ticket & a berth; & I think that is only reasonable although I had planned to go there on a coach train with reserved seat, on which, it then turned out, you can make reservations only to California. – I am very much looking forward to my visit with the Carnaps. I have no recent news from them, incidentally.

You may recall that the Brunswicks had invited me to spend the first week of August at their "Camp" [a large, elegant estate with many buildings in a beautiful location, 6 miles from here]; but it has turned out that the work on degrees of confirmation, in which I am engaged here with Oppenheim, cannot be pushed as far as we had planned unless I continue here until I leave for Santa Fe; & so, the arrangements have been changed today. Mr. Brunswick (who leaves tonight for N.Y. to bring up his wife plus some patients on Sunday) was very nice about it, & I find it in fact more satisfactory to stay here with the Oppenheims: First, the work is really interesting this time (largely as a consequence of a few suggestions made by Gödel during my stay in Princeton) & I like to continue; & second, living with the O's has worked out very much more satisfactorily than I had anticipated. I have a fairly strictly organized "Tageseinteilung" every day, and I think that that is psychologically very good for me at present. – As to our work: We have several (!) explicit definitions of dc (S_1,S_2) – degree of conf of S_1 with respect to S_2 – which satisfy all the probability postulates as presented, say, in the Hosiasson article in J.S.L.[229] *However*, the definitions refer to a much too simple linguistic structure: lower predicate calculus of *one*-place predicates only, & they have various characteristics of the typical a-priori type of probability definition. While this bothers my empiricist conscience, I find the formal theory quite fascinating; in fact, some of its mathematical aspects have become so complicated that Olaf is going to be consulted on them. He will come up here about 8/11 – hence unfortunately only after my departure – for the rest of the month. Oppenheim asks that you keep this

227 The journal *Philosophy and Phenomenological Research* hosted a Symposium on *Meaning and Truth* in three different parts. The first part was published in volume 4, No. 2, Dec., 1943, and was made of articles by C. I. Lewis, Gustavus Watts Cunningham, and "Verification, Meaning, and Truth" by Felix Kaufmann. The second part was published in volume 4, No. 3, Mar., 1944, and gathered four contributions from Curt John Ducasse, Alfred Tarski, Wilbur M. Urban, and Norman Dalkey. Finally, the third part was published in the volume 5, No. 1, Sep., 1944, and contained one article by Nagel ("Truth and Knowledge of the Truth" [Nagel 1944a]), a reply by Kaufmann, and a counter reply by Nagel ("Rejoinder to Mr. Kaufmann's Reply" [Nagel 1944b]). The third part of the Symposium also hosted contributions from Charles A. Baylis, C. I. Lewis, Kaufmann, Roy Wood Sellars, and finally Andrew Ushenko.

The Russell contribution to which Hempel refers is Nagel's essay for the Schilpp volume *The Philosophy of Bertrand Russell*, entitled "Russell's Philosophy of Science" (Nagel 1944d).

228 Stuart C. Dodd (1900–1975) was an American sociologist and educator, working mainly on mathematical sociology and scientific polling.

229 Hosiasson-Lindenbaum (1940).

news for yourself, please; simply for the reason that he has not yet informed Kaplan[230] of this recent development; and he had asked the latter to help him with the study of quite a different phase of his "system", and Kaplan had agreed at least in principle. Now if the news about Olaf should get to him incompletely, a misunderstanding might occur. But O. is going to write Kaplan sometime fairly soon.

Thank you for your advice re Mind and the Mat. Article! In the case of the latter, I was surprised simply by the fact that I have never received any confirmation of the arrival of the ms from the editor; that the decision about its acceptability would take some time, perhaps even several months, would not seem surprising to me.

I had wondered why I had not been assigned any reviews for J.S.L. lately, & I had felt at the same time that that was a fortunate coincidence in view of the various things that were on my mind during the past months. And now it turns out it was no coincidence in the first place! Thank you very much, Ernest! And also for your suggestion of waiting until the fall, which is most welcome to me!

Well, I *did* cut down considerably on my preparation time for Til Brunswick's[231] German lessons. But I still have to refresh my memory, or to learn entirely ab ovo, on matters of German grammar. I find these 2 hours every other day quite dull; but the girl is making some progress, & I am fairly confident she'll pass that exam.

Today, I received an invitation for Eva for final citizenship hearing to be held on 8/25. How glad Eva would have been! – I had informed the Naturalization office of her death a long while ago;[232] but apparently, the letter was mislaid. I found the incident somehow saddening, although it is of no great objective significance. – Now I wonder whether my hearing will come up soon, too. I hope it won't be scheduled during my trip West, in which case I would ask for a later date, a possibility explicitly provided for in the form letter giving the date of the hearing. –

What are Edith's ideas about teaching at Q.C [[Queen's College]] in the fall?

The Goheens seem definitely to have decided to stay in Flushing this summer.

Please give my best regards to the Hooks! Sidney had apparently given instructions that those copies of the New Leader containing his articles on St. John's be sent to me;[233] I read them with very great interest & took them with me for the Carnaps & possibly other people who might be interested.

Tomorrow, Eileen should arrive at your house. Please give her my very best.

My love to both of you!

Peter

The Oppenheims send kindest regards!

We have visited Einstein again & on Sunday are invited to see the Pauli's,[234] who live a few miles from here on another lake.

---

[230] Presumably Abraham (Abe) Kaplan (1918–1993), an American philosopher. Born in Odessa, he emigrated to the USA in 1923 and became an American citizen in 1930. After receiving a PhD at UCLA, he taught there for almost 20 years before moving to the University of Michigan and then to the University of Haifa in Israel. He worked mainly in the fields of aesthetics, ethics, public policy, and the methodology of social sciences.

[231] Til Brunswick (n/a), was a social worker (with a degree from Berkeley) who had grown up in Vienna, "where her mother, Ruth Mack Brunswick, one of the few Americans in analysis with Freud, was a pioneer in analytic research." Information taken from Sklenicka (2019, 238).

[232] Eva died in January 1944 shortly after giving birth to their child, Peter Andrew.

[233] Hook published a two-part article on St John's College in the magazine *New Leader*, called "Ballyhoo at St. John's"; the first part is entitled "Education in Retreat", while the second part bears the title "The 'Great Books' and Progressive Teaching", published respectively May 27 and June 3, 1944.

[234] Wolfgang Pauli (1900–1958) was an Austrian physicist. A pioneer in the study of quantum physics, he received the Nobel Prize for Physics in 1945. From 1940 to 1946 he lived in the USA, working at the Institute for Advanced Study in Princeton.

4.1.42 Hempel to Nagel, July 25, 1944

[Handwritten]

July 25, 1944

Dear Edith:

Just a line to wish you many happy returns, and lots of luck and happiness for the coming year! I am writing you on the train to New York: I received word that I am to be naturalized on Thursday! I was very pleased about the great event coming to pass as well as about the fact that it happens before my departure West. I expect to stay in N.Y. until Friday or Saturday; various things have to be attended to, including discussions with Olaf on some of those mathematical problems I mentioned in my last letter. I also expect to see the Goheens; I shall be staying at their house tonight. –

Mrs. Richmond, who is now staying in a cottage at Lake Spofford with PA & her mother-in-law, wrote me that the drops Dr. Otis prescribed have relieved his symptoms, & that he is liking his latest summer residence very much. Mrs. Gabree, who went to Brattleboro to pick up the mail, found some very nice gifts from the Hooks for P.A. on the doorsteps. I am writing Ann & Sidney today; I assume their address is the same as yours.

Please give Ernest my thanks; and also those of Oppenheim, who was very pleased – for the reprint of his Moore article,[235] which I am reading with great interest on this trip.

So much for today! Love to both of you.

Peter.

4.1.43 Hempel to Nagel, August 22, 1944

[Handwritten]

Santa Fe, N.M., August 22, 1944
c/o R. Carnap
P.O.B. 1214

Dear Edith and Ernest:

There are so many things of interest to tell you that I am afraid I'll just have to make a selection and leave the rest for oral report.

Thanks very much to both of you for your letter of July 30, which I received after my return from New York. The two days I spent there were extremely hot and sultry, but I got quite a few things done and thus was rather pleased with my trip. Among others, I saw the Goheens twice, met several friends of theirs, had a long and worthwhile discussion with Olaf, saw Eileen for an hour – she just returned that evening –, dropped in at our apartment and found again that it was being kept in fine shape. The main event, the awarding of citizenship in Brooklyn, was, I am sorry to say, an extremely dull & uninspiring procedure, which consisted of 3 ½ hours spent standing without interruption; 5 minutes of these were used for clerical details, about 20 by a speech of the Judge plus some general information; the rest consisted in just waiting in line. Well, I am very happy to be a citizen now!

The trip to Santa Fe, though rather long, was very fascinating; it showed me a really very impressive cross-section of this enormous country: The big industrial centers, gigantic and terribly ugly and slummy, the simply endless flat agricultural acres of the Great Plains, the

[235] Presumably Nagel (1944c).

Kansas oil fields, the ranch lands of New Mexico, and finally the high mountains. The college of the Carnaps has a superb location atop a mountain 2 ½ miles outside and well above Santa Fe; it commands a magnificent view in all directions, over rows & rows of mountain chains, over the whole picturesque and quiet city of Santa Fe, and 20 miles down toward the Rio Grande; the scene is quite unbelievable. The climate is magnificent; warm and sunny in the daytime, cool at night; & the air is almost always dry. The starry sky at night is simply overwhelming; the colors of sky, clouds, & mountains at any time of the day are simply wonderful, & I seem never to tire of admiring them. The vegetation is poor, according to the altitude (>7000 feet), with thousands of cacti; some needle trees & other evergreens, & plenty of sagebrush. There are most fascinating animals around; wonderful instances of mimicking in insects; then there are horn toads, attractively ugly; they look exactly like a fat toad with a lizard's tail and legs and a cock's comb, really red, on their heads. It is very quiet up here, & on my regular walks with Carnap, I've never yet met a soul.

These walks, which I insist on extending gradually, take 1–1 ¼ hours in the morning, and another hour in the afternoon. C. always wears his brace when he is up, & on walks he uses a cane. He walks still a trifle hesitantly, but does remarkably well; two days ago, when we visited a ranch nearby, he followed me up to the top of a cliff which I had never expected him to make. He is in his usual well-balanced, confident & pleasant mood & is working rather steadily (although my presence seem to have reduced the rate of his writing). Ina looked *very* much better than I had anticipated; but her health is by no means perfect, & running the house & attending Carnap is a very tough job, which certainly does not make things easy for her. The C's are planning on leaving here about Sept. 10 or 15 or so; they have found it impossible to be referred to even a single apartment by any of the [n/a] 4 real estate agencies they contacted, & they are now seriously considering the purchase of a house; they plan to stay at a hotel first & then try to find something. The housing situation in Chicago really looks like a very tough proposition; the Ch'go papers have lately been carrying ads offering a reward of $75 to $100 for being referred to an available apartment! –

I have had several talks with C about his work in probability and degree of confirmation, and I am also reading those parts of the ms which are already typed (about 500pp, & that is less than half of the total work, which C feels is his most important book to date.) His ideas on the subject seem to me to be very fruitful & important, & I hope to discuss them with you in some detail when we meet again. He had obtained many of the results that had come out of my work with Oppenheim this summer; and in many respects, he has of course gone far beyond them; but there are also some points which are, as it were, "original" with us (or with Gödel).

I have had good news from Mrs. Richmond about Peter Andrew; the only trouble is that he continues to have wheezing and coughing spells, especially in damp weather – It would be very nice if you could have a look at him again; but I do not know when Mrs. R. expects to take him back to Brattleboro; at present, she, her mother-in-law, & the baby are staying at Lake Spofford, N.H. (their address is the Brattleboro one). I expect to return to N.Y. on Sept 10 or so & suggested that Mrs. R. return with P.A. on Thursday, Sept. 14, after I have straightened out things in the apartment a little. When do you expect to come back?

Many thanks for Ernest's suggestion to Krik about sending me a copy of the Naturalism & the Human Spirit volume;[236] I appreciate that very much & am looking forward to getting the book!

[236] Hempel refers to the 1944 book, edited by Yervant H. Krikorian, *Naturalism and the Human Spirit* (Krikorian 1944). Krikorian (1892–1977) was a philosopher who taught at the City College of New York. Born in Turkey, he left the country in 1915 to go to the USA, where he received a PhD at Harvard University. *Naturalism and the Human Spirit* is a collection of essays by some of the most influential American philosophers of the time (John Dewey, Sidney Hook, Ernest Nagel), which aimed at presenting a distinctive form of naturalism.

Finally, Ernest, I am honestly distressed & have been quite concerned about those expressions of your dissatisfaction with your own work, & when I came to Santa Fe, I found the Carnaps feeling very much & very sincerely the same way about it; we all feel that this attitude is certainly without rational justification, & that in particular the book[237] you are now working on is going to be important and valuable, & I feel that it may well exert a far greater influence than many of Carnap's writings. All this does not mean that I could not well appreciate the emotional depression about your work – which does not wait for a rational justification –; in fact, I think I can well realize what it is like, & I am very sorry you have to contend with these difficulties; but I do hope very much that they won't persist, and that you won't allow yourself to be discouraged by them!

I had expected to do various kinds of reading & perhaps writing up here, but I find that the time is simply flying, and I have not got half the things done I had counted on. But I think that the talks with C, walks, little trips, and the general physical & psychological atmosphere here will prove beneficial in their way, anyhow.

The Carnaps ask me to convey to you their very best, & also to say that there is absolutely no hurry about returning the ms on extension[238] [n/a]; that may well wait until after you get back to N.Y.

I do hope you've both been well! Is Edith going to teach at Q.C. this fall?

Love to both of you, and my best regards to the Hooks if they are around!

Cordially,

Peter

4.1.44 Hempel to Nagel, January 13, 1946

[Handwritten, Box 3, Folder 3]

Flushing, N.Y., Jan. 13, 1946.

Dear Edith and Ernest:

many thanks for the charming gift you gave Peter Andrew; he unwrapped it with glee this morning and is delighted with his blocks; and I know he will get a great deal of fun out of them, because Mrs. Richmond told me he found some similar blocks at the house of her mother-in-law in Greenfield recently and played with them a very great deal. Peter A. has been pulling his little cart with the blocks in it all over the house, and he proudly showed his new toy to some student visitors who came over this afternoon. –

The Kipferl Ernest's mother sent us are delicious and really not even a trifle stale; we took a more-than-fair sample of them this morning at breakfast. –

In a few days, it will be two years since Mrs. Richmond came to bring Peter Andrew home from the hospital and to take care of him here; and I think that in retrospect I can appreciate even more fully than at that time what an inestimable help you and other friends have given me by making this arrangement possible. I need not tell you – for you know it – about Peter's wonderful physical and psychological development under her care, and the reassurance and comfort it has given me to watch this process; nor do I have to mention the various practical ways in which I have benefited by Mrs. Richmond's keeping the house for me in her extremely efficient manner; but there has been yet another important consequence of her staying at the house, which is perhaps less obvious: Mrs. Richmond seemed to bring

[237] Presumably *The Structure of Science* on which Nagel worked for more than a decade.

[238] In the mid 1940s, Carnap was working on what became his first materials on modal logic, especially Carnap (1946, 1947a).

with her an atmosphere of quiet sanity and confidence which was a very great help for me in the first difficult months and also later in occasional periods of discouragement and depression.

I am sure you know how very deeply I appreciate all the help you have given me in these past two years – help which has gone far beyond the arrangement with Mrs. Richmond – but I would like to take this opportunity to express to both of you again my very, very warmest thanks!

With every good wish for all three of you,

Peter.

4.1.45 Hempel to Nagel, July 25, 1946

[Handwritten, Box 3, Folder 3]

7/25/46

Dear Edith and Ernest:

I am scribbling this – without the aid of alcohol – on the train to Boston; I am going to see John Goheen, who is teaching at Harvard, and with whom some departmental matters have to be taken up. And I hope to have a talk with Van, too, and maybe to pick his brains on some logical issues which have [n/a] in my work with P. Oppenheim (on explanation, emergence, etc. I can't tell yet whether it'll lead anywhere).

First of all, I want to send Edith special greetings and warmest wishes for many very happy returns, with Sandy thriving and happy, and Ernest doing the same and enjoying his work (I'd almost said, his overwork).

It was so nice to see you some weeks ago; I hope Sandy got by without the mumps! Here is a copy of one of the duo pictures I took; the other one (with Sidney, Ann, Katherine Goodman)[239] is quite poor.

I was glad to hear that Ernest has agreed to serve as the representative of the Ass. Symb. L. with the A.A.A.S. It'll mean new work for him, again, but he is of course, *the* man for this job.[240]

I have a copy of the ms of Carnap's article which contains his reply to Goodman's query.[241] C. asked for your summer address, which I sent him, and I hope that he has sent you a copy, too. What do you think of it? The loci of potential difficulties seems to me to lie in his completeness requirement and in his notion of simple property, which, as far as I see, he does not define in the paper. What he says about the [n/a] of complete evidence seems to me quite clarifying, on the other hand. –

W. Hole has been quite pleasant so far, except for the occasional psychol. strain that has to be expected of this kind of arrangement. Somehow I have not got around to any work of my own, which irks me; but there has been so much correspondence (also in connection with parcels f. Europe, etc) that, with some time out for weeks [n/a], I simply don't know how to

[239] Katherine Goodman was the wife of Nelson Goodman.

[240] Nagel served as representative of the Association for Symbolic Logic on the Council of the American Association for the Advancement of Science for the year 1946. The list of officers and members of the Association for Symbolic Logic for the year 1946 can be found in *The Journal of Symbolic Logic*, Volume 11, Number 4, December 1946. Nagel also served as President of the Association for the years 1947–49.

[241] Carnap (1947b).

arrange. I'd like to talk to both of you in the fact concerning the whole question of working with P.O. – I read, and liked very well, Hospers' article on explan. in the Journal.[242]

I've met some interesting people here, and – just a few days before my departure from N. Y., so to speak – a very nice woman in N.Y. (divorced; secretary of a physician), who I hope will spend her short vacation in the end of August in Woods Hole, and I am much looking forward to it.

Can you read this at all? It is almost an imposition. I do hope these lines will find all of you well!

Love to all, —» Over

Peter.

Do you happen to have Morty White's summer address? –

The Goodmans wrote me briefly from [n/a] where they were on a house hunt. Don't you think Katherine is a particularly nice woman?

4.1.46 Hempel to Nagel, August 11, 1946

[Handwritten, Box 3, Folder 3]

Woods Hall, Mass. 8/11/46
Hotel Breakwater

Dear Ernest,

many thanks for your cards and the preceding interesting letter.

First, as to the Carnap ms on confirmation: I am afraid that I won't find the leisure to read it for quite some time to come, and since I think I have read most of it when I was in Santa Fe, I feel it will be better if you return it to Carnap. I received from him a copy of his planned reply to Nelson Goodman's and some of your criticisms and discussed it in great detail with Nelson Goodman, when I visited in Rockport, as an addition to my trip to Cambridge. Both Nelson and I have sent long comments to Carnap. I feel that the article is helpful in stating explicitly the (or some of the) assumptions which are presupposed in applying either Carnap's or our system of confirmation theory. But on the other hand, I think these assumptions go extremely far and certainly raise serious problems; thus, e.g., Carnap uses the notion of *simple property* (mind you: not "primitive predicate" or "designatum of a primitive predicate") quite essentially in meeting Goodman's objections, and the very meaning of the idea seems to be quite problematic.[243] C. uses it in what seems to be an absolutistic ontological sense; but Nelson has pointed out that it may be possible to give it a more acceptable meaning by presupposing a suitable constitution system in which the primitives would then denote simple properties; at the same time, the primitives would have to be chosen in such a way as to denote projectible properties; and the decision as to the choice of suitable primitives of that kind would apparently have to be made on intuitive

[242] See Hospers (1946). John Hospers (1918–2011) was an American philosopher and political activist. He was educated at the University of Columbia, where he studied in 1941 with G. E. Moore. He was chairman at the philosophy department at the University of Southern California for many years. He was a good friend of Ayn Rand for years; in 1972, he was the first presidential nominee of the new Libertarian Party.

[243] Carnap uses "simple property" in "On the Application of Inductive Logic" (1947b), to which Goodman has replied in "On Infirmities of Confirmation Theory" (1947). Possibly what they discussed here was Carnap's then- unpublished reply that would come out next year as Carnap (1948).

grounds, more or less – despite all the great difficulties which have arisen, I feel that through his intense occupation with the matter of inductive logic I have learned a great deal, and that I have become sensitive to certain (logical) facts which I had overlooked entirely before. –

My visit to Cambridge + Rockport was extremely enjoyable and stimulating. I saw John Goheen – he and his family in Tiverton are very well –, John Cooley, Van, briefly also Morris Weitz,[244] Henry Aiken,[245] and then Donald Davidson,[246] who is finishing his work on a thesis on Plato and has been assisting Van, and who, I hope, will come to Queens in Sept. for a year to teach mainly C.C. and perhaps a little philosophy (all this is not yet official). Van seemed in good spirits, but did not look too well; the subject of his wife and his (probable) divorce was not mentioned. I attended a session of Van's very well attended "elementary" logic course, where he proceeded with lightning speed after a longish discussion of a tricky exercise in symbolic paraphrasing. The exercise: "A boy is a boyscout in good standing only if he has been duly admitted before and has done a good deed every day thenceforth". The answer: "(x)[x is a boy » <(t)[t is a day »] x is a boyscout in good standing at t » (Ey)[y is a day. x has been duly admitted on y. t is later than y. (z) [z is a day. z is [n/a] and t » (Eu)[u is a good deed. x does u on z]]]>]. – so Davidson told us (incl. Van), the students have a word for problems as hard as this, they call them Quinish. –

I had some good talks with Van, who also spoke a little about his present work which [n/a] the development of a nominalistic system of arithmetic, and then perhaps of further branches of math. This implies: no use of variables for classes etc., in fact for anything but individuals. – In the coming spring term, he expects to give a course in the philosophy of language. – Rockport proved to be a very lovely place again, and I decided to go there for a few days towards the end of this month. I am looking upon my "achievements" this summer (except perhaps for the economic side of it) with considerable misgivings; somehow, I have not got around to doing anything worth mentioning besides the work with Oppenheim (and, at times, with Felix, who during his presence here, discussed some problems of his more recent work on the logic of law with me). The work with Paul Oppenheim includes much material which I do not consider as significant, and often I feel that these discussions are just so much waste of time. Now and then, something interesting turns up; such as, quite recently, a certain suggestion, which came from O., as to how one might define the degree of systematization conferred upon a mass of singular sentences by a general theory. I followed this up and was eventually led to a concept which might be characterized as the dual analogue of the concept of degree of confirmation. But apparently, the measure I have been using for the content of sentences is the dual analogue to the measure of their ranges (which is allright) – but as chosen by Wittgenstein (which is not so good). Now I'll have to see whether there is any way of saving the basic idea in a manner which does not entail the difficulties connected with the various approaches to confirmation. –

Olaf and Lou wrote that they got 2 divorces and one marriage. They are on their way to California, where Olaf is going to join the research staff of – Douglas Aircraft, where they offered him an even higher salary (> 7000) than the Navy. He got his release from his contract with the latter. –

---

[244]Morris Weitz (1916–1981) was an American philosopher. He was professor of philosophy at Brandeis University, where he worked mainly in the field of aesthetics.

[245]Henry Aiken (1912–1982) was an American philosopher. He received his PhD from Harvard University, where he also taught for almost two decades before moving to Brandeis University. His main fields of research were ethics and aesthetics.

[246]Donald Davidson (1917–2003) was an American philosopher. After receiving his PhD from Harvard University with a thesis on Plato's *Philebus*, he taught at the University of Stanford, Princeton University, and the University of Chicago before being appointed Slusser Professor of Philosophy at the University of California, Berkeley. He wrote extensively on philosophy of language and philosophy of action.

I am really glad you are continuing work on your book, Ernest. I think I can imagine some of the psychological obstacles only too well; but I am so convinced the book is going to be first rate, and I feel so strongly it is needed that I am happy to hear it is progressing. Of course, not the least reason why I am so glad about it is that I know how much satisfaction it will give you to have accomplished the job!

The news about the Nesses[247] interest us all here vividly. As Oppenheim and I talked some vagueness and precision some months ago, and O. is quite fascinated by the subject, we would be interested to hear what kind of approach Ness is using, if this can be indicated *briefly*, and when you have time for a post card some time. *No hurry at all*, may also wait until N.Y.; the O's are going to Brussels for 3 months on 9/15, as I may have told you. –

I hope Robert Finkel is allright again;[248] it must have been quite a worry for his parents. If they are still around, please give them my very best.

I was much interested in your account of your work on convention programs, etc., but my, I really pity you!

Enough now for this time. Best regards from the O's, incl. Felix.

Love to all three of you!

Peter

4.1.47 Hempel to Nagel, February 5, 1950

[Typewritten, Hempel 1999, Box 28, Folder 1]

February 5, 1950.

Dear Ernest:

I have just finished reading your essay on reduction in the natural sciences, and I want to tell you how very much I like it.[249] I think it is a really wonderful piece of analysis, extremely lucid and convincing, and it has this additional merit that as one approaches the end, a number of disparate issues all fall into place and appear in a clearer light. Something I appreciate quite particularly is the inclusion in your essay of your rather detailed analysis of the "definition" of temperature in terms of statistical mechanics. Just a few weeks ago, in my seminar, Bob Cohen[250] (whom you met at Harvard) raised the question as to the status of the identification of temperature with mean kinetic energy of the molecules; I argued that it had to be considered as an empirical law rather than as a definition; but if he wished to build up a kinetic theory of matter and define in it a concept of temperature by means of that equation, then the usual definition, or partial determination of meaning, of the temperature concept by ref. to thermometric procedures or even to the Carnot process would become empirical laws.

---

[247] Arne Naess (1912–2009) was a Norwegian philosopher. A member of the Advisory Committee of the Encyclopedia of Unified Science, he is best known for his book *Interpretation and Preciseness* (1953). Naess was also very much involved in the environmental movement and wrote extensively on deep ecology.

[248] Possibly the son of William and Irene Finkel, see letter 62.

[249] "The meaning of reduction in the natural sciences" (Nagel 1949).

[250] Robert S. Cohen (1923–2017) was a philosopher, physicist, and historical sociologist of science. Trained as a physicist, he served as professor of both philosophy and physics at Boston University from 1959 to 1993. His influence on the philosophical debate and landscape was achieved through the founding of the Boston University Center for Philosophy & History of Science and the launch of the book series *Boston Studies in the Philosophy of Science*.

I wish I had been able to refer the class to your essay at that time; I will surely call Cohen's attention to it right away. And when I give my course on concept formation again, this essay of yours will be required reading!

Your criticism of Laplace's super-intelligence seems to me very well taken, and it was all the more striking to me as I had never thought of applying this kind of consideration to Laplace's idea.[251] Incidentally, the formulation of your objection in the first few lines of p. 133 gives rise to a vexing little puzzle: For strictly speaking, it is possible, of course, to deduce from a given sentence a consequence which contains expressions not occurring in the given premise; for ex., by means of the schema: S_1 —» S_1 v S_2. On the other hand, the import of your critique is undoubtedly correct, and I subscribe to it entirely; but I am afraid I would be unable to state the intended criticism quite generally and in purely logical terms.

It was awfully nice of you, Ernest, to have presented me with this whole volume,[252] and I am very glad at some of the other pieces, there seem to be a number of interesting studies among them. Many thanks!

I hope you have gotten over your miserable cold by now. Diane[253] has had a particularly bad one, with some fever, and then Toby got it, but both seem to be over the hum, and Peter, who had been going around with a stuffed-up nose and pale cheeks, has been looking a little rosy again since yesterday. Now I hope I won't come down with the same thing!

A few days ago, I was asked by Dean Hall of Washington Univ., St. Louis, Mo.,[254] to comment on a statement of tentative plans they have developed there for a series of faculty seminars, which are to extend over 2 years and are meant to further inter-discipline understanding, cooperation, and integration on various levels (undergrad. general education, graduate work, and faculty work). It sounded good to me, although I had certain reservations based on the impressions I had gained in the Columbia Seminar,[255] and I wrote him a fairly detailed statement based on that experience, and suggesting certain precautions which I thought would have to be taken to ensure success. As you were one of the outside consultants who are to be invited to help with the seminars if the plan materializes, I hope they have submitted the whole plan to you for comment; but in order to take no chances, I suggested this strongly to Hall, on the ground that in my opinion you and Lazarsfeld[256] had been the two people who had done most to make the Columbia Seminar serve its purpose. I also suggested that Lazarsfeld might be very useful as a consultant in a seminar, provided for in their plans, on the interrelations of the social sciences – Are you giving a seminar with

[251] "Laplace was thus demonstrably in error when he imagined a divine intelligence that could foretell the future in every detail on the basis of knowing simply the instantaneous positions and momenta of all material particles as well as the magnitudes and directions of the forces acting between them. At any rate, Laplace was in error if his divine intelligence is assumed to draw inferences in accordance with the canons of logic, and is therefore assumed to be incapable of the blunder of asserting a statement as a conclusion if it contains expressions not occurring in the premises" (Nagel 1949, 132–133).

[252] Presumably the volume that contained Nagel's essay as well, *Science and Civilization* (Stouffer 1949).

[253] Diane Perlow was Hempel's second wife. Miranda TobyAnne (1949–) is Diane and Carl Gustav's daughter.

[254] Thomas S. Hall was the Dean of the College of Liberal Arts at the Washington University, St. Louis.

[255] The Columbia Seminars were launched in 1944 by President Frank Fackenthal, based on the idea of Professor Frank Tannenbaum and the former President Nicholas Murray Butler that groups of Columbia professors and experts should be created to explore matters no single department had the resources to study.

[256] Paul Lazarsfeld (1901–1976) was an Austrian born sociologist who emigrated to the USA in 1933. Very well versed in quantitative methods, his most important studies concerned the social effects of unemployment, mass communication and voting behavior, and higher education.

Lazarsfeld this year? I wrote Hall you were; and I hope I am not wrong in this assumption. How is it working out?

I wish I could come in tomorrow, but I am behind with my work, and now even a little more so because Diane's cold, so I shall have to let it go. – I got the notice of the prospective termination of the Circle;[257] it is a pity, but seems surely the right step to take.

I hope Edith and the children are well! We've had such a mild winter this time; let's hope it will stay mild to the end.

Love to you all from Diane and myself!

4.1.48 Nagel to Hempel, February 6, 1950

[Typewritten, Hempel 1999, Box 28, Folder 1]

Professor C. G. Hempel
64 Anthony Street
New Haven
Connecticut

Dear Peter:

I told Lazarsfeld about your unfinished manuscript on concept formation[258] for the encyclopedia. He was very much interested in it and since in the joint seminar he and I are giving we shall be discussing the subject at some length, he wondered whether you would care to have your manuscript mimeographed for student use. It seems to me a very good idea and might enable you to get some detailed comments and criticisms. The mimeographing would be done in Lazarsfeld's office. Please let me know what you think.

In haste,
Ernest Nagel
Professor of Philosophy

4.1.49 Nagel to Hempel, March 7, 1950

[Typewritten, Hempel 1999, Box 28, Folder 1]

March 7, 1950.

Dear Peter,

There are so many things I want to write you about, but I shall have to stick to pressing matters for to-day. I owe you at least a half-dozen letters, and yet I cannot get organized and find the time for them.

As you doubtless heard from Frank, there is to be a meeting of the Unity of Science in Boston over the week-end of April 28th. We plan to have four sessions, two on semantics and theory of meaning, one on the sociology of science and one (I think) on probability and induction. I agreed to organize the first two sessions, and it is in this connection that I am writing.[259]

[257] There is no indication of which Circle Hempel had in mind.

[258] Hempel wrote a monograph for the Encyclopedia of Unified Science entitled *Fundamentals of Concept Formation in Empirical Science* (Hempel 1952).

[259] For the program see letter 52.

We would like to have a profitable discussion on the verifiability of meaning, and if possible to get a clear formulation of outstanding difficulties and conflicting views. In this way we might prepare the ground both for a settlement of at least some of the problems involved, and for directing further research along needed lines. We plan to have two speakers at this session, allowing a half hour to each, with ample time for discussion. And now for the question: would you be able to be one of the speakers? I have already written to Reichenbach, and I imagine he will consent to be the other one. I do not think this would place any burden on you except for the trip to Boston. It seems to me that you could base your paper (or talk) on the one you are printing in the REVUE INTERNATIONALE DE PHILOSOPHIE.[260] The fact that you gave a talk on this subject at Harvard earlier in the year is no obstacle, for it will not be the same audience and in any case different people will take part in the discussion. I need not tell you how desirable I think you should take part in this symposium, and hope you will say yes without hesitation.

I am authorized to say that the IUS [Institute for the Unity of Science] is prepared to cover at least the substantial part of the expenses of invited speakers. Specifically, the IUS can pay the cost of a round-trip coach ticket, plus $15. to defray the cost of hotel and meals during the meetings.

I wrote Carnap sometime back inviting him to speak at the session on semantics, but he declined saying that he could not come east. I don't quite know whom to ask in his place (you would be the obvious person, except that I would prefer to have you on the verifiability theory symposium). Unless I get a better idea, I'll write to Church inviting him, and perhaps either Van or Nelson may agree to present an opposing standpoint.

Lazarsfeld is having your Concept-formation ms. mimeographed, and I'll send you more details when the job is ready.

Please let me know as soon as you can about your decision. I'll be anxious until I hear from you.

Our love to you all,

Ernest

P.S. I saw Tarski for a short while last week. He was on his way to Europe for some lectures. But I'll tell you more another time.

4.1.50 Hempel to Nagel, March 10, 1950

[Typewritten, Hempel 1999, Box 28, Folder 1]

Dear Ernest:

Many thanks for your very nice and interesting letter! Today, in haste, just a note to say that I shall be very glad to speak at the Unity of Science meeting on what seem to me some crucial problems in the testability criterion of meaning and indeed perhaps in the whole concept of empiricism. I suppose that the exact title can be agreed on later? I shall plan to send you a definite suggestion within a week. – The whole program sounds very good to me; it would be splendid if Church and Van or Nelson could speak on the semantics issues!

Incidentally, are these meetings going to be open only to members of the Institute (if there is such a status), or may anyone who is interested attend? I am asking because Oppenheim had heard about the plan for a meeting from Reichenbach, who had written Oppenheim that he was coming East to attend it, and Oppenheim was eager to know who can attend.

[260] "Problems and Changes in the Empiricist Criterion of Meaning" (Hempel 1950a).

I hope the burden of your work is not wearing you down too much! I heard a rumor the other day that you were toying with the idea of taking a leave next year to travel and work in Europe; is there something to it? If so, I surely hope your plans will work out well! I suppose that in that case you might think of getting a Fulbright grant?

Love to you all from Diane and myself!

As ever,

[Hempel]

4.1.51 Hempel to Nagel, March 26, 1950

[Typewritten, Hempel 1999, Box 28, Folder 1]

March 26, 1950.

Dear Ernest:

Please forgive my delay in sending you the title of my talk for the IUS meeting. I still do not feel quite clear about how I am going to organize my material nor in fact just what I shall want to include; but the title, *On Meaning and Confirmability* should at any rate give an approximate characterization and yet leave me sufficient leeway.

I am sure you have been exceedingly busy these past months; I only hope you are not overdoing it! How is the preparation of the IUS meeting coming along? I wonder whether Popper will be speaking in one of the sessions; I certainly hope so.

About a week ago, a strange and alarming rumor circulated among our graduate students here, to the effect that Hao Wang[261] at Harvard had discovered a contradiction in something like elementary number theory or a similarly innocuous fragment of mathematics, and that a number of specialists were checking his ms. Had you heard about it? I finally wrote Quine and asked for information, and was both relieved and disappointed to hear that while Wang had indeed come quite close to proving the sort of thing that rumor ascribed to him, it was now known why the proof could not be clinched; in fact, Wang himself had never been very sanguine about it.

I was very glad to receive from Lazarsfeld a parcel of mimeographed copies of my ms the other day; would you be good enough to give him my greetings and thanks when you see him again?

I hope you, Edith, and the children are all in good health. We have all been well, but Toby is teething again and seems to be in considerable discomfort quite often, which means that she is fussy and cries a lot, and that Diane and I do not nearly get an adequate amount of sleep. I sense very acutely and painfully the sort of intellectually numb state that gets me into on certain days, but there is simply no remedy to this at present. Not even acoustical insulation is feasible in our present place. We hope that will all be better in our house, which is well along now, so that we may (*perhaps* etc etc) be able to move in some time in April. – Oh, I must tell you how much pleasure Toby has been getting out of the soft little

[261] Hao Wang (1921–1995) was a Chinese-American mathematician, logician, and philosopher who obtained a PhD with Quine at Harvard in 1948. Wang is known for his work on Gödel's logic, his interpretation of Wittgenstein's philosophy of mathematics, and for various technical-mathematical results along with computer programs that proved theorems from Russell and Whitehead's *Principia Mathematica* back in the 1950s. The manuscript, mentioned in this letter, was possibly what became Wang (1952)—there the contradiction hardly appears (although Wang mentions it as a motivation for the paper) and he argues that certain simple reasonings cannot be formalized even in very strong systems.

dog you and Edith sent her; the other day, when nothing would calm her, I gave it to her, and promptly she started to gurgle and laugh, and really played herself to sleep with it; the same thing has happened several times – quite apart from the fun she has with the little dog when she is feeling well.

Love from all of us to all of you!

4.1.52 Nagel to Hempel, March 27, 1950

[Typewritten, Hempel 1999, Box 28, Folder 1]

March 27, 1950.

Dear Peter,

I have not answered your question about the meeting of the IUS in Boston, because I did not know the answer. But yesterday I saw Frank, who came to N.Y., and we worked out all the fundamental arrangements. The program is now complete, and the tentative schedule is as follows:

Friday April 28, 3 P.M.: Science as a Human Enterprise
P. Frank and Abe Kaplan
Friday evening: Business meeting of the board of governors
Saturday April 29: 9:30: Theory of Meaning:
Hempel and Reichenbach
Saturday 2:30: Semantics
Nagel (Introd. remarks), Church and Quine
Saturday evening: social gathering
Sunday April 30, 9:30: Science of Man
Morris, Sheldon[262] and Kluckhohn[263]

Anybody who cares to come will be welcome, but we are going to charge a registration fee of $1.00

Oppenheim called me up Saturday, very much distressed over the possibility that he might not be welcome. Of course, I reassured him. But I think he is a bit of a fool, and for the first time since I've known him he left a bad taste in my mouth.

Last night I also saw Popper, at the home of Boschan,[264] whom perhaps you know: he is a statistician, badly crippled, who knew Popper in Vienna. As a matter of fact, the evening was a sort of "old Vienna" affair – in addition to a lot of people whom I did not know, there were present Lazarsfeld, Wald,[265] and Frank. It was quite pleasant, though naturally there

[262] William Herbert Sheldon (1898–1977) was an American psychologist who created the field of somatotype psychology, correlating body types with temperament. He was known also for directing the controversial project of making nude photographs (front, back, and side) of all incoming freshmen (and freshwomen) at the Ivy League universities to make statistical and medical data of certain ostensible and visible medical conditions in the population (such as rickets, scoliosis).

[263] Clyde Kluckhohn (1905–1960), an American anthropologist and social theorist, known for his work among the Navajos. He studied, for example, in Oxford and Vienna, and obtained his PhD from Harvard in 1936, where he remained as a professor for his entire career.

[264] Paul Boschan (1903–1985) was a German-American statistician. He worked at such places as the Econometric Institute and at the Guardian Life Insurance Company of America, New York.

[265] Abraham Wald (1902–1950) was a Hungarian-born mathematician and statistician. He was influential in the development of operational research.

was next to no talk of philosophy. But Popper is going to give a talk at NYU on April 17, and perhaps I'll learn then just what he is doing.

Have you received the mimeographed copies of your CONCEPT-FORMATION ms.? I wonder whether it would be too great an imposition on you to ask you to send a copy of the bibliographical references. We'll make mimeographic duplicates of it, and then distribute it to the students who bought the Concept Formation.

I don't think I know yet how the course with Lazarsfeld is going. I've been doing a lot of the talking – it is hardly a seminar, for we have close to a 100 students in it – but Lazarsfeld's turn is coming soon, and then I expect to learn much more. But it is fun doing, and if we continue with it over a number of years it may work up into something both interesting and valuable.

At the last meeting of the University Seminar Lazarsfeld gave what I thought was a brilliant exposition of his work in latent attribute theory, and he is to continue with it at the next meeting (April 10). I think you would have enjoyed it. But some of the discussion and comment was fantastic, especially by the economists. Wald is now a member of the group, and he comes fairly often.

There is of course no basis in the rumor that I am planning to go abroad next year. I am taking a sabbatical, to which I am entitled, and there is a chance that I may take the whole year off (if the Guggenheim Found. gives me enough money to cover the salary I would *not* be getting). But whether it is a half-year or a full year, I expect to stay in New York, and try to work on the book.

I saw Tarski last week, on his return from England. He is the same as ever, though I did not have too much chance for serious talk. He has been urging me to go to California if I take the whole year off, but there are too many difficulties in the way and I am not sure I would be able to work as well away from my own desk and room.

Will you be getting any reprints of your article in the Revue Internationale? I would be grateful for one, and so would John Cooley. I've read the essay through once, and of course enjoyed it very much. But there are a number of questions I would like to ask you about it, and perhaps the opportunity for this may come before very long.

We are all pretty well, though Bobby is teething and tends to be cranky. It is a hard life for Edith, and she is not looking forward very eagerly to Vermont – at least in New York she gets some relief when her mother comes, but in the country there is practically no break in the fixed routine.

How are things with you and yours? Does Diane get as tired as Edith does? I suppose you are almost ready to move into the new house. I know you'll be too busy to write, but I certainly expect to hear about all your doings the latest at the end of April.

Our love to you all.

In great haste,

Ernest Nagel

4.1.53 Hempel to Nagel, April 6, 1950

[Typewritten, Hempel 1999, Box 28, Folder 1]

April 6, 1950

Dear Ernest:

Many thanks for your nice letter. Two days ago, I received from Boston the tentative program for the meeting and was very pleasantly surprised to find Feigl listed as a speaker; and the news was confirmed the same day by a letter from Feigl, in which he suggests that you, he, and I stay at the same hotel so that we'll have a chance to see something of each other. I like that idea a great deal, and yesterday wrote Nelson Goodman to ask whether he knows any suitable hotel not too far from the Academy; I also suggested that he might want to stay at the same place. If I hear anything from him, I'll let you know in the hope that you'll be able to put up at the same place. To be on the safe side, I have also asked the secretary listed at the bottom of the Tentative Program to send me information about accommodations if he has any. The program looks really very enticing to me, and I am looking forward esp'y to the Sat. pm session. I shall plan to get to Boston in time for the Friday pm meeting and for the evening meeting of the Board.

I am sorry to have delayed my answer concerning the bibliographic references in my ms;[266] the point is that I do not have a typed list of the material; it is on cards, and these had to be picked out of a larger collection and also had to be gone over to assure at least a certain degree of uniformity in the arrangement of the references. I got around to that yesterday and today, and have just left the cards with our departmental secretary, who hopes to start typing them up on Monday. As soon as she is through, I shall mail the cards to Lazarsfeld's office so that they may be mimeographed. I hope this is not too late for your purposes. The number of students in your joint seminar is breath-taking! Do you ever have any seminar with just about our number of students (between, say, 6 and 25) in it? The Faculty Seminar looks much better now than it was in the beginning, and I surely am sorry not to be able to hear Lazarsfeld. Also Wald's contributions should be very much worth hearing.

Diane and I are keeping our fingers crossed for the extra funds from the Guggenheim to come through so that you may take a year off and perhaps even finish the book; what a demand there is for a good work of that kind! I recently saw two book mss on philosophy of science that a NY publisher wanted to have a professional reaction to, and they were both so poor that it seemed to me they shld not be published; one of them esp'y seemed to be the work of an eager dilettante, who was confused and had no first-hand knowledge of any of the fields under discussion. So please get your book finished, Ernest, it will pay big dividends in acclaim and dough!

I myself certainly sympathize with your decision not to go to Europe; I am sure I wld not get any work done that way. California is a different thing if you cld sublet your ap't; from the experience of the Goheens it seems that you wld stand a good chance of having a year without colds for the whole family, and that is not to be sneezed at. And you cld perhaps just settle down at some nice place not too far from a library and would then really be in a setting conducive to work, while Edith and the children would be out on the beach. How does that strike you and Edith?

You would long have received a reprint of my article, Ernest, if it had not been for a special arrangement with the publisher, or rather editor, under which Carnap, F[ei]gl, and I get a crate full of copies of the whole issue; these arrived quite recently in Minneapolis and are being cleared by the customs officials now, and then Herbert Feigl is sending them out to the various people whom he, Carnap, and I would like to send reprints. So you will get the whole thing, with Russell and Barzin thrown into the bargain.[267]

We have all been well, but tired – just as tired, Diane asks me to report, – as Edith is feeling. So we have great sympathy even for her hesitations about the summer in Vt. Toby has been teething, and for months she fell back into the habit of requiring a feeding each night; then Peter got a cold, and so there was one thing or another which interfered with our

[266] See the previous letters, concerning Hempel's manuscript of *Fundamentals of Concept Formation in Empirical Science*.

[267] Hempel refers here to the *Revue Internationale de philosophie*, Vol. 4, No. 11, 1950, where Hempel's "Problems and Changes in the Empiricist Criterion of Meaning" was published alongside Carnap's seminal article "Empiricism, Semantics, and Ontology", Russell's "Logical Positivism", "The Mind-Body Problem in the Development of Logical Empiricism" and a "Selected Bibliography of Logical Empiricism" by Feigl, and Marcel Barzin "L'empirisme logique".

sleep, and Diane and I often walked about like zombies. I find it especially trying when I have to teach or do any serious thinking. But we feel lucky that the children have been well otherwise; and both are lively and very charming; Toby is simply a delight, and she can wrap me around her little finger any time, esp'y since a few days ago she suddenly and quite spontaneously started feeding me a cookie.

Our house has been slower in getting finished than we expected, and we have been obliged to set back the date of our moving in several times. Now it looks as though we may be able to move in about 4 or 5 weeks from now; so on your way up to Vt, at the latest, you shld be able to drop in and see for yourself what it is like. Unfortunately, the architect has miscalculated the cost of the house no less than the date of its completion, and we find it will cost ca 33 percent more than the estimate said it wld. Well, I can do nothing about it at this point, and I do think it will be a most pleasant place to live in. But it will oblige me to do outside work for some time, and as you know I expect to be going to California this summer, although formal arrangements with Rand[268] have not been made as yet.

Love and best wishes from all of us to all of you!

4.1.54 Nagel to Hempel, April 8, 1950

[Handwritten, Hempel 1999, Box 28, Folder 1]

April 8, 1950.

Dear Peter,

Morty White invited me to stay at his house during the Boston meeting, and I've accepted – otherwise I would have liked very much to put up in the same hotel as you and Feigl. By the way, Frank promised to arrange for reservations of invited speakers, and he will probably write you about this matter before long.

There have been further oddities to the program – G. Bergmann[269] and Max Black[270] have accepted invitations to appear as "discussants" in the "Meaning" and "Semantics" sessions respectively. The program is beginning to look a bit too crowded!

I heard this week from the Guggenheim Foundation: I am being granted a Fellowship, and though it will be a tight squeeze financially I have decided to accept it and take the whole year off. I hope I'll be able to get something done – I feel it is next year or never.

I am sorry that the bibliography put you to any trouble – had I suspected that you would have to make it up especially – I would never have asked you.

I am looking forward to seeing you in Boston, and hope there will be time for some good discussion – there may be so much to do that the most important reason for the meeting may

[268] Rand Corporation was created in 1946 by the United States Army Air Forces, based on an idea from the Douglas Aircraft Company. The mission of the Corporation was to develop economic analysis and research in the fields of military strategy, health industry, aerospace, universities, and so on.

[269] Gustav Bergmann (1906–1987) was an Austrian-born philosopher who emigrated to the USA in 1938. He was closely associated with the Vienna Circle and taught philosophy and psychology at the University of Iowa. Bergmann is mostly known for coining the phrase "linguistic turn". His major works are *The Metaphysics of Logical Positivism* (1954) and *Realism* (1967).

[270] Max Black (1909–1988) was a Azerbaijani-born British philosopher who moved to the USA in 1940. He taught at the University of Illinois at Urbana–Champaign and was then appointed professor of philosophy at Cornell University. He is best known for his influential work on metaphors, *Models and Metaphors: Studies in Language and Philosophy* (1962).

be ignored. I expect to arrive for the business meeting in the evening on Friday, but doubt that I can make it in time for the afternoon session.

A happy Easter and our love to you all
Ernest

4.1.55 *Nagel to Hempel, April 20, 1950*

[Typewritten, Hempel 1999, Box 28, Folder 1]

April 20, 1950.

Dear Peter,

You are ever so much more prompt in making acknowledgement than I am, for I should have written you some time ago to thank you (and Carnap and Feigl as well) for the copy of the Revue Internationale de Philosophie.[271] I am very grateful for it, and I wanted to have it very much. I enjoyed your paper immensely, and have already referred a number of students to it. But I think I don't quite see eye to eye with you on several points, and I am fairly convinced that the alternative you propose – of translatability into an empirical language to be constructed along the lines you suggest, as the criterion of empirical meaningfulness – is bound to be a blind alley. But this is a large question, and this is not the occasion for going into it. I hope we shall be able to go into it at Boston, and I am looking forward eagerly to seeing you there. I will not be able to arrive in time for the Friday afternoon meeting, but certainly expect to make the business meeting in the evening at Frank's home.

I was naturally pleased that you found the piece I did for the Dewey volume of some interest.[272] I did not press my difficulties with Dewey's views – partly because I did not think it would have been appropriate for such a volume, and partly because I do not have too clear ideas on how to solve them. Like yourself, I was glad that finally a more or less explicit statement of the content of the "revolt" against the distinction between "analytic" and "synthetic" is finally available in Morty's piece.[273] But I must confess that I have not been impressed by it, and still am not sure what the shooting is about. What on earth is being demanded by a "criterion" of analyticity in a natural language? Since no answer in terms of a constructed language will do, and since natural languages are admittedly vague, I suspect the issue is a pseudo-problem. But again, this is a large question, not to be discussed here. However, I've been tempted to write out a reply, though I am not likely to do it – the dissertations are now coming in almost in car-loads, and I have barely time to breathe.

Last Monday night I heard Karl Popper give a talk – a critique of analytic philosophy – at N.Y.U.[274] It was most disappointing, and rather boring, and I do not think a single member of his audience was impressed. He dragged in things by the hair, most of it irrelevant, and used the results of Tarski and Goedel in what seemed to me a most slippery and almost scandalous fashion. He said he was going to give a talk at Yale before he sailed, so that perhaps by now you have heard him give the same talk. I'll be eager to learn what you thought of it.

Our love to you all. In great haste,
Ernest

[271] See note above in letter 53.

[272] Nagel wrote an essay entitled "Dewey s Theory of Natural Science" (Nagel 1950a).

[273] Morton White's essay "The Analytic and the Synthetic: an Untenable Dualism" was published in the same volume about John Dewey (White 1950).

[274] See letter 52 above.

4.1.56 *Nagel to Hempel, May 17, 1950*

[Typewritten, Hempel 1999, Box 28, Folder 1]

May 17, 1950.

Dear Peter,

After you hung up, I found Irwin Edman;[275] he said he would call you back right away and explain more clearly the situation. But I had to leave, and do not know whether he managed to reach you.

But in any case, he said that if you cared to teach only one semester, that would be quite all right – he suggested that in that case, perhaps the spring semester would be preferable, but agreed that the question should be left to you to decide.

As to salary, he also agreed that the minimum which should be offered you is $1,000. for a one-semester course, although he did not feel entirely sure that he could get that from the administration. He said he would ask for at least that.

Perhaps this information may help you to come to a decision.

I am sorry that Yale is not able (or willing) to grant you an immediate promotion; but I guess you are right in deciding to remain at Yale in spite of the lower salary and rank. It seems to me that they are rather stupid, or perhaps just irresponsible in taking the chance of your leaving them – where would they be able to get anyone a quarter as good as you are to replace you?

Does your decision not to go to the Rand this summer mean that you will be doing some work with Oppenheim? I hope not. You have had the patience of Job in working with him so many years.

I can barely hold on till the end of this semester – I feel pretty draggy, and just sick of the rush and endless chores. I was down to Bryn Mawr last Saturday for a meeting of the Fullerton Club. It was quite pleasant, and enjoyed meeting some old faces, and especially Nelson Goodman.

I am rushing off now to my class at the New School.[276] Perhaps we can see you before we go off for the summer. Meanwhile, our love to you and Diane.

Ernest

4.1.57 *Nagel to Hempel, July 3, 1950*

[Typewritten, Hempel 1999, Box 28, Folder 1]

South Wardsboro, Vt.
July 3, 1950.

Dear Peter,

I suppose you received the message last Tuesday that I was in North Haven and tried to see you. I am terribly sorry that I was unsuccessful, but circumstances were such that I could

[275] Irwin Edman (1896–1954) was an American philosopher. After receiving his PhD from Columbia University, he taught there for most part of his career. He wrote extensively on many different Western philosophers and thinkers (Plato, Machiavelli, Schopenhauer). His main field of research was the philosophy of art.

[276] Nagel lectured on "Philosophy of Science" and "Logic in Practice" during the spring of 1950 at the New School (*New School Bulletin*, January 2, 1950, pp. 51–52) Seemingly Nagel first taught at the New School in 1945 a course on "Logic in Practice" (*New School Bulletin*, September 1945, p. 87); later on, he lectured on the same topic in 1947 and 1948.

not drop you a card in advance to tell you that I would try to stop at your place on my way back to New York from Vermont. Indeed, we've had quite a peck of troubles since I talked to you on the phone two weeks ago. I left with the family on June 19th as we planned, and the trip up went smoothly; but when we arrived we found that the hot-water system in the house was ruined – though I went through the usual routine last fall in draining the water out, I must have been careless and some must have been left in the hot-water tank and heating coils, with the inevitable result – tank and coils burst wide open. We immediately ordered another appliance from a local man, and he promised to get it in within a week. But two days afterwards Bobbie came down with an illness, and ran a temperature of 104.8 – this happened at 9:30 P.M., and I was frantic and went into a tail-spin. It turned out that Bobbie had nothing more serious than tonsillitis, and by Sunday he was almost normal. Then on the following day I had word that the local man could not get a heating system, and did not expect to do so indefinitely; and so I had to make sudden decision to return to New York, and in addition to bringing up the rest of our stuff try to buy the appliances there and have them shipped up. I therefore had to leave in a hurry, and did not have time to drop you a card. When I got to North Haven, I inquired at the Post Office where Cooper Road was, but the directions were so complicated that I got lost. I would have persisted, however, and would even have tried to find you in your office if the car had not decided to act up – the radiator leaked badly and over-heated, and I thought it best to take no chances but get back to New York before it got too late and have things attended to. I had to hobble along on the Parkway, and got home fairly late as it was.

So you see we've had our hands full, and I hope the jinx will let up now and allow us to have a decent summer. The children are now in good condition, though the hot-water system is not yet repaired – the unit will not arrive until later this week, and I don't know when it will be finally installed.

I had a pretty busy time in New York, and still feel the effects of the rush. I saw a couple of doctors – Mack and also my eye specialist, and got a pretty clean bill of health. Mack found me underweight, and recommended that I eat and play more, but otherwise the ear is almost healed up. And I think that if new worries do not arise I ought to get back into decent physical shape in a week or two.

I finally decided not to go to California. It was very tempting to say "yes", but even apart from the problem of what to do about Edith being left alone for at least two weeks I think it was a wise decision to remain here. I've already lost a month of good time because of my miserable body, and I know that I would have been eventually quite miserable if I had squandered another month out of my life. I have something like a half-dozen books for which reviews are long overdue – I had hoped to get this out of the way during May and June; and once this chore is done I'll have to strain every muscle to get down to the real job before me.

I would have so liked to see you and Diane and the children, and have a good talk – we really haven't had one for more than a year. On the trip back to Vermont I had Meyer Schapiro's wife for a passenger, and had to make time to get her to Brattleboro to make a bus to her own place; and so I could not stop in New Haven. Maybe in the fall things will be less hectic, and I could try again to see you in your new place.

You know perhaps that the Hooks are in California this summer – Sidney is teaching at Santa Barbara, though he took a week out to fly to Berlin. The Munitzes[277] have their place here, and it is nice for us to have them as neighbors. I talked to Albert Hofstadter[278] on the

[277] Milton K. Munitz (1913–1995) was an American philosopher. He received his doctorate in philosophy from Columbia University and then taught at New York University before moving to CUNY. His work was mainly devoted to discussing the philosophical implications of cosmology.

[278] Albert Hofstadter (1910–1989) was an American philosopher. He taught at Columbia University, the University of California of Santa Cruz, and the New School for Social Research. He was a Heidegger scholar and translated some of his works into English. He also wrote extensively on the philosophy of art.

phone while in New York last week – I forget whether I wrote you that he is coming to Columbia permanently – and he is very happy. They are buying a house in Larchmont, and though there will be financial headaches it will solve lots of problems for them, including the schooling question for their boy Mark.

Edith is planning to resume work next fall, if things go all right. It will be a half-time position, doing some library research for one of the research teams at Columbia, so that her program will be flexible and she will not feel that she is neglecting the boys. Sandy is starting in kindergarten at the New Lincoln School, which will take him out of the house from 9 to 3, and Bobbie will be going to the Columbia nursery school from 10 to 12. The tuition for Sandy is something out of this world – the tuition alone, without lunches and bus-transportation is $585. for the first year and the rates go up in the higher grades – and I don't quite know how I'll make ends meet, especially if there are going to be more unexpected doctor's bills to pay. But Edith is looking forward to the fall and to the chance for doing adult work once more. But her salary will just about cover the additional expense of having some one come in and take care of Bobbie while she is away.

I'll be starting to take some pictures of the boys pretty soon, and will send you one if successful. If you have one of your family lying around, please don't forget us.

I'll write soon again, when things have settled down. Meanwhile, our warmest love to you and Diane.

Ernest

4.1.58 Hempel to Nagel, July 6, 1950

[Typewritten, Nagel 2020, Box 3, Folder 4]

July 6, 1950

Dear Ernest:

Diane and I regretted very much having missed your visit by such a narrow margin; but we regretted even more all the adverse events which finally prevented you from coming! You really had an awful lot of bad luck and of scares of one sort or another, and we both hope very eagerly that the rest of the summer will compensate you for all the trouble by all-around enjoyable, relaxingly and profitable stay in Vermont. And please, do try to drop in on us at the next possible opportunity, we are most anxious to see you all; and of course, I would thoroughly enjoy a chance for some short talk with you on that occasion.

We are enjoying our house a good deal, although it is still unfinished and continues to be the source of ever new aggravation, especially through the irresponsibility – if not outright dishonestly of our architect. Last Friday, for example, we finally received the sliding doors for our many closets, for which we had been waiting for months (the architect kept telling us they were on order, but when I finally checked with the firm in NY, they did not even know his name, and much less had any orders from him on file. Well that was 5 or 6 weeks ago); and now it turns out they are all one foot [[?]][279] short, and the architect tries to convince us that this was intended [[?]] we should simply close up permanently the top part of the closet. Now we have to fight this out with him, which means another probable long delay, a prolongation of the period during which the house looks disarrayed inside, and also of the time before the bank will give us our next mortgage payment. We still have wires sticking out of the ceiling in the living room, where some special fixture is to be installed, we still have to put up towel racks etc., and I am reluctant to spend too much time on these things.

[279] [[The right margin of the letter is missing; thus some words are unreadable.]]

Some of the other members of the group seem to devote a great deal of their time to the completion and improvement of their houses, but I do not want this sort of thing to interfere too much with my regular work, and surely do not want it to become my major project for the summer. Since so much about the house still needs being done, I have begun going to the office again rather than stay at home and be interrupted by phone calls, visitors (who drop in all the time, including students), the children, etc. Eventually, of course, I hope to be able to use the study regularly at least at night. – The location is simply wonderful with a view ranging over miles of hills and valleys, and with a pleasant breeze even on hot days. Our land in front of the house, and also for some 100 feet behind it, has been landscaped and grass has been seeded, which has just begun to come up, so that the outside of the house now looks pleasant, though somewhat austere, in the absence of flowers and trees. I did transplant one fairly large dogwood tree with the help of one of my students, and it seems to be pulling through after a period of alarming withering. I have also built an outdoor incinerator out of cinderblocks and hope it will be possible for us to burn our garbage in it; it is collected once a week but in the summer develops a terrific smell and attracts flies and finally teems with maggots, so that we'd rather get rid of it quick. We now wish we had installed a disposal, as many of our neighbors did; but it seems now too late for that since it calls for a larger septic tank. We still have to learn about the care of septic tanks, f. ex., that no water containing detergents rather than soapsuds must be allowed to run into septic tanks because it would destroy the particular bacteria (if that is what it is) which exercise some sanitary function in the things. On the other hand, it is allright to pour detergents down the drain that is connected with the drywell for our kitchen. We have also made the discovery that one diaper, flushed inadvertently down the toilet can put the two bathrooms completely out of commission and we have had to pay a flat 30,- to have it removed from the drain pipe by the roto-rooter people after our plumber had spent 10,- worth of his time trying in vain to clear the drain!

The children are both very well, and especially Peter is having the time of his life at the new house. There are about 20 lively and bright children in the project (5 houses in use, 3 a-building), and Diana and I think that his relations with them mean much more than merely his having a good time etc.; he learns how to get along with other children and is getting less timid; he also goes swimming (or rather wading) with the children of one colleague up there, who takes the children to a nearby lake. The school, in which he spent some 3 weeks before it closed for the summer, is not only very close by (3 min walk), but is the largest, entirely modern, school near and far, and has competent instructors and a very pleasant atmosphere. The children get their lunch in school, at 25 cents a meal, and the meals are quite good, including occasionally turkey sandwiches, and often ice cream etc. Diane and I appreciate all this doubly in view of the exorbitant cost of the Lincoln School you mention in your letter. I suppose your public schools are not satisfactory? This is generally said also of New Haven schools, and many of the faculty members at Yale send their children to private schools, but for the time being, we do not seem to have to contemplate this course. And North Haven is now planning the construction of a similarly modern high school not far from the public school PA is now attending. Toby is just delightful and looks a picture of health, somewhat like the peasant children in 15th century Flemish paintings. She is very lively, but still does not walk, and even in standing she holds on to the railing of her crib. Her resemblance to me – and even more to my mother, esp'y in my mother's baby pictures – is amazing, and even more so is the fact that she sometimes makes faces, when she tastes something new, which on such occasions; it is almost eerie to see that, because it is such a very characteristic facial expression, and I don't have it, so that there seems to be an inheritance of some subtle trait. There is one form of immortality almost, isn't it?

You must have been terribly worried, you and Edith, about Bobbie's sudden illness; we know how such things take it out of one, and we are glad it is all a thing of the past now. – Good that you seem to be getting into a more decent physical condition; it was really high time, and I hope you'll allow yourself some rest before you get at all those reviews and things. I can't help feeling that the decision not to go to California was well-advised both for you and for me. It seems to me that the preparation looked intelligent and thorough, but much of the success would now depend on how conscientiously every participant will live up to his obligations; and from talks with Rulon Wells,[280] I gathered that there was to be a comprehensive exchange of ideas, through circular letter, prior to the conference, and that also every participant was to prepare a list of his own publications, of which then all the other participants were expected to read as much as possible (this does not seem to me a very good idea). Gilpatric[281] called me once more after I had written him I could not accept the invitation, to see whether I might not change my mind; but it seems clear to me that even a free trip to Calif, plus a probably stimulating and enjoyable week, does not justify so much of an expenditure of time and energy if one has so many other things to do. My progress is excruciatingly slow; I am still working on a written version of my Boston paper; whenever I get a chance for work of my own, I seem to develop serious mental blocks, and during the academic year, when teaching takes first place, I always think that if only I had a few weeks, I could write lots of worthwhile things. Also, I have been seeing students rather regularly; among other things, I have had many long talks with Grünbaum[282] on his thesis, which he is completing this summer; and while it is interesting, it gets to be a nuisance to have to spend time on these things during the vacation.

Many thanks for your latest reprints, Ernest (the memorial address on Kaufmann and the Philo Sci contribution);[283] I am sorry to say that I did not order any reprints of my own note in Ph Sci;[284] I think I kept waiting for information about the possibility of getting reprints of the whole symposium, but I do not even know whom I had asked about it. Well, no great loss, to be sure.

In a few weeks, I hope to get busy with the revision of the concept formation monograph. You have already given me a number of very helpful comments and suggestions on it; if there should be any further ideas that may have come to your mind during the past term, I would be much interested in them. Did Lazarsfeld have anything in mind that I had better consider? Perhaps I should drop him a note to ask him? If you think this advisable – could you by any chance tell me where he is this summer?

Off now or I'll miss my bus home!

Love from all of us to all of you!

Peter

---

[280] Rulon S. Wells III (1919–2008) was an American linguist and philosopher. He received his PhD from Harvard and was appointed professor of linguistics and philosophy at Yale.

[281] Chadbourne Gilpatric (1914–1989) was an official of the Rockefeller Foundation. He served there as associate director of humanities and social sciences.

[282] Adolf Grünbaum (1923–2018) was a German-born philosopher who emigrated with his family to the USA in 1938. He wrote a PhD thesis on Zeno's paradoxes under the supervision of Hempel. The dissertation was later published as a book called *Modern Science and Zeno's Paradoxes* (1967). Grünbaum was the Andrew Mellon Professor of Philosophy of Science at the University of Pittsburgh, where he founded the Center for Philosophy of Science. He worked extensively on the philosophy of space and time and wrote an influential book on *The Foundations of Psychoanalysis* (1984).

[283] Nagel wrote a "Memorial Address for Felix Kaufmann", which was published in *Philosophy and Phenomenological Research* (Nagel 1950b) and was originally read at the New School for Social Research on January 4, 1950. The "Philo Sci contribution" is "Science and Semantic Realism", published in *Philosophy of Science* (Nagel 1950c).

[284] Hempel participated in the same symposium as Nagel, which was devoted to Feigl's influential paper on "Existential Hypotheses". Hempel published his article entitled "A Note on Semantic Realism" (Hempel 1950b). The other contributors are C. West Churchmann ("Logical Reconstructionism"); Philipp Frank ("Comments on Realistic versus Phenomenalistic Interpretations"); A. G. Ramsperger ("On Feigl's "Existential Hypotheses"); and Herbert Feigl ("Logical Reconstruction, Realism and Pure Semiotic").

Nelson and Katherine Goodman stopped by on their way to Boston; they are leaving for Europe soon from Canada. The Quines are in Europe already. Have you seen Quine's new logic book?[285] I think I'll use it for an introd. graduate course next fall. Olaf Helmer expects to spend some weeks in August in New England, then to attend the Math Congress at Cambridge. We also hope to see Feigl soon. I knew about Hofstadter, and was genuinely delighted when I heard about his joining your department. – Aren't the Korea events bitterly discouraging in many ways?[286] And what do you think will the outcome of it all be? Is this "merely" another Spain?

4.1.59 Hempel to Nagel, September 5, 1950

[Typewritten, Nagel 2020, Box 3, Folder 4]

September 5, 1950.
9 Cooper Road
North Haven, Conn.

Dear Ernest:

As usual, the summer has gone by much too fast for my liking; two weeks from tomorrow, our classes start here at Yale, and a week later, I am to begin my course at Columbia already!

We have had a good summer and have enjoyed our house, although the progress on its final stages has been very slow indeed; in fact, it is not completed even now: we are still without some lighting fixtures, floor covering in some rooms, and closed outdoor storage space. But gradually I have got used to the slow pace and don't get too excited any more. The cost has become pretty staggering, and there is still no end in sight: we will have to have the four outside doors weather stripped, then some sort of quasi-drainpipe has to be put on the roof to keep the water from coming down just over our entrance door, etc. Still, we would be reluctant to sell the place, for it is comfortable to live in, affords maximum privacy for all of us, is beautifully located, and it has all the advantages which result from the particular character of this community: all the people in the group (three houses are still a-building; five are in use) are very congenial and cooperative without being obtrusive; the children (there will be 20 of them when all the houses are occupied) are all very pleasant, and for them it is an ideal setting; Peter plays outdoors almost all day; sometimes he plays in one of the other houses or brings his friends into ours; the children get lunch wherever they happen to be; there is none of the conspicuous consumption which we noticed among the children on Anthony Street, where Peter had to have guns and cowboy hats and Good Humor sticks and what not because all the other children in the neighborhood had it – while now, none of his friends totes a gun or even boasts a cowboy hat, and no Good Humor man puts in his appetite-stifling pres-supper appearances.

We had a lot of visitors during the summer, including Felix Oppenheim and his wife, Olaf Helmer and his prospective wife Maggie (who seems to be a very pleasant person and rather warmer than Lou), Herbert Feigl alone and also with Maria (he enjoyed his work at Columbia a great deal and also the stay in New York; Maria seemed to be in fine shape and is as slender as ever; the two drove up to Quebec for a little vacation before returning to Mpls), and the Lamonts with their little daughter Tessa – they are old friends of Diane, and I like them very much. They lived in NY until some two years ago and are now in Easton, Pa. – We heard from the Goodmans, who were enjoying their trip enormously. I suppose they ought to

[285] Quine published his book *Methods of Logic* in 1950 (see Quine 1950).

[286] North Korea invaded South Korea on June 25, 1950; military events escalated quickly between the Soviet Union-supported North Korea and the United Nations Command-supported South. As the UN consisted mainly of American military personnel, the States entered the war later in September 1950.

be back soon. – No news from the Carnaps, I hope he has not had a recurrence of his back trouble. No doubt you have seen his new probability book;[287] I find it very impressive indeed. – John Goheen wrote a number of letters from Japan, where he had gone with a group of colleagues from Stanford to give lectures and seminars on recent American philosophy. He was enthusiastic in many ways. By now, he must be back in Palo Alto, where they have found a very comfortable house and a beautiful garden.

My work, especially during the first half of the summer, went painfully slowly; I often felt myself completely blocked in trying to write; but finally, I got the revised text of my Boston paper written (wonder when and where the proceedings are going to be published; I am quite hazy about this) and then started revising the ms on concept formation. I found I had to rewrite a good deal of it, but at least it seems to me that it is now in a much better shape than the first version. I am almost through with the job, but then still have to check a long list of memoranda to see whether the reminders I wrote down for myself at various times have been taken into consideration, and then I will have to type my handwritten text before a clean copy can by typed (I hope by our departmental sec'y). – Even during the summer, I was kept busy now and then with a thesis – Adolf Grünbaum is finishing his dissertation and kept bringing it in to me in installments for comments and suggestions; that took up quite a few afternoons and evenings. – I have not yet received the ms by Bertalanffy[288] which I am to discuss at the Philosophy meeting in Toronto.

Rulon Wells told me about the semantics meeting in California, and I think I have no reason to regret not having attended it. Not as if Rulon had given an unfavorable account; but I sense that was not too exciting, and he said upon my question that no ideas had been put forward by anyone which the speaker had not previously advanced in speeches or in writing. I would be very eager to get Morty's side of the story. Well, as you know, reports of the proceeding are in the process of being written up, and we will probably see tho[n/a].[289]

May I bother you with a few questions concerning my course at Columbia? It would help me a lot in my thinking about it if I knew just about how many students I have to count on (hence: whether it will be more or less a lecture course or whether it is to have the character of a seminar), and what the preparation of the regularly enrolled students is likely to be; also on what evidence the course marks are based (is it customary to have mid-term written exams, for example? Term papers? Finals?); and what steps I should take in order to have certain books put on reserve in the library (I suppose there is the system of reserved bookshelves for the different courses?). [[n/a]] it customary to require a text – and how much money can one expect a [[n/a]] student to spend for the purpose? (I want them all to have your monograph, which I also use in the same course here at Yale; beyond that, it is hard to think of a book of which I would use a sufficiently large chunk to justify the expense for the students. Reichenbach,[290] Carnap, Kneale,[291] Keynes,[292] Williams,[293] Mises,

[287] Carnap's book *Logical Foundations of Probability* was published in 1950.

[288] Ludwig von Bertalanffy (1901–1972) was an Austrian biologist. He was one of the founders of general systems theory. His most important book is *General System Theory: Foundations, Development, Applications* (1968).

[289] The side of the letter is missing here, and later in the next paragraph.

[290] Reichenbach (1949), Carnap (1950), von Mises (1939).

[291] William Kneale (1906–1990) was an English logician and historian of logics. He was appointed White's Professor of Moral Philosophy at Oxford. He is best known for his book *The Development of Logic* (1962) write with her wife, Martha Kneale. For his book on induction, see Kneale (1949).

[292] John Maynard Keynes (1883–1946) was an English economist and philosopher. He studied at Cambridge and then became a fellow of King's College. He is best known for his books *The Economic Consequences of the Peace* (1919), *A Treatise on Probability* (1921), and *A Treatise on Money* (1930).

[293] Donald Cary Williams (1899–1983) was an American philosopher. He received his PhD from Harvard University, where he also taught from 1939 to 1967. He published a book on *The Ground of Induction* (1947) and was influential in renewing debates on analytic metaphysics, particularly through his two articles "On the Elements of Being", I and II, published in the *Review of Metaphysics* in 1953.

Kolmogoroff[294] – I want to refer the students to all of these books, and specifically to certain sections in them, but of course, I do not follow any one of them. What have *you* done in the past?) And if you shld have any other suggestions, I shall be very grateful for them all!

Diane and I have thought and spoken of you all many times; we hope you have had a pleasant summer, too, and that your health, Ernest [[n/a]] is now much improved! How long are you planning to stay in Vermont? Are we going to see you all on your way back? We would enjoy it so much!

Love and best wishes from all of us to all of you!

Peter

4.1.60 Nagel to Hempel, September 7, 1950

[Typewritten, Hempel 1999, Box 28, Folder 1]

South Wardsboro, Vt.
Sept. 7, 1950.

Dear Peter,

I feel very much ashamed of myself for letting the whole summer go by without writing you. But somehow time has just galloped by, and I haven't been able to do a fraction of the things I had hoped to. It is hard to realize that we'll be on our way back to New York in a week from to-morrow at the latest.

It has been a pleasant summer, even though we did have a long streak of bad luck. The latest is that during the terrific downpour of a week ago our roof developed leaks all over the place, and we collected rain water in the living room. It is not something that I can fix myself – I have patched up small leaks, even though I do not feel sure-footed even on the low-pitched part of the roof – and the man who promised to come and do it has not appeared thus far. I don't know whether the job will be done before we leave. But the children have been fine, except for Bobby's illness early in the summer, and they have been quite happy. Sandy is eager to get back to New York, for he is starting kindergarten at the New Lincoln School. (I've paid the first half of the tuition, lunch and transportation charges, and it is a terrific drain!) Bobby is quite a boy, and now is beginning to talk in sentences. None of the snap-shots I took this summer have been successful, otherwise I would send you one. Edith, too, has found the country a great relaxation, though at the moment she has some trouble with her sinus. It is going to be a hard year for her. Before we left New York, she made arrangements with one research group at Columbia to go part-time work for them, of the kind she did a year ago for the Bureau of Standards – it was a convenient arrangement and quite flexible as far as time is concerned. But 3 weeks ago she had a letter from the chairman of the physics dept at Barnard, asking her whether she should be interested in doing part-time teaching at Barnard; and after much calculation on our part she finally agreed to do so. The hours are still convenient – most of them are in the forenoon, when Bobby will be in nursery school and Sandy in kindergarten – but obviously not as flexible as in the case of the other job; and if one of the children becomes unwell, we'll have to count on getting a sitter for a few hours. But it seemed to us like a great opportunity to get a footing in Barnard, and this job does not exclude her doing a certain amount of research work in connection with the other one. Edith wants very much to return to physics in a serious way, and at the same time

[294] Andrej Nikolaevič Kolmogorov (1903–1987) was a Soviet mathematician. He worked extensively on the theory of probability; in his book *Foundations of the Theory of Probability* (1933) he set out the modern axiomatic foundations of probability theory.

she also feels that her main responsibility is to the children. So it will be tough going for a while, and you must wish us luck.

I've had only a moderately successful summer. June was killed because of my illness, and most of July was taken up with getting rid of the pile of book-reviews that had mounted up. I've begun working on the book since early August, but progress has been depressingly slow. I've been at work on a very long chapter on explanation, and the end is not yet insight; the stuff I had written some years ago is just deplorable, and I am afraid what I am doing now is not much better: it is neither fish nor fowl, and says nothing that I haven't seen better stated in any number of places. So I have plenty of hard work ahead of me, but whether by the end of the year I'll be anywhere near finished is becoming progressively more doubtful.

I was interested in your brief account of the California meeting. I had no regrets that I turned the invitation down, and what you report certainly does not change my feelings. But I'll be eager to see that full report of what went on, and I expect that Morty will be able to give us lots of interesting asides.[295]

I hope we'll be able to see something of you when you come down to New York this year, in connection with your course at Columbia. The course, if I remember correctly, is listed as a "one-hundred" course (the actual number is, I think, Philos. 107), and one-hundred courses are intended to be primarily lecture-courses, though the individual instructor has a completely free hand in running it anyway he likes. My own feeling is that you ought to do what is easiest for yourself, compatible with the size of the class. Unfortunately, it is not possible to tell in advance how many students you'll have, though my guess is that there will be *at least* 30 registered students plus a goodly number of auditors. If my guess is right, it might prove to be too difficult to run the course as a seminar-discussion group, though I've never hesitated to invite questions and discussions even in classes of 80. Most of your students will be straight philosophy graduate students, with perhaps a small contingent from the mathematics department and a handful of advanced undergraduates. Some of them, of course, will have a considerable background both in logic and mathematics; but most of them will have no previous familiarity with the materials of the course, and I think it would be wise to assume very little as far as the equipment of the students is concerned. I have always asked the class at the first meeting to fill out a slip with information about previous training, and I have tried to adjust the rest of the course to what I find out. Books are placed on reference in the Philosophy Library, situated in the (new) Butler Library; you ought to draw up a reference list, and send it to the Librarian, Philosophy Library, Butler Library, mentioning the fact that you are giving the course this year. It is not usual to ask graduate students to buy any texts – at least not in my department – but there is no reason for you to be bound by precedent if you want them to have a text. I imagine the chief difficulty in this connection would be to find a text that covers at least approximately the materials you plan to discuss, and also one which is not prohibitive in price – the cost of both Reichenbach and Carnap books seems to me to make them beyond the reach of most students. I myself have never used a text in a graduate course, and have simply referred students to the relevant literature. Instructors vary in the basis they use for grading in graduate courses, especially since one-hundred courses can be taken for so-called H or attendance credit, in which case the student is supposedly not responsible for anything except attendance (which no one checks). (In two and three-hundred courses, some written work is required). Several devices are widely used: oral reports (which I do not like, for they are usually a waste of time, especially in larger classes), final examinations based on the lectures and assigned reading, and a written term report on some subject agreed upon. I have tended to use the latter method (except when the class is so large that I can get a reader), partly because in the case of an inhomogeneous group it is difficult to set a final examination which is not either too easy or too hard, and partly because having to do a term paper

[295] See letter 59 above.

compels students to do some independent reading and to become familiar with literature that otherwise they might not encounter. It seems to me that you ought to save yourself unnecessary wear and tear and to give the course pretty much in the way you have done it previously. I am quite sure it will go over with a bang.

By the way, you ought to have an office, and I see no reason why you should not use mine. I don't know when Irwin Edman will be back, but one of the dept. secretaries will be able to take care of this.

I have been reading in the Carnap book, and as you say it is quite impressive, even though I have serious reservations about the whole approach.[296] I am curious to see the reviews – I will *not* review it myself – and wonder also just how things will work out in the second volume. I don't know whether I already wrote you that I did a short review of the Reichenbach, and a somewhat longer one of the Kneale.[297] I found the R. in some ways ingenious, but just pig-headed; and I'll be surprised if he does not send in a reply. Contrary to my first impression, the Kneale seemed to me quite an interesting and suggestive book, even though there is much in it that is not convincing. Have you seen Feller's book on probability?[298] I've merely had it in my hand; it is a technical development of the subject, but the first chapter seemed to me to have general interest and I'll have to read it when I get back to New York.

We were very much interested in your account of your house, and are eager to see it. But it is problematic whether we can stop on the way down – I should say the chances are definitely against it – and we'll have to make a special occasion for a visit. We haven't seen Diane for years now, and of course Toby not at all; and we hope to do both before very long. But if we should not see you before you make your first trip to Columbia, will you give us a ring and perhaps you'll even find a few moments to drop in.

Meanwhile, our love to all of you.

Ernest.

4.1.61 Hempel to Nagel, September 15, 1950

[Typewritten, Hempel 1999, Box 28, Folder 1]

September 15, 1950

Dear Ernest:

Thanks lots for your interesting letter and your most helpful information and advice on my course at Columbia. The librarian obligingly anticipated my wishes and spontaneously sent me a form to list the books I want to have put on reserve, and I also had a communication from a Miss Peters in your Department concerning texts that could be purchased by the veterans under the GI bill; here, I gave them your monograph. – I should be very happy if you would let me use your office; many thanks for offering me hospitality! I shall follow your suggestion and get in touch with one of your secretaries – maybe Miss Peters – about it. But, come to think of it, don't *you* want to use the office, or can you study undisturbed at

[296] See Carnap's *Logical Foundations of Probability*.

[297] Nagel reviewed William Kneale's *Probability and Induction* and the 1949 English translation of Reichenbach's 1935 book on probability—entitled *The Theory of Probability. An Inquiry into the Logical and Mathematical Foundations of the Calculus of Probability*—in *the Journal of Philosophy*, see Nagel (1950d, e).

[298] William Feller (1906–1970) was a Croatian-American mathematician, known mainly for his work on probability. For his book, see Feller (1950).

home (as I can not)? I shall let you know, at any rate, just at what hours I would presumably be there (it depends somewhat on the forthcoming winter schedule of the New Haven RR), but it would be for about an hour before and possible a short while after my class, which is scheduled for 4.10–6 p.m. on Wednesdays.

Your reviews of Kneale and Reichenbach have not appeared as yet, I suppose?[299] At least I have not come across them so far, and I am very eager to see them. I had not heard about Feller's book[300] at all and am curious to see it.

Diane and I want to congratulate Edith most sincerely upon her teaching appointment at Barnard. We can imagine how exciting the prospect must be for Edith, and we hope very much that it will work out to her fullest satisfaction. Perhaps I will hear about Edith's first impressions when I give you a ring on my first visit to Columbia; and of course, I hope to have a chance to drop in on you before very long for a little chat.

I hope you will have a good year, Ernest, and won't be drawn into lots of University matters with which you should not be bothered during your leave. I am eager to hear more about your book, about which you always write so disparagingly and in such a discouraged spirit; well, wait until it is out, to see the reception it will get, and see how you'll be rolling in dough, too! By the way, did you notice that James Newman included the Cohen and Nagel book in his list of 20 outstanding works in scientific method and in popularized science (in the September issue of the Scientific American).[301]

The Goodmans stayed with us for a shortish overnight visit; they seemed very pleased indeed with the trip to England and appeared to be well rested and chipper. – I heard that Van Quine's going to England for the Spring term is not quite sure now because of the peculiar dates of the British semesters which conflict with the Harvard semesters.

If you can manage, please stop over and see our house! And the children, too! Toby is just beginning to walk (quite late, but she is strong and very active, and gets about a lot by sliding on her fanny with amazing speed and skill.); P.A. has started in 2nd grade and already reads quite well. Love to you all from us all,

4.1.62 Nagel to Hempel, September 1, 1951

[Typewritten, Hempel 1999, Box 28, Folder 1]

South Wardsboro, Vt.
Sept. 1, 1951.

Dear Peter,

It was good to get your letter, though it made me even more shamefaced for not having given any sign of life all summer. I mislaid your first letter from en route, and did not remember when you would be back in North Haven – for some reason I thought you would not return until the end of July, and then one thing after another kept my mind away from correspondence. And now the end of our stay here is in sight – we plan to leave around the

[299] See letter 60 above.

[300] Feller (1950).

[301] James R. Newman (1907–1966) was an American mathematician and historian of mathematics who served as member of the Boards of Editors for the magazine *Scientific American*. In the September Issue of 1950 he published an article entitled "BOOKS: the 50 years and the popularization of science: a listing of 20 good examples": Cohen and Nagel's book *Introduction to Logic and the Scientific Method* is described as "a masterly textbook, but actually much more than a textbook" (1950, 103).

12th – fall weather is upon us, and I feel very remiss in letting the summer go by in so silent a manner.

Your trip to [P]oints [W]est sounds wonderful, and makes me want more than ever to see something of this country before very long. The trouble is that I could not afford to take the whole family, and I could travel alone if at least part of the expenses were covered by some institution; and I think I would not consider the latter unless I could take everyone along too. So it is not likely I'll see the Gold Coast for a long while. Your brief report of the state of the Carnaps makes me feel quite sad, and it is a great pity for a number of reasons that they are located in such a remote part of the country. I'll be eager to read the monograph you announce – when is it to appear, and is it a substitute for Vol. II?[302] But perhaps I'll have a chance to see you in New York sometime this fall, and you'll be able to tell me more about it.

The Rand group must be quite stimulating. I know Girschick[303] – he is a former student – and heard a good deal about Blackwell;[304] and it is too bad G. turned down the Columbia offer (about which I had heard), for our statistical dept. is now in a very bad way – Wolfowitz[305] accepted an offer from Cornell to replace Feller (who went to Princeton), and what remains of the department is at best a bunch of second-raters. With Wald's death and Wolfowitz's leaving I feel that I have lost not only good friends but important channels for ideas. I am not sure that you ought not to consider seriously Olaf's invitation to spend a half-year at the Rand. Your position at Yale is quite secure, and taking a leave to act as a consultant for the Rand can certainly do you no harm – on the contrary. The only thing I would suggest is that you go over your calculations as to finances carefully – by the time one counts costs of travelling, contributions to the annuity fund (which I assume you would want to make even while on leave from Yale), property tax, etc. even a superficially attractive offer to visit usually begins to take on another appearance. A year ago I was asked to visit at the U. of Illinois for a half year, and if I remember rightly, the salary was to be $6000. for the period – but I found that I would be out quite a sum if I took it, despite the fact that the pay at first seemed fabulously high.

It is a shame that your Encyc. pamphlet will not be out early next year. I am eager to see it in its revised form, and expect to profit much from it.

We too have had a very pleasant summer, though I've done very little serious work on the book. The first two weeks here Edith and I spent painting floors in the living room, kitchen and bath-room, while the two boys were away at a farm near Kingston, N.Y. Edith needed the rest from them, and things worked out pretty well; but we were glad to have them back with us. They've flourished here, and Edith has been quite happy. We went into

[302]Presumably Carnap (1952), that was meant to be part later of the conceived volume II of Carnap's studies on probability.

[303]Meyer Abraham Girshick (1908–1955) was a Russian-born statistician who emigrated to the USA in 1922. He studied at Columbia, where he also received his PhD under the supervision of Harold Hotelling. After a career in government service, he was appointed Professor of Statistics at Stanford University in 1948.

[304]David Harold Blackwell (1919–2010) was an American statistician and mathematician. He received his PhD from the University of Illinois at Urbana-Champaign. An African American, he worked at several historically black colleges and universities before being hired at the University of California, Berkeley. He made significant contributions to the fields of game theory, theory of probability, and Bayesian statistics.

[305]Jacob Wolfowitz (1910–1981) was a Polish statistician who immigrated to the USA in 1920. He received his PhD from New York University before joining the faculty of the Statistics Department at Columbia University, where he worked with Abraham Wald. In 1951 he accepted an offer from Cornell University.

partnership with the Hooks and the Finkels (my sister and brother-in-law)[306] in getting a small garden tractor, which can mow the hay-fields, cut brush, and also plow. It is now next to impossible to get any help around here, and so we decided that we would have to become independent and do as much of the necessary labor to keep the places looking decent as we possibly could. It is fun to run the machine, but it is also time-consuming; and I often wish I did not have property for which one develops a sense of responsibility and so eliminate a constant source of distraction. I started out bravely enough in trying to continue the rigid routine I had established for myself in New York. I was in the midst of a longish chapter on the social sciences when we left the city, and turned to rewrite my first draft of a section on economics. But what I had seemed to me in awful shape, and so I began to read in the subject; however, I still have not been able to make the stuff jell. I spent weeks trying different ways of organizing the material, but had to throw all of it away. Finally, I decided to postpone this section, and return to it later. But all I have been able to do is to rewrite a section on history, so that I have very little to show for my pains this summer. I feel I need to read much more in economics literature, but whether my resolution to do so during the coming year can be kept is an open question. And so what I have as a result of the year off is just Part I of the book (except for the section on economics); this now runs to about 190,000 words, and this is obviously too long if I am also going to include (as I feel I must) the material I have for Parts II (roughly, concept-formation) and III (probability and induction) – these latter parts are in first draft, but badly needing rewriting. So even after I get Parts II and III into order (the lord knows when) there will be the job of cutting. I am afraid I will never see the finished book.

I've read the other [day] the galleys of a book on SCIENTIFIC INFERENCE AND ITS PRESUPPOSITIONS by John O. Wisdom[307] (a colleague of Popper at the U. of London, not the Cambridge Wisdom), which is appearing in England and which Wiley is considering publishing here. It is not a very good book, though it may be useful because it brings together some of the current discussions (in England) on induction and probability. I did not write a very enthusiastic report for the publishers, and do not know whether they will market it here. I've also just received the English translation of von Mises' book on Positivism[308] – it makes very attractive reading, and should make a good text for a course in the philosophy of science.

Early in July I spent three days at Harvard, attending a conference on science and philosophy organized by the Harvard Summer School.[309] The conference was not much, but it was pleasant meeting some old and new faces, including John Cooley, Steve,[310] and Butterfield[311] (the English historian who published a very good little book on the history of

[306] William L. and Irene Finkel, to whom Nagel (1979) was dedicated. William L. Finkel was a linguist and professor at the Speech Department of City College, originally with a PhD from Columbia.

[307] John O. Wisdom (1908–1993) was an Irish philosopher who wrote extensively on philosophy of science. There is no such book of Wisdom as *Scientific Inference and its Presuppositions*. But in 1952, Wisdom published his *Foundations of Inference in Natural Science*. See Wisdom (1952).

[308] von Mises (1951).

[309] The Harvard Summer School was held on 9, 10, and 11 July 1951, and consisted of three conferences—"Science and Society", "Science and Philosophy", and "Science and History". The Science and Philosophy conference was directed by Philippe Le Corbeiller, who was professor at Harvard.

[310] Later Nagel and Hempel refer by "Steve" to Charles Stevenson, who also attended the Harvard Summer School.

[311] Herbert Butterfield (1900–1979) was an English historian, Regius Professor of History at the University of Cambridge. The book mentioned by Nagel is likely *The Origins of Modern Science, 1300–1800* (1949).

science). Bridgman[312] read a paper, on the whole pretty poor, and Black and Stace pitched into him very vigorously. I felt that something should be said on the other side, so that I went out of my way in emphasizing the stimulating quality of much of Bridgman's analysis of scientific concepts. Black felt that I had not been altogether upright in doing so; but I still feel that Bridgman deserved some defense before such an audience.

The Hooks left about a week ago, and Sidney and Ann are now on the ocean for about a month's trip to Europe – in connection with the work of the Comm. on Cultural Freedom. So we are pretty lonely now on this hill. I wish we could drop in on you on our way home, but this is still too difficult with Bobby along – we want to reach N.Y. in time for his bed-time. But I do hope we can manage a visit soon. Meanwhile, our warmest love to you and Diane

Ernest

4.1.63 Hempel to Nagel, November 23, 1951

[Typewritten, Hempel 1999, Box 28, Folder 1][313]

Dear Ernest,

I am very sorry I never thanked you for your nice long letter you wrote us from Vermont early in September.[314] Diane and I enjoyed very much getting your news, and we both hope that your stay in Vt. was pleasant to the end, and that you all have been well during the two months that have passed since your return. I have not been in New York since mid-July – when I simply came through on my way to and from Princeton, and I really regret very much seeing so little of you all. Diane does get in from time to time, but – as you can imagine – is then always quite tied up with her family, and especially her mother, whose health is deteriorating slowly but irreversibly. We all have been well; just today, Diana began complaining of some pain in her lymphatic glands on the neck, and Peter seemed to be under the weather; I hope this will pass soon without any more serious consequences.

I have rather enjoyed my work here, including even the lectures in introductory logic twice a week to an audience of 230 students;[315] in lieu of the blackboard, which is almost invisible from the back rows of the dimly lit lecture hall, I am using a very special projector with something like a Schmidt lens system in it, which I have on my desk only two feet from the screen behind me, and which enables me to project what I write on a special kind of paper coated with black wax, while I am writing it. It works out well, except that when the thing has been functioning for a while, it still gets so warm, (despite a fan I have to out-talk) that the wax tends to melt and my writing to disappear. – What irks me is that my routine jobs take so much of my time that I get very little reading and less writing done on the side; but that is my old and by now somewhat stale story. I did turn in the Encyclopedia ms and hope the Chicago Press won't be too slow in getting it out.[316] Some weeks ago, I received a number a copies of the proceedings of the first conference held at the Inst. of Unified Science in Boston in 1950; if you have not received a copy of it, will you just drop me a post card,

[312]Percy W. Bridgman (1882–1961) was an American physicist who received the Nobel Prize for Physics in 1946. Bridgman is also known for his work in philosophy of science, as he developed an influential account of scientific concepts called *operationalism* in his book *The Logic of Modern Physics* (1927).

[313]We are grateful to Sander Verhaegh for bringing this missing letter to our attention.

[314]See letter 62.

[315]At the time, Hempel was teaching at Yale.

[316]It will be Hempel (1952).

and I'll send you one; I would send one without asking, but I want to avoid duplication since I received only 25 copies of the thing (and no separate reprints of my paper).[317]

Many thanks for the reprint of your article on the causal character of modern physical theory;[318] I have read it with the greatest enjoyment and think it is really a masterpiece of lucidity and cogency. It will be a help for me in my philos. of science course (although I suspect it may be to[o] hard for the students as an assignment; the benefit to them must come via my increased enlightenment on the subject) and on the graduate level, where it should provide an excellent basis for discussion.

I hope to attend the Bryn Mawr meeting.[319] The Goodmans have asked me to stay at their house; prob'ly the Whites and Van will be staying with them, too. The programs in Philos and in Symb.Log. look quite intriguing and promising, and I hope we'll be having two really good meetings.

I gave two lectures away from Yale this fall, one at U of Conn and the other at Amherst, where I had unusually lively and very gratifying discussion (my topic was: Science – Unlimited?).[320] I was recently asked to speak at the first meeting of the Sigma Xi club at Brooklyn College, but decided I was getting too involved in extraneous activities and declined. As perhaps you know, I did write Gregg[321] that I'd be willing to speak at the biolog[ical] colloquium at Columbia; it seemed too tempting to decline, although I felt as if I was going into some trap, since Gregg mentioned that you had so far been reluctant to commit yourself to a talk before that group. I wish I cld get some pointers from you on that matter!

Today, I received a request from the Guggenheim Foundation for statements on three candidates, Carnap, Myhill[322], Pap[323]. I surely am in favor of all of them receiving fellowships. Myhill, in my opinion, is doing well here and is really an extremely able man. How well he has established rapport with his undergraduate students (he has three discussion sections in connection with that elementary lecture course I wrote about) I simply have no way of telling. But Fitch[324] and I are auditing his grad. course on recursive functions, Gödel

317 See Volume 80, No. 1 of *Proceedings of the American Academy of Arts and Sciences*, special issue on "Contributions to the Analysis and Synthesis of Knowledge, published in July 1951. See Hempel (1951a).

318 See Nagel (1951).

319 The 48th meeting of the Eastern Division of the APA was held at Bryn Mawr College in Bryn Mawr, Pa., December 27–29, 1951. According to the program, Hempel did not give a talk. See "Eastern Division", *Proceedings and Addresses of the American Philosophical Association* 25 (1951–1952): 69–77.

320 A similarly titled paper of Hempel was published much later as Hempel (1973).

321 John Gregg (1916–2009) was an American biologist and zoologist. After graduating from the University of Alabama, he received his PhD from Princeton University. He worked at Columbia University and later became a professor of zoology at Duke. His book *The Language of Taxonomy. An Application of Symbolic Logic to the Study of Classificatory Systems* (1954) was reviewed by Hempel in the *Journal of Symbolic Logic*; see Hempel (1956).

322 John Myhill (1923–1987) was a British mathematician. He studied with Quine at Harvard, where he received his PhD, and worked at SUNY Buffalo.

323 Arthur Pap (1921–1959) was a Swiss philosopher who emigrated to the USA in 1941. He enrolled at Columbia University and then at Yale University; he received his PhD from Columbia under the supervision of Ernest Nagel. His most important work are *The a Priori in Physical Theory* (1946), *Elements of Analytic Philosophy* (1949), and the posthumous *An Introduction to the Philosophy of Science* (1962).

324 Frederic Fitch (1908–1987) was an American logician, working on combinatory and modal logic. He obtained his PhD from Yale, where he taught between 1937 and 1977.

and Church theorems, etc. Unfortunately, I find that I would have to take time out to work on my notes to get maximum benefit out of that course; it has gotten quite technical and I sometimes just get the general points, but miss some of the technical details.

Enclosed a carbon of a letter to Feigl, *re* the Carnap-Rbch volume.[325]

Our very best to you all!

As ever,

[Hempel]

4.1.64 Nagel to Hempel, February 28, 1952

[Typewritten, Hempel 1999, Box 28, Folder 1]

25 Claremont Avenue,
New York 27, N.Y.
Feb. 28, 1952.

Dear Peter,

I've been very remiss in writing you, and can only plead the stock excuse – a sense of being overwhelmed with the endless chores that seem a permanent feature of New York life. But I look forward eagerly to seeing you next Tuesday here at Columbia – John Gregg told me that you would be able to stay for dinner and I plan to join you then as well as go to your lecture.

I imagine however that there will not be much chance to talk about some private things, and in particular about the Yale situation and your plan to try to take off next year. Your problem has been very much on my mind, and I've tried to think of ways for solving it – but thus far with no results. I dropped discreet hints to Sidney Hook that you might consider an offer from N.Y.U.; and I also talked to my brother-in-law, Willie Finkel, who was President Wright's[326] ear at City College, suggesting that he mention your name to Wright – since as I think I told you, Wright would like to strengthen the Philosophy Dept at the College by making a major appointment before he retires. None of these seeds has begun to sprout as yet, and I have thought of no way of planting others.

I am of course very eager to learn whether you have heard anything yet from the Guggenheims about a renewal. I don't know whether Moe[327] has any fixed policy about renewals, and in any case Moe has not consulted my opinion in your case.

Thank you ever so much for the reprint of your Bertalanffy critique[328] – I think your points were excellently made and quite decisive. John Cooley told me the other day that you

[325] Still at the end of 1951, there was a Schilpp-volume planned for Carnap and Reichenbach to celebrate and evaluate their philosophies. But following Reichenbach's death early 1953, the volume focused only on Carnap and was published in 1963 as *The Philosophy of Rudolf Carnap* (edited by P. A. Schilpp, LaSalle Ill.: Open Court).

[326] Harry Noble Wright (1881–1969), a mathematician, who was the sixth president of the City College of New York between 1941 and 1953.

[327] Henry Allen Moe (1894–1975) was the principal administrator of the Guggenheim Memorial Foundation between 1925 and 1963.

[328] Nagel is referring to Hempel's paper on "General System Theory: A New Approach to the Unity of Science" (see Hempel 1951b), originally read at a symposium. In addition to Hempel's essay, the symposium was made up of three texts of Bertalanffy ("Problems of General System Theory", "Conclusion", and "Towards a Physical Theory of Organic Teleology: Feedback and Dynamics), and by two other essays, one by Hans Jonas ("Comment on General System Theory") and the other by Robert E. Bass ("Unity of Nature")

sent him a copy of your final draft of the Concept Formation monograph, and he is most enthusiastic about it. He is using it, so that he has not been able to lend it to me. Have you any extra copies, and if so can you spare one if only for a short period so that I may read through it?

The boys have been having the usual series of minor illnesses, but though there has been nothing seriously wrong with either of them even the minor things are drain on time and energy, as you of course know. Sandy is now undergoing a series of scratch tests, to determine what are the things to which he is allergic; and Bobby seems always teetering on the edge of running high temperatures and being dosed with terramycin. We both hope that the spring is not far away and that it will bring relief.

But I must be brief, and stop now. Our love to you and Diane.

Ernest

4.1.65 Nagel to Hempel, October 8, 1953

[Typewritten, Hempel 1999, Box 28, Folder 1]

25 Claremont Ave.,
New York 27, N.Y.
Oct. 8, 1953.

Dear Peter,

I feel quite ashamed that I haven't written before this, especially since I am writing to get some specific information from you. I had meant to get in touch with you during the summer, but just never got around to it; and since we've returned to New York it has been pretty hectic.

Let me come to the point right away. I got a letter from Ledger Wood[329] renewing the invitation of two years ago to come to Princeton next fall as a Hibben Fellow; and he makes clear that the only obligation on my part would be to give a graduate seminar, and to be around for consultation. Now I am strongly tempted to accept, on the hypothesis a) that I would have considerable time left to do some work of my own and b) that I would suffer no financial loss. Since you've been through this experience, I wonder whether you could give me a brief but clear picture of what the conditions really are like at Princeton in these respects. I have only a vague recollection of the few things you told me about your stay at Princeton – somehow we never found time to talk about it at any length – and I hope you can refresh my memory and perhaps add to what you already told me.

I would like to know, specifically, whether you were able to give most of your time while at Princeton to your writing. I would see no point in going there myself if I could get as little done there during the semester as I can in New York (which is just about nothing). Does preparation for the seminar involve a good deal of labor? Are social obligations heavy? Are there many students around who consume a good deal of time? Also, Ledger Wood writes that though no salary is stipulated, "we make every effort to provide a stipend which will involve no sacrifice for the holder of the fellowship". I suppose that this means I would be paid the equivalent of my Columbia salary during the Fellowship. But since, if I accepted, I would leave Edith and the boys here in New York, I would of course come home for week-ends. Now did Princeton pay you for travel expenses, or your living expenses while at Princeton? If I should go, the cost of keeping up the New York home would not be

[329] Ledger Wood (1901–1970) was an American philosopher who got his doctorate from Cornell and taught at Princeton University between 1927 and 1970.

significantly diminished, certainly by not enough to make up for the cost of my living and eating away from home. And I would like to know whether such differential cost can be legitimately charged to Princeton. I would be very grateful for any information or suggestions you could send.

I assume you will not be coming to the meetings of the N.Y. Philosophical Club this year, at least not regularly, so that I will see less of you this year than ever. Are you enjoying Harvard – students and faculty? And do you find time for work of your own? There is so much I would have to tell you, and get your advice and opinion; but I'll save it for another day. Edith and the boys are fine, though everybody is terribly busy, and quite fatigued by the end of the day.

I hope you and Diane and the children are flourishing. Edith joins me in sending our love to all of you.

Ernest

4.1.66 *Nagel to Hempel, October 11, 1953*

[Typwritten, Hempel 1999, Box 28, Folder 1]

Oct. 11, 1953.

Dear Peter,

Thank you every so much for your letter – I am terribly grateful to you. Unless some hitch occurs at the last moment (whether administrative or financial), I shall most likely accept the invitation. Will write you more fully later.

Meanwhile, our love to you all,

In haste,

Ernest

4.1.67 *Hempel to Nagel, November 7, 1953*

[Typewritten, Hempel 1999, Box 28, Folder 1]

Dear Ernest:

No doubt you have, like myself, received a confidential letter from Johns Hopkins asking for suggestions as to how to replace Boas[330] and Hammond[331] upon their retirement by men of promise or even of eminence. I have not really given the matter serious thought so far, but I am sure I could benefit a lot from your own ideas; you will know much better the total group of likely candidates. Machlup,[332] who wrote me, indicated that they might want to appoint one full prof about 50 years old and an assistant or associate professor less than

---

[330] George Boas (1891–1980) was an American historian of philosophy. He studied at Harvard and received his PhD from the University of California, Berkeley, before being hired by Johns Hopkins University.

[331] Albert L. Hammond (1892–1970) was an American philosopher who was professor of philosophy at Johns Hopkins.

[332] Fritz Machlup (1902–1983) was an Austrian economist who emigrated to the USA in 1933. He taught at the University of Buffalo, Johns Hopkins University, and Princeton University.

40 years old; one of the two men should include symbolic logic among his fields of competence. It occurred to me that Pap would fill the bill for the second position very nicely, provided that whatever personal peculiarities he is supposed to have do not interfere with the cooperation and interpersonal harmony of the department. Do you have any feeling on this matter? An alternative that occurred to me is Rulon Wells in our Dept at Yale, who may have a long while to go at Yale before he reaches a full profship – he still is an assist. prof. Linguists here have spoken to me very favorably about him, and I myself consider him as thorough and competent, if not especially powerful. Does this check with your reaction? My opinion of Wells is surely favorable in the sense that I feel he would make a very good professor at Johns Hopkins, and I myself would like to see him stay at Yale, but in view of the crowding we have at the top level I thought I might mention him to Machlup's committee. – For the top position I had wondered whether they wld be willing to try Popper – although that prospect also raises the question of personal frictions. – I have heard very fine things about Mates,[333] and his recent historical study in logic seems a very scholarly work, but I heard a rumor to the effect that the frail health of some member of the family ties him to the San Francisco area. This, however, should perhaps be no reason for me not to mention him to J.H. – What wld you think of Black in the top place at J.H.? – Shld one recommend Suppes[334] as a possibility for the lower of the two posts? Kaplan for the top? (Had you heard that after his return from summer teaching at Harvard, Abe K had some sort of *spontaneous* pneumothorax and was quite sick for a longish time? He seems to be getting better, though. I was very sorry to hear of it.) Well, this has been pretty rambling, but I was simply thinking out aloud on my typewriter.

In a fairly recent letter, Ledger Wood wrote as if it were now definite that you are going to Princeton as a Hibben Fellow next fall. I hope that was no overstatement, and I will be very happy if you go, and hope it will work out as well for you as it did for me.

I was shocked to read of Felix Cohen's[335] death. I had hardly known him personally, but had heard about him from you and had read some of his published work. From the Times I gathered that you spoke at his funeral in Washington; it must have been a terribly hard blow for you to lose this friend.

Has life been reasonably kind to you all on other counts at least? Are you, Edith, and the children well? (Have you heard that the Princeton public schools are excellent – oh but that is of no interest to you, with the boys staying in NY) We are all well, we have had our bouts with colds, I just got over one and still am a bit tired. The social life here is rather intense, with many invitations, which is very pleasant of course but also a bit strenuous. The philos. colloquium, an informal affair just for members of the Dept and guests they may invite, has been revived, and we meet abt once a month in the evening at the house of one of the

[333] Benson Mates (1919–2009) was an American philosopher. He received his PhD from the University of California, Berkeley with a thesis on stoic logic, which later became his first book *Stoic Logic* (1953). He also worked extensively on the philosophy of Leibniz, to which he devoted the book *The Philosophy of Leibniz. Metaphysics and Language* (1986).

[334] Patrick Suppes (1922–2014) was an American philosopher. After studying at the University of Chicago, he received his PhD from Columbia University. He taught at Stanford University. His work has been very influential in the fields of the philosophy of science, the philosophy of measurement, psychology, and educational technology.

[335] Felix S. Cohen (1907–1953) was an American lawyer and a legal philosopher. He received his PhD in philosophy from Harvard University before attending Columbia Law School. His most influential work was "Transcendental Nonsense and the Functional Approach" (1935). Cohen was also extremely influential in American Indian policy making, to which he contributed his book *Handbook of Federal Indian Law* (1942).

members of the Dept. A few days ago, I spoke on the Churchman[336]-Braithwaite[337] ideas – based on Wald's theory of decision functions – concerning the valuational presuppositions of science. I found that it has all sorts of intriguing aspects, this problem; I came across it in thinking about the use of probability statements (statistical ones) in explanations.

It was good to hear you will be attending the AAAS[338] etc meeting here – will you be going to Rochester at all? Margenau[339] asked me to read a paper on the operationism symposium, and I have agreed. What I have heard of the program sounds quite interesting.

Best to you all from us all, I do hope this finds you all well.

As ever,

4.1.68 Nagel to Hempel, January 3, 1954

[Typewritten, Hempel 1999, Box 28, Folder 1]

25 Claremont Ave.,
New York 27, N.Y.
Jan. 3, 1954.

Dear Peter,

It was good to get your nice letter, and I am grateful for your congratulations. I got the news when I returned home from Boston – Edith had received a telegram from James Gutmann[340] who had been at Rochester. I suppose I ought to be pleased over the honor, but at present all it means to me is that I have another chore to do. I had no official statement, but Gutmann reported that Baylis is the VP, and that you had been elected to the Executive Comm. I have no idea whether the latter meets except at the annual sessions of the Association; but if it does, perhaps it will provide an opportunity to see you during the year.[341]

---

[336] C. West Churchman (1913–2004) was an American philosopher and theorist of management. He studied philosophy at the University of Pennsylvania, where he obtained his PhD. He began his career at the University of Pennsylvania and then he became professor at the University of California, Berkeley. He was editor of the journal *Philosophy of Science* and also served as the secretary and program chairman of the American Philosophy of Science Association.

[337] Richard B. Braithwaite (1900–1990) was an English philosopher of science. Mostly associated with the University of Cambridge, he worked mainly in the field of philosophy of science. His most important book was *Scientific Explanation* (1953).

[338] The 120th meeting of the American Academy of Sciences was held in Boston, from 26 to 31 December 1953.

[339] Henry Margenau (1901–1997) was a German physicist and philosopher of science who emigrated to the USA in 1922. He received his PhD from Yale University, where he was appointed professor of physics. His field of expertise was the theory of microwaves, but he also wrote some important books on the nature of science and became the president of the Philosophy of Science Association between 1951–1957.

[340] James Gutmann (1917–1988) was an American philosopher who worked at Columbia and who was the Director of the University Seminars between 1970 and 1976.

[341] Nagel served as President of The American Philosophical Association, Eastern Division (1954–1955); in that role, he delivered his famous lecture, "Naturalism Reconsidered" (1954/1956a).

I found the Boston meetings quite pleasant, but as usual it is seeing friends which justifies going to such gatherings. Some of the sessions could have been improved, so I think, by having fewer prepared papers and fewer invited discussants. But I suppose it is difficult for any program committee to resist the temptation to make a program look attractive on paper – though at the cost of cutting down on the live discussion.

I am sorry I missed the trip to Skinner's[342] workshop – I would have liked to see him at closer range. He seems terribly intense, as well as bright. When we next meet you must tell me about the details.

The boys received the cards announcing the subscriptions you sent. Both of them are terribly pleased, and want to thank you all. Bobby has been quite jealous of Sandy's monthly mail of Jack & Jill, and now he will be able to hold his own in the s[i]bling rivalry.

I would have liked to attend the Business Meeting of the IUS.[343] The publications business is quite crucial – our present arrangements with the Academy do not seem to me entirely satisfactory. I also think (and I talked to Herbert a bit about this) that Frank and the rest of us ought to look around seriously for someone who would eventually take his place as chief director. My feeling is that it would be desirable to find a practicing scientist, who has some prestige among other scientific workers. But this is not easy. Herbert is already tied up with his own Institute, and will probably continue to be indefinitely. Neither you nor I are cut out for administrative work, I think, and would not enjoy it. Bergmann would probably rub people the wrong way, though I imagine he would like to be a king-pin in an important organization. From what I know of him, I think Adolf Gruenbaum would be an excellent choice, but have no notion whether he has enough interest to take on such work. Other names occur to me, but these matters are best discussed orally; and perhaps the next time we meet we can compare ideas and impressions.

I saw the news about Victoria Weiss in the Times, and wrote Paul.[344] It is a terrible experience to live through.

Apparently Morty White did not go to Rochester, because he came down with the flu. I had reports both from Sidney and from my colleagues at Columbia about the Boas-Hook symposium[345] – from all accounts it was a very heated and quite unpleasant session. Personalities entered into the discussion, and much hard feeling was produced. Perhaps political and semi-political issues should not be raised at the meetings. In any case, I am glad I was not there.

---

[342] Burrhus Frederic Skinner (1904–1990) was an American psychologist. He got his PhD from Harvard University, where he also taught for most of his career. A proponent of radical behaviorism in psychology, he promoted the study of human behavior in a scientific way. Some of his most important books are *Behavior of Organisms* (1938) and *Science and Human Behavior* (1953).

[343] The Institute for the Unity of Science was established in 1947 under the directorship of Philipp Frank. It was first financed by the Rockefeller Foundation (it began in 1950, thus for 3 years not much had happened). The Institute organized several conferences in the Boston-area, held essay contests for students and early career scholars, and published various volumes and special issues. After the initial legal difficulties, the American Academy of Arts and Sciences provided useful legal and institutional help for Frank and provided offices for the Institute in the 1950s, and also publication forums and rooms for conferences.

[344] On 1 January 1954, the New York Times announced that Victoria Brodkin Weiss, wife of Paul Weiss, had died on 31 December 1953 after a long illness. She was 47 years old. Paul Weiss (1901–2002) was an American philosopher. He received his PhD from Harvard University, and he taught at Yale for most of his career. Weiss wrote several books on metaphysics, but he is best known for his work as editor of the *Collected Papers of Charles Sanders Peirce*.

[345] Boas and Hook discussed the questions of the ethics of academic freedom at the annual meeting of the American Philosophical Association's Eastern Division (University of Rochester, December 28–30, 1953). See their papers in White (1953).

It was so refreshing to see you and Diane again – I wish we could repeat the experience more frequently. Here is hoping that the new year will make it possible. In any event, our very best wishes to all of you for the year.

We all send you our love

Ernest

4.1.69 *Hempel to Nagel, February 26, 1954*

[Typewritten, Hempel 1999, Box 28, Folder 1]

February 26, 1954

Dear Ernest:

The immediate occasion for this note is a letter which I received yesterday from a former graduate student at Yale, who recently received his PhD there, Leon Goldstein.[346] He expressed a great interest in an opportunity to talk to you on questions in the methodology of the social sciences, and especially of anthropology, and asked me if I would write you a word about him. Let me say right away that I would appreciate it greatly if you could see him. Goldstein has had a quite a number of courses with me, and I have taken a considerable interest in him, but I am also concerned for him because he has had difficulty in finding a job (this is *not* what he wants to see you about) and in finding recognition of his somewhat unusual interests. He came to philosophy from a study of ancient languages or something of the sort, and at first did not do very well; if I recall correctly, he got a C in symbolic logic from me. But then, as a result of his steady and intensive work, he improved all the time, and in the last two courses he took with me, he reached the Honors level. He has had considerable work in anthropology, and his thesis deals with the use of the concepts of form, function, and structure in anthropological theory. I have seen large sections of his thesis (which was completed only during the time I was away from Yale), and thought them substantial; and I was especially impressed with the very wide reading knowledge of which his writing and also his discussion gave evidence. But he is not an especially brilliant person, and his range of interests is somewhat limited, and so you can imagine that he has not exactly had an easy time getting recognition. He is now working on several articles – only some of which are pieces of his thesis –, and I hope they will help him professionally. Well, all this just as background; and since he is a little unimpressive at first sight I am all the more eager to help him to broaden his outlook and give him some encouragement.

I gave a talk at the Bell telephone Co project at the U of Pa[347] recently (they seem a good group and were quite responsive) and on that occasion stopped over in Princeton and saw the Carnaps and the Oppenheims. Carnap is greatly improved and goes for half-hour walks by himself and sits up for longer times, though he still does his reading and writing in bed. The UCLA matter was still hanging fire, with no final offer having arrived when I was there.[348] Ina has a full time job as a psychiatric social worker for the State of NJ and handles her complex duties with what seems to me amazing efficiency.

---

[346] Leon Jay Goldstein (1927–2002) was an American philosopher of history. He was professor of philosophy at Harpur College, Binghamton University. His most important book is *Historical Knowing* (1976).

[347] It is unknown what talk Hempel gave there.

[348] After Reichenbach's death in 1953, UCLA managed to hire Carnap to join the Department of Philosophy, though he had previously turned down the offer of a Flint Professorship in the early 1905s because of the loyalty oath requirement.

I am having a good deal of work here, but nothing that compares with your load. I hope you are feeling well, physically and psychologically, and that all the work is not getting you down too much!

We are all well; Diane has just returned from New Haven, where she went to look at damage caused by a blow-back of our furnace; fortunately, it is slight, and covered by insurance.

All our best to you all!

Very cordially,

4.1.70 Hempel to Nagel, May 24, 1954

[Typewritten, Hempel 1999, Box 28, Folder 1]

May 24, 1954.

Dear Ernest:

Enclosed are Scriven's[349] notes on the Princeton Conclave, which I received this morning. I don't think they help much in this form, and I do not know how I can add anything useful without spending a considerable amount of time and effort; and that expenditure might be useless since I do not know just what points would be of special importance for the Minnesota group. I wish that Carnap's ideas on fusion had been stated in the notes; by now, I am rather hazy on the details although I am sure I learned something much worthwhile from that phase of the discussion.

I was sorry to be unable to hear more from you about your plans and about your work the past term. Several people talked in terms of high praise of your Vanuxem lectures;[350] about these, too, I wish I could have heard from you. This must have been an excessively hard academic year for you, and I am glad to think you can now think forward to a peaceful stay in Vermont soon. I very much hope you all will have a really good summer there.

John Goheen visited us for a short time; he had to come to NY on university business and then flew on to Boston to see some of the Japanese students here who are somehow under Stanford's benevolent sponsorship. He seemed very well (as did Arnold Isenberg,[351] who dropped in one evening about three weeks earlier) and spoke with great enthusiasm about the work of his department and of the value study project which is being carried on at Stanford under Ford Foundation sponsorship.[352] He wanted to know whether you would be visiting Berkeley the year after next, and I had to tell him I did not know. Will you?

---

[349] Michael Scriven (1928–2023) was a British philosopher who grew up and was educated in Australia. After receiving his PhD in philosophy from the University of Oxford, he spent almost his entire career in the USA. A major figure in evaluation theory, he is also known for being one of the harshest and most persistent critics of Hempel's account of scientific explanation.

[350] Based on its website, "the Louis Clark Vanuxem Lecture Series was founded in 1912 with a bequest of $25,000 under the will of Louis Clark Vanuxem of the Class of 1879 for the purpose of supporting public lectures at the University on subjects of current scientific interest" (https://lectures.princeton.edu/lectures/louis-clark-vanuxem-lectures).

[351] Arnold Isenberg (1911–1965) was an American philosopher, whose main areas of research were ethics, aesthetics, and value theory. He studied at Harvard, where he also got his PhD in philosophy. He taught at Queen's College of City University of New York, Stanford, and Michigan State. The volume *Aesthetics and the Theory of Criticism* (1973) collects a selection of Isenberg's essays.

[352] Stanford University received a grant of $23,425 from the Ford Foundation to run a two-year project about value theory and decision-making at the Department of Philosophy. While Isenberg was running the program, John D. Goheen was the chairman of the research committee.

Along with Scriven's material, I also received quite a desperate letter from Leon Goldstein, who said that he had not even been interviewed for any teaching position so far, and that he doubted whether Hendel[353] was actually going to do anything for him. I am at a complete loss as to what to do; I put him in touch with the Princeton Department earlier, and I recommended him to Northwestern. I will ask him whether he has registered with the APA; I do not know whether anything ever comes of this, but perhaps he might get a start. I also thought of whether he could go into highschool teaching, at least for a while, but I am not sure that he won't have to meet all sorts of additional requirements to do so. I feel awfully sorry for Goldstein, and the experience reinforces my conviction that one must [[be]] very cautious indeed about encouraging anyone to choose philosophy as a vocation.

A note from the Carnaps told me that C had announced two public lectures at the Institute, which were to be given last Th and Friday; I think this is excellent news.

As I think I mentioned to you in Princeton, I have again been using your probability monograph as a required book in my course;[354] and it again proved very useful. You will also figure in at least two places on my final exam, of which I will send you a copy.

I hope you, Edith, and the children are well! Our very best wishes to you all!

As ever,

4.1.71 Nagel to Hempel, May 27, 1954

[Typewritten, Hempel 1999, Box 28, Folder 1]

May 27, 1954.
Professor Carl G. Hempel
Department of Philosophy
Harvard University
Cambridge, Massachusetts

Dear Professor Hempel:

I have been asked by the Chairman of the Program Committee to assist in the preparation of a session to be held at the annual meeting of the American Historical Association in New York on December 28–30, 1954.[355] The theme of the session is a critical examination and evaluation of history as a discipline by scholars whose own work is principally in other fields.

From the viewpoint of developments and perspectives in their own fields, a philosopher, a sociologist, and anthropologist will be asked to comment in over-all terms upon the values, deficiencies, or unrealized potentialities they observe in historiography. Searching critiques of this nature, it is anticipated will stimulate discussion among historians concerning the past, present, and future theory and practice of their discipline, as well as make a contribution towards the breaking down of interdisciplinary barriers. One or two historians will be asked to comment formally upon the three papers, but since it is also anticipated that the papers will be general enough to have wide appeal, time will be provided for discussion from the floor.

[353] Charles William Hendel (1890–1982) was an American historian of philosophy. He studied at Princeton University, where he received his PhD in philosophy in 1917. He taught at Princeton University, Williams College, McGill University, and Yale. Hendel was best known for his work on David Hume; his main book—*Studies in the Philosophy of David Hume* (1925)—helped to revive interest in the Scottish philosopher.

[354] See Nagel (1939).

[355] The 69th annual meeting was held on December 28–30 at the Hotel Commodore in New York.

We would be delighted if you could accept this invitation to present such a critique from the viewpoint of philosophy. Paul Lazarsfeld of Columbia University has accepted an invitation to present a similar paper in terms of sociology,[356] and an equally distinguished anthropologist is expected to participate. To allow maximum time for discussion, it would be desirable, if possible for the main papers to run from twenty to twenty-five minutes. Unfortunately, the American Historical Association has no funds with which to reimburse participants for expenses incurred in connection with its programs.

I would appreciate learning at your earliest convenience whether such a session seems desirable to you, and whether it would be possible for you to participate in it. If any point needs clarification, please feel free to call upon me to supply it.

Sincerely yours,
Ernest Nagel, per H.F.

4.1.72 Hempel to Nagel, May 31, 1954

[Typewritten, Hempel 1999, Box 28, Folder 1]

May 31, 1954
Professor Ernest Nagel
Department of Philosophy
Columbia University
New York 27, N.Y.

Dear Ernest:

It really flatters me very much that you should have thought of me as a possible representative of philosophy on the panel planned by the American Historical Association. The program for the discussion seems to be unusually interesting, and I would very much like to participate. But you know that, being a slow worker, I would have to set aside a considerable amount of time in order to prepare an appropriate paper. Now I have already allowed myself to be sidetracked from working on my book by several other commitments, and I have felt increasingly concerned about this. To take on the assignment for the American Historical Association would mean to aggravate the situation, and, though with great regret, I feel therefore I should not accept your invitation. I do hope you will understand!

Two or three days ago, I received your reprint on Symbolism and Science and your essay on functionalism' many thanks![357] I have read the former with great interest and pleasure; several of your observations struck an especially responsive chord because I had recently been thinking along the same lines (e.g., on formal analogies). Your analysis of functionalism could not be more welcome: As I may have mentioned to you, I had to give a talk on

[356] Paul Lazarsfeld gave a talk entitled "Some Possible Applications of Recent Social Research to Historiography" on 29 December.

[357] The essay "Symbolism and Science" was first published in the volume *Symbols and Values: An Initial Study*, edited by L. Bryson, L. Finkelstein, R.M. MacIver, and R. McKeon, pp. 39–71, and later reprinted as Nagel (1954/1956b). Regarding the "functionalism paper", Nagel's most famous "A Formalization of Functionalism" was originally prepared for the *Planning Project for Advanced Training in Social Research* (Case Studies in the Formalization of Discursive Material) at Columbia University in 1953, although it was published only in *Logic Without Metaphysics* in 1956. Another option is Nagel's "Teleological Explanation and Teleological System", which also covers the discussion of functions and appeared in 1953.

the subject before the Social Relations Colloquium[358] here, and I based it to a large extent on a study of Merton's long essay,[359] which seems to me by far the best thing available on the subject. My analysis had not gone into the formal details of a study based on Somerhoff's[360] machinery; I recall that you suggested this procedure in an earlier article,[361] which my students had to read this term for a comparative methodology course; and I am very glad now to have the details. I have so far had time to look at your essay only superficially, but will study it with great care soon.

This is an extremely busy time of year, and I am sure you must be snowed under with work, too. I do not know what thesis or essay to read first, and I want to get busy with my Carnap essay.[362]

Cordially,

4.1.73 Nagel to Hempel, August 30, 1954

[Typewritten, Hempel 1999, Box 28, Folder 1]

South Wardsboro, Vt.
August 30, 1954.

Dear Peter,

I've been very remiss in not writing to you before this, but it has been a busy summer, even though I've little to show for it. I felt pretty tired when we came up, and shortly thereafter, and until about a week ago, we've had Edith's mother and my mother in succession as visitors. I put in plenty of time at my desk, but writing came terribly slowly, and I have not scratched the surface of the pile of writing obligations I had hoped to complete before the summer is over. At the moment, I am beating out my brains trying to think of some appropriate theme for a presidential address for next Christmas. My feeling is that one ought not be technical in any sense on such an occasion, and deal with a large subject with a certain amount of eloquence and imagination. Thus far, no idea is stirring, and I haven't much time left to fool around with possibilities. I haven't even begun the article for the Carnap volume, and will turn to it as soon as the immediate problem is solved. But I hope the matter is not too pressing, and that Carnap himself will not be ready to write his comments for some time.

I had hoped to visit the Carnaps' before they left for the west, but the pressure of obligations was too great; and so I will have to wait until (if and when) I get out to California myself. We saw Steve at his place in Wilmington yesterday, and he told us that he has tried in vain to have you visit them. Too bad you could not make it. But perhaps this year you will be

[358] In 1946, the Department of Social Relations was established at Harvard, after a long campaign of behaviorist social scientists who took up practical wartime and policy-related work.

[359] Hempel is probably referring to the essay by Robert K. Merton "Manifest and Latent Functions", published in *Social Theory and Social Structure* (Merton 1949).

[360] See Sommerhoff (1950).

[361] Nagel refers to Sommerhoff (1950) in the essay "Teleological explanation and teleological system" (1953).

[362] See Hempel (1963b).

coming down to New York in connection with the Philosophy Club,[363] so that I hope to see something of you. (By the way, I notice from Albert's circular that you are scheduled to read a paper next March! I look forward to hearing it.)

We plan to-morrow to go to Hanover for a short visit. I owe a return visit to Mandelbaum,[364] and Lazarsfeld is summering there and I would like to see him before he leaves for California for the coming year.

I don't envy your job of reviewing Reichenbach's monograph.[365] I've gone through it rather hastily, and am not impressed. Assuming that there is a genuine problem (which I doubt), he seems to me to "solve" it by his usual high-handed methods, plugging in holes in an ad hoc fashion, and labelling a difficulty by calling it some new "principle". This is, of course, a bit harsh and unfair, but I think not too far off from the truth.

The rest of us have been fine, and the boys are flourishing. We had a local man with a bull-dozer dig out a little pond at the bottom of our field, and the prospects are good that even in dry seasons there will be water there. The boys enjoy it, and all the neighboring kids come around for some "dunking". But it has been an unusually cold summer up here, and there have not been too many days for such things. Edith enjoys the country very much, and though she works hard keeping the house in running order she gets a great pleasure out of puttering in the garden. I hope sometime soon you can all pay us a visit – Diana has never seen our place, as I recall.

The Hooks are here this summer, and are in good form. Sidney is working on a book of some sorts on Marx, though he is sick of the subject.[366] They are going into Brattleboro this morning, and I must stop now if this is to go along with them to the post-office.

Our love to all of you.

Ernest

4.1.74 Nagel to Hempel, September 30, 1954

[Typewritten, Hempel 1999, Box 28, Folder 1]

2 Dickinson Street,
Princeton, N.J.
Sept. 30, 1954.

[363]The New York Philosophy Club "met during the academic year in the men's faculty club at Columbia University." Members included mainly philosophers from Columbia, but it was open to a few outsiders too (some from Yale). As Hempel (2000, 33) recalled, "there was one very curious system in these meetings at Columbia, namely after the paper had been presented, there was a discussion. But the discussion was not your usual free form, i.e., whoever has something to say raising his hand. Rather the discussion went around the table and everybody was expected to say something no matter whether he knew anything about this subject, or even if he knew about it and had anything to say that was worth mentioning".

[364]Maurice Mandelbaum (1908–1987) was an American philosopher. He received his PhD in philosophy from Yale and then taught at Swarthmore College, Dartmouth College, and finally Johns Hopkins University. He is best known for his work in the philosophy of history and historiography. His most important books are *The Problem of Historical Knowledge* (1938) and *The Anatomy of Historical Knowledge* (1977).

[365]Reichenbach's book *Nomological Statements and Admissible Operations* was published in 1954, a year after his death. Hempel reviewed Reichenbach's book in "The Journal of Symbolic Logic"; see Hempel (1955).

[366]Sidney Hook published the book *Marx and the Marxists: The Ambiguous Legacy* in 1955; see Hook (1955).

Dear Peter,

It is a long time since I've written you, and even longer since I've seen you. But I have been busy (even though not very profitably), and I've found it hard to summon up energy for correspondence.

This is almost the end of my first week at Princeton, and I shall be going to New York to-morrow. I had a preliminary meeting with the seminar on Sept. 22, but did not move down here until last Monday. I decided, as I think I told you, to devote the whole semester to probability and induction, partly in order not to overlap seriously with the material you covered two years ago, and partly because I thought it would be profitable both for the students and me to concentrate on a narrow field. After meeting the group, I am not sure this was a wise decision. The actual registration for the seminar is quite small (just 3 students), although there are a fair number of auditors; and while some of those who've come have had some exposure to the material, most of them I am afraid have little background and will not be able to read what I had planned very easily. I thought I'd spend the major part of the semester going over the Carnap material (parts of the Logical Foundations, and the whole of The Continuum),[367] plus four or five chapters in Braithwaite.[368] But as I see it now, and also in consequence of the small number of registrants, I'll have to lecture on much of the material myself, instead of having the students carry the main burden of the work. And so I am beginning to wonder whether I will have much more time for myself after all to do some writing, and the whole idea may turn out to be a bad mistake.

Ledger Wood managed to find for me a suit of unfurnished rooms, which happened to have in them a bed, chest of rooms, a rickety table and a chair. I have to supply my own linen, etc., and of course eat all of my meals out. Although I often thought it might be nice to live the life of a monk, I am getting a bit more than I bargained for. The bare rooms are a little grim and I do miss the family. Add to this that at the beginning of an academic year I feel as nervous as a wet hen, and do not at all feel sure of myself, and you will understand why at the moment I feel rather discouraged. But I am hoping things will pick up before long, and that I will finally be able to get some worked turned out.

My immediate headache is to write a presidential address for the Xmas meeting. I just wasn't able to get down to it this summer, and now the time seems very short. Part of my trouble is that I don't quite know what to write about. I don't think a technical paper (even if I were able to cook up something) would be appropriate; and I feel as juicy as a baked brick as far as fresh general ideas are concerned. The only thing I've been [able] to think of as a possible theme is a sort of a credo what the world looks like to me, what the function of philosophy might be, with some pot-shots at a standpoints I dislike. I've started to get something down on paper, but haven't managed to fill a single page.

I still have done nothing about the Carnap piece, and am simply ignoring Schilpp's[369] pleading notes. But once I get the presidential address out of the way, I think I'll be able to turn to that, and perhaps get it off before the end of October. Pray for me!

I've seen some of the younger men in the dept. here, but not at great length. A few of them are going to meet bi-weekly, to discuss Wittgenstein and perhaps some other recent book, and I'll join them if and when I can. I haven't seen Oppenheim, but expect to give him a courtesy ring on the phone before very long.

---

[367] See Carnap (1950, 1952).

[368] Presumably from Braithwaite (1953).

[369] Paul A. Schilpp (1897–1993) was a German-born philosopher who studied at Northwestern University and the Garrett Theological Seminary. He received his PhD in philosophy from Stanford University. He taught at the College of Puget Sound, the College of the Pacific, Northwestern University, Southern Illinois University at Carbondale, and the University of California at Santa Barbara. Schilpp is best known for the series *The Library of Living Philosophers*, inaugurated in 1939 by a volume on John Dewey.

Edman's death is a serious loss for the department in many ways. Whether any replacement will be made I do not know, nor what sort of a person my colleagues will agree upon. I expect to be consulted, but serious differences divide some of us, and I doubt very much that most of the department would be interested in strengthening the logic and philos. of science offering. But Edman's death is also a great personal loss; for despite our differences in interest, I was always able to talk to him, and he had an unusual sane and realistic grasp on things. I returned to N.Y. from Vermont for the funeral, and was asked to make a talk, which I did. I am beginning to feel like a professional funeral orator. But I suppose one ought to become reconciled to the fact that if one lives long enough apparently permanent faces in the environment vanish.

Hillary Putnam[370] told me he received a letter from the Carnaps full of cheer and delight over California. He also told me of a paper C. had circulated, in which he replied (according to HP very effectively) to Quine's attack on the analytic-synthetic distinction.[371] I suppose you've seen it, and I'll try to borrow it from HP. I'll be interested in hearing what your own reactions to it are.

Will you be attending the NY Philos. Club this year? I hope you can come down at least occasionally, and if you do perhaps we can make a date for some private talk before the meeting, on some Friday or other.

Affectionate regards to you and Diane,
Ernest

4.1.75 *Hempel to Nagel, October 9, 1954*

[Typewritten, Hempel 1999, Box 28, Folder 1]

October 9, 1954

Dear Ernest:

Two days ago, I received a letter from Jerry McGill,[372] in which he informs me of his dismissal from Hunter and describes some of the details of the investigation which led to this result. It certainly sounds extremely unfair, although I suppose the Board of Higher Ed. has one formal ground to stand on, namely that he perjured himself many years ago before the Rapp-Coudert Committee.[373] I do not know McGill too well, but I am pretty sure he is not a threat to higher education, and I am very sorry to see him in an almost hopeless professional impasse. He wrote to say that if I knew of any possibility for him to receive even a very modest grant that might enable him to do some research, he would like to know about it; but I feel virtually certain that no university, no college, and no foundation is going to give him a chance in the present climate of overheated anti-communist sentiment. I just had to write him that at present, I could make no constructive suggestion, but that I would surely let him know if I shld hear of any likely opportunity. You mentioned to me some time ago that you had

[370] Hilary Putnam (1926–2016) was an American philosopher. He graduated from the University of Pennsylvania and got his PhD in philosophy at the University of California at Los Angeles, under the supervision of Hans Reichenbach. Putnam taught at Northwestern University, Princeton, MIT, and Harvard.

[371] See Quine (1951). Presumably Carnap (1955).

[372] Jerry McGill was Vivian J. McGill, introduced in letter 10 above.

[373] The Rapp-Coudert Committee was a committee established by the New York State Legislature to investigate the Educational System of the State of New York.

rendered him help by a statement – I think to Shuster[374] – and as I recall there was a chance that you might review McG's book.[375] I just want to say that if you {would} know of any way in which one cld help McGill, I would be glad to do what I can for the purpose. At present, the only thing I was able to add was that I would like to be approached in case a collection would be made towards a fund intended to help him meet the expenses involved in the appeal that he is planning.[376]

One other thing: Would it be possible to purchase from Columbia Univ. one or two copies of your "Formalization of Functionalism"?[377] I would very much like to use your essay in my current seminar on explanatory procedures; I have already included in my reading list your earlier piece that is reprinted in Feigl and Brodbeck.[378] – Where would one send the order?

I very much hope Princeton is beginning to look a bit better to you, and life a little more cheerful!

With many good wishes,

4.1.76 Nagel to Hempel, October 11, 1954

[Typewritten, Hempel 1999, Box 28, Folder 1]

2 Dickinson Street,
Princeton, N.J.
Oct. 11, 1954.

Dear Peter,

I've returned this morning from a week-end in New York, and found your letters waiting for me. Thank you for writing so encouragingly and kindly. I am sure I'll get adjusted to a new environment pretty soon (I *am* sensitive to change of locale), and hope eventually to get some work done. At the moment I am getting over one of those colds that seem to drain all one's energy. And I guess I'll have to solve the seminar problem simply by having to do lecturing half the time, if not longer. Some of the material I am reading or rereading I have to go over anyway, so it will not be a total loss.

[374] George N. Shuster (1894–1977) was an American educator and journalist. After receiving his PhD in English literature from Columbia University, he taught at Notre Dame. He was an influential figure in the Catholic cultural scene in the United States during the 1930s and, as a delegate to the Conference on International Education, he helped to found UNESCO.

[375] In 1954 McGill published the book *Emotions and Reason*. It does not seem that Nagel reviewed McGill's book.

[376] McGill found a position through Mortimer J. Adler's Institute for Philosophical Research and worked in San Francisco after 1956 and eventually became a professor at San Francisco State College.

[377] As noted above in letter 72, "Formalization of Functionalism" was prepared originally as a case study for Columbia's *Planning Project for Advanced Training in Social Research*, but got published only in 1956, after all.

[378] Four of Nagel's papers are published in Feigl and Brodbeck, *Readings in the Philosophy of Science* (1953). It is likely that Hempel refers here to "Teleological Explanation and Teleological Systems" (on pp. 537–558).

The McGill case is sad, but like you I cannot think of anything that can be done about it.[379] I doubt very much whether he'll get another teaching job, unless his appeal turns out successfully, which is not likely. Although I am quite prepared to believe that he has really broken with the C.P., I have heard that those who have seen the record of his testimony have found it curiously evasive – but of course, this is just hearsay, and strictly *entre nous*.

I am not sure just how to go about getting additional copies of the Functionalism piece. As you perhaps know Lazarsfeld is in California for the year (at the Ford center for soc. sci.), though he will be back in New York for short visits a few times (at the end of this month, definitely, to take part in a panel discussion in connection with the Columbia Bicentennial). Apart from writing to him directly, the only suggestion I have is to write to his secretary at Columbia. (The piece is not on sale at the C.U. Bookstore, as far as I know). Thus far, I have not been getting into N.Y. early enough on Fridays to be able to get in touch with her, and I do not recall her name. Why don't you address a letter to: Secretary to Prof. Lazarsfeld, and see what happens. And I'll try next week-end or the Friday after to make a contact myself.

It is too bad you will not be coming to the Club meetings this semester – not because the papers read are worth the trip, but simply because I wish there were more frequent opportunities to see you. But should you come down to Princeton, please let me know.

Meanwhile, all good wishes and regards,

Ernest

4.1.77 Nagel to Hempel, December 6, 1954

[Typewritten, Hempel 1999, Box 28, Folder 1]

2 Dickinson St.,
Princeton, N.J.
Dec. 6, 1954.

Dear Peter,

I have a very embarrassing favor to ask, but in order to ask it I must first explain the circumstances.

Three weeks ago, my mother was suddenly taken to the hospital (while I was here at Princeton) in a very critical condition; Mack Lipkin[380] gave here a fifty-fifty chance to pull through. It was apparently a "slow" stroke, complicated with a generally failing heart. She has managed to recover, and was discharged from the hospital Saturday. Mack does not think she can, or ought to, live by herself any longer; and for a while at least she is staying with us. Mack's prognosis is not very cheerful; but as long as she can take care of minimum bodily needs and Edith's strength holds out, we'll be able to manage. After that, we don't know – my sister has a small apartment, and cannot possibl[y] make room for her. As you can imagine, the medical bills have been staggering, though we hope that they will diminish now that she is out of the hospital.

And now for the request. Would you mind if we did not exchange Christmas gifts for the children this year? Quite apart from the fact that at this end we are very much harassed, with little time to take care of shopping that is not immediately necessary, we have to cut down on expenses in some way, and even though the costs of gifts are small, every little bit will help

[379] See letter 75 above.

[380] Mack Lipkin (1907–1989) was an American physician and clinical professor of medicine at the University of North Carolina's Chapel Hill School of Medicine. He served as an attending physician in New York.

for a while. I wish I did not have to write in this vein, but I feel driven to it, however petty all this may seem.

Edith is having a rough time this semester, with me away most of the week. Princeton is pleasant, and I think I would really enjoy it were it not for the fact that I've been feeling guilty in running off and leaving so much of the burden on her shoulders. I haven't got much work done, except for finally finishing the presidential address – a terribly concoction of which I feel ashamed, though I feel lucky that my having a job does not (I hope) depend on anything I say. At least two of the graduate students are excellent, and one of them really first-rate. But the situation being what it is, I as well as Edith am looking forward to the end of the semester, and the resumption of normal life in New York. Partly as a consequence of Irwin Edman's death, and partly because of my mother's condition, I've had to write to Pepper[381] at Berkeley that I could not accept an invitation to go to California next year as a visitor (you may remember, that was the plan last year, to see whether I would like the place sufficiently to consider staying on permanently). Obviously, they cannot keep that vacancy open indefinitely, so that I expect the whole matter is closed as far as I am concerned.

Are you planning to go to the Goucher meeting Christmas week?[382] The program doesn't look inviting to me, and if I did not have to, I certainly would not go. It also involves an expense that I cannot at present afford, but I am afraid I have little choice in the matter.

Hope you are flourishing. My love to you and Diane.

Ernest

4.1.78 Nagel to Hempel, December 18, 1954

[Typewritten, Hempel 1999, Box 28, Folder 1]

25 Claremont Avenue,
New York 27, N.Y.
Dec. 18, 1954.

Dear Peter,

It was good to get your letter in Princeton (I am home now until after the Xmas recess); but the letter you say Diane sent to Edith must have gone astray. About a week ago, we did get a card announcing that you are sending JACK AND JILL to Bobby, so that my previous letter to you was sent too late. We are terribly grateful to you and Diane, but as you can imagine the situation is nevertheless embarrassing. Will you forgive us if we do not reciprocate this year? My finances are in a very bad way, and the trouble is that I do not at present know how to improve them. My fixed expenses far outrun the take-home pay. This obviously cannot go on indefinitely, and I'll have to find ways next year of adding to the income by taking on some extra work if I can find it. There appears to be little chance of a salary raise at Columbia; and I may have to double my load at the New School next year if the New School is willing, and perhaps teach during the summer (though this is out for the summer of 1955, since all arrangements have already been made).

[381] Stephen Coburn Pepper (1891–1972) was an American philosopher. He studied at Harvard University, where he received his PhD in philosophy. He taught for most of his career at the University of California at Berkeley. His most important books are *The Basis of Criticism in the Arts* (1945) in the field of aesthetics and *The Sources of Value* (1958) in the field of value theory.

[382] From 28 to 30 December, Goucher College hosted the fifty-first annual meeting of the American Philosophical Association Eastern Division. In 1954 Nagel served as president of the Eastern Division of the APA, while Hempel was in the executive committee.

I have had to turn down the chance to go to Berkeley as a visitor next year, partly as a consequence of the plight of my dept. because of Edman's death, and partly because of my mother's condition. I wrote Pepper that he will have to make his plans for his dept. without counting on me, so that I am sure that this chapter is definitely closed. What they will do I am unable to say, but I would not be surprised if they made you an offer. Would you go if asked?

There is a small possibility that Princeton might make me an offer (but please keep this strictly confidential). Ledger Wood said he would approach his administration, but he acknowledged that there would be difficulties; and in any case, I don't expect to hear anything definite for some time. (My presidential address will not, I am sure, improve my chances.)[383] But if a reasonable offer is made, I will consider it seriously, and use it for all it is worth at Columbia. For even if the salary offered should not be larger than my Columbia one, the effective income would be larger: the public school at Princeton is by all accounts excellent, which would itself mean a saving of at least $1200. a year; there is a group insurance plan at Princeton, which would enable me to drop one of my life-insurance policies; and there is no New Jersey state income tax. I think life could be pleasant in Princeton, though I would not consider making a transfer if the chance were given me, unless Edith could also find something to do there in physics.

Although I would very much like to see you, quite frankly I would not go to Goucher if I were you. The meetings promise to be quite awful, to say nothing of time lost and the expenses. I don't think the Exec. Comm. meeting is sufficiently important to warrant your taking the trip. And I'll look forward to seeing you sometime this coming spring.

I am far behind with my obligations, and still haven't done anything on the Carnap paper.[384] I have at least one chore to do before the New Year which will take what free time I'll have during the Xmas holidays, so that I do not know just when I'll get around to keeping my promise to Schilpp. Life has become a mess.

All good wishes to you and yours for the holidays. We send you our love.

Always yours,

Ernest

P.S. My mother is much improved, and is living with us – whether permanently or not is still unclear. This is not the ideal solution, and is especially hard on Edith.

E.

4.1.79 Nagel to Hempel, No Date

[Handwritten, Hempel 1999, Box 28, Folder 1]

2 Dickinson St.
Princeton, N.J.

Dear Peter,

I have a vague recollection that about a dozen years ago you reviewed a book in German on various eminent mathematicians, which included a chapter on Frege and also contained a

[383] Nagel's presidential address delivered before the annual meeting of the Eastern Division of the American Philosophical Association at Goucher College (see letter 77 above) was entitled "Naturalism Reconsidered" and was published in the *Proceedings and Addresses of the American Philosophical Association* (Vol. 28, 1954–1955, pp. 5–17), and then republished in *Logic Without Metaphysics* (1954/1956a).

[384] Nagel contributed to Schilpp's volume on Carnap with the essay "Carnap's Theory of Induction" (1963d).

photograph of the man.[385] Do you remember the author and title of the book, and when you reviewed it? Please drop me a post-card with the data if you can spare a moment.

I am trying to make up my mind whether to accept an offer from Princeton. I wish I could talk to you about it – it can't be discussed easily by mail or phone. But I'll write you more fully later.

Meanwhile, my love to you and Diane.

In great haste

Ernest

4.1.80 Hempel to Nagel, January 13, 1955

[Typewritten, Hempel 1999, Box 28, Folder 1]

January 13, 1955.

Dear Ernest:

I do not recall having reviewed a book of the kind you describe, I am sorry to say. If it should be a picture of Frege that you are interested in, I happen to have one that seems quite good, and I'd be glad to lend it to you for any length of time.

I am very happy to hear that Princeton made you an offer; I hope that no matter how you decide, it will be to your advantage. In particular, I am sure that Columbia will make a big effort to keep you – your departure would mean an enormous loss to the Department and to the University. I do wish we had a chance to talk about the pros and cons; I am very favorably disposed towards Princeton, as you know; somehow I see you there under less stress and strain, and in a setting and living conditions that would no doubt be very attractive to your family; the children would have an excellent public school and would be practically living in the country, and still you would have all the attractions of New York nearby. One of the difficult questions for you and Edith might be whether Edith would be able to do some work in physics that would be to her liking; and I do not know what the prospects might be in Princeton. I would much like to hear whether you have found the members of the Department there congenial; my impression is that if you were to accept, you would be the most influential man in the Department and could counteract effectively any trend in their policies that you might consider unsound.

On the day of the next meeting of the NY Philos Club – and indeed even one Friday after that, I still have to meet my seminar; otherwise I would have hoped to talk these things over with you at the next meeting. I doubt that my own reactions to the problem you are facing can be very important for your own decision; but in case there should be some aspects of the matter that you would want to air orally with me, I would try to come to NY for the purpose towards the end of January.

Best wishes to you and your family, also from Diane!

As ever,

[385] In letter 80 below, Hempel mentions that he did not write such a review.

4.1.81 Hempel to Nagel, January 27, 1955

[Typewritten, Hempel 1999, Box 28, Folder 1]

January 27, 1955

Dear Ernest:

It is a pity I won't be able to attend tomorrow's meeting of the NY Philos. Circle – I suppose you will be going there, and there are all sorts of things I'd like to talk over with you. None of them are of any immediate urgency, though and so I am hoping for a good talk with you at the latest on the day of the February meeting of the Club.

I am glad to have a little breathing spell between classes; apart from meetings of the Department and of the full professors (repeatedly from 8 pm to midnight) and apart from giving and grading exams, I am trying to use the time for preparing a series of three lectures I am to give early in March at Carnegie Tech before an audience consisting of teachers and students of philosophy and of people working in other fields who are connected with various institutions in the Pittsburgh area.[386] The talks are to deal with the nature and scope of scientific knowledge and understanding, and they will have to be no more than slightly technical at the most. During the Easter vacation, I have been invited to visit, and talk at, Feigl's Institute in Minneapolis, and that is an occasion that I am looking forward to a great deal.[387] I understand you are going to see Feigl at a meeting in Phila soon; I hope he will be able and willing to return to Mpls via New Haven. I have suggested that to him; perhaps you can add your weighty voice to mine?

During the past term, I had to give an undergraduate course in inductive method, and I used your monograph in full and found it very helpful. I also referred the students to your critique of Reichenbach in *Sovereign Reason*;[388] but that was just suggested rather than required reading.

I suppose you are just about ready to pack up and return to New York, I hope that as you look back over your term at Princeton, you feel reasonably satisfied with your decision to accept the invitation, despite the various difficulties – some of the unforeseeable, after all – that arose during your absence from New York, and despite the fact that you had to live in rather uncheerful quarters. During my stay in Princeton, I lived at the Oppenheims', and I think that had a good deal to do with my feeling more at home there. Have you seen anything of the Oppenheims, by the way? When I spoke to Paul some time ago, he expressed regret that he had only met you at some meeting, and he was hoping that you would come to see them at their house. If you did, I know he'd be very pleased, and maybe you'[d] enjoy it, too.

I wonder whether the fate of Princeton *vs.* Columbia is still in the balance; I should think it must be quite a difficult decision to make, and I very much hope that whichever course you take, it will prove to have been a felicitous choice that will be satisfactory and rewarding to you all! – Please give our love to Edith and the children, and our very best wishes to your mother.

As ever,

[386] On 2-3-4 March 1955, Hempel gave three lectures in the "Lectures in Modern Philosophy Series", on the topic "On the Nature and Scope of Scientific Understanding", at the Carnegie Institute of Technology. The three talks were entitled "Explanation and Prediction in Natural and Social Sciences", "The Status of Theories and Constructs", and "Science and Valuation".

[387] On 4 April 1955 Hempel gave a lecture entitled "On the Status of Theoretical Constructs in Science" at the Minnesota Center for Philosophy of Science.

[388] Presumably "Probability and the Theory of Knowledge", Nagel (1939/1954).

4.1.82 Nagel to Hempel, April 3, 1955

[Typewritten, Hempel 1999, Box 28, Folder 1]

April 3, 1955.

Dear Peter,

The contents of your letter were hardly a surprise, for I did not doubt for a moment that you would accept the Princeton offer. I think it was a wise decision, and as you know I thought you had not made the best choice a few years ago when you turned down a similar offer. I am sure you will be happy at Princeton, and I think you'll be in a congenial department. Ledger is a sweet person, and Gregory Vlastos[389] has impressed me as both a fine man and a good philosopher. In addition, I'll be terribly surprised if Abe Kaplan is not a member of the Princeton department before the year is out. And so I am convinced everything is set for a very bright future for you. In any case, our warm congratulations and all good wishes!

As for myself, I do *not* know whether I choose wisely in deciding to remain at Columbia. I did get (or rather, think and hope I will get) a handsome salary raise out of the affair, but I am not sure this will compensate for other things that are perhaps the inevitable consequence of being in a department that is set up and constituted as mine is. My decision to remain was controlled by a number of extraneous factors that seemed, and still seem, very weighty, though whether my weighting of them was realistic I have no way of deciding. I suppose the tantalizing thing is that, in the nature of the case, I'll never know whether or not I made a mistake. But however this may be, I've made my bed and I'll have to lie on it.

Bobby has been home from the hospital for a week, and everything has turned out well in that respect. He took the various ordeals like a man, and enjoyed the attention he received at the hospital from patients and nurses. He liked the card you sent him, but I think that what he enjoyed most was seeing Sandy again, whom he worships and adores but also fights with and teases.

I have ordered, but not yet received, the Lazarsfeld Reader,[390] though I expect it will arrive by the time of the next Philosophy Club meeting. I look forward to seeing you then.

Meanwhile, warmest regards and best wishes,

Always yours,

Ernest

4.1.83 Nagel to Hempel, May 16, 1955

[Typewritten, Hempel 1999, Box 28, Folder 1]

May 16, 1955.

[389] Gregory Vlastos (1907–1991) was a Turkish-born philosopher. After graduating from Roberts College in Istanbul, he received his PhD from Harvard University. He taught at Queen's University in Ontario, Cornell University, Princeton University, and finally, after his retirement, at the University of California at Berkeley. His area of research was Greek philosophy, and he wrote important essays on Plato and Socrates.

[390] Lazarsfeld and Rosenberg (1955).

Dear Peter,

I saw Sidney Ratner[391] at the New School conference yesterday, and he told me that you had been in an auto accident last week, while at Princeton. S.R. could not tell me how it happened or how bad it was. I am hoping that you haven't been hurt, and that the incident was a minor one; but in any case I was shocked, and am writing to find out how things really are. Drop me a card if you can, though I expect to see you Friday.

The Conference was not profitable, and I managed to kill a beautiful day. Frank charming as usual, but Lewis Feuer presented some dubious historical materials; and Henry Aiken was just a bore, taking an hour and a quarter to read his paper. Abe Kaplan gave a pleasant talk, though not much in it.[392]

In haste,
Ernest

4.1.84 *Nagel to Hempel, December 3, 1955*

[Handwritten postcard, Hempel 1999, Box 28, Folder 1]

Dec 3. 1955

Dear Peter,

I am most sorry I've missed seeing you and hearing your talk. Edith and I had made a previous date, and there was no help for it. But it is too bad there are such long intervals between our getting together. Our last meeting in Princeton was *so* enjoyable – though I am afraid I've never thanked you and Diane properly. I am flying next Wednesday to Minneapolis, so feel more rested than usual, I do not look forward to the trip, and besides I do not have anything to say which is not a rehash of old stuff.

I hope to see you at the next Philos. Club meeting. Meanwhile best to you and yours
In haste
Ernest

4.1.85 *Nagel to Hempel, December 19, 1955*

[Typewritten, Hempel 1999, Box 28, Folder 1]

25 Claremont Avenue,
New York 27, N.Y.
Dec. 19, 1955.

---

[391] Sidney Ratner (1909–1996) was an American philosopher. He studied philosophy at the City College of New York and at Columbia, where he also received his PhD in the history department. He worked as an economist in the US State department before joining the faculty of the history department at Rutgers University.

[392] The 37th semi-annual meeting of the Conference on Methods in Philosophy was held on May 15, 1955, at the New School for Social Research, New York City. The conference consisted of five talks by Philipp Frank ("Sociological Aspects of Science"), Lewis S. Feuer ("The Hedonistic Ethics and Seventeenth Century Science"), Donald Bell ("Social Power and Scientific Research"), Abraham Kaplan ("Intention and Expressiveness in Art"), and Henry D. Aiken ("Critical Interpretation and Artistic Intentions").

Dear Peter,

Though you certainly missed nothing by not hearing the paper at the Philosophy Club meeting last Friday, I am sorry you could not come, for I would have enjoyed the opportunity for a bit of talk, some serious some gossip. I guess I will not see you again until the next year, for even if you are going to the Boston meeting I will not be there[393] – I find such meetings exhausting and not very profitable, to say nothing of the expense. So this must be my occasion for wishing you and yours a good holiday season and a happy year to come.

As I wrote you, I went to Minnesota last week (flew both ways, on the whole pleasant enough, though the descent in New York from an altitude of about 21,000 feet was excruciatingly painful to my ears). I enjoyed my stay, despite the bitter cold, and though I hardly distinguished myself either with the prepared paper for the Conference on Development or the informal lecture before the Center + Philosophical Society,[394] I did not completely disgrace myself. It was fun seeing Herbert and exchanging points of view and prejudices; and I took in several discussions with him, Sellars,[395] Scriven and Meehl[396] – while I do not think I learned anything, I found Sellars more congenial than his writings made him appear, and Meehl especially interesting and impressive. I envy Herbert the opportunity for continued and intensive discussions – this seems just about impossible to arrange here in New York. He has probably written you by now that he is going to be Matchette lecturer in Brooklin College next Spring, and I am hoping to see something of him again at that time.

I spent a pleasant few hours with Mehlberg[397] two days before my trip, and he mentioned that Diane's mother died recently. I can well imagine that Diane has mixed emotions about her loss, but in any case please convey to her our warm sympathy. I find Mehlberg most agreeable and interesting, and I wish I knew how to help him find a position in a less remote and isolated part of the world. Do you get much chance to talk with him?

Though I've bunched my classes into two days a week, I find as little free time for work of my own as ever. Part of the reason is that I am giving a new course this semester, in the philosophy of law, for which I feel badly equipped and reading for which takes up both energy and leisure. In addition, I am running a seminar in the theory of knowledge, and I made the bad mistake of to choose Price's[398] Thinking and Experience as one of the major books for discussion. I have to read it, find it slow going, in addition terribly dull and

[393] It is likely that Nagel is referring to the fifty-second annual meeting of the American Philosophical Association Eastern Division, which was held in Boston from 27 to 29 December 1955. It seems that Hempel did not attend it either, as his name does not appear in the program.

[394] Nagel lectured on "Determinism and Development" that was published in 1957; see Nagel (1957a).

[395] Wilfrid Sellars was an American philosopher. Son of Roy Wood Sellars, he studied at the University of Michigan, University of Buffalo, and Oxford. Though he never completed a PhD degree, he taught at the University of Iowa, University of Minnesota, Yale, and finally at the University of Pittsburgh. His major works are "Empiricism and the Philosophy of Mind" (1956) and "Philosophy and the Scientific Image of Man" (1962).

[396] Paul E. Meehl (1920–2003) was an American psychologist and philosopher. Trained as a psychologist, he received his PhD in psychology at the University of Minnesota, where he came into contact with Herbert Feigl, and contributed thus to the foundation of the Minnesota Center for Philosophy of Science.

[397] Henryk Mehlberg (1904–1979) was a Polish-born philosopher. He studied at the Jan Kazimierz University in Lviv, from which he also received his PhD. He emigrated to the US after World War II; he taught philosophy at the University of Chicago.

[398] See Price (1953).

unrewarding. Indeed, partly as a consequence of the sense of harassment from having to keep abreast of things not quite up my alley, I've been in low spirits – though this is nothing unusual or noteworthy. If I may indulge in a profound statement, life is not a bed of roses.

My department is having lunch with I. Berlin[399] this Wednesday, and I expect that will be fun. Have you read B's recent HISTORICAL INEVITABILITY?[400] It is atrociously written, and in spots disappointingly poor though on the whole correct enough.

I must stop now. Once more, all good wishes and love from us all.

Ernest

4.1.86 Nagel to Hempel, January 7, 1956

[Typewritten, Hempel 1999, Box 28, Folder 1]

25 Claremont Avenue,
New York 27, N.Y.
Jan. 7, 1956.

Dear Peter,

Edith and I, returning from a shopping trip to Macy's this afternoon, had a pleasant encounter with Diane and P.A. in the subway, as you probably have heard. Despite the noise and crowd we had a bit of a chat, all too brief, until we had to get off the train at our station. (Edith had not seen P.A. since that summer in Vermont, and she is very much taken with him – he is certainly a fine boy). It is a pity our respective families see each other so infrequently: after all, in terms of travel-time Princeton is less far away than parts of Brooklyn, Queens and the Bronx! But it is an effort to get things organized for a visit, and I am acutely aware how busy all of us are.

I am nevertheless writing in the hope that perhaps we can arrange for some talk (and/or discussion), on the model more or less of the trip the two Johns (Cooley and Gregg) and I took to Princeton a few months back, and that you might be persuaded to spend a few hours in New York. John Cooley has repeatedly urged me to write you about this, and to have dinner with him and me and John Gregg (if the latter would be available). I haven't written before, because I thought I would see you at the Faculty Club and make arrangements that way. This hasn't turned out the way I had expected; and Diane mentioned to-day that it is unlikely you would come to the next meeting because the announced paper would not make it worth your while (indeed, I concur in your estimate, and I would not attend either if I had to travel any distance). I do not know your own schedule at Princeton, so I cannot suggest a time that would be clearly convenient for you. John Cooley usually goes off to his Connecticut house Friday afternoons and does not return until Sunday evenings, so the week-ends would not be good if your coming is contingent on having a discussion in which J.C. is present. Tuesdays and Thursdays are very full days for me from early morning until 4 P.M., but the times thereafter would be fine for me as well as on the remaining days of the week (though occasionally I am tied up with various University chores). It also occurs to me that you might have both the inclination and the free time during the inter-term period – at

[399] Isaiah Berlin (1909–1997) was a Russian-born philosopher and historian of ideas who spent most of his life in Britain. He studied at Oxford where he taught for his entire career. Berlin wrote extensively on many different topics: he is best known for his work in political philosophy (thanks to his essay "Two Concepts of Liberty"), and in the history of the Enlightenment and Romanticism.

[400] Berlin's lecture "Historical Inevitability" was delivered at the London School of Economics in 1953 and was published as a volume a year later, see Berlin (1954).

Columbia the exams begin on Jan. 23, and the new term does not begin until Feb. 8th. But all this is just a "feeler", and if there is the slightest chance of your being able to come to the city, I am sure we can work out a convenient time at this end.

I hope you got some rest over the Xmas holiday, despite your trip to Boston. Judging from reports, I did not miss much by playing hookey, though I would have liked to meet Father Bochenski[401] again.

Have you heard from Dean Jensen[402] of Dartmouth yet? I suggested that he get in touch with you, since I thought there was a chance (even if small) that you might be interested in spending the summer in Hanover. I had a brief talk with him here in New York. For a short spell I played with the eye of accepting his offer to take part in their summer program, but on second thought decided against it: I could not leave Edith in Vermont with the boys and her mother, especially since Edith does not drive; and it would be too complicated to try to find suitable living quarters for so many of us – and senseless too, since we are fairly comfortable in South Wardsboro and the boys are happy there. But I thought that quite apart from the remuneration you might want to get away for the summer.

Did you get a letter from B[u]rchard[403] of the Amer. Acad. of Arts and Sciences, about the contemplated conference in honor of Bridgman and Frank?[404] I think it is a happy idea to have one, and I sent my acceptance to participate. I feel certain that you have also been asked, and I hope you will say 'yes'.

I had some pleasant talks with John Myhill and Pat Suppes when they passed through New York during the holidays. I did not know until one of them told me that Wang was going to Oxford for good. Quite a surprise, and it seems that Oxford is importing quite a number of foreign scholars to make up for its own weaknesses.

Well, so much for to-day! But let us try to get together before very long.

Love to you all,

Ernest

4.1.87 Hempel to Nagel, January 27, 1956

[Typewritten, Hempel 1999, Box 28, Folder 1]

January 27, 1956

---

[401] Józef Maria Bocheński (1902–1995) was a Polish Dominican and philosopher. He studied philosophy at the University of Fribourg in Switzerland, where he received his PhD. He then studied at the Pontifical University of St Thomas Aquinas in Rome. He taught at the University of Fribourg for most of his career.

[402] Arthur E. Jensen (1903–1998) was an American scholar, working on Victorian literature, who joined Dartmouth in 1937, became a professor in 1946, and Dean in 1955.

[403] John E. Burchard (1898–1975) was an American historian. He studied at the University of Minnesota and at MIT, where he also taught for most of his career. Burchard served as President of the American Academy of Arts and Sciences from 1954 to 1957.

[404] After their retirement from Harvard in 1954, Philipp Frank and Percy W. Bridgman were honored by a conference, organized mainly by Gerald Holton, on May 5–6, 1956, at the Harvard Faculty Club. Participants included Frank, Bridgman, Georgio de Santillana, Henry Guerlac, Harcourt Brown, Jerome S. Bruner, J. Robert Oppenheimer, I. I. Rabi, W. V. O. Quine, Charles Morris, Howard Mumford Jones. Nagel was chairman, but Hempel did not give any talk or took any official role. The lectures were published in *Daedalus* (number 87, issue 1).

Dear Ernest:

John Cooley called me last night from his mountain retreat and told me that this coming Tuesday he would normally be in the country and asked what about next week. Monday Feb. 6 is not good for me because of a Faculty meeting in the afternoon, but on Tuesday I could come in after my morning classes. Now you mentioned in your letter that Tues is a heavy day for you until ca 4 pm, I hope it is not too much for you then to get together with John and me (and perhaps John Gregg?) afterwards? John Cooley seemed to think that Tuesday would be allright with you and said it would suit him; so I wonder whether we could make it, say, between 5 and 5.15 pm at the Faculty Club or at some other place you might wish to suggest; I just thought of the Club because it seems that there would be lounges and corners where we might go into a huddle and chat undisturbed. John Cooley said that he'd return from Connecticut on Monday, Feb 6. at any rate. – Could you drop me a postcard, some time at your convenience, to let me know whether this play is allright?

Ernest, you never told me that the John Dewey chair, which you mentioned some time ago as a vague possibility, has actually been established and given to you. My very, very warmest congratulations, in which Diane joins enthusiastically. I can assure you very sincerely that in my judgment no other philosopher, however distinguished, deserves that honor as much as you do. I very much hope that you will have many professionally happy and rewarding years as John Dewey Professor at Columbia.

Diane and I have not changed our minds about Princeton; we are very satisfied with our decision to come here, and we like living and working here. There are, of course, some minor reservations about some matters concerning the Department if and when there is an opportunity to speak just entre nous.

I hope we'll be able to meet on Feb 7, and I also am planning to attend the next Phil Club meeting and hope perhaps to have a chance for a little personal exchange with you then.

In the meanwhile, best wishes to you all from us all –

4.1.88 Nagel to Hempel, January 28, 1956

[Typewritten, Hempel 1999, Box 28, Folder 1]

25 Claremont Ave.,
New York 27, N.Y.
Jan. 28, 1956.

Dear Peter,

Tuesday, Feb. 7th will be fine for me, especially since classes for the spring term do not begin until the 8th – so that I am entirely free on the day you are planning to come in. (I hope postponing your trip from the day you had originally suggested, Jan. 31st, is not too great an inconvenience for you. John Cooley had hoped the present arrangement would be possible, because he wished to be in Conn. during the coming week.) One other thing occurs to me: perhaps you and I might meet in advance of meeting with John, I would love it – our apartment is as good a place as any, although I could meet you downtown (e.g. at the Penn. Station) if this would interfere with any other plans of yours. Any time you suggest would be just right for me. Just drop me a post-card. Meanwhile, I'll try to find out from John whether he wants to have dinner at the Faculty Club or prefers some other place – I wish I could ask you all to have dinner at our house, but this is rather difficult at present because of Edith's schedule at the lab plus the complication of my mother's special diet. I'll write you as soon as I get [a] word from him.

Thank you for your congratulations on the Dewey chair. To be entirely frank, I feel somewhat embarrassed over the title, for a variety of reasons; and the real point behind the designation is the financial arrangements that had been worked out in connection with it. I'll tell you about it if we meet privately.

Anyhow, I am looking forward eagerly to seeing you; and meanwhile, warmest regards to all of you.

Ernest

4.1.89 Hempel to Nagel, January 31, 1956

[Typewritten, Hempel 1999, Box 28, Folder 1]

Jan 31, 1956

Dear Ernest:

Many thanks for your note. My schedule on Tuesdays does not make it possible for me to leave here before 2.36, a train which is due at Penna St at 3.36. I have a few errands to do which I have put off for some time; they may not take me more than an hour, or possibly just a bit less. I would be very glad indeed if we could have at least a brief chat before we got together with John, and I will plan to give you a ring at your house when I am done with my errands. We can then decide whether there is any point in my coming to your house or whether it is too late for that. In order to give our little private meeting a better chance of materializing, we might think of putting the date with John off until, say, 5.30 or 6. There is no need to trouble with replying to this or telling me about John Cooley's plans as long as Tues pm suits you both; you can tell me on Tuesday when I call you just what the details of the arrangement with John Cooley are.

All our best to you all!

4.1.90 Nagel to Hempel, April 22, 1956

[Typewritten, Hempel 1999, Box 28, Folder 1]

Dear Peter,

In the flurry of talking to so many people last Friday, I somehow forgot to ask you about your more specific plans for going to Boston two weeks hence. It occurred to me that if I knew what train you were taking, I would try to board the same one, so that we might spend part of the trip in pursuing a couple of the issues we got started on at Princeton. I have not consulted train schedules thus far, but I expect to take a train sometime in the afternoon on Friday, May 4; and if you care to let me know which train you'll be (as soon as you know) perhaps I can get on it also in New York.

I am sorry you took the time to come to New York Friday for so poor a paper – but at least I did warn you. On the other hand, some of the comments on it seemed to me just irrelevant; and though I doubtless deserved many of the harsh criticisms, my opinion of several of our colleagues has not improved. I also have the impression (though this may be quite unfair) that Oppenheimer tends to throw his weight around, and that even in quantum physics things are not quite as settled as he appears to hold – but I admit that this comment may have as one of its determining conditions my wounded vanity. I did not get around to replying to your own criticisms – obviously not because I did not think them important, but simply because did not get the time for it. I would very much like to discuss them with you.

Please thank Diane for me once more for all her kindness and trouble last Thursday. Despite her disclaimer, it meant a lot of extra work for her – the one serious fly in the ointment of our visit. I know I am speaking for all three of her guests in saying how much we enjoyed the hospitality. And it was just wonderful to have a chance to talk with you.

My love to you both.

Ernest

4.1.91 Hempel to Nagel, April 25, 1956

[Typewritten, Hempel 1999, Box 28, Folder 1]

Wednesday, April 25, 1956.

Dear Ernest,

Thank you very much for your letter, which must have crossed mine in the mails. I was very glad to see how closely we were in agreement in our reactions to last Friday's meeting – with one exception, and here I was right, of course, and that is the quality of your paper. It was a good thing I wrote you before receiving your note how much I had liked it; I think it was a *superb* job! I certainly would like to talk with you about it some more, and would also enjoy having an opportunity to continue our discussion of some of the points that came up in our talks here last Thursday. It would be great fun to go up to Boston with you; thanks for the suggestion. I had thought of three possible trains: The 2 pm or the 4 pm from Penn Station or the Advanced Merchants Limited, 4.45 from Grand Central. Now the former two have the advantage for me that I need not change stations in NY; my trains from Princeton are due at Penna at 1.36 and 3.36 respectively; I would take the latter also if we were to agree on the Adv. Merchants Limited. Now the trains from Penna Station come from Washington (they don't stop at Princeton Jctn) and are likely to be more crowded. On the Adv. Merchants Limited, we would, I think, be able to get seats together if we met at the information booth of Grand Central in good time, say at 4.15; that train is due in Boston at 8.45; the 4.00 pm from Penna Station is due at 8.30. Hence I would suggest that you make a choice – if this suits you – between the Adv. Merchants and the 2.00 pm from Penna Sta (due in Boston at 6.45); in the former case, we would agree to meet at 4.15 at Inf. Booth, Grand Central; in the latter case, I would come up as fast as possible to that gate where the 2.00 pm leaves; and if by the time the train is announced and the gate opened I am not there, I'd suggest strongly that you go down and hold two seats and I'll try and find you on the platform outside the train or in your coach. If one of us should be too late to make the train agreed upon, the other should not wait, of course, but proceed to Boston. Just drop me a line at your convenience and tell me whether one of the two suggested trains would suit you. (Adolf Grunbaum writes me he is flying from Newark – looks impressive: 65 min; but Diane prefers me to take the train, and those low-flying planes are so much more conducive to air-sickness.)

Bernays[405] is speaking tonight, and we are having a dinner for him at the Peacock Inn before his lecture.

I started reading Urmson's[406] book on analytic philos between the wars and would like to talk to you about it.

Diane joins me in sending you all our love.

[405] Paul Bernays (1888–1977) was a Swiss mathematician and philosopher of logic and mathematics. He taught at Zürich, Göttingen, and later again at the ETH when he emigrated after the Nazis took power.

[406] J. O. Urmson (1915–2012) was a British philosopher in Oxford, working on linguistic and ancient philosophy. He is known as the editor (with G. J. Warnock) of the papers of J. L. Austin and for his mentioned book on the history of analytic philosophy, see Urmson (1956).

4.1.92 Nagel to Hempel, October 24, 1956

[Handwritten, Hempel 1999, Box 28, Folder 1]

The Presbyterian Hospital
Bway. 168 St
Oct 24, 1956

Dear Peter,

I hope you can make out this writing, but it is the best I can do on my back, and I do not want to note to say how pleased I was to get your letter. Since I could not have typed it myself it is some sort of argument for the existence of the external world.

I think I'm beginning to see a bit of daylight – at least since this evening (it is now 10 PM) – though for a while things looked grim, and I was already resigned to go through another operation on my back. But this evening the doctor in charge examined me, found definite progress and ventured the prediction that in about 5 days I might be able to stand up. So I feel relatively cheerful, even though I cannot oppress entirely the sense of having wasted a month already, of being a terrible burden, and of having to remain pretty useless for an indefinite time. So many things left undone! I was in the midst of typing my paper for the Carnap volume (at long last), and now still some more delay! It is not a good paper, and certainly is not work waiting for.

The first volume of the Feigl series[407] arrived when I was on my back at home, and I haven't had a chance to do more than glance through it. Even so, I can't get steamed up over Sellars, who is certainly overrated by Herbert.

But there is no sense in such casual remarks. I hope I'll see you before very long, and have a good talk. Meanwhile, remember me to Van.

Love to you and Diane,
Ernest

4.1.93 Nagel to Hempel, January 21, 1957

[Typewritten, Hempel 1999, Box 28, Folder 1]

25 Claremont Ave.
New York 27, N.Y.
Jab. 21, 1957.

Dear Peter,

John Cooley called me by phone this evening from Connecticut, to ask whether it would be possible to arrange for a meeting with you and Van either sometime this week or next. I told him that I saw you last Friday, and that you indicated you would be too pressed until the end of the month because of the beginning of the new semester and would prefer some later week. John then suggested that I write you to find out what days of the week would be most convenient for you, assuming that some week early in February is not ruled out. John finds that Monday, Wednesday and Thursday are his best days and these would be fine with me too – I think both he and I would give a slight edge to Thursday. [John Gregg appears to be

[407] The Minnesota Studies in the Philosophy of Science was launched in 1956. The first volume was edited by Herbert Feigl and Michael Scriven and was entitled *Foundations of Science and the Concepts of Psychology and Psychoanalysis* (see Feigl and Scriven 1956).

out of commission at the moment – according to John Cooley he is down with the mumps. This can be rather mean at his age. I'll find out more about his condition within the week.]

Could you drop me a post-card about possible days? Even if John Gregg is not back in trim, the two other two of us could take the train.

It was good to see you after such a very long interval. Hope Diane is quite recovered. My best to you both.

Ernest

4.1.94 *Hempel to Nagel, January 31, 1957*

[Typewritten, Hempel 1999, Box 28, Folder 1]

January 31, 1957

Dear Ernest:

I am sorry to have been so laggard in answering your note; but first there was some uncertainty in my schedule for February, and then Van was out of town. Finally reached him this morning. He likes the idea but has to prepare a lecture on his recent philos. work (to be given at the Institute on Monday Feb 18) and would like to keep his schedule free until then. Wednesday Feb 20 and Mo Feb 25 would suit him fine; I would have a preference for Monday since Wed is "ladies' day" on the Penn RR with a reduced fare that Diane often takes advantage of to go to NY. So what about Monday Feb 25, and you with one or two Johns come here, and we have a simple dinner at our place; Diane is looking forward to seeing you, and sends warm greetings. I am sorry to hear John Gregg has the mumps; I do hope it will all go well, and that he will be able to come along.

Don't bother to write now: I expect to see you on Feb. 8/9 at the NYU meeting.

I hope you are feeling a bit stronger every day; it certainly was a relief to see you up and around (and looking quite well) again!

Our best to you and your family!

As ever,

4.1.95 *Nagel to Hempel, October 28, 1957*

[Typewritten, Hempel 1999, Box 28, Folder 1]

25 Claremont Avenue,
New York 27, N.Y.
October 28, 1957.

Dear Peter,

This is really just a note to say hello, and to report that we got home all right from Minneapolis (in fact, in what for me is a record time, of which I feel boastful – just 3 days!). At the moment we are all flourishing, though we've had some bouts with various bugs – I was in bed ten days ago with the flu, whether Asiatic or not I don't know. I've finally settled down to my long grind, and feel quite disheartened at present over a) the unbelievably poor quality of some of the chapters I wrote seven years ago and which I had fondly believed would require only minor revisions at best, and b) the snail-like pace at which the work proceeds – I'll have to live to be as old as Methuselah to ever finish, and there is fortunately no danger of that. My one rational excuse and source of melancholy comfort is that one cannot make a silk purse out of a sow's ear.

There is one question I would like to raise, but I don't expect you to answer it unless you have time hanging on your hands, which I do not think is likely. You may remember that at Minneapolis we briefly discussed the distinction between fundamental and derivative laws, and that to my query whether in deducing the latter from the former it is permissible to use other than instantiation derivations – you answered, if I recall aright, that the intent of the distinction is to permit such other derivations (e.g., by adding other premises to the fundamental laws). But if so, then it seems that many statements which one would not wish to call laws at all, turn out to be derivative laws, and hence laws. For example:

All the screws in my car are rusty

can be derived from the presumably fundamental law

All iron exposed to oxygen rusts

when coupled with

All the screws in my car are iron, and
All the screws in my car have been exposed to oxygen.

And I think a plausible case can be made for maintaining that Nelson G's famous example: All the coins in my pocket on VE-day are silver,[408] is also a law, by a parallel argument (if, for example, you allow that All US coins of a denomination greater than 5 cents are silver is lawlike).

Please don't bother answering, for we may be able to discuss the point when I next see you. By the way, had a note from John Cooley from Maine. He's had some back trouble, but seems to have got over it. He plans to come down to New York before long, and expressed the hope that we might get together with you.

Best to you and Diane, from all of us.
Always yours,
Ernest

4.1.96 *Hempel to Nagel, November 1, 1957*

[Typewritten, Hempel 1999, Box 28, Folder 1]

Princeton, N.J.
November 1, 1957

Dear Ernest:

It was good to hear from you and to learn that you and your family had a good (and amazingly fast) trip back (I wonder whether having Henry Mehlberg trail you proved difficult or not), and that you are all reasonably well. We have had bouts with the flu, too; all of us except for Diane got the bug, and I even despite a vaccination two weeks earlier; but perhaps I had a lighter case as a result, I was in bed for only three days, and had a temperature on only two of them.

I sympathize with your sense of discouragement, Ernest, but I know also that it is greatly exaggerated. Just a couple of weeks ago, I reread your pieces on Toulmin's and

[408] See Goodman (1955, 118).

Braithwaite's books and again found them really excellent and very helpful as a preparation for discussing parts of the books in the senior seminar;[409] and I will be very happy when your book – or vol.I of it in case it is more than one volume – gets out and hits the bookstores and the eagerly waiting readership; it will be an important contribution and a great success both professionally and financially, I am quite sure!

It will be good to see you again and have a long talk with you on theoretical matters and on all sorts of things of a more personal character. And it would be great fun to get together with John Cooley, too. I have a notion that I won't be coming in for the next meeting of the Philos. Club either; I am quite busy and I have a feeling that I would not get enough out of Tillich's talk to tear myself away from my work here. I like Tillich and I have enjoyed listening to him, but it is really too far removed from my interests, and the prospect of simply *having* to contribute a 3-minutes' remark does not make the occasion any more attractive. I have thought however of coming in December for Stace's talk, and of bringing Hare[410] (if he is still here) or Smart[411] along as a guest.

I gave two talks in Sidney's graduate seminar at NYU and enjoyed it; there are several quite bright students in the group. The one thing that makes me feel uneasy is that there is too little time for discussion, so that issues cannot be thrashed out nearly as well as in a graduate seminar, and so I think the students must be left with a sense of frustration. I wonder what your impression will be. – Early in October, I gave a "College Lecture" to some 500 students at Mt. Holyoke[412] on, if you can believe it, the nature of logical truth, and I wound up with a general and brief description of the decision problem and of Goedel's and Church's results. It met with considerable interest, but I cannot tell, of course, whether it was really well understood.

As for derivative laws, I am not clear about the issue myself; but it would seem to me that instantiation pure and simple would not suffice to derive certain laws that we would surely want to qualify as derivative laws, such as Kepler's. Suppose, for ex., the law of gravitation is formulated as saying: For any physical body x and for any physical body y, if the distance between them is R, then the central force etc. is m(x). m(y). R^{-2}. c, where c is a constant. Now to infer Kepler's laws or similar consequences for a planet, say, the Earth, moving about the sun, we would have to substitute 'the Earth' for 'x' and 'the sun' for 'y', but now we would need an additional set of premisses of singular form stating what the masses of the Earth and the sun are, and giving their distance and relative velocity at some particular time. Thus, there is involved here some process of instantiation in a wide sense, which however, I do not know how to explicate. One thing that might be useful here, and that would serve to rule out your examples, would be the requirement that a derived formula, to count as a derived law, must not only be syntactically of universal form, but must be lawlike in the sense pointed out by Nelson Goodman,[413] namely, that we would use it to support counterfactual conditional and hypothetical conditionals. And this we would clearly do in the case of Kepler's laws when formulated for the sun as the central body: we would apply it to any new planet, to any super-sputnik, etc., whereas we would not say of a shiny new iron screw that if it were put into your car it would be rusty. But here I am at the end of wits.

---

[409] See Nagel (1954, 1957b).

[410] R. M. Hare (1919–2002) was a British philosopher who worked mainly on moral questions, especially on meta-ethics. He graduated from Oxford University and worked there for decades. Hare visited Princeton in 1957.

[411] J. J. C. Smart (1920–2012) was a British-Australian philosopher who worked mainly on metaphysics and the philosophy of mind. He graduated from Oxford and was Chair of the Philosophy Department at the University of Adelaide between 1950 and 1972. He visited Princeton in 1957–8.

[412] Mount Holyoke College is a private liberal arts women's college in Massachusetts.

[413] See Chap. 1 in Goodman (1955).

Will you be going to the Harvard meeting at all, or will you proceed directly to Berkeley? Adolf Grunbaum wrote me recently that he has decided to attend both occasions, flying to Berkeley after a day or two at Cambridge. What energy! (Apparently Tarski wired him and urged him to come.)

Diane joins me in sending warm greetings and best wishes to you all.

Cordially,

4.1.97 Nagel to Hempel, November 15, 1957

[Typewritten, Hempel 1999, Box 28, Folder 1]

25 Claremont Avenue,
New York 27, N. Y.
Nov. 15, 1957.

Dear Peter,

Albert Hofstadter, whom I just met a few minutes ago, told me that you will not come this evening (as you indicated in your letter of a week or so ago, for which many many thanks). And so I am writing you in compliance with John Cooley's suggestion – he was in New York for the past two days – about the possibility of the three of us getting together for a bit of talk sometime soon, either in Princeton or if that is possible for you in New York. Clearly the coming week is out, and John also has dates for Dec. 2 and Dec. 9, as well as the Thanksgiving week-end. He would also prefer some mid-week afternoon and/or evening. Would you care to suggest some time convenient for you, whether in Princeton or New York?

Since John Gregg is no longer around, if we come to Princeton it will have to be by train. So we would have to leave early enough to catch the last shuttle from Princeton (as I recall that is around 10 P.M.); and in order to get in some discussion, perhaps we could meet you a bit earlier than last year. Also, John Cooley raises the question whether you would approve having Sidney Morgenbesser[414] replace John Gregg.

Needless to say, I hope that you can find time for this; and I think also in order not to place extra burdens on Diane, it should be understood that if we meet in Princeton we will eat out somewhere.

John Cooley is in good form, his back is behaving, though he has some trouble with one of the knees.

My own work is moving along at snail's pace, and it continues to be discouraging. But I have no alternative to keeping at it, even if days go by when I have nothing to show for the work I put it.

Have you seen the new book by Dray, Wm. "Laws and Explanation in History" (Oxford).[415] I am in the midst of it. A good deal of the material I've read thus far is devoted to a critique of deductive model view, and your paper is almost always at the center of the stage. Some of the points are well-made, and I don't think you would dissent from them. But

[414] Sidney Morgenbesser (1921–2004) was an American philosopher. He originally studied at the Jewish Theological Seminary and was ordained as a rabbi. He also studied at the City College of New York, the University of Pennsylvania, where he got his PhD. He then taught at Swarthmore College, the New School for Social Research, and Columbia University. He did not publish very much but was very influential as editor of anthologies and volumes.

[415] William Dray (1921–2009) was a Canadian philosopher. He studied at the University of Toronto and at the University of Oxford, where he received his PhD in philosophy. He taught at the University of Toronto and at Trent University in Ontario. His main field of research was the philosophy of history and historiography. *Laws and Explanation in History* (1957) is his most important work.

thus far I have no idea what his positive analysis is of historical explanations. I'm sure you would find the book interesting, and I think it would be a good subject for discussion if we should meet.

The rest of the family is pretty busy too, and for the moment in good health.

Warmest regards to you and Diane,

Always yours,

Ernest

{P.S. Judging from the recent letter from Schilpp (enclosing a copy of a letter from Carnap) all of the contributions for the Carnap volume – including one from Gödel[416] *– must now be in. Is this really so? When I next see you, I must tell you about the dénouncement of the affair – Gödel (re the [N/A] by Newmann and myself) – quite fantastic!*

E.}

4.1.98 Nagel to Hempel, November 29, 1957

[Typewritten, Hempel 1999, Box 28, Folder 1]

25 Claremont Avenue,
New York 27, N. Y.
Nov. 29, 1957.

Dear Peter,

After talking to you this evening, I got in touch with John Cooley. Sunday, Dec. 15 will be just fine for him. I haven't been able to contact Sid Morgenbesser thus far, but I had talked to him before calling you, and he had indicated that this date would most probably work out well for him too, since he was expecting his mother to make up her mind about the surgery sometime this coming week. So tentatively at least let's assume that Dec. 15 will be the day.

What time would you like us to come? John said that he would plan to come down from Hartford the very same day, so that he probably will not arrive in New York much before noon. Could you drop me a post-card suggesting a convenient time for you – it didn't occur to me to ask while we were on the phone.

I hope the moving will not be too strenuous, and that you will feel settled and comfortable before very long. I still recall our last moving operation some seven years ago as a nightmare.

I've decided not to go to the conference in Berkeley during Christmas week, nor even to the Association meeting at Harvard[417] – though I would very much like to hear your comments on the two papers which I have seen in the Journal but have not had the time to read carefully. Maybe there will be time when we see you in Princeton to rehearse your remarks to us.

Best to you and Diane,

In haste

Ernest

[416] Gödel's paper has appeared only decades later in the collected works (see Gödel 1995).

[417] The fifty-fourth annual meeting of the American Philosophical Association, Eastern Division took place at Harvard University, December 27–29, 1957.

4.1.99 Nagel to Hempel, December 28, 1957

[Typewritten, Hempel 1999, Box 28, Folder 1]

25 Claremont Avenue,
New York 27, N.Y.
Dec. 28, 1957.

Dear Peter,

By the time you get this note, you will have had your fill of meetings, and I realize I am writing at a very unpropitious time. However, I had promised John Cooley that I would get in touch with you about the possibility of arranging another discussion with you perhaps sometime in January and perhaps in New York. What do you think? If you could come to New York we could dine at some good restaurant, as well as discuss a number of issues. According to John, anytime in January would suit him. Drop me a line when you can.

I did not find a chance to talk to John in private on his last trip, to find out what were his plans in relation to Columbia. My chairman (Gutmann) told me that John had talked with him, and asked for another leave of absence for the following year. This sounds to me very much like the beginning of the end, and that John is not likely to return. I cannot tell you how desolated I feel. To be sure, there is still hope, for after all he has not resigned thus far. But John disliked New York and disliked most of my department (I am not far behind him in this latter respect), and under the circumstances he would be doing the reasonable thing by cutting his ties with Columbia.

As usual, I enjoyed listening to Stace at the last meeting of the Club. Smart was present, and he made a better impression on me than on the previous encounter. But the whole discussion did not amount to much; and you certainly were wise in not bracing that utterly fearful weather that hit New York for the occasion.

Our best wishes for the New Year to you and Diane and the children.

Always yours,
Ernest

4.1.100 Hempel to Nagel, January 13, 1958

[Typewritten, Hempel 1999, Box 28, Folder 1]

January 13, 1958.

Dear Ernest:

I am sorry to be so late answering your letter; but I knew my schedule for the first three weeks in January would be quite crowded, and after that, Diane had tentative plans for going up to Boston for a few days, which would limit my mobility and availability to some extent. But now the prospects are definite enough for me to make a suggestion.

First of all, if you, John, and Sid don't mind the trip to Princeton, Diane would be *delighted* to fix some supper for us and to catch a brief glimpse of you; and I think we have here a pretty good chance for quiet conversation. Would Friday, Jan. 31 or Monday Feb.3 or possibly Tues. Feb 4 suit you? I will keep these days open for the time being. I expect to come to the next meeting of the Philos. Club this Friday; perhaps we can make somewhat more definite plans then?

I am planning to bring Smart to the meeting on Friday; it will be his last chance since he and his family will be leaving for the West Coast fairly soon thereafter.

The Harvard meeting was rather good, it seems to me, as meetings go, although two of the sessions I attended struck me as disappointing. The Determinism symposium came off rather well,[418] I thought, especially thanks to a lively and rather disciplined talk by Wiener.[419] None of the central issues gets really thrashed out, of course, but then, I don't expect that; and there were no digressions into irrelevant matters and no crackpot remarks. I am more and more impressed with the difficulty of the rather unrewarding task of the program committee. – You probably have heard of the Randall[420]-Sellars incident. I was not present, but I heard many expressions of strong disapproval and dismay concerning Randall. No doubt many things are eating him, but even that realization does not make the incident less painful.

It will be good to see you Friday! Many good wishes to all of you from all of us!

As ever,

PS Maurice Mandelbaum called from Cambridge last night and mentioned you had suggested Shimony[421] as a possible candidate for the Hopkins vacancy. At Maurice's request, I got in touch with Shimony and to my very pleasant surprise he was definitely interested and was going to get in touch with Maurice. If this materialized, I think Maurice will really have a first-rate man, and a prince of a person. I am glad you made that suggestion!

4.1.101 Nagel to Hempel, January 21, 1958

[Typewritten, Hempel 1999, Box 28, Folder 1]

25 Claremont Ave.
Jan 21, 1958.

Dear Peter,

Heard from John and just finished speaking with Sid Morgenbesser. Their preferences are, in that order, for:

Tues. Feb. 4 {*«– yes*}
Mon. Feb 3
Fri. Jan 31

[418] The Determinism symposium was held as the first annual conference of the New York University Institute of Philosophy at Washington Square, New York, on February 9–10, 1957. For the proceedings, see Hook (1958).

[419] Norbert Wiener (1894–1964) was an American computer scientist, mathematician and philosopher. He was known as a prodigy (who graduated from Harvard at the age of 17) and is seen as the originator and most influential researcher of cybernetics. At the "Determinism" symposium the speakers were Wiener, Adolf Grünbaum, and Michael Scriven, with Hempel as commentator and W. V. O. Quine as chairman.

[420] J. H. Randall Jr. (1899–1980) was an American philosopher who graduated from Columbia in 1922, and spent his whole career there. At the Harvard meeting Sellars was a speaker (with Rogers Albritton) at the "Substance and Form in Aristotle" symposium, where Randall was a commentator (with Joseph T. Clark), with Gregory Vlastos as chairman.

[421] Abner Shimony (1928–2015) was an American physicist and philosopher, with a philosophy PhD from Yale (1953) under Carnap's supervision and a second PhD in physics from Princeton (1962) under Eugene Wigner. Shimony worked on quantum theory and developed important theories and ideas.

So unless you veto the idea, lets assume that we'll come down on Feb 4th. According to the time-table I have (which may be out of date) there is a train from Penn Station at 3 P.M., to arrive at Princeton at 4. P.M. Where should we meet you? In your office? And is the time all right?

It was good seeing you last Friday. But that was a queer paper by Ducasse.[422] I had a pleasant day in Washington, saw Alfred Tarski and Kleene, as well as some of the history of science group (including your colleague Chas. Gillispie,[423] whom I like a good deal).

All good wishes meanwhile,
In haste
Ernest

4.1.102 Nagel to Hempel, April 3, 1958

(On behalf of The Tamiment Institute and Library)
[Typewritten, Hempel 1999, Box 28, Folder 1]

April 3, 1958
Professor Carl Hempel
Department of Philosophy
Princeton University
Princeton, New Jersey

Dear ~~Carl~~ {*Peter*},

On behalf of the Advisory Committee of Tamiment Institute, it gives me great pleasure to invite you to attend a seminar on the philosophy and practice of education which the Institute is sponsoring in the first week of June next.

The seminar is the first of an annual series sponsored by the Institute and aimed at dealing with problems that lie in the forefront of the nation's intellectual concerns. The program for the seminar is as follows:

THE CHALLENGE TO AMERICAN EDUCATION

1. The Aims and Content of Education

   Sidney Hook
   George N. Shuster

2. Education in the Sciences and the Humanities

   Douglas Bush[424]
   Ernest Nagel

---

[422] Curt John Ducasse (1881–1969) was a French-American philosopher who obtained his PhD from Harvard in 1912. He worked mainly on aesthetics and philosophy of mind. He was also president of the Philosophy of Science Association between 1958 and 1961.

[423] Charles C. Gillispie (1918–2015) was an American historian of science, with a PhD from Harvard (1949). He established the History of Science program at Princeton during the 1960s.

[424] Douglas Bush (1896–1983) was an American literary critic. He studied at the University of Toronto and received his PhD from Harvard University, where he taught for most of his life. He was an expert on Renaissance literature.

3. Education and the American Scene

   Arthur Bestor[425]
   John L. Childs[426]

4. Education and the World Scene

   Hans J. Morgenthau[427]
   Reinhold Niebuhr[428]

Each of the persons named above is preparing a paper which will be distributed in advance of the seminar to those who attend. The papers will not be read at the panel sessions; instead the authors will summarize their contents briefly and comment on the paper of their opposite number. This will be followed by a general discussion, in which all the participants will be invited to join.

We are planning upon an attendance of twenty persons in all. Thus the seminar should be intimate, informal, and conducive to a thorough exploration of different viewpoints.

The meeting is to take place at Tamiment-in-the-Poconos, Pennsylvania, which is about eighty miles west of New York City. The setting of the newly erected cultural center is most appropriate for a scholarly meeting. The seminar sessions will be held on June 3rd and June 4th. We are inviting participants to arrive on June 2nd, and they will be able to depart on June 5th or, if they choose, the preceding evening.

If it is possible for your wife to accompany you, Tamiment Institute will be happy to welcome and accommodate her. The Institute will, of course, bear all the travelling expenses; and your wife and you will be our guests at Tamiment.

I should appreciate your letting me know at your earliest convenience whether you are able to accept this invitation. The members of the Advisory Committee are aware that you are busy with many commitments. But this occasion for the discussion of educational problems is, they feel, of primary importance at the present time.

I do hope that you will be able to accept this invitation. I think you and your wife can look forward to an intellectually stimulating and physically enjoyable time.

Sincerely yours,
{Ernest}
Ernest Nagel
for the
Advisory Committee

[425] Arthur Bestor (1908–1994) was an American historian. He studied at Yale University, where he received his PhD in history. He was a critic of Progressive education.

[426] John L. Childs (1889–1985) was an American philosopher. He studied journalism at the University of Wisconsin before moving to China. He went back to the USA to get his PhD from the Union and Teachers College of Columbia University, where he was appointed assistant professor. Strongly influenced by Deweyan pragmatism, he worked extensively on the philosophy of education.

[427] Hans J. Morgenthau (1904–1980) was a German-born political theorist. He studied philosophy at the University of Frankfurt and jurisprudence at the University of Munich. He received his doctorate law degree from the University of Frankfurt. He moved to the USA in 1937; he taught at Brooklyn College in New York City, the University of Kansas City, and the University of Chicago.

[428] Karl Paul Reinhold Niebuhr (1892–1971) was an American theologian who studied at Yale's Divinity School. He was appointed Professor of Christian Ethics at the Union Theological Seminary in New York City. His most important book is *Moral Man and Immoral Society* (1932), in which he laid the foundations of his Christian realism.

4.1.103 Hempel to John Cooley, April 12, 1958

[Typewritten, Hempel 1999, Box 28, Folder 1]

Princeton, April 12, 1958.

Dear John:

Many thanks for your note. Friday, April 25 suits me very well; the other two days are taken, alas. So I do hope we can get together on Friday. Since that will be the time when spring should be just at its best here in Princeton, Diane and I want to urge you to have the session here, though I much appreciate your invitation and will of course be glad to come to New York if you prefer. But though the train trip is a nuisance and an expense for all of you, please do consider coming out here: Diane will be delighted to have you come for dinner; you know that we have a reasonable quiet place for talking shop both before supper and after; and then, there should be the charm of Princeton in the spring. I am writing Ernest in the same sense; would you please drop him a line with your reaction: I expect to see him at the Philo Club meeting on Friday, April 18, and he can tell me then about you, and about Sid Morgenbesser.

I have some questions concerning rules for induction, and their justification, and would be glad if we could air those at our next meeting.

Very cordially,

P.S. I find that you have given no return address, and I hope I am doing the right thing in sending this note to Hartford.

4.1.104 Hempel to Nagel, April 12, 1958

[Typewritten, Hempel 1999, Box 28, Folder 1]

April 12, 1958
Professor Ernest Nagel
Advisory Committee
The Tamiment Institute
7 East 15th Street
New York 3, N.Y.

Dear Ernest:

I am very much honored by the invitation of your Committee to attend the forthcoming seminar planned by the Tamiment Institute. To my sincere regret, I find that at the time of the meeting, I will be in the midst of several groups of examinations here in Princeton, so that, unfortunately, I will be unable to attend this time.

With renewed thanks, and with best wishes for the fullest success of the seminar.

Cordially yours,
Carl G. Hempel

4.1.105 Hempel to Nagel, April 12, 1958

[Typewritten, Hempel 1999, Box 28, Folder 1]

Princeton, April 12, 1958

Dear Ernest:

Diane and I were greatly touched by the invitation to the Tamiment seminar; the plan sounds very attractive indeed! Unfortunately, I will be in the midst of our examinations here and won't be able to get away, and I have just written you a formal note to that effect; I should add informally that also I am so far behind schedule with all sorts of work that it makes me feel quite depressed, and I want to use every bit of available time to catch up. Many thanks for asking me!

Diane has gone through a long series of thorough examinations and had been pronounced to be in excellent shape; the intestinal trouble, which seems to be gradually subsiding, is attributed by the physicians to a thyroid deficiency, and she has been put on an increased dose of thyroid medication.

I got a note from John Cooley yesterday, saying that he would write you, too, and proposing April 23, 25, or 26 for a get-together. Of these days, only April 25 is free for me, and it suits me very well; I do hope that it will fit into your plans and Sid Morgenbesser's. Since at that time, spring should really be at its best here, Diane and I want to urge you to have the session here rather than for me to come into New York. You know Diane is always delighted to have you come! I am aware that the train trip makes a session in Princeton a rather costly pleasure for you, and if you wish, I'll come to New York; but I would definitely prefer to see you come here, where we also have a place for undisturbed discussions. I'll write John Cooley today to the same effect; and I will ask him to let you know his reactions. Then you need not bother to write me in case you plan to come to the next meeting of the Philosophy Club: you can tell me then. – As for our topics, I would like to bring up a question concerning the justification of induction.

I heard a rumor that Wilfrid Sellars is going to Yale and am eager to learn from you when I see you what there may be to this. He would no doubt make an efficient and energetic chairman when Hendel retires.

Ledger Wood had sent out letters proposing May 9 and 10 for the Advisory Council meeting without finding out whether other people in the Dept had any previous commitments for that date and I have: I am to give two talks {*at Yale*}, in Directed Studies, and before the Grad Philos Club, on May 8 and 9, and thus would have to miss the Friday evening part of the occasion. Ledger is now trying to shift it to the following weekend, though this would interfere with the NY Phil Club meeting; but he wants to get first the reactions of the chairman and the secretary of the Council before suggesting the new date to anyone else. – I hope you're all well! Best wishes to you, Edith, and the children! As ever,

4.1.106 Nagel to Hempel, April 14, 1958

[Typewritten, Hempel 1999, Box 28, Folder 1]

25 Claremont Ave.,
New York 27, N.Y.
April 14, 1958.

Dear Peter,

I was relieved to learn from your letter that there is nothing seriously wrong with Diane, and that your worries were happily groundless. I had been tempted several times to call or write you, but feared to obtrude.

Of course I'll see you this Friday, especially since I am host. I haven't had a chance to talk to Sid M. since your letter arrived this morning; but I expect to see him to-morrow

(Tuesday), and in any case he too will be at the Club Friday as my guest. So we'll be able to arrange for a discussion meeting then.

I've also heard the rumor about Wilfrid Sellars and Yale – in fact, from Herbert Feigl in a letter I had from him from Hawai[i]. On the other hand, the story around here (I don't recall from whom I heard it) is that W. is going to Yale just for a year – perhaps as a trial. I wonder what all this is going to do to Arthur Pap's chances to stay on at Yale.

Too bad you and Diane did not see your way clear to go to the Tamiment seminar. I've never been there myself, but from all accounts it is quite a place; and I thought that a few days there, without any serious obligations for you and without any expenses, is just what the doctor ordered. Not too wisely, I agreed to get a paper ready. I haven't even begun to think about it; but I am sure it will cost me blood, sweat and tears, plus lost time on the book.

I forgot to ask you when I saw you at the NYU meetings what you decided to do about the presidency of the Symb. Logic. Assoc. Hope you were able to say yes. Also, I forgot to mention to you that I've heard Carnap was most definitely hurt by my contribution to the Schilpp volume,[429] and that he will probably deal sharply with me. As I told you, I had great difficulty in doing that paper, in large measure because of my rather negative appraisal of C's inductive logic; and I regret that my worst fears have apparently been realized. My paper is of course full of very complimentary and genuinely felt comments; but it is true that if I am anyway near correct in some of my criticisms, the paper is hardly the sort of thing one contributes to what is essentially a Festschrift. I have not heard from Carnap for years, and I regret more than I can say that I have apparently hurt him.

We have an increase in our family since last Saturday! The boys nagged us to have a dog, and we finally gave in. Edith mentioned the matter to her brother, who turned up last Saturday with a 7 weeks old beagle pup – very cute, still quite a baby, and like a baby a bit of a nuisance that one does not mind too much. The boys are of course in seventh heaven.

I did not mean this to be so long. Will be seeing you Friday, and meanwhile our best to you and yours.

Always yours,
Ernest

4.1.107 Nagel to Hempel, July 11, 1958

[Typewritten, Hempel 1999, Box 28, Folder 1]

South Wardsboro, Vermont
July 11, 1958.

Dear Peter,

I meant to write long before this, just to give you some sign of life; but it apparently required the card I received to-day from Herbert Feigl to jolt me into overt activity. The card invites me to join the conference he is organizing at Princeton Sept 5 to 12, and asks me to let you know post-haste whether to make reservations for me at one of the Princeton inns. But I really do not know what to say. Needless to say, I would very much like to be at the meetings. On the other hand, there is the usual difficulty – I cannot leave Edith and the boys here, since Edith does not drive and without a car they could not get along; and to have them go back to the city at such an early date (we usually do not leave here before mid-September) would also be most uncomfortable for them. Furthermore, I do not think I could afford to take at least 8 days for a conference this year – there are too many things undone which I

[429] See Nagel (1963d).

hope to get at least partially completed before the summer is gone, and I'll need those 8 days very much. I suppose I ought to know my own mind and make a definite decision right away; but if it is at all possible, I should like to keep the matter open for a while at least, even though (as I suspect) I will feel impelled in the end to decline Herbert's kindness. Perhaps the best way to put it is this: If it puts you (or Herbert) to any inconvenience to have me procrastinate, then just count me out; but if an immediate decision is not essential, then please put me down tentatively as a possibility.

We have been here some three weeks now. It has been pleasant, especially after the unbelievably hectic month in New York previously, which when compounded with the noise of constant blasting on Claremont Avenue (Barnard is building a new library) was quite an ordeal. Every one of us is busy with something or other, and the days fly by with tremendous speed. I had hoped to get one part of the book done before I resume teaching; but I am now so far behind schedule that only a miracle could help me out. Although I am now revising stuff some of which has been published, it just takes me ages to get anywhere; and the final chapter (roughly on the social sciences, including history), in its present form some 75 typed pages long, will require radical rewriting, which I know will cost me plenty in sweat and agony. I have, as usual, some additional writing chores (book-reviews and such) that I must get off before the end of this month; and so I cannot complain of having nothing to occupy my mind.

Do you know what decision, if any, has been made about Arthur Pap at Yale? You may remember my telling you that Hendel asked me to report on Arthur's new book.[430] I sent what I think was on the whole a strongly favorable recommendation, and had only a somewhat laconic note of acknowledgment from Hendel, saying that a decision would not be made until the latter part of June. (Incidentally, he mentioned that two other men had been requested to comment on Arthur's book – had Hendel given me this information originally, I would certainly not have undertaken the thankless job.)

You must also have seen the letter I sent to Ledger Wood at his request, giving my estimate of Hilary Putnam. It was not an easy letter to write, though I could not refuse to write it. I'll be curious to learn what your impression of it is, and what effect if any it is likely to have.

Do you know a chemist at Yale by the name of Cassidy?[431] I read a manuscript of a book by him before leaving New York (on science and the humanities). Well intentioned, with some sensible things in it; but I had so many reservations to make that the publisher decided not to take it. I also read the ms. of a new book by Bridgman, which I did recommend strongly for publication, despite some questionable things in it.[432] There is a home-spun honesty about almost everything that Bridgman writes for which I am a push-over, even when some of his dicta (on logic as well as on matters of social policy) seem to me just hilariously funny.

How is Nelson Goodman? Is he still in Philadelphia, or has he been permitted to go to one of his country homes? Thus far, we have seen nobody in these parts, except our neighbors the Hooks (who, by the way, are going to Asia next month for a 5–6 months stay) and my sister (also close by). I haven't heard from John Cooley since we saw him in Connecticut in May, though I expect to get word from him before long, and hopefully at least a brief visit. As far as I know, Steve is still out west – you've doubtless heard that Louise underwent a major operation last spring.

---

[430] Pap (1958).

[431] Harold G. Cassidy (1906–??) had a PhD in chemistry from Yale, and stayed there for his career, becoming a professor in 1958, and retired in 1972. Besides his work on chemistry, he was known for his work on science and the humanities. The book mentioned was possibly the manuscript that became Cassidy (1962).

[432] Quite possibly Bridgman (1959).

Sidney told me that you decided not to contribute to his psycho-analysis volume, because of the pressure of other commitments.[433] I am sorry you had to decide this way, for I would be most eager to profit from an explicit statement of your views on the subject. However, I know how much time such "casual" contributions can take (at least judging from my own experience), and you were probably very wise in conserving your time for more important things.

Will you be getting away from Princeton this summer? Drop me a line when you can and feel so inclined. Meanwhile, love to you and Diane from all of us.

Ernest

4.1.108 Nagel to Hempel, July 30, 1958

[Typewritten, Hempel 1999, Box 28, Folder 1]

South Wardsboro, Vt.
July 30, 1958.

Dear Peter,

There is not much point in my procrastinating any further about my attending the conference at Princeton, especially since the discussions will center around the contributions to Vol. III of the Minnesota Studies.[434] There is absolutely no chance of my getting anything ready for this volume, and I think it is best therefore that I decline the opportunity to be a sort of fifth wheel at the meetings and that I cultivate my own garden instead. I'll write to Herbert sometime soon, explaining my decision; and meanwhile you can just cross off my name from the list of invitees, and so save yourself one headache less.

It was good to learn from your letter that you will be taking a vacation next month with the whole family. It will be an excellent thing for you, despite the pressure of work to be done, for I am sure you will make up for the apparent time lost by greater efficiency as a consequence. It is too bad, though, that the visit from Eva's father will come just at that time and cut short your period of complete relaxation.

It is an excellent idea for you to apply for a Fulbright, despite the fact that it is probably impossible to make a reasonable estimate of what the likelihood is of getting one to England. Apriori my guess is that your chances are excellent, for I find it difficult to imagine who would receive priority over you. But in this connection let me complete the story about the Ford Center[435] which, as you remind me, Frank interrupted. I wrote to Tyler[436] sometime in May (I believe), asking whether the rain check he gave me some four years ago was still good. He wrote back that there would be a place for me in 1959'–60, or during any of the succeeding three years; and I accepted the invitation for 59–60, though the formalities have not yet begun. (I want very much to get away from New York for at least a year, partly for

[433] See Hook (1959).

[434] The third volume of the Minnesota Studies in the Philosophy of Science, entitled *Scientific Explanation, Space and Time* (1962) was edited by H. Feigl and G. Maxwell; see Feigl and Maxwell (1962).

[435] Nagel had in mind Stanford's Center for Advanced Study in Behavioral Sciences, an interdisciplinary research center that was established in 1954 by the Ford Foundation.

[436] Ralph W. Tyler (1902–1994) was an American educator, working mainly on educational evaluation and assessment, often for government-funded bodies. He was the first director of Stanford's Center for Advanced Study in Behavioral Sciences between 1954 and 1966.

family reasons, – just the 4 of us would go, and my sister would have to take care of my mother – and partly because I want to get out of sight and sound of some of my colleagues at Columbia). I've already talked to my chairman (Gutmann) and dean (Barzun),[437] and they raised no objections, despite the fact that I've been on leave this year; and so unless some unforeseen hitch arises, we are hoping to be in California a year from next September. But to come to the really relevant point: When Tyler wrote me, he asked me whether I could suggest men whom I would like to be at the Center during my own stay; and I sent him your name (together with those of some social scientists). I gather from your letter that you have not heard from Tyler thus far. But neither have I since I accepted his informal invitation – I've been told that the formalities do not begin until the fall. And so my point is this: I think you ought to apply for a Fulbright; but I also think you ought to consider the Stanford Center as another possibility, and have both irons in the fire. If both possibilities should become actual, you could easily get a postponement of the Ford Center invitation.

I was in New York last week for a day, to take part in a half-hour television discussion with – of all people – A.J. Ayer,[438] who was in the city for a short stay to attend the marriage of his daughter. The groom, Ayer told me, is a student in my department working for an A.M. in the philosophy of religion, and I had the impression Ayer was not too enthusiastic about the man.

I also had a letter from Arthur Pap, in which he mentioned that no decision has [been] made about him as far as a permanent tenure is concerned, but that his present contract has been extended to 1960. He apparently knows where the opposition comes from, but declined to enter into details.

It is sad news about Louise Stevenson,[439] but we hope all will go well with her. I don't know whether they are back in Wilmington already, but we'll try to get in touch with them before long. About ten days ago I saw Nelson Goodman in the street in Brattleboro, but I was driving the car in the midst of heavy traffic and was unable to stop. The brief glimpse I had of him gave me the impression he lost some weight. I do not know whether he is still in Vermont, and we have been too busy to try to get in touch with him. No word yet from John Cooley.

Warmest regards from all of us to you and yours, and all good wishes for a wonderful holiday at Folly Cove.

Ernest

4.1.109 Hempel to Nagel, August 23, 1958

[Handwritten, Nagel 2020, Box 3, Folder 4]

8-23-58

Dear Ernest, many thanks for your letter. I sympathize with your decision not to come to the Princeton meeting, though I am very sorry that we won't have your reactions to the

---

[437] Jacques Barzun (1907–2012) was a French-born historian who studied at Columbia University, where he also obtained his PhD and taught for most of his career. In March 1958 he was appointed Dean of Faculties and Provost of Columbia University,

[438] Alfred Jules Ayer (1910–1989) was an English philosopher. He studied at Oxford and then taught at the same university and at UCL, before being appointed Wykeham Professor of Logic at Oxford. Ayer is best known for his highly influential book *Language, Truth, and Logic* (1936), which helped to popularize logical empiricism in the English-speaking world.

[439] Louise Stevenson (1098–1963) was the wife of the philosopher Charles Stevenson (see letter 33).

various contributions; I'd very much have liked to hear what you think of some of the questions, for example, that I encountered in grappling with statistical explanation. And I know Herbert will be very disappointed when he learns you're not coming. – We are all enjoying our Rockport vacation. Ray Goodman had found us a comfortable cottage on a large tract of land (thus much privacy), and with a fine view; and Rockport has its attractions for each of the four of us.

It was very good of you to suggest my candidacy to Tyler! It wld be great fun to be at the Center with you. So far, I have heard nothing from Tyler; and I plan to put in for a Fulbright by Oct 1st, provided that [n/a] can get Dean Brown's[440] OK for a leave for me. – Diane and I watched you and Ayer discussing and really thought you brought the obviously difficult job off *very* well (you looked very impressive, by the way!).

Warm greetings and many good wishes from us all to you all!

Peter

4.1.110 Nagel to Hempel, September 23, 1958

[Typewritten, Hempel 1999, Box 28, Folder 1]

Sept. 23, 1958.

Dear Peter,

Just a hasty line to let you know that I've sent off the material to the Fulbright committee. I suppose there will be a long wait before you hear about the decision – I am of course most optimistic, but there is nothing like a bird in the hand, and I know how unsettling it can be to wait for an indefinite date.

We are at sixes and sevens at home, for the painters are coming in, and what with classes starting Thursday I feel in a daze.

Hope to see you before long. Meanwhile, love to you and yours from all of us.

In haste,

Ernest

4.1.111 Hempel to Nagel, November 8, 1958

[Typewritten, Hempel 1999, Box 28, Folder 1]

Princeton, Nov. 8, 1958

Dear Ernest:

Many thanks to you, from Diane and myself, for sending us an inscribed copy of *Gödel's Proof!*[441] As you can imagine, I have read it with the keenest interest and enjoyment; and I think it is a truly wonderful job of lucid exposition of an enormously complicated argument! When Nelson Goodman stopped into my office some weeks ago, I showed the book to him, and he thought right away that might be something to discuss in a special problems course

[440] Presumably James Douglas Brown (1898–1986), who was an economist and educator, known mainly for his work on social security. He was the dean of Princeton between 1946 and 1966.

[441] Nagel and Newman (1958).

for seniors; two days ago, I saw him again in Philadelphia, and he told me, to my great satisfaction, that he has decided to discuss the book this semester with his seniors; he liked it very much and appreciated the introductory discussion of axiomatic procedure etc.; in fact, he wishes you had included even more on Non-Euclidean geometry. I also have shown the book to Kemeny,[442] who was here recently, and I feel tempted to read it with some good undergraduate students, or in the context of a graduate course.

Please forgive me for having been so laggard in thanking you; I had meant to write you much sooner, but I have been kept quite busy, and then there was a lecture for the U. of Pa's Benjamin Franklin lecture series which I had to prepare (on Science and Human Values),[443] and which cost me a good deal of time and headache. I gave the lecture last Wednesday and am not sure whether I really established contact with the audience; sometimes I feel very clearly that my audience did go along, and this time, that feeling wasn't there at all.

I have filed my Fulbright application, and some time later received an invitation from Tyler to the Stanford Center. I explained to him about the Fulbright, and my reasons why, after a 20 years' absence from Europe, I thought a year there might be profitable, but I asked also whether by any chance it might be possible for me to be kept on their list tentatively, and to my very pleasant surprise he said that that could be done. So I hope now that my leave of absence for next year will be granted, and that I will be spending the year either in England or at Stanford, in your company!

I learned yesterday morning that you will be coming here next Friday and Saturday for a meeting of the Hum. Council;[444] Oates[445] called me to ask whether I could attend {Dick Brandt[446] is expected, too} (I am not on the committee now, but am to substitute for Gregory [Vlastos] next semester), and told me then about this meeting of which I had known nothing. I am so glad to have a chance to see you then, but I am very glad also that the Council will have your reactions to this large-scale project it is planning under Ford Foundation auspices; I am eager to talk to you about it. And then lets make plans for a day when you, John Cooley, and Sid will come out here for one of our confabs!

Diane and I send all our best to you all.

Peter

---

[442] John G. Kemeny (1926–1992) was a Hungarian-born philosopher and mathematician. He studied at Princeton University, where he also received his PhD in mathematics. He taught philosophy at Princeton before moving to Dartmouth College as professor of mathematics.

[443] The essay resulting from this conference was published under the title "Science and Human Values" in the book *Social Control in a Free Society* (see Hempel 1960), edited by Robert E. Spiller. It was then reprinted, under the same title, as Chap. 3 of Hempel (1965a).

[444] The Princeton Humanities Council was established in 1953 with the mission of supporting teaching and research in the humanities.

[445] Whitney Jennings Oates (1904–1973) was an American classicist who studied at Princeton, where he also received his PhD in classics and taught for most of his career. Oates was responsible for the establishment of the Council of the Humanities and served as its first chairman.

[446] Presumably Richard Brandt (1910–1997), an American philosopher educated at Denison University, Cambridge, and the University of Tübingen. He received his PhD in philosophy from Yale and taught at Swarthmore College and at the University of Michigan. Brandt was an influential advocate of utilitarianism in moral philosophy.

4.1.112 Nagel to Hempel, November 16, 1958

[Handwritten, Hempel 1999, Box 28, Folder 1]

25 Claremont Ave
New York 27, NY
Nov. 16. 1958

Dear Peter,

This is apropos of your remark about the statement in Gödel's Proof (p. 78) that one expression can be an initial part of a second *if and only if* the Gödel # of the former is a factor of the Gödel # of the latter.

As I understand the difficulty (our parting words were rather hasty), the difficulty is illustrated by the following example.

a)	The Gödel # of:	$p \supset p$
	is	$2^{=2} \times 3^3 \times 5^{=2}$
b)	But the Gödel # of:	p
	is	= 2 (as per p. 71)

which is not a factor of the above.

And so "if and only if" should be replaced by "if".

I was aware of this difficulty when writing the stuff. But I thought it could be outworked by distinguishing between i) the Gödel number of a single sign and ii) the Gödel number of single sign *when that sign is regarded as a "sequence of signs"* (containing just one member). With the help of this distinction, what has to be said is that *one sequence of signs* is an initial sequent of a second sequence if and only if etc. (I believe Carnap[447] distinguished between the Gödel # of a *term*, the Gödel # of a *sequence of terms*, and the Gödel # of a *sequence of sequences of terms*.) I decided not to mention this in the exposition for fear of confusing the readers; but perhaps this was poor judgment.

I wonder whether you think this gambit takes care of the trouble, and whether the difficulty arises in any other case than when the initial expression contains only one sign. But *please don't take the time to reply to this* – I'll ask you when I next see you.

I am not sure whether anything valuable will come out of Project Osgood. But I found it worthwhile if only because I had the chance to see you again, though I wish there had been more chance to talk to you at quarter length. The hour was getting too far along for me to stay any longer, as it is I did not get home till 5:30, and rather tired at that.

I've sent a note to John Cooley about Dec. 12. Will write you as soon as I hear from him. Meanwhile, my best to you both – please tell Diane how sorry I am not to have seen her yesterday.

As ever
Ernest

4.1.113 Hempel to Nagel, November 23, 1958

[Typewritten, Hempel 1999, Box 28, Folder 1 and Nagel 2020, Box 3, Folder 4]

[447] Nagel is referring, perhaps, to Carnap (1937b, §§19ff).

Princeton, November 23, 1958

Dear Ernest:

Many thanks for your letter. I very much enjoyed seeing you, and your presence was a great help in sustaining my morale at the meeting, about whose objectives I feel at least as skeptical as you, though I hope that the project of having survey studies written may yield some worthwhile results.

As for *Goedel's Proof* your procedure surely does take care of the possible objection your mention; but my example is somewhat different: the formula, or sign sequence, '~p' has the Gödel number 2^1 x 3^{11x11}, and this is a factor of the Gödel number of '(pvp) ⊃ p', though the former sequence is not an initial segment of the latter. I think therefore that your criterion may have to be reformulated so as to say that a sequence s_1 is an initial part of a sequence s_2 if and only if the prime factorization of the Goedel number of s_2 in the form 2^a x 3^b x 5^c x ... contains the total prime factorization of the Goedel number of s_1 as an initial segment. And this condition can clearly be expressed with the help of the arithmetical concepts you are presupposing, so your argument goes through without a hitch.

I very much hope our plans for December 12 will work out. Diane and I look forward to seeing you, John, and Sid Morgenbesser here.

Our very best to you, Edith, and the children!

As ever,
Peter

4.1.114 Nagel to Hempel, December 1, 1958

[Typewritten, Hempel 1999, Box 28, Folder 1][448]

25 Claremont Avenue,
New York 27, N. Y.
Dec. 1, 1958.

Dear Peter,

Thank you so much for the note on the error in the Gödel book.[449] I am afraid I was using "being a factor of" in a most peculiar sense, for I have a vague recollection of thinking that if *x* and *y* are Gödel numbers, and *x* is a "factor" of *y*, then the quotient obtained on dividing *y* by *x* will contain no prime in its decomposition that is presented in decomposition of *x*!! That passage in the book (and doubtless many others) certainly needs rewriting, though I doubt whether the opportunity to do so will ever arrive.

I meant to acknowledge your comments much sooner – but I have been immersed in chores up to my neck, and thought that I would wait till I heard from John Cooley (whose reply I had expected last week) and so save my thanks until I could report on his response. A note came from him this morning, saying that Fri. Dec 12 was just fine for him to take the trip to see you; and of course both Sid Morgenbesser and I have left that date open. Is it still convenient for you to have all three of us come down on that day? There are trains from Penn Station at 3 and at 4 P.M. (the first arriving at Princeton at 4, the second at 5). If the day is still all right with you, have you a preference as to the train we take?

This Friday I am going up to Amherst to give a talk, though I feel so weary at present that I am not looking forward to the trip. I've had a confirmation of the invitation to the Ford

[448] We are grateful to Jason Rampelt at Pittsburgh for helping us out with this letter.

[449] See Nagel and Newman (1958).

Center for next year (I was beginning to feel jittery over not having had any further word since last spring), and I can't tell you how much I anticipate going out to the coast for a whole year. And though I know you would prefer to go to England, it would be wonderful if you too were at Stanford during my stay. I am of course fully confident that you will be going abroad, but it is still nice to think of what is just a logical possibility.

My love to you both.

Ernest

4.1.115 Hempel to Nagel, December 6, 1958

[Typewritten, Hempel 1999, Box 28, Folder 1]

December 6, 1958.

Dear Ernest:

I am so glad you can come; please take the 3 pm from Penn Station! I will be waiting for you in my office, and we can then have some coffee before starting our proceedings.

One thing more, which I hope can be said between friends without being taken amiss. I have sometimes hesitated to suggest Princeton as our meeting place because, though most convenient for me, it requires of the three of you not only the investment of travelling time but also considerable travelling expenses, both of which I save. Diane and I have therefore felt all the more troubled in our consciences to see you arrive carrying all the sorts of delicious but also very precious gifts. This really makes the imbalance of burdens even more unfair; it really makes the meetings a sort of luxury for all three of you. So please don't Ernest, honestly! And if the thought should cross your mind that we are putting ourselves out, let me say that if anyone here owes a debt of gratitude for much help in more ways than one, it is owed by me to you and Edith.

By now your Amherst talk is over and I hope you found it a pleasant experience after all! But it is good news to hear your arrangements with the Stanford Center are now complete – I so much empathize and sympathize with you!

All the best to you, Edith, and the children, and greetings to your mother!

As ever,

4.1.116 Hempel to Nagel, January 24, 1959

[Typewritten, Hempel 1999, Box 28, Folder 1]

January 24, 1959

Dear Ernest:

I am awfully sorry to be so late in writing you, but the past week has been unusually busy. Now I want to suggest some dates for our next meeting here in Princeton: Tues, Feb. 10; Friday, Feb. 13; Tues, Feb 24, Friday Feb 27. The intervening week is not so good because on Monday, I will be attending the Zubin psychiatric workshop in NY,[450] and on Friday Feb. 20, there should be the next meeting of the Philosophy Club. I hope that one of

[450] The paper that came out of the conference was Hempel (1961).

the dates will suit you, John, and Sid jointly. I am certainly looking forward to our next get-together!

Ernest, do you know of any book and article that addresses itself to the status and function of the theory of evolution as a scientific theory? I would be interested, for example, in seeing a discussion of what and how the theory could be said to explain; whether explanation goes together with prediction here, and so forth. I have some ideas on the subject, but would certainly like to be apprised of any publication that deals with these questions systematically. The reasons for my interest are twofold: One, at present I have to supervise a senior thesis that is to deal with this kind of issue; two, I sense that whatever explanation the theory of evolution provides for a great variety of data is not quite of the order of the explanation of planetary motions by Newtonian celestial mechanics. At a first level, there seems to be something like a genetic-historical account of how the evolution of the various existing and extinct species took place; and then, on a higher plane, there are some more nearly explanatory principles, apparently statistical in character, that account, at least in part, for the changes in the course of that evolution: mutation, and some sort of survival of the fittest principle.

Another question: I seem to recall that someone like Arrow[451] had a fairly long essay, perhaps in some collective volume, on the logic of measurement, something, if I recall correctly, [a] bit along the lines of S.S. Stevens' earlier pieces on various types of scales;[452] but more sophisticated. Would you know whether I am right, or what at any rate there exists of this character? I am interested in something covering also various quantitative methods in psychology and sociology. No rush about this.

I wish your paper on the cognitive status of theories[453] were available in mimeographed form: I would use it in my graduate course this term; it is really an excellent job!

Austin[454] was here and read "Three ways of spilling ink" last night and conducted a discussion group this morning; both very interesting indeed; one feels that here is a subtle and perceptive mind, even if the problems seem to have little relevance to what one normally expects of philosophy.

Best to all!

4.1.117 Nagel to Hempel, January 27, 1959

[Typewritten, Hempel 1999, Box 28, Folder 1]

25 Claremont Avenue,
New York 27, N.Y.
Jan. 27, 1959.

---

[451] In a fairly long paper, Paul F. Lazarsfeld and Allen H. Barton, right after Kenneth J. Arrow's essay in the same volume, have discussed the logic of measurement and referred to Stevens' papers and ideas. See Lazarsfeld and Barton (1951).

[452] Hempel is probably referring to Stevens (1946).

[453] The sixth chapter of *The Structure of Science* (1963) is entitled "The Cognitive Status of Theories".

[454] J. L. Austin (1911–1960) was a British philosopher. He studied at Oxford, where he spent his entire career. A major proponent of the philosophy of ordinary language, he is best known for his extremely influential book *How to Do Things With Words* (1962). For the paper mentioned by Hempel, see Austin (1966).

Dear Peter,

It was good to get your letter. I've written to John mentioning the two Friday dates you suggested (the Tuesday ones are difficult or impossible for Sid or myself), and will let you know which he prefers as soon as I get word from him.

I am not sure that I know exactly what you are looking for in connection with the logic of measurement, but perhaps one of the following is what you have in mind:

S.S. Stevens, Chap. 1 of the HANDBOOK OF EXPERIMENTAL PSYCHOLOGY

Coombs, Raiffa and Thrall, "Mathematical Models and Measurement Theory" in DECISION PROCESSES (ed. by R.M. Thrall, C.H. Coombs, and R. L. Davis)

Patrick Suppes, "A Set of Independent Axioms for Extensive Quantities" in PORTUGALIAE MATHEMATICA, Vol. 10, 1951

Dana Scott and Pat. Suppes, FOUNDATIONAL ASPECTS OF THEORIES OF MEASUREMENT, Technical Report No. 6, Apr.1, 1957, Applied Mathem. and Statistic Lab., Stanford University

Nor am I familiar with anything that answers the questions one wants to ask about the character of the explanation offered by evolution theory. With the possible exception of the first item on the following list, you are doubtless familiar with rest, but I mention them on the chance that they have slipped your mind:

Morton Beckner, THE BIOLOGICAL WAY OF THOUGHT (just published by the Columbia Univ. Press, this is a dissertation which John Cooley, Albert Hofstadter and I supervised, and I think it is a useful book, with a chapter on genetic (including evolutionary) explanation; I think you will find it generally stimulating.)

L. von Bertalanffy, PROBLEMS OF LIFE, has two chapters on evolutionary theory. (1952)

T. Dobzhansky, EVOLUTION, GENETICS AND MAN. Informative, and descriptive, but with little concern for methodological problems.

G.G. Simpson, THE MAJOR FEATURES OF EVOLUTION. Though this too is not methodological[l]y oriented, it has lots of important and not generally familiar material.

A.J. Lotka, ELEMENTS OF PHYSICAL BIOLOGY. Has some useful comments on evolution.

R.A. Fisher, GENETICAL THEORY OF EVOLUTION. Some incidental comments of methodological interest.

I had lunch to-day with Yourgrau,[455] and it turned out much better than I had dared hope. You will probably meet him this week, for Y. is going to Princeton at Oppenheim's invitation. For the moment, my department has nothing to offer him for next year – he was obviously disappointed, though when I wrote him I tried to make as explicit as I could that the dept was unable to commit itself in any way at this time. However, something may turn up later, despite the tight budget. Despite the fact that my salary will be available next year, it doesn't seem as if we were going to have enough money to invite any visitor to replace me.

But I must stop now to read exams and term papers! I look forward to our next talk, and meanwhile our love to you and Diane.

Ernest

[455] Presumably Wolfgang Yourgrau (1908–1979), who was a German-American physicist, with interests in chemistry, biology, and philosophy. He earned a PhD from Berlin on color blindness. He traveled extensively and worked as the head of the Department of Logic and Scientific Methods at the School of Higher Studies in Jerusalem after the war; later he went to South-Africa, became professor of philosophy of science at the Minnesota Center in Minneapolis between 1959 and 1960, and worked also at Smith College and in Denver, primarily as a philosopher of science.

4.1.118 Nagel to Hempel, April 8, 1959

[Typewritten, Hempel 1999, Box 28, Folder 1]

25 Claremont Avenue,
New York 27, N.Y.
April 8, 1959.

Dear Peter,

Quite apart from the contents of your letter, it was so good to hear from you. I have been so miserably busy during the past six weeks or so, that I haven't had a chance to write you as I so often wanted to. Even now, I don't know whether I am coming or going (except for an acute sense that I have been misspending my time on silly chores, some of which never get done); and I am dashing this off, simply because your letter requires a prompt answer.

I really feel very pleased that you want me to undertake a monograph on a critical appraisal of work in the philosophy of science during the past 30 years or so; and I admit the proposition is tempting. But as you know, though I work hard enough at it, writing is a painfully slow business with me, and things I should have been able to finish years ago still continue to be in limbo, and will probably remain in that state during the remainder of my present incarnation. I had hoped to get the concluding chapter of the first part of the book on the philosophy of science out of the way months ago; and though I shall try to turn to it before going to the Center, in my sane moments I've begun to wonder. And even if this expectation should be realized, I've made promises that will surely require much more time than the year at the Center to fulfill. Accordingly, I think it would be irresponsible if not dishonest on my part to yield to the temptation you hold out, since I would then be delaying indefinitely the completion of Project Osgood. For reasons of elementary decency I must therefore say no.

One of the things I had wanted to write you about was my meeting with Schlatter,[456] whom I found quite agreeable. He asked my opinion of the plan to have studies of the sort you mention. As I recall, I told him that I thought such studies would really make a contribution in those areas of philosophy in which there is a cumulative process of advance (I distinctly remember citing logic, the philosophy of science and the theory of knowledge, and possible ethics), but that in other areas where such a process does not obtain (and I thought that the history of philosophy, esthetics and metaphysics were in this category) I had some doubts whether except for a small group of specialists such studies would do much for the humanities. I also expressed the view that if it were at all possible to do so, it would be desirable to bring out in the proposed studies in what way developments in other areas of thought influenced the direction of research in given sectors of philosophy, and conversely what impact if any professional philosophy has had upon the thinking of those outside the profession. And finally, I also raised the point (I think we discussed it at Princeton last October) about the desirability of giving philosophers in cross-disciplinary areas the opportunity of obtaining a more solid training in the disciplines (such as physics or law) in which they had an interest. What I *don't* know is whether these comments to Schlatter have made life more difficult for you – something which I need hardly say I did not intend and would never think of doing.

This *precis* of my talk with Schlatter is a necessary preliminary to any comments I can make on your question concerning the metaphysics monograph. What sorts of things are

[456] Presumably Fredric W. Schlatter (1926–2016) who was a Jesuit priest with a doctorate in classics from Princeton. Schlatter taught classics for decades at Gonzaga University.

likely to be covered in such a study? I suppose not only the kinds of things that Whitehead[457] and Dewey did, but also Broad[458] (and McTaggart)[459] and maybe even Alexander.[460] But one cannot stop here, and so the existentialists, and the Thomists, and Paul Weiss, and Hartshorne, and Wild,[461] and so on and so on – in short stuff that turns my stomach. Unless I misconstrue the situation, in this area each man is a prima donna pretty much on his own, and except for the inevitable polemics, there is no cumulative outcome. If one is to be highly selective of the material to be discussed (so that a top-notch man will examine themes that are currently also treated in the theory of knowledge or philosophy of science) the outcome will not be representative of the great bulk of material now written under the rubric of metaphysics. But if the aim is to be genuinely representative, my fear is that the kind of person whom one could get to do the job will also be one whose monograph neither you nor I will want to read. Richard Taylor[462] would have done an interesting job. As for Abe, who certainly is keen and gifted, I find it hard to know what to expect, for he is also erratic and sometimes too clever for his own good. On the whole, I think I would prefer Arthur Murphy[463] to Abe, since I don't really know Shaffer.[464] Perhaps C.D. Broad might be interested in the project – one can count on his doing a pretty solid job. In short, I really do not know what to say to your query, perhaps because I can't get very enthusiastic about the subject.

---

[457] Alfred North Whitehead (1861–1947) was a British-born philosopher and mathematician who studied at Cambridge. He then taught at the University of Cambridge and the University of London before being appointed professor of philosophy at Harvard in 1924. Whitehead is best known for his collaboration with Bertrand Russell, which resulted in the publication of *Principia Mathematica* (1910–1913), and for his metaphysical *opus magnum Process and Reality* (1929).

[458] C. D. Broad (1887–1971) was an English philosopher. He studied at the University of Cambridge and then taught at the University of St. Andrews, Dundee, Bristol, and Cambridge. He is best known for his influential book *Mind and Its Place in Nature* (1925).

[459] J. M. E. McTaggart (1866–1925) was an English philosopher. He studied at the University of Cambridge, where he spent his entire career as fellow and lecturer. Although he was an important Hegelian scholar, he is now best known for his argument against the reality of time formulated in his 1908 essay, published in *Mind*, "The Unreality of Time".

[460] Samuel Alexander (1859–1938) was an Australian philosopher. He studied in Melbourne before moving to Oxford; he then went to Germany to study experimental psychology. Alexander became professor at the Owens College in Manchester. His major work is *Space, Time and Deity* (1920).

[461] John Daniel Wild (1902–1972) was an American philosopher. He studied at the University of Chicago and Harvard University, and received his PhD from the University of Chicago. He then taught at Harvard, Northwestern University, and Yale. In his work he tried to find a composition of realism and existentialism.

[462] Richard Clyde Taylor (1919–2003) was an American philosopher. After completing a BA in zoology, he studied philosophy at Oberlin College and received his PhD from Brown University, where he was appointed professor of philosophy. He went on to teach at Ohio State University, Columbia University, and the University of Rochester. Taylor published a book entitled *Metaphysics* in 1963.

[463] Arthur Edward Murphy (1901–1962) was an American philosopher. He studied philosophy at the University of California, where he also received his PhD and got his first job as an instructor in philosophy. Murphy taught at many universities in the United States. He is best known as a proponent of objective relativism.

[464] Perhaps Jerome Arthur Shaffer (1929–2016), an American philosopher who graduated from Cornell, and earned a PhD from Princeton in 1952. Between 1955 and 1967 he taught at Swarthmore College. His main research interest were in philosophy of mind and epistemology.

I also heard that Sellars is remaining at Yale, and that Arthur Pap has been given assurance of being kept on. Rosa Rand[465] called me last week when she was in New York, though I had no free time that day to see her as she wanted me to; I hope you will not let her occupy too much of your time and attention.

It is too bad you've still not had word from the Fulbright, though I am willing to bet my bottom dollar that when you do it will be favorable. Sid Morgenbesser was turned down by the Guggenheim – it is too bad, for he deserved a grant in my judgment, and certainly needed the encouragement it would have given him. I imagine that Hilary must be receiving one, though the list has not yet been published.

There is lots more I'd like to gossip about, including the plans for an international congress in logic and philosophy of science to be held at Stanford in August 1960.[466] I saw Pat Suppes and Alfred Tarski about ten days ago. But I'll have to save the rest until I next see you. Bobby went to the hospital to-day to have his tonsils out – I think it went well, and we expect him home to-morrow. If I get through this semester with all my wits still with me, I'll believe in miracles.

All our best to you and Diane.

Always yours,

Ernest

4.1.119 Nagel to Hempel, May 14, 1959

[Typewritten, Hempel 1999, Box 28, Folder 1]

May 14, 1959.

Dear Peter,

I talked with John Cooley Monday night on the phone, and mentioned to him that you thought the pressure on you might be lifted toward the end of the month or early in June, and that you suggested we might then visit you in Princeton. John was of course delighted at the idea, and asked me to inquire of both Sid M. and you whether May 25 or 26 might be feasible. He also said that June seems out of the question for him, presumably because he has already made arrangements to go sailing.

I talked to Sid to-day, and May 25 would be fine for him, but since he has a PhD late afternoon on the 26th this latter date is impossible for him. So unless I misunderstood John, Monday 25th is the one day this season that is suitable for him. That day is also fine with me.

Of course, I don't know whether so early a date fits into your schedule. While I hope it does, however, I also think that you must not hesitate to suit your own needs and convenience, and to say 'No' if you will still be too busy to have the additional burden of visitors. So please let me know how matters stand, and I'll communicate with John (at the moment I believe he is somewhere in Maine).

---

[465] Nagel is referring to Rose Rand, an Austrian-American logician and philosopher. She studied at the University of Vienna, where he encountered the Vienna Circle. As a Jew in Vienna, she could not apply for a job in her own country. She moved to England and, after the Second World War, to the United States. Rand was an important translator of German, Polish, and Russian logicians.

[466] The first international Congress for Logic, Methodology and Philosophy of Science was held at Stanford University from August 24 to September 2, 1960, under the auspices of the Division of Logic, Methodology and Philosophy of Science of the International Union of History and Philosophy of Science. The proceedings of the conference were published in Nagel et al. (1962).

Should May 25th be impossible for you, perhaps something will turn up that will give me a chance to see you before we go off to Vermont – the date is quite indeterminate, and we will not start until after Edith finishes her experiment in the lab, and this might not be until July. We expect to start the big adventure of driving across the country to Stanford either the first or second week in August.

I was disappointed not to get in the dittoed copies of Carnap's replies Herbert sent me, the comments on your paper. I read most, though not all the others, but in the case of the original papers I have not seen it is difficult to follow some of the detailed discussion. It must have been an enormous labor for Carnap to go through the material, and there is no doubt the volume is going to be a most valuable one. I have not had the chance to look up my own paper, to see how badly Carnap damaged the points I tried to make. I do know, however, that Carnap did not consider all my points, understandably enough. On the whole, I am relieved that Carnap was not as displeased as I feared he would be – after all, I rejected pretty much in toto the work on which he has spent a larger fraction of his life than on anything previous; and he would have to be inhuman not to show some signs of disappointment with me. It was a difficult piece for me to write, just for this reason.

Hope your trip to Dartmouth was a pleasant one, and not too wearying.

Our best to you and Diane,

As ever,

Ernest

4.1.120 Hempel to Nagel, October 11, 1959

[Typewritten, Nagel 2020, Box 3, Folder 4]

Oxford, October 11, 1959
93 Woodstock Road

Dear Ernest:

I have often thought of you and your family since we last saw each other in Princeton; I hope you had a pleasant time in Vermont and a good trip to Stanford, and that you find your house as well as the Center to your liking and are all comfortably settled by now. Are you and Edith satisfied with the schools for your boys? Are the people and the general atmosphere, at the Center pleasant and stimulating? And do you have a chance to do reasonably uninterrupted work in your office?

As for us, we had a terribly hectic time getting ready for our departure; during the last week, Diane slept only some three of four hours each night, and I got along with my minimum; and still even on the last morning, there were things undone; but fortunately one of our neighbors jumped into the breach and offered to tidy up the last odds and ends for our tenants from Michigan, who are not the Copis or Frankenas, each of whom is coming for one term only, but a specialist in Japanese history. – We had a wonderful crossing on the Queen Elizabeth, and the children enjoyed their end of it as much as we were glad to have some chance to relax and catch up on our sleep. Our house in Oxford, which belongs to St. John's College, and whose regular occupant is a Prof. of Turkish who is spending this year in Istanbul with his family, is quite small both in number and in size of rooms; but this, of course we had known. It lacks also many of the standard conveniences of home; so far we had little occasion, fortunately, to try out our diverse heating devices, which include gas as well as electric "fires" and so-called paraffin stoves. We did feel let down, however, by the equipment and general condition of the house; for the lady of the house had written several letters which gave us the impression that the place was filled with what were in effect museum pieces, and was kept in immaculate condition; whereas actually, it is all quite ordinary stuff downstairs, and in the bedrooms, much of the equipment – chairs esp'y and a sofa – are so rickety and filthy that one hates to sit down on them; nor do I dare try carry them into the basement, lest they disintegrate when picked up; Toby has unspeakably dirty rug in her room that cannot have been cleaned in 15 years or so. Well, but we have things under control and find the house quite livable; and we are very lucky to have found one in Oxford and in such a location, a 15 minutes' walk from the center of town, where the Colleges and the large stores are. We all find Oxford a fascinating place with its old streets and buildings, the tradition-heavy colleges with their beautiful garden – Magdalen, where Ryle[467] is a Fellow, has its own deer park with some 25 or so well-fed deer; with pleasant old inns, and with a lot of cultural activities; 2 theaters, for example, and the most expensive ticket for, say, a performance of the Royal Ballet is far below the cheapest you could get in New York. For ordinary plays, you can get a seat for 35 cents, and coffee served in your seat for 12 cents. On our trips by rail we have several times treated ourselves to dining car snacks or meals, which by our standards are very reasonable; Diane, for example, had a complete tea (pot of) with several courses of sandwiches, toasted tea buns with marmalade, then raisin cake, then still something else for 42 cents, and the waiters solicitously ask whether you want more. Diane has taken the children to Stratford-on-Avon and individually to London; she and I have been to several of the interesting "Houses" that have been turned over to the National Trust by the nobility that once owned them; in one case, we were taken around by the former owner, who still lives in one of the annexes; he is Hare's father-in-law and made a most charming guide; he also showed us the rooms of Florence Nightingale, his aunt, who had lived in the House for many years. – In London, I've seen the Rosetta Stone and the Magna Charta, to mention two for me awe-inspiring items. – The children are both in private schools and like them reasonably well, but both miss their old friends in Princeton a great deal. The schools are good, but we think that in America, the English school system is perhaps somewhat overrated, or perhaps it is that the Princeton Public School System is really very good, for the level here seems by no means higher than at home. – The Grices[468] and the Hares have been very nice and helped a lot in showing us the ropes; I have also had lunch with Ryle at Magdalen, and he was very pleasant. The courses for Michaelmas term start tomorrow, and I expect I will sit in on an occasional lecture and some sessions of Hare's seminar with Hinton[469] (on Moral Problems); but I want to avoid getting involved too much. Two weeks ago I attended the annual meeting of the Brit. Soc. for the Philos. of Science at Cambridge[470] and was pleased to see Braithwaite and Woodger and Hutten[471] again and

---

467 Gilbert Ryle (1900–1976) was a British philosopher who studied and taught at the University of Oxford, where he was appointed Waynflete Professor of Metaphysics. A proponent of ordinary language philosophy, he is best known for his *opus magnum*, *The Concept of Mind* (1949).

468 Paul Grice (1913–1988) was a British philosopher. He studied at Oxford University, where he taught until he moved to the University of California, Berkeley. A highly influential philosopher of language, he wrote important essays on the theory of meaning and the theory of implicature.

469 John Michael Elliott Hinton (1923–2000) was a British philosopher who lectured at Oxford from 1958. Hinton worked mainly on epistemology and perception and is known for his disjunctive theory of perception.

470 The Fourth Annual Conference in Philosophy of Science was held in Cambridge, 25–27 September 1959.

471 Ernest H. Hutten (1908–1996) was a German-British physicist-philosopher who obtained his physics-PhD in 1932 in Berlin. He worked as physicists in European laboratories, then after 1941, he taught philosophy of science in Chicago, at the City College in New York, and then in London. He was one of the founders of the Philosophy of Science Group in England. For Hutton's talk see Hutten (1960).

interested in meeting some people I had heard of (J. O. Wisdom, Dingle,[472] Agassi[473]). The quality of the symposia was uneven, and in part rather stimulating. Es wird aber auch hier nur mit Wasser gekocht! Polanyi[474] read a paper on Knowing and Being which I found strangely wooly; Socrates[475] was on a symposium with Hutten on Biology and Physics, but he talked almost exclusively on the formal logic of hierarchies with some illustrations from genetics, and I was pleasantly impressed by Hutten who managed to make it seem relevant to the topic of the symposium by comparing biological hierarchies with the radio-active decay series which have a somewhat similar structure. – I had a note from John Gregg reminding me to make up my mind re the Woodger Festschrift,[476] and after thinking it over I suggested one topic to him on which I might write a short piece, but which seems not very close to Woodger's work nor to his interest; it has to do with a logical inconsistency in many standard formulations of rules of induction, such as Reichenbach's, Black's, etc. Nothing very weighty either, so I have asked him for his candid opinion as to whether this might be suitable; I don't want to take time out now to think about a new topic so as to be able to write a contribution. – What will you be writing on?

Have heard nothing from John Cooley, to whom I wrote weeks ago asking for a reprint of his review[477] of Toulmin's[478] book. Did he make the crossing to Ireland? And have you heard how he is now?

I was shocked to hear of Arthur Pap's death; it was totally unexpected, and it must be dreadful for his wife, who now must raise three or four young children!

I received a letter from Ducasse asking me to become a member of the Editorial Board of *Philos. of Science* and also of some Board headed by Margenau that is to direct policies etc. I declined the latter because it would demand too much time and effort of me and I have no talents in this direction anyway; but did accept the former, but I hope I won't get drawn into it during this year; I had written Rudner,[479] who had asked me to do some refereeing, that I

[472] Herbert Dingle (1890–1978) was a British physicist and philosopher of science. He studied at the Imperial College, London. A Fellow of the Royal Astronomical Society, he taught at the Imperial College and at the University College London.

[473] Joseph Agassi (1927–2023) was an Israeli philosopher of science. He studied at the Hebrew University of Jerusalem and received his PhD at the University of London. He taught at several universities in North America, Israel, Europe, and Asia. Agassi was an advocate of Popper's rationalist philosophy.

[474] Michael Polanyi (1891–1976) was a Hungarian chemist and philosopher of science. He studied at the University of Budapest (where he also received his PhD in physical chemistry) and at the University of Karlsruhe. He taught at the University of Manchester and at Oxford University. Polanyi's most important work is *Personal Knowledge* (1958).

[475] "Socrates" refers to Joseph Henry Woodger who talked about "biology and physics" at the conference.

[476] The published Festschrift (Gregg and Harris 1964) did not have any papers from Nagel, after all.

[477] Presumably Cooley (1959), reviewing Toulmin's *The Uses of Argument* (Toulmin 1958).

[478] Stephen Toulmin (1922–2009) was a British philosopher. After his undergraduate studies in mathematics and physics, he earned both his MA and his PhD in philosophy from the university of Cambridge. He taught at several universities both in England and the USA. Toulmin wrote extensively about the nature of rationality.

[479] Richard Rudner (1921–1979) was an American philosopher. He studied under Hempel at Queens College in New York City and then received his MA and PhD from the University of Pennsylvania. He taught at several American universities. He is best known for introducing the argument of inductive risk in his 1953 article "The Scientist Qua Scientist Makes Value Judgments" and acted as editor of *Philosophy of Science* between 1959 and 1975.

would be prepared to do so but only after my return. Are you on either of the two Boards? And do you find it burdensome, or is it too early to tell?

For my work here, I miss having an office or study away from home, but I have no alternative to the present set-up, and I do have a nice study here in the house, and so I will have to make the best of it.

If you should see the Goheens, would you give them our warmest greetings? And have you seen anything of Arnold? How is he? Our best to him, too! I hope to write them all, but I do not quite know how soon it will be.

Diane sends her best to you all; she is enjoying her stay here a great deal and is making the most of it; I think she knows London better than some natives, and she has been to several of the very picturesque little towns in the vicinity of Oxford. We all spent a pleasant weekend in the country when we visited Eva's half-sister Ruth who lives in a small Sussex village with her (research-)psychiatrist husband and two children.

Does Edith like Stanford? Do you have a car there, and have you been to San Francisco and to the Giant Redwoods?

I have received the announcement of the big Ph Sci meeting at Stanford; it sounds enormously impressive and should be very stimulating; as things look now, however, we won't be back in the U.S. until the end of the congress. Are you much burdened by the preparations for it?

But enough for this time. I hope this letter will find you all in good health and good spirits!

All my best to you all –

Peter

4.1.121 Hempel to Nagel, January 13, 1960

[Type- and handwritten, Nagel 2020, Box 3, Folder 4]

Oxford, 13. January 1960
93 Woodstock Road

Dear Ernest:

Diane and I were terribly sorry to hear of your and Edith's accident, but we are glad that you escaped without serious injury. We know what it is like, for we had a potentially quite serious collision in 1955, with the kids in the car; and although no one was hurt in either of the cars, ours was pretty badly damaged, and the psychological shock wore off rather slowly. We do hope that by now you are pretty much over it in every respect! You mention the financial aspects of the damage you suffered, and we wondered whether that was not covered by insurance, at least to a considerable extent? – Over here, we have so far been without a car (we sold our faithful old Plymouth to the Vlastoses) and have not missed it, though when spring comes, we will no doubt be more aware of what one could do with a car. We returned recently from a week's visit to Paris, which, though a bit tiring, was quite successful, also with the kids. The weather was at least dry, and we got around a lot, mostly walking, and saw a good deal that one would not see on a guided tourists' tour. For about April 1st, I have been invited to come to the U. of Jerusalem under the sponsorship of the U.S. Educational Commission there, and we plan to go all four, though that is going to be quite expensive; but I am glad Diana will have this chance to see Israel (which I am eager to see, too), and we think that for the children it will after all be a rare and long-remembered opportunity. Now the U.S. Commision has written me that it is making inquiries to find out whether Turkey, Greece, Italy, and Germany would not be interested in having me come for a few days each, again on Fulbright terms; but even if the answer is in the affirmative, I do not think I shall want to accept it all, however tempting it may be: it is just too strenuous and cuts to[o] much into my time, here, which has slices cut off it in various other ways anyhow. Among other things, Ryle has asked me to give a class (=grad. sem) on historical explanation with him during the eight-week term beginning next week, and has assured me that apart from one or two papers which I would have to give, I won't have any obligations except for being there and participating in the discussion; and as I am just working on hist. expl., I accepted, esp'y since the Fullbright people in London informed me that this sort of invitation (without any compensation) is sometimes extended to selected people as a courtesy. Diane had strong misgivings about the thing, and I hope she won't be right this time.

Yesterday, Val Vlastos and her son Stephen arrived here, by ship, and tomorrow, Gregory is to come by plane. Steve is now attending Magdalen College School with our Peter.

Have you seen the Gardiner volume, *Theories of History*?[480] I find it very useful and well arranged. – In the fall, I read a good part of Toulmin's *The Uses of Argument* and found it pretty disappointing, especially the part on probability, and I sympathized with John Cooley's review.[481] – The stir about the Gellner[482] book and Ryle's refusal to have it reviewed in *Mind*[483] had now died down; I can empathize with Ryle, who seems to have been full of righteous indignation at Gellner, but I think it was unwise to take the step he took.

Please give our love to Edith! We hope that this note will find you all in good spirits, and that 1960 will prove to be a very good year for you!

As ever yours,

Peter

{I think of both of you often.

Peter has told you how sorry we both are about the accident. I myself am simply *not* one of Nature's Born Drivers and am appalled by Oxford's (England's generally?) drivers and traffic. The *very!* first thing Val Vlastos did after one night in Oxford was to rush out to buy an electric blanket. Please write us about the sunshine so we will not lose faith that such does exist somewhere. It will be vast fun to meet after [n/a] all back, and share impressions – I've plenty! Have a good year – all of you

Love

D.

4.1.122 Nagel to Hempel, October 12, 1960

[Typewritten, Hempel 1999, Box 28, Folder 1]

[480] Patrick Gardiner (1922–1997) was a British philosopher who studied and taught at Oxford University. A scholar of continental philosophy, he also published influential works on the philosophy of history and historiography; see Gardiner (1959).

[481] Toulmin (1958) and Cooley (1959).

[482] Ernest Gellner (1925–1995) was a Czech-British philosopher and anthropologist. Born in Paris and raised in Prague, he earned his MA at Oxford and received his PhD in anthropology from LSE. His first post was as Professor of Philosophy at the London School of Economics; he was then appointed Professor of Social Anthropology at Cambridge University. He also taught at the Central European University in Prague. The book to which Hempel refers in the letter is *Words and Things* (1959), subtitled "An Examination of, and an Attack on, Linguistic Philosophy".

[483] Ryle was the editor of *Mind* between 1947 and 1971.

25 Claremont Avenue
New York 27, N.Y.
October 12, 1960.

Dear Peter,

I feel a bit embarrassed in writing you, for I have been such a bad correspondent during the past year and still have on my conscience the fact that I never answered any of your letters from Oxford. But I'll apologize and try to explain some other time. This is just a hasty note (for I continue to be in a whirl with endless things to do and none of them getting done) to send my greetings, and to express my hope that all is well with you and yours and that it will be possible to see you before long.

Although I enjoyed the Center very much, we are glad to be back on home territory, and this despite my nostalgic longing for the leisure I had in Stanford. We got back early in July, and it was fortunate that we did not remain in California (as we had originally planned) until the Stanford Congress, for I was able to see my mother before she died in mid-August. She was in a pretty poor condition when we returned, and Mack Lipkin made it clear to me that she might go any moment. Even so, her sudden death (it happened in Vermont) has been a severe blow, and the fact that I was "intellectually prepared" for it has not done much to soften it, however much I tell myself that she was no longer getting any joy out of life.

Have you heard that John Cooley has resigned his post at Columbia, and is now a *Privatgelehrter* living at his home in Connecticut? I've seen him once briefly in Vermont at the Stevensons, and he said that he would be coming down to the city fairly often, so that I can [look] forward to seeing him now and then. But those of us at Columbia who admired him, miss the almost daily contacts we had with him when he was formally connected with the department – and even frequent but discontinuously spaced meetings with him will not be a substitute for the things to which we had become accustomed.

There is a great deal else (seriously *sachlich* as well much gossip – not all as lugubrious as the items mentioned above) about which I would like to write – but I will save it for another time. Meanwhile, warmest regards to you and Diane,

Always yours,
Ernest

4.1.123 Hempel to Nagel, December 2, 1960

[Typewritten, Hempel 1999, Box 28, Folder 1]

Princeton, December 2, 1960.

Dear Ernest:

Please do forgive me for not having written you sooner; but several times when I was planning to take a closer look at your letter to Dick Rudner and at some of the literature on the Goedel theorem, something came up that obliged me to put it off.

I find your reply fully convincing; in fact, your section (b) might have mentioned also that your footnote 28 not only explicitly says that Goedel did not prove statement (1), but in addition states what he did prove. I am especially struck by the inconsistency between Hilary's assertion[484] that you fail to prove the non-demonstrability of G and his admission that you prove that the demonstrability of G entails that of ~G. The only question I am not quite clear about is what position you take concerning the converse proposition, namely, that

[484] See Putnam (1960). Eventually Nagel and Newman (1961) published a reply to Putnam.

if ~G is demonstrable then so is G. Does the last paragraph in footnote 28 still stand? And would the proof require the assumption of omega-consistency for the system?

It is too bad that Hilary is often so rash and at the same time so irritatingly apodictic in his critical reactions; I know how you must feel, and I am terribly sorry about this so [u]nnecessary aggravation. I wish there were a way for you to reply to him in public. Would not his review in conjunction with the forthcoming article by Myhill[485] which you mentioned give you more than ample justification to publish a reply, not in the form of a letter to the editor, but in that of a note or short article, say, in the *Journal of Philosophy*, or, perhaps better, in *Philosophy of Science*? What do you (and James Newman) think?

Diane and I were delighted to hear that Edith is up and about again and has even begun to go back to her laboratory. Please give her our love and best wishes; we hope she is feeling better every day and that she will steadily gain strength!

We got the good news about Edith's recovery from Meyer Schapiro whom we met with his wife for a most pleasant dinner at the Szathmary[s][486] before Meyer gave a very stimulating and enjoyable public lecture here on philosophy in painting.

I hope your trip to Fort Worth was interesting and not too strenuous! Did you go by jet? – Have you had any news from John Cooley yet? – John Gregg, whom I saw at Duke a few weeks ago, sends his very best to you. He seemed in very good shape and very pleased with his position there. I stayed overnight at his house and thus met his wife and his children, and I enjoyed it a great deal.

If Sheldon reads a paper at the December meeting of the Philosophy Club, I am afraid I won't be coming in, though I would so very much like to see you and talk to you about many things!

Donald Davidson just called he is in the East – among other things to look for potential candidates for specialized work in logic and the foundations of mathematics at Stanford, for which they have built up a very impressive department. He will be visiting us a week from Monday.

A representative of Harcourt Brace was here recently and told me that your book will be out in a few days.[487] I look forward to it with the keenest interest, and the representative, Mr. Bremer, said he would rush a copy to me as soon as it comes off the press. It must be a wonderful feeling to see this big job done!

Best wishes to you all!

As ever,

4.1.124 Nagel to Hempel, December 30, 1960

[Handwritten, Hempel 1999, Box 28, Folder 1]

25 Claremont Ave
New York 27. NY.
Dec. 30. 1960

[485] John Myhill (1923–1987) was a British mathematician with a PhD from Harvard with Quine. He taught logic, philosophy, and linguistics at SUNY Buffalo. The article in question is probably Myhill (1961).

[486] Presumably Arthur Szathmary (1916–2013), who was a Hungarian-American philosopher, with a PhD from Harvard. He was working at Princeton between 1947 and 1986. His field of interest included the philosophy of art, music, and literature.

[487] Nagel (1961).

Dear Peter,

I've just heard this evening about your election to the presidency of the Eastern Division. Warmest congratulations on the obviously well-deserved distinction, though really it is the APA which has done honor to itself by this act.

I've also heard that you gave a remarkably graceful introduction to John Wild's presidential address[488] – my reporter said that your remarks were the best thing that happened during the entire set of meetings. I'm sorry I missed it.

But I've been acting as nurse for Edith is ill again. She's been in bed for the last ten days, running a fever from a continuing infection whose course is still undetermined. She may have to be hospitalized again, but probably not for an immediate surgery, though this will eventually have to be done as soon as the infection quits down. I may know more to-morrow, when the surgeon is expected to examine her here at home.

Anyway, our very best to you and Diane for the coming year!

As ever

Ernest

4.1.125 Hempel to Nagel, January 3, 1961

[Typewritten, Hempel 1999, Box 28, Folder 1]

January 3, 1961.

Dear Ernest:

Thank you so much for your good wishes à propos of my election to the presidency of the Eastern Division. I am all the more touched by them because you wrote under such trying circumstances: Diane and I are so terribly sorry to hear Edith has this troublesome infection, and that some more surgery seems to be ahead; how hard this must be on you all! I had an impulse not to write but to call you tomorrow instead, after the surgeon had been there: but you will probably have many solicitous inquiries, and I don't want to add to them just then, but I'll try to reach you a few days later. In the meantime, Diane and I will be thinking of you and Edith and will be telepathing all our good wishes for a speedy and thorough recovery!

Your book has come, and what I have read of it so far impresses me enormously. I was acquainted, of course, with the views you express in some of the chapters, having read your articles or heard you read papers on the subject – but there is so much new material, for example in chapters 7, 8, and 9; and here as well as in the chapters on the social sciences I am struck by the wealth of illustration and documentation in addition to the extremely thorough and lucid presentation and analysis of all the issues. There is absolutely no doubt in my mind but that this book is going to be the standard work in the field, and I hope that it will be as long-lived in its general use as the Cohen and Nagel,[489] of which I regularly use some chapters in my undergrad. courses in logic and in the philosophy of science this spring term, and some other chapters – especially 4, 5, and 6 – in my grad. course on selected topics in the philos. of science; in the latter, I will also use your probability monograph.[490] During the fall, Paul Benacerraf[491] and I have been giving a senior seminar on explanation, using several of

[488] Wild's presidential address delivered before the Fifty-Seventh Annual Meeting of the Eastern Division of the American Philosophical Association at Yale (December 27–29, 1960) was about "The Exploration of the Life-World". See Wild (1960–1961).

[489] Cohen and Nagel (1934).

[490] Nagel (1939).

[491] Paul Benacerraf (1931–2025) was a French-American logician and philosopher. He spent his entire career, both as a student and teacher, at Princeton University, where he studied under Hempel. Benacerraf wrote extensively on the philosophy of mathematics.

your articles on reduction, organismic theories, functional analysis, etc.; I hope we will be able to offer the seminar again next fall, and in that case, we will surely use the corresponding chapters of your book, and also that on historical inquiry. My warmest congratulations, Ernest! And best – but entirely unpressurized – wishes for the progress of your work on the next volume!

I confess I was very pleased that you had had a favorable account of my introduction of John Wild at the New Haven meeting. I felt rather awkward at the prospect of introducing a man with whose ideas I was not well acquainted at all and I had been urged by various people to say something in a light vein; and I am relieved to think that my remarks were well received. John Wild himself seems to me a very nice person, but rather humorless, as far as I can tell, and I do not know what he thought of it all, though he cannot have been in doubt as to the good-natured character of what I said.

It was a well-attended meeting this year, and I was glad to see lots of old friends and acquaintances. Stevenson was there, and Frankena[492] and Henle;[493] Dick Rudner and Bob Callaghan;[494] also Gustav Bergmann, to whom I tried in vain to sell one or another of our PhD candidates. – Diane came on the second day in hopes of seeing various old New Haven friends in addition to the very fine Yale Art Gallery; but then, the next morning, it started to snow, and so we left earlier than we had intended, but we had made the gallery and had seen a few friends. I also went to a huge cocktail party given by Paul Weiss, who had his own paintings hung in all the rooms; I must say I was quite fascinated by some of them, though I think there is a lack of technical mastery of the medium.

We had a rather laconic card from John Cooley, and we do hope that before long, things having taken a decided turn for the better, you and Sid and John will be able to come out here for one of our little talk fests!

Please give our love and warmest good wishes to Edith!

As ever,

4.1.126 Nagel to Hempel, January 6, 1961

[Typewritten, Hempel 1999, Box 28, Folder 1]

January 6, 1961

Dear Peter,

I talked to John on the phone two days ago about the possibility of visiting you. Is Wednesday January 24 still a good day for you – it happens to be very convenient for the three of us, although we could also make it on Thursday the 25th? And if the day is all right, at what time would you like us to arrive and where should we meet you?

[492] Presumably William K. Frankena (1908–1994), an American moral philosopher who taught at the Department of Philosophy, University of Michigan, and acted as its chairman between 1947 and 1961. After his death, the University created the chair for the "Carl G. Hempel and William K. Frankena Distinguished University Professor".

[493] Paul Henle (1908–1962) was an American philosopher. He studied at Harvard University, where he received his PhD. He taught at Harvard, Smith College, the University of Michigan, and Northwestern University. Henle wrote several texts on logic and the notion of validity.

[494] William Callaghan (1912–1986) was an American philosopher. He earned his BA and MA at Harvard University, while he received his PhD in Philosophy from Columbia University. He taught at Queens College in New York City and Michigan State University. Callaghan was a specialist in Greek philosophy.

Although I wrote Paul Meehl Dec. 1 asking him whether he would accept the nomination to serve on the Board of Governors of the Philos. of Science Association, he replied only yesterday saying that he is unable to do so. And so we must find another candidate. When we were discussing the question in November, Ducasse as you will recall mentioned 4 men in the social sciences in this country (Furfey,[495] Catholic University of America; Hartung,[496] Wayne; Price,[497] Chapel Hill; and Trend, Wayne), but he said he knew nothing about any of them. However, I vaguely recall that Hartung has been interested in the sociology of science and that he once held some kind of office in the PSA; and so I would be inclined to pick him from Ducasse's list (although I do not find his name on the membership list in the Jan. 1961 issue of PHILOSOPHY OF SCIENCE). But another sociologist occurs to me: Llewellyn Gross,[498] Univ. of Buffalo, the editor of *Symposium of Sociological Theory*, to which you contributed; and my preference is to ask him first. Could you send me your suggestions? I am also writing to Ducasse.

Freddy Ayer read us a paper Thursday on PROPER NAMES AND DESCRIPTIONS;[499] the paper was puzzling and not too clear (perhaps because it was read very fast), but talking to Freddy is always lots of fun. I saw him again last night at a dinner for Krik[orian], who will be 70 this month and is retiring from City College at the close of the present semester. It was a very pleasant affair, but I find it hard to realize that Krik is that old.

Edith will be returning to the hospital as soon as a bed becomes vacant (probably this coming week), and if the X-rays which will be taken are favorable she will undergo what we are hoping will be the final operation (to close the colostomy). We are keeping our fingers crossed.

Love to all of you,
Ernest

4.1.127 Hempel to Nagel, August 21, 1961

[Handwritten, Nagel 2020, Box 3, Folder 4]

8-21-61

Dear Edith and Ernest: I can't tell you how happy we are to know that you were able to go up to Vermont for a spell, and that it is doing all of you good; we very much hope you will have especially fine weather for your whole stay, and that you will all get a good rest!

Sandy must have earned his vacation especially this summer after his course work (I hope it was interesting and worthwhile for him!). Peter, who will be starting freshman

---

[495] Paul Hanly Furfey (1896–1992) was an American clergyman and sociologist.

[496] Frank Hartung (1908–1985) was an American sociologist. He studied at the City Colleges of Detroit and at the University of Michigan, where he received his PhD. He taught for most of his career at Wayne State University. His field of expertise was the sociology of crime and deviance.

[497] Daniel O'Haver Price (1918–2012) was an American sociologist. He studied at Florida Southern College and the University of North Carolina-Chapel Hill, where he also received his PhD in sociology. He taught at UNC, the University of Texas-Austin, and UNC-Greensboro.

[498] Llewellyn Z. Gross (1914–2014) was an American sociologist. He received his PhD from the University of Minnesota and taught for most of his career at the University of Buffalo/SUNY Buffalo. He was the editor of the *Symposium of Sociological Theory* (1959), to which Hempel contributed the essay "The Logic of Functional Analysis", later reprinted in Hempel (1965a, Chap. 11).

[499] See Ayer (1962).

year here at Princeton in September, has been holding down a number of lawn-mowing and similar jobs and earned himself some money and an amiable suntan. I have been working steadily and have made some progress; but boy, how slowly!

You herewith have a rain check, and indeed a warm invitation, to visit us whenever you feel like it after your return; do try to make it while the weather is still pleasant! There will be lots of news to exchange!

Love to you all from Diane + myself!

As ever yours,

Peter

{Isn't it sad to hear of Bridgman's death?}

4.1.128 Hempel to Nagel, November 17, 1961

[Typewritten, Hempel 1999, Box 28, Folder 1]

November 17, 1961
Professor Ernest Nagel
Department of Philosophy
Columbia University
New York 27, N.Y.

Dear Ernest:

I will be very glad to serve on the Nominating Committee to nominate a successor to Professor Bridgman on the Board of Governors of the Philosophy of Science Association. I do not know what procedure the Committee is to follow, but perhaps its business can be transacted by mail – if necessary with the aid of some telephone calls.

It was very good to see you a week ago, and I very much appreciate your having kept me company up to the departure time of my bus, despite all the things on your mind. I hope that early next year you will be feeling in the mood to come out for a good relaxed talk, perhaps with Sid and John; and of course, Diane and I hope you will have dinner with us after your seminars whenever possible!

Last Tuesday, I gave my talk in the Delaware series;[500] the audience seemed small in the large-ish auditorium, perhaps 80 to 100 people; but there was quite a lively discussion afterwards, with many interested questions, and a few that I thought good and incisive. – I was shocked to learn from Baumrin[501] that Hanson,[502] on his return from a lecture to a Dupont group in Hershey, Pa, had crashlanded his own airplane a short distance from his home airport because he had run out of gasoline. He was in critical condition, but after

[500]The Delaware Seminar in the Philosophy of Science took place at the University of Delaware between 1961 and 1964. Materials of the seminar were published as Baumrin (1963a, b). Hempel talked about "Explanation and Prediction by Covering Laws", see Hempel (1963a).

[501]Bernard Baumrin (1934–2018) was an American philosopher who obtained his PhD from Johns Hopkins University in 1960 and later a doctoral degree in law in 1970 at Columbia. He was an assistant professor and Chairman of the Delaware Seminar in the Philosophy of Science at the University of Delaware from 1961 to 1964.

[502]Norwood Russell Hanson (1924–1967) was an American philosopher of science, known for his so-called post-positivist views on science, theory-ladenness of observation, and historical orientation. He was educated at Chicago, Columbia, Oxford, and Cambridge, and founded the Department of History and Philosophy of Science at Indiana University. Hanson died in 1967, in another airplane-accident, when his airplane crashed in fog on his way home.

several operations in brief succession is off the critical list, and there are rumors that he may be teaching next semester. – Incidentally, I did the U. of Del. an injustice: I thought they had offered me a honorarium of 100,–, but it turned out that it was 200,–.

Everybody is inquiring about Edith, and I wish all the good wishes expressed and all the warm feelings for her could help speed her recovery proportionately – she would be up and around tomorrow! Please give her my love, and Diane's.

Cordially,

4.1.129 Hempel to Nagel, January 9, 1962

[Typewritten, Hempel 1999, Box 28, Folder 1]

January 9, 1962

Dear Ernest:

Wednesday, January 24, suits me very well, and Diane and I are much looking forward to seeing you, John, and Sid then! I'd suggest you try out the bus to get an idea of how it would suit you when you come out here for your seminar during the spring term. It leaves from the Port Authority Bus Terminal from Gate 8 on the lower level; tickets can be purchased on the street level at windows 1,2,3,4; and indeed, you have to have the tickets in order to be allowed on the bus. Round trip fare is cheaper, and the buses run every half hour during the day, once an hour during the evening. How would it suit you to take the 2.00 bus, arriving at Palmer Square (practically opposite the Firestone Library) at 3.30? I'll await you in my office, and we might start by having some coffee. If for some reason you prefer to come by train, I'll be in my office from 3.00 on and shall be delighted to see you!

Llewellyn Gross does seem to me a much better choice than any of the other candidates for membership on the Board of Governors of the Ph Sci Assoc., and I hope that Ducasse will go along, and that Gross will accept!

Diane and I send our love to Edith, and warm good wishes for a prompt and happy completion of her long and hard surgical treatment; how wonderful it will be for you all to have all these troubles over and done with!

All our best to you all!

4.1.130 Nagel to Hempel, May 14, 1962

[Typewritten, Hempel 1999, Box 28, Folder 1]

25 Claremont Avenue
May 14, 1962

Dear Peter,

I too regret that we did not have a chance to talk after the NYU meetings and do some post-mortems, and especially to tell you how fine and thoroughly convincing your analysis of Dray's paper seemed to me. But I could not stay any longer, for I felt increasingly uncomfortable because of my back. I've been lying down during the past two days (I've had no classes today), but there is no improvement; and though I hope to meet my scheduled classes for the rest of the week, I am afraid that I will have to resign myself to having another protracted bout with my miserable backbone. Should my fears be well-founded so that I will be unable to come to see you with John and Sid I will let you know; and in any event, I will send you a report on the students in my seminar as soon as I can get the papers read (within the next couple of days).

On the whole, I found the NYU conference quite disappointing; and though, like you, I learned something from it, I am not sure whether the price in terms of time and energy does not come too high. Paul Lazarsfeld told me over the phone that he was appalled at the quality of most of the discussion during the Saturday forenoon session, and was in fact incensed at the way Sidney Hook chaired the meeting. But I must save further comments for another time, since it is difficult for me to type now. For the moment, all the best to you and Diane,

Always yours,

Ernest

4.1.131 Nagel to Hempel, February 27, 1963

[Handwritten, Hempel 1999, Box 28, Folder 2]

Feb. 27 1963.

Dear Peter, I've been so busy and tired that I have not been able to drop you a line sooner (as I had intended to), to tell you how much I enjoyed your [n/a] at the Club. It was so characteristically and admirably clear, stimulating, and solid; and if argument has any persuasive form, your various critics (including Mike Scriven!) ought certainly be persuaded by your rejoinders to stop making silly mistakes. At any rate (though this is no news to you), I agree with just about everything you said. However, I would have liked to discuss with you some of the finer points – especially your "good reasons" requirement, but also the "predictive" power of extraordinary theory – about which I do not feel myself to be sufficiently clear, and I hope that some day soon both of us will have time for this. Meanwhile, love to you and yours,

Ernest

P.S. I finally received proper proof of my piece for the Carnap volume, so that it looks as if the volume will be published during my lifetime after all!

4.1.132 Nagel to Hempel, May 9, 1963

[Typewritten, Hempel 1999, Box 28, Folder 2]

May 9, 1963

Dear Peter,

I saw both John and Sid today, and mentioned the list of dates you suggested as convenient for you to see us. We decided on Wednesday, May 22, and I hope that is still possible for you. When I see you at the Club next week, you can tell me whether it is, and also the time of day you will be free to have us call.

It will be good to see you again. Meanwhile, love to you and Diane.

As ever

Ernest

4.1.133 Hempel to Nagel, July 15, 1963

[Typewritten, Hempel 1999, Box 28, Folder 2]

July 15, 1963
Professor Ernest Nagel
South Wardsboro Vermont

Dear Ernest:

I hope that the above address of yours is not just wishful thinking on my part, that you are actually in South Wardsboro, that you had a good trip, and that all four of you are enjoying the much-deserved and badly needed stay in the country. It should be very pleasant in Vermont now, and the thought of sunny days but cool nights and of the smell of woods and of hay seems very attractive in the rather hot and very dry weather we have been having here. It seems to make every effort a bit more strenuous and I really face with misgivings the task of getting things ready for the move to Stanford. We have decided to take the Southern route, after all, so as to be able to see, and to show the children, Santa Fé and some of the Indian country, and to visit the Grand Canyon. I just hope it won't be too frightfully hot! Pete will come along, which will be useful also because he can help drive the car, and he is eager to make the trip; but he will then soon have to turn around to get back to Princeton for the start of the term. He has currently a job as a waiter in Commons, the Univ. dining hall, where during the summer, Prudential Life Insurance executives and similar groups are fed while they participate in special summer programs.

This morning, I received the Wesleyan Conference volume on Induction[503] that you had sent me; it was a great surprise, and a most pleasant one. I have read your introductory essay, which I think is very helpful, and I was much interested in Kyburg's[504] prefatory account of how you had handled the presentation of the discussions; to judge from the various parts I have so far been able to read, your procedure has been very successful. I find that there is a sense of relevance and coherence as one reads through a discussion, and stylistic awkwardnesses have been filed out, so that the reading is pleasant and holds one's attention, which I cannot say of some similar efforts in other recent volumes, especially the Delaware one, where too much irrelevant material and sheer "noise" has been allowed to be taken over verbatim into print. I am especially glad to have this new volume because almost anywhere I look I find things of immediate interest to me, and I have already started marking some of the passages on the margin. I am very grateful to you for having thought of me, Ernest – thanks a lot! And Diane joins me in sending greetings and her appreciation!

I have heard nothing from John Cooley; I sent him a note some time ago, to his P.O.B. address, and I hope he is by now all over the operation and in good shape. It was such fun seeing you and Sid in June, and I wish John could have been along. I wonder whether Sid is off to Europe already. Paul Benacerraf with wife, baby, and poodle are flying off tomorrow for a five-week visit to France, where mother lives. And virtually everybody else seems to be in Europe. The Cohens (he is in the Rutgers Law School – remember?) are off to Scandinavia, Russia, Yugoslavia, Italy; the Helmers came through and are travelling in Europe (Olaf told me that Albert Wohlstetter is leaving his top-level job at Rand, has a one-year profship at Berkeley this coming year and I think some grant for writing a book for another term or year, but no fixed plans thereafter). Feigl has huge travel plans: Mexico City in September, and from next June on, Europe and Australia. – Are you thinking of going to

[503] Kyburg and Nagel (1963).

[504] Henry E. Kyburg (1928–2007) was an American philosopher and computer scientist who worked on probability and logic, and is known for his "Lottery Paradox". For Nagel's mentioned paper, see Nagel (1963a).

the Jerusalem Congress? Feigl wrote that the Carnaps have to leave their house after all, but he does not know what they have decided to do; I feel very sorry for them, this is a problem, with their difficulty in making decisions – with which I can very much empathize.

Did I tell you that Saul Kripke,[505] having received permission from the Senior Fellows, will spend the spring terms of his second and third years as a Junior Fellow teaching and doing research in our department as a visiting lecturer with the rank of assistant professor (and without PhD)? This means the spring of 1965 and 1966.

All the best from Diane and me to you, Edith, and the boys!

As ever,

4.1.134 Nagel to Hempel, July 18, 1963

[Typewritten, Hempel 1999, Box 28, Folder 2]

South Wardsboro, Vt.
July 18, 1963

Dear Peter,

This is just a hasty note to tell you how good it was to get your letter, and to keep you abreast of a few items of personalia. The last Sunday in June, Sid M., Arnold Koslow[506] (whom I think you know – he is an instructor in my Department) and I visited John in Bakersville. He looked very well, but his knee was still healing, and though he was off the crutches he had been using he was not able to move around very much. Since then I talked to him on the phone a couple of times (last week most recently), and he is much improved – he plans to visit Chas. Stevenson (in Wilmington, Vt.) and perhaps us next week, and if all goes well to do some sailing in August. So I think things are going as well with him as can be expected.

As for Sid, he is teaching in the first Summer Session at Columbia, will be through by next week, and promised to visit us before he sails for Europe sometime in August.

We have been here since the 2nd of the month, and find the country wonderfully pleasant despite a longish drought – it's lucky we have a driven well, so that we have not had a water problem. Edith continues to be tired, but with the two boys around she has considerable muscle at her disposal for whatever digging and weeding that needs to be done in her flower garden. She is scheduled to give an advanced Electricity and Magnetism course in the fall, and she has been saving most of her energy to working out the problems in the textbook she will be using. On the other hand, I am not of much use to anybody, for I've had a good deal of trouble since I last saw you. In the middle of June I had another siege with my back, and thought for a while that I would have to be hospitalized again – but I managed to get complete bedrest for a week at home, and the attack quieted down sufficiently to enable me to get on my feet again. However, the day after we got here, I unexpectedly passed out while I was shaving, and the experience was repeated two days later. So we had a local doctor in, he diagnosed my condition as "exhaustion", and prescribed some more bedrest and

[505] Saul Kripke (1940–2022) was an American philosopher and logician who did not have a PhD, but revolutionized modal logic, when he was still in high school. He taught at Rockefeller University, Princeton, and the City University of New York. Kripke made unique contributions to logic, metaphysics, philosophy of language, philosophy of mathematics, and set theory.

[506] Arnold Koslow (1933–2024) was an American philosopher. He studied at Columbia University, where he also received his PhD in 1965. He then taught at Columbia University before becoming a professor at the City University of New York.

sedatives. I am up and about again, but my head does not feel quite right and I have no ambition to do anything but sleep. So the summer is flying by without my having gotten a single thing done. If I don't pick up within another week I'll get in touch with Mack Lipkin to see whether he has anything to recommend. Maybe my "time has come" – though quite apart from my wanting to get some things done before it does, I was hoping that I'd be around long enough to see the boys through college.

Toulmin had written me asking whether I would contribute to a symposium on history at the Jerusalem Congress, and I accepted the invitation. I confess that the temptation to visit the near east, with a side trip to Greece, is very strong. Are you planning to go? I hope you will – it seems like a wonderful chance not only to travel at relatively little expense (Toulmin wrote that travel expenses will be paid!), but also to meet people not otherwise accessible. But don't bother replying now to this query – you must have your hands full with preparations for your transcontinental tour, and perhaps you'll write me from California.

Meanwhile, all the best to you and Diane from all of us, and affectionate regards,

Ernest

4.1.135 Hempel to Nagel, July 23, 1963

[Typewritten, Hempel 1999, Box 28, Folder 2]

July 23, 1963

Dear Ernest:

What a shame that you have not been feeling quite right. Diane immediately made the diagnosis: "It's lack of sleep", she said; and I think she has something there. I really don't know how you have been managing for so many years to get along on so little sleep, and very little other relaxation; and I just hope that you can just make yourself slow down for a while and get lots of rest; this is a relatively optimal time, and *other things will have to wait*. I also think it would be a very good idea to call Mack Lipkin just to get this advice; you know, don't you, that after 9 you can call anywhere in the U.S. for one dollar for the first three minutes (I have heard it said that from then on, the rate goes up to the standard one!); but just possibly, this bargain is limited to dial phones. We have used it to call Carnap, for example. (That was quite some time ago, and he and Ina were well; recently Feigl wrote that they have to vacate their house after all – as I mentioned in my last letter. We have been meaning to call them again, but have not done so as yet.)

It was good to hear that John Cooley is a-mending; let's hope he will be able to get some sailing in before the fall. I wonder whether he will continue to teach for your Department, or whether he now wants to retire again.

I read your article on Assumptions in Economic Theory[507] with pleasure and find it clarifying and helpful; it makes me want to discuss some of these issues in my seminars, as I should long have done. Many thanks! I wish I could have heard the discussion.

You know, three days ago I had a letter from von Wright[508] asking me to give an invited address, perhaps on the present status of confirmation theory, at the Jerusalem meeting. I feel tempted, of course, but I am also concerned that, the way I am, it will take a considerable amount of time and worrying to get something together for the occasion (though

---

[507] Nagel (1963b).

[508] Georg Henrik von Wright (1916–2003) was a Finnish philosopher who succeeded Wittgenstein upon his retirement at the University of Cambridge in 1948. He is known for his work on philosophical logic (especially modal and deontic logic), the philosophy of Wittgenstein, and the philosophy of language and action.

v.W. indicated that I might choose another topic in the general area; there'll also be a symposium on rational belief with Black, Kyburg, and perhaps Braithwaite, and I feel I'd rather be asked to participate in *that*). Now the news that you are going makes me especially eager to accept, but I am still wavering. Oh for that decision machine whose desirability for the likes of us you have more than once very eloquently indicated!

Now Ernest, please do get a rest, and do get some really informed medical advice; you have an iron constitution, but you just shouldn't tax it too hard!

Warmest good wishes to all of you from Diane and me –

4.1.136 Hempel to Nagel, September 24, 1963

[Typewritten, Hempel 1999, Box 28, Folder 2]

Tuesday, September 24, 1963

Dear Ernest:

Before Diane picks me up at the gate by the bottom of the hill, I want to send you a greeting from here and tell you that things have gone very well with us so far. The trip across the country proved to be pleasant and of course very interesting; Pete was a great help in doing a lion's share of the driving and enjoying it hugely; and we all found that the tiresome aspects of the trip were not nearly as trying as we had anticipated.

The Jeffreys' house, for which we swapped ours, is pleasant, if rather smaller than ours, and it has a nice garden whose ingenious layout makes it seem larger than it is (or rather, larger than it might otherwise seem). But it's five miles from the Center, and so Diane has the chore of driving me and then collecting me again in the evening; and on top of that, she takes Toby to school early in the morning, and has been picking her up, though we expect that soon she will get into the habit of walking the 1 mile home from Cubberly High, where she was transferred after two days in Junior High, when it turned out that her so-called fast group in Princeton had been ahead of the corresponding grade; so she technically skipped a grade and went into 10th; this will (or should) enable her to rejoin her group upon her return to Princeton. – The other day, Toby got badly clawed by a black cat against whom she tried to protect our own cat; this happened during our absence, and fortunately, our very nice neighbors took her to the hospital, where she had seven stitches put into her right hand. But it healed well, the stitches are out, and Toby is in good spirits, though from time to time she complains that she'd rather be back in Princeton with her friends. – Pete stayed for two weeks and enjoyed it all; Diane took him twice to San Francisco, where I have yet to go (though I was there once ten years ago), and also to such places as the Almaden winery. We hope he'll have a good and profitable year at Princeton. For Xmas, he expects to visit us.

The Center has come up to the most glowing advance billings, and has actually exceeded them; my only question is how one can live up to and justify the expectations that must be presupposed by this kind of thoughtful organization and eager service. I come up here every day and sweat over my ms and slowly seem to gather some momentum, but the result is still far from what I'd like it to be. At any rate, I have so far declined all invitations to join the Visiting Committee for MIT, and so forth, except for accepting an invitation to be a commentator on v. Wright's paper on the paradoxes of confirmation at the Xmas meeting of the APA, Pacific Div.[509]

---

[509] The thirty-seventh annual meeting of the American Philosophical Association's Pacific Division was held between December 27–29, 1963, at the University of California at Los Angeles. There von Wright spoke on the first day about "The Paradoxes of Confirmation" with Hempel as official discussant.

Bar Hillel[510] was here, and we had lunch up here at the Center, and he said that I would get an official invitation to the Jerusalem Conference, at which I had agreed to read a paper upon v. Wright's request. And I was delighted to hear from B-H that Carnap has agreed to attend the Jerusalem meeting as well!

How are you, Edith, and the boys? We've often thought of you and we hope that you had no recurrence of those fainting spells, that you got a good rest, and that you are feeling better. If and when you feel like it, pls drop me a post card, not more, and let me know how you are.

With many good wishes, to you all,

As ever,

Peter

Any chance of your coming West this year? I decided not to attend the meeting with the Advisory Committee of our Dept in Princeton this weekend.

4.1.137 Nagel to Hempel, October 14, 1963

[Typewritten, Hempel 1999, Box 28, Folder 2]

25 Claremont Avenue
New York 27, N.Y.
Oct. 14, 1963

Dear Peter,

It was good to get your letter of Sept. 24, and to learn that you are enjoying the Center (as I had no doubt that you would – I suspect also that like myself you will be spoiled for life by the freedom from distraction, the climate, and the beautiful physical setting); and I am sorry that it has taken me so long to give you any sign of life. Let me tell you briefly about the family's state of health. The summer was a poor one for both Edith and myself. Although I had no recurrence of the black-outs, I could pump up no energy whatsoever, and the continued back-trouble (not entirely gone) did not help in this respect. I got absolutely no work done, except for the review of the recent W.H. Watson book for SCIENTIFIC AMERICAN,[511] over which I sweated for a good three weeks. (Indeed, my difficulty in getting words down on paper is beginning to verge on the pathological.) I had a complete check-up on our return to the city, and Mack Lipkin found everything normal; so there is no physical excuse for the lack of drive from which I have not recovered or the depressed spirits which are its consequence. However, I've had no time since we returned from Vermont to think of myself, so there is perhaps a silver lining to the hectic pressure of odd jobs that fill my days. So much for that! During August, Edith began to feel unwell, her temperature began to climb, and finally she spent a week in the hospital nearby until the temperature could be brought back to normal. Mack Lipkin has found no indications that the diverticuli are starting to cut up again; and Edith is back teaching physics at The City College. But she continues to be extremely tired, and is full of apprehensions (not surprising) that the cycle of infections would start up once more. But for the moment at least we are carrying on. The trouble is that there is no cure for old age, especially in my case. As for the boys, they are

[510] Yehoshua Bar-Hillel (1915–1975) was an Israeli philosopher, mathematician, and linguist. He was born in Vienna and lived in Berlin, and after the Second World War he obtained a PhD in philosophy from the Hebrew University. He was in close contact with Carnap, and they worked together on logic and information theory.

[511] Nagel (1963c).

flourishing. Sandy is enjoying his second year at college thus far, and Bobby is kept quite busy with a heavy program he is taking at the Bronx H.S. of Science.

I was down at Princeton for the meeting of the Advisory Committee of your department, and had a brief chat with Stuart Hampshire[512] who is as charming as ever. But I was sorry that you were not there. You really have a fine department, and Gregory is a wonderful person – though I confess that Walter Kaufmann[513] is not quite my dish of tea. Feinberg[514] wasn't present, but last Thursday he read us a paper here at Columbia – not as good as the previous ones I heard him give, but he is undoubtedly a man of great promise. As you probably know, we have a new member in our department: Richard Taylor. I haven't seen much of him this fall, and hope he works out well and brings something vital to our midst. However, I recently finished his little book on metaphysics in the Beardsley series, and found it disappointingly thin.[515] Art Danto[516] told me that his book has been accepted by the Cambridge Univ. Press.[517] I recall your telling me that you were the referee for it in this country (Danto does not know this), so that I expect the book is reasonably good, even though I did not get much from that long piece of his that appeared in HISTORY AND THEORY.[518] I am about half-way through Is Scheffler's book,[519] and I think some of it is very good, despite the excessive and to me irrelevant Goodmaniana in it. And this morning I received a copy of Adolf's book[520] – it looks very solid and will not be easy reading. I am beginning to find it extremely difficult to keep up with the outpouring of the literature, especially since I feel a professional obligation to keep up with the latest wrinkles, even though I do not often profit much that I read. By the way, have you seen Mary Hesse's MODELS AND ANALOGIES IN SCIENCE[521] which came out this summer in England (Sheed and Ward)? It is quite small, but I think worth reading, and you may find it useful for your own work.

I am delighted that you are going to the Jerusalem Congress, which I am also planning to attend (on the assumption that my expenses will be paid). I've been told that Carnap was at

---

[512] Stuart Hampshire (1914–2004) was a British philosopher. He studied at Oxford, where he also taught in three different occasions, before and after World War II. He also taught at the University College London and Princeton University. His major work is *Thought and Action* (1959).

[513] Walter Kaufmann (1921–1980) was a German philosopher, emigrated to the USA in 1939. He studied at Williams College and received his PhD in philosophy from Harvard University. He taught for his entire career at Princeton University.

[514] Joel Feinberg (1926–2004) was an American philosopher. He earned his BA and MA from the University of Michigan, where he also received his PhD. He taught at several American universities. He published extensively in the fields of moral philosophy and philosophy of law.

[515] Taylor (1963).

[516] Arthur Danto (1924–2013) was an American philosopher. He earned his BA degree from Wayne State University and received his MA and PhD in philosophy from Columbia University. A towering figure in the philosophy of art, he also published important works in the philosophy of mind and the philosophy of historiography.

[517] Danto (1965).

[518] Presumably Danto (1962).

[519] Israel Scheffler (1923–2014) was an American philosopher and theorist of education. He studied at Brooklyn College and received his PhD from the University of Pennsylvania. He taught at Harvard for his entire career. The book to which Nagel refers in this letter is Scheffler (1963).

[520] Presumably Grünbaum (1963).

[521] Mary B. Hesse (1924–2016) was a British philosopher. She studied at the University College, London, and at the Imperial College of Science and Technology, London, where he also received her PhD. She taught at the University of Leeds, the University of London, and the University of Cambridge. *Models and Analogies in Science* is her major work. See Hesse (1963).

the Mexico meetings, but have otherwise heard nothing about what went on. But apparently Carnap is in good shape, and I am looking forward to seeing him if he does go to Israel. Will you be visiting him in L.A. sometime this year?

We had a picture card from John Cooley from Maine, and he seems to have recovered completely from his operation. I also had a letter from Sid Morgenbesser from England, perhaps a little lonely but in reasonably good spirits.

I'll try to write again before long, and hope meanwhile you and Diane will continue to enjoy California. If you see Pat Suppes, please remember me to him.

Love from both of us to you and Diane,

As ever,

Ernest

P.S. I am afraid I never acknowledged the copy of Jocelyn Davey's A TOUCH OF STAGEFRIGHT[522] you sent me to Vermont. Thank you most warmly though belatedly for it – it was fun reading, and also surprising to find Morty White mentioned in it. Philosophy does bake bread, after all! Many, many thanks.

E.

4.1.138 Hempel to Nagel, November 18, 1963

[Typewritten, Hempel 1999, Box 28, Folder 2]

November 18, 1963

Dear Ernest:

Thanks so much for your very nice long letter; I appreciate it all the more because I know how busy you are, and also how pre-occupied with many things.

What a miserable summer you and Edith have had! I do hope you are both feeling better now; I can understand only too well how depressing it must have been for Edith to have that fever spell, and for you to be so totally exhausted. I think I know what should be invented for you: A substantial fellowship for a year off with the one string attached that you must use it to do whatever you please except for doing any writing, and that at least six weeks must be devoted to lazying in some attractive place with a good climate (how's about Stanford?). Which reminds me that I recently saw an application for a grant in support of a man with a very high annual salary who wants 2/9 of it for three summers to do research on a philosophical problem; and with overhead and secretarial help this came to such a staggering amount of money that I could not help saying in my statement that I thought the request was out of proportion. Maybe you have seen the same thing; I had to comment on it for NSF. There would have been no reason why you should not have applied for the same terms in order to write whatever you wish to write, and I am afraid that you need it more. It seems to me this system is getting out of hand.

I read your review of Watson's book as soon as I got my copy of Sci Am, which was before your letter came, and I found it again a truly-first-rate job. You give one such a clear idea of what a book is all about before you proceed to your criticisms, and I thought what you said by way of critique was entirely convincing.

Thanks for calling my attention to Mary Hesse's little book: it is indeed relevant to what I am doing; I ordered it immediately from Blackwells and read it with interest. I am dubious about the value of her discussion of the logic of analogical reasoning, but I find some of her general ideas about models suggestive. I am not sure, though, that I would be able, even after

[522] Davey (1960).

re-reading the book a couple of times, to give a precise account of what she means by a model etc.; all her definitions or non-definitional characterizations are couched in fairly general and sometimes imprecise language. I'd very much like to see a review of the book by you! (Perish the thought) She certainly writes pleasantly and engagingly.

Adolf's book, of which I received a copy, looks rather overwhelming to me, and while I am sure it is very substantial I am afraid its readership will be rather limited, perhaps less because the subject matter is too difficult than because he writes in a quite involved style; I feel that I have to read every sentence twice or three times before I understand it. And I was really taken aback when I first opened the book and started right at the beginning: how many readers will follow him beyond the first page, seeing that he seems to be starting right in the middle of things and thus presupposes an awful lot that one would like to have presented explicitly to begin with? I do not know whether he realizes that, and I am surprised that Knopf did not urge him to change the organization and presentation. Or am I taking a too extreme view of these matters?

I have had some really astonishing experiences with The Voice of America. I recorded my talk for them last June in Princeton, having sent my ms long before to Sid. In mid-September I received a long-distance call here in Stanford from their Washington office saying that Sid had not sent them my ms before leaving for Europe, and his sec'y was unable to locate it; would I send them another copy. So I had my copy xeroxed and mailed it to them and after some time received an acknowledgment and the news it would be printed in the Philippines. A few days ago, another call from Washington: the ms had been given to a sec'y for duplication, and the sec'y had since left the country and the ms had vanished: could I send them another copy!

The Carnaps did spend some weeks in Mexico and are now back at their old address but under constant threat that the house they live in may be demolished to make room for a larger building. They are reluctant to buy something now since Carnap expects to be in Europe in the spring – that Institute in or near Vienna, and some other places – and later in Jerusalem, if he does go, which now seems a bit more dubious to him. We have not yet visited them, alas, but hope to do so at Christmas time or rather right afterwards, when I have to attend to APA meeting in L.A.; I agreed to be a commentator on von Wright's paper; and Carnap is to read a paper – I understand somewhat like the one he presented at the Stanford Congress – as well, and I want to hear that, of course.[523]

I have read a little in Russ Hanson's book on the positron[524] and found it pleasant and lively, but when one looks at the detail of the argument, it does not always seem so satisfactory nor do I even always get the relevance of what he is saying to the issue he is discussing. In the second part of the chapter on explaining and predicting, he says some interesting things that seem to me perhaps to provide some ammunition for an anti-Hesse (or anti-Campbell)[525] position, though I am not sure he'd want it to be construed that way.

I learned today that Paul Oppenheim is in NY Hospital for a prostate operation, probably this week. He is in very good general condition, so I hope it will all go well.

---

[523] The 37th Annual Meeting of the Pacific Division of the American Philosophical Association was held between 27–29 December 1963 at the University of California Los Angeles. Von Wright spoke about "The paradoxes of Confirmation" and Hempel was discussant. Carnap spoke about "Rationality in Inductive Thinking" and was also discussant in a Symposium about "The Scope and Significance of Higher-Order Languages".

[524] Hanson (1963).

[525] Norman Robert Campbell (1880–1949) was an English physicist and philosopher of science, known for his books on the philosophy of physics and the nature of science.

Have you heard from Sid? Had a nice note from Nozick[526] today, he's in Oxford and enjoying his stay, but critical of some of the philosophizing.

All the best from Diane and myself to you, Edith, and the boys!

As ever,

4.1.139 Hempel to Nagel, November 23, 1963

[Typewritten, Hempel 1999, Box 28, Folder 2]

November 23, 1963

Dear Ernest:

We are all shocked and stunned by Kennedy's assassination,[527] Diane is depressed, and the Fellows up on the Hill seem, like myself, sort of paralyzed. Diane came up to the Center after she had gotten the news, and we just sat in the lunchroom over coffee with Toby, who was very upset; then Savage[528] drifted in and despondently sat down at another table just by himself. On an impulse, I went over and asked him to join the three of us, and was he feeling as depressed as we were; and he said he had cried at the news and had rushed down to his boy's school to make sure he would be the first to break the news to him. Savage is not only an enormously sharp man, but an incredibly gentle and sweet one. His wife did not come with him; we understand that in fact she is in a sanatorium; his older son is a student at Michigan, and he is here all alone with his 9-year old, who is said to be very very bright, but who has reading difficulties. Savage then wondered out aloud, in a detached way, about the extent to which, to his surprise, he had been emotionally involved even though he was a-political and would not have been willing or able to argue pro or con Kennedy the day before. It is indeed striking to find that one feels Kennedy's death as a personal loss – I was reminded of the way I felt at the news of Roosevelt's death.[529]

I received a letter yesterday from Bar-Hillel extending a formal invitation to me to read an invited paper, as previously informally arranged with von Wright, the chairman of this session. B-H also enclosed the information about the travel arrangements, and I must say this is certainly not what I had expected when I was invited. Once one goes over to Europe one would like to be able to stop over elsewhere, say in Greece, and not be tied to a rigid schedule; also I might prefer to stay less long in Israel. But what is especially on my mind is the thought that preparing a paper for an hour's lecture will take a good deal of time; and I have two contractual obligations to meet: the essays for the Free Press, and the little book on

[526] Robert Nozick (1938–2002) was an American philosopher. He studied at Columbia University and went to Princeton University, where he received his PhD in philosophy, to work under the supervision of Hempel. He taught at Princeton, Rockefeller University, and Harvard University. He wrote seminal books on political philosophy, ethics, and the theory of rationality.

[527] John F. Kennedy, 35th president of the United States was assassinated in Dallas, Texas on November 22, 1963, by Lee Harvey Oswald.

[528] Leonard J. Savage (1917–1971) was an American statistician and mathematician. He studied at Wayne State University and the University of Michigan, where he also received his PhD; he taught at several American universities, including Columbia University. His masterpiece, *The Foundations of Statistics* (1954), advocated a subjective conception of probability.

[529] Franklin D. Roosevelt was the 32nd president of the United States who passed away just before the end of the Second World War on April 12, 1945.

philos. of science for Prentice Hall.[530] The essays would not take much time if I had not decided, much earlier, to write a long new piece for the volume; and I have found that even where I thought I could use previously published material at least in part or in substance, actually I leave hardly a line unchanged; for example, I have modified my discussion of statistical explanation a great deal as compared with the position in the Minnesota volume.[531] So I have the uneasy feeling I will be able to use my time very well to complete my obligations instead of embarking on something new, and I am wondering now whether I should not beg off. I have the impression that it would not be too difficult to get somebody else to present a paper on the problem of induction, though of course I'd hate to let down von Wright. Why am I bothering you with this? Just to ask whether you might drop me a brief line to say how you feel about the travel arrangements (and the subsistence allowance of 83.33 for ten days), and if you have any reaction to my quandary. Please don't trouble to write extensively; a brief note would be most welcome, if you can spare the time.

I have read Pat Suppes's article in Synthese, Sept. 1960,[532] arguing that the concept of model in all its uses in empirical science coincides with the logical or mathematical concept: I find this quite unconvincing and also very difficult to follow his argument at some places.

Warm greetings to you and Edith.

As ever,

4.1.140 Nagel to Hempel, November 26, 1963

[Typewritten, Hempel 1999, Box 28, Folder 2]

25 Claremont Avenue
New York 27, N.Y.
Nov. 26, 1963.

Dear Peter,

Although your Saturday letter arrived this morning, I left the house before the mail was delivered and did not get home till late, so that I finished reading it only a few minutes ago (it is now 9 pm). It is curious how much comfort you gave me just by telling me that you have been devastated by last Friday's events as much as I have been – the sense of sharing a sorrow with a friend somehow manages to reduce its intensity. I still find it hard to believe that I haven't been having a nightmare. In a way Kennedy's death has been a more poignant experience than Roosevelt's, at least in retrospect. Roosevelt was well along in years, he had worn himself out, and his death was "natural"; but Kennedy was in the prime of life, full of vitality and grace and full of promise, and his assassination without rhyme or reason. It just happened that I was in Washington last Friday, to attend what turned out to be a nonsensical committee meeting of the Nat. Science Teachers Assoc. in which I had foolishly agreed to participate. I had a seminar from 9 to 11 am, caught the noon plane to Washington, arrived there around 1:30, overheard someone at the airport saying that Kennedy had been shot, found nothing to support this in the newspaper headlines at the stands and thought I must have misheard, but when I got to the NEA building at 2 pm there could no longer be any doubt, and we stood around the radio until Kennedy's death was officially confirmed. Certainly the horror and distress of that afternoon have not been diminished by subsequent events; and I confess to a continuing suspicion, admittedly unfounded in any evidence, that

[530] Hempel (1965a, 1966a).

[531] Hempel (1962).

[532] Suppes (1960).

something more than meets the eye is involved in the assassination – not a happy thought, and one that will undoubtedly linger for a long time.

But I must turn away from this sorry subject. I have not myself heard from Bar-Hillel (all my communications about the Congress have been with Toulmin), so that I do not know just what travel arrangements are being made for invited participants (when Toulmin invited me to read a half-hour paper, he said only that travel expenses would be paid). On the other hand, Tarski wrote me (in my capacity as president of the Philosophy of Science Association) that the US National Committee of the IUHPS[533] has applied for a grant to the NSF for travel expenses, and asked me (in addition to other impossible requests) to appoint some member of the PSA residing in the vicinity of Berkeley to help him administer the distribution of the grant when made. I wrote him saying that in my judgment the grant should be administered with the advice of the PSA representatives on the US Committee, and that I do not feel myself empowered to supersede the views of those representatives by making such an appointment. (I sent a copy of my reply to Tarski to the three representatives: Feigl, Grünbaum, and Salmon.[534]) Last week I heard from Tarski again, suggesting among other things that I should nevertheless designate someone nearby who could advise himself and Leon Henkin[535] in "the selection of scholars in the philosophy of science" for receiving travel grants to attend the Congress. I haven't had a chance to write Tarski again, and was on the point of asking you whether you would be willing to serve in an *unofficial* capacity to *advise* them on this matter – for I continue to think that final decisions in the distribution of travel funds to philosophers of science should remain with the U.S. Committee. I really hate to ask you to do this, and certainly do not want to urge you to take on this chore, for you should not be subjected to unnecessary distractions. However, the names that occur to me of qualified members of the PSA who are residents of the Bay Area are so completely dominated by Tarski's views on what is significant work in the philosophy of science (e.g. Pat Suppes), that designating one of them to advise Tarski would in effect leave the choice of suggested recipients of travel funds entirely up to Tarski himself. But in any case, the information in this paragraph may be useful to you – for it may well be that *in addition* to travel funds about which Bar-Hillel wrote you further funds may be available from the NSF.

Without any more knowledge about funds than the above, I had assumed myself that my own expenses would be fully covered, and that barring unforeseen contingencies I would be able [to] take Edith and the two boys with me and besides seeing Israel also visit Greece if not other parts of the eastern Mediterranean. But in view of what you write about the meager subsistence allowance that is being offered (unless it is supplemented by NSF funds), I am now uncertain whether I could do this; and I must now wait until I get definite word (from Toulmin or from someone else) just how much money will be available – and I expect that I will have to revise my plans.

As to the hour's address you agreed to give, I have the strong feeling that a) you should not permit its preparation cut into your leisure at the Center and interfere with your set

---

[533] The International Union of History and Philosophy of Science (IUHPS) was founded in 1956 by the federation of the International Union of History of Science and the International Union of Philosophy of Science.

[534] Wesley Charles Salmon (1925–2001) was an American philosopher. He studied at Wayne University and the University of Chicago. He received his PhD in philosophy from the University of California at Los Angeles, where he worked under the supervision of Reichenbach. He then taught at several universities, including the University of Pittsburgh. Salmon is mostly known for his work in the theory of explanation.

[535] Leon Henkin (1921–2006) was an American logician and mathematician. He earned his BA degree in mathematics and philosophy at Columbia University, following Nagel's courses in logic, and received his PhD in mathematics from Princeton University. He taught at Princeton, the University of Southern California, and the University of California, Berkeley.

schedule; and b) you should not take the hour talk too seriously, so that instead of preparing a brand new paper you should either talk from notes or use the material which you are currently writing. On a) I would add that I think it is more important that you work on (and hopefully finish) your present project than postpone it for something less pressing – what you have to say in the two books that will appear under your own name will be more widely read than what appears in the proceedings of a congress, and the chance to work undisturbed which the Center provides is much too rare to worth a gamble. As to b), I *know* that some of the hour addresses given at the Stanford Congress in 1960 were presented from notes, and were written up only subsequently. Moreover, I do not see why you should not use material that you already have at hand – after all, the decision as to what you include in your address is entirely up to you; and I also do not see why you should worry whether what you say meets the imagined expectations of von Wright or anyone else. My hope is that you will give the invited address, but that you do not begin to think about it until after you finished your current two projects (even if this means that you start worrying about the congress only after you left the Center).

I also have read Pat's Synthese paper, and have not been convinced by it. I do not recall the details now, but I do remember that I did not profit much from it, since he lumped too many things together that cannot be accommodated to the rather simple analysis he offers. (By the way, an F.B.I. agent visited me last week, to inquiry about Pat, and he told me that Pat was being considered for a Presidential appointment but did not know in what capacity. However, this is now water over the dam.)

At the request of the Dean of the college, Charles Gillispie talked to our committee on the philosophy of science (appointed to get work on the subject started at Columbia, and I am a member of it), and told us about the expanded program at Princeton. I had already heard that Tom Kuhn[536] had accepted an appointment to Princeton, and Gillispie said further appointments were in the offing. Do you know Tom? He will be a lively addition to your staff, and though I disagree with a good deal in his recent book on revolutions in science,[537] I find him stimulating. I would like to compare notes with you some time about Gillispie – some of the things he's published seem to me excellent, (but only when he writes on what is at bottom social history), while rather poor when he deals with scientific *ideas*.

There has been no word form John Cooley, though he had written to my chairman that he would be in N.Y. sometime before the end of this month. I haven't heard from Sid Morgenbesser for some time, but according to Isaac Levi[538] (who is a better correspondent than I am) Sid has had his fill of Oxford and talks of returning to N.Y. before the New Year.

Sandy is coming home to-morrow for Thanksgiving. We haven't seen him since early September, and we expect that his presence here will help dispel the gloom of recent events. The rest of us are in reasonable form (not quite spry but I suppose in a state that one most now accept as normal), with much too much to do and in my own case with discouragingly little to show for the time I devote to trivia. Our love to all of you,

As ever,
Ernest

[536] Thomas Kuhn (1922–1996) was an American historian of science. Trained as a physicist, he studied at Harvard, where he also received his PhD. He taught at Harvard, the University of California, Berkeley, Princeton (in 1964), and MIT. His masterpiece *The Structure of Scientific Revolutions* (1962) radically changed the field of the history of science, by introducing the notion of paradigm.

[537] Kuhn (1962).

[538] Isaac Levi (1930–2018) was an American logician and philosopher. He earned his BA from New York University and received his MA and PhD in philosophy from Columbia University, where he studied under the supervision of Nagel. He taught at Western Reserve University, the City College of New York, and Columbia University. Levi worked extensively in formal epistemology, with a particular emphasis on the idea of belief revision.

4.1.141 Hempel to Nagel, December 3, 1963

[Typewritten, Hempel 1999, Box 28, Folder 2]

December 3, 1963

Dear Ernest:

It was very good to get your letter, and I appreciate it all the more because I know how busy you are!

Concerning the matter of advising Tarski, I fully agree with your sentiments, and I do not understand why Tarski wants to have someone in the Bay area when the other three people can readily b[e] reached by mail or, in a pinch, by phone. If it should seem desirable to you, however, that you accede to his request, I would be willing to act as an informal adviser. I can't say that I'd be glad to get involved with Tarski, and especially not in a capacity which gives me no formal vote in the matter but still, if you think it might help, I'll do it.

The passage in Bar-Hillel's communication concerning the funds suggests that what he refers to are the funds that can be expected from the NSF. Here's what he says:

A list of all the invitees from the U.S.A. is being sent to Prof. Leon Henkin…., the chairman of the Nat'l Subcommittee of DLMFS/IUHPS, who asked the NSF and other American Institutions to subsidize the expenses for these invitees. It is by no means assured that the funds that will be put at the disposal of the subcommittee will allow it to cover all regular travel expenses to their full extent. The subcommittee will try to arrange for group flights from the USA to Israel and back. These flights shld arrive on August 23 and leave on Sept, 6. It is not excluded that invitees will be asked to use these group flights or receive (in cash) their monetary equivalent – which wld amount to something like 500,-, sld they prefer to make their own arrangements. Let it be stressed that the exact details are still to be worked out and that even drastic changes in these preliminary arrangements are not excluded.

Another passage says that each participant will receive the equivalent of 83.33 to cover expenses for room and board for the ten-day period of the Congress, and that this shld be adequate for room and board in the dormitories of the University.

It does not sound too encouraging, but of course, perhaps the NSF will prove to be more generous than expected. B-H says that "part of the financial difficulties in which the Organizing Committee finds itself are due to the fact that the number of invitees has turned out to be much larger than originally envisaged (close to 60 rather than 45)."

Despite your sound and sensible advice, I became so panicky during the past week about the prospect of not getting my planned work completed and then having to carry it over into the next academic year, when I will have little free time, that I wrote von Wright and asked him to release me from my earlier acceptance of his unofficial invitation, assuring him however that if he shld find it impossible to make suitable alternative arrangements, I would honor my commitment. At moments when I can look at myself from the outside, I realize that my panic level is much too low to be reasonable, but there it is; near-panic seems to become something like a habit with me.

Yes, I have met Tom Kuhn, he came to the Center for a long talk about Princeton after he had been there to talk things over with Charles Gillispie, Gregory Vlastos, and the administration. I liked him, and I find his book very stimulating and intelligent, though no doubt extreme and thus not fully tenable. Shapere[539] at Chicago sent me a preprint of a review of

[539] Dudley Shapere (1928–2016) was an American philosopher. He studied at Harvard, where he earned his PhD in philosophy. He taught at several American universities, including the University of Chicago and Wake Forest. Shapere is known for his critical works of Kuhn, and for his historicist philosophy of science.

the book which is to appear in Phil Review, and I was very much impressed by his sober but pertinent critical observations.[540] He and Kuhn will be on one of the symposia of the AAAS this December; Adolf Grünbaum sent me the program, it looks so interesting that I am sort of sorry not to be able to go.[541] I have just rec'd von Wright's paper on the paradoxes of confirmation[542] and am rather puzzled by what seem to me flaws in his reasoning. But I have to think some more about it before I can be sure.

Diane and I dropped in on the Wohlstetters the other day, when we went into San Francisco to meet the Baumols[543] for a cup of coffee: Bill is lecturing in the L.A. area and brought his family along to see the Coast! The Wolhstetters live in a stunningly located apt on Lombard Street; Albert is rotating Ford Prof of Politics (Govt?) at Berkeley this year. He must have had an enormously influential position with Rand and seems to have met all the top brass in the Eisenhower and even the Kennedy administration. It is still quite incredible that Kennedy is gone. But Johnson seems to be rising to the demands of his office, doesn't he?

Paul Oppenheim is back in Princeton after an astonishingly successful prostate operation at NY Hospital; he said it was done with spinal anesthesia and that he had virtually no after-pains; and 8 days after the operation, he was allowed to go home. This is really remarkable, judging by the experience of other people I have talked to.

I hope you have all had a pleasant Thanksgiving! It must have been nice to see Sandy again. We are looking forward to Pete's Christmas visit. He has apparently settled down to work now, and while his grades perhaps will not show much immediate change, we are very glad that his attitude seems to have changed a great deal.

Greetings, and many good wishes, also from Diane, to you all!

As ever yours,

4.1.142 *Hempel to Nagel, January 21, 1964*

[Typewritten, Hempel 1999, Box 28, Folder 2]

January 21, 1964.

Dear Ernest:

I got a letter from Paul Benacerraf today in which he had enclosed a clipping from the New York Times with the story of your having received the Great Teacher award.[544] Congratulations Ernest; I am *delighted* at this recognition of your achievement as a teacher – it seems to me you have more close students of yours in our profession than any colleague could claim and you certainly have powerfully stimulated the interests and attitudes of able people with many different special aptitudes. Some weeks ago, I looked again at J. Turner's[545] pieces on Maxwell and felt sure this work had derived inspiration and help

---

[540] See Shapere (1964).

[541] The 130th AAAS meeting was held in Cleveland from 26 to 30 December 1963.

[542] Appeared as von Wright (1965).

[543] William J. Baumol (1922–2017) was an American economist. He studied at the City College of New York and at the London School of Economics. He taught at Princeton and New York University.

[544] The article was published in the New York Times, January 16, 1964, on page 71.

[545] Joseph Turner earned his PhD in philosophy from Columbia University. He worked for *Science* magazine, for the Office of Science and Technology in the federal government, and for educational research projects at the school and college level. As for the papers, see Turner (1955).

from you. – Diane was here when I got the mail, and she asked me to send you warm greetings and her congratulations as well!

Since I wrote you last, I have been to Los Angeles to attend the meeting of the Pacific Division, which I found more lively and enjoyable than I had expected. Von Wright's paper on the confirmation paradoxes came up the first afternoon, and though it was quite technical – he had two backboard covered in advance with probabilistic formulas – an audience of I suppose 200 or more people followed him with interest and stayed on through the discussion and was in fact still willing to continue at 6.10, when Kalish[546] called a halt. In my comments I disagreed on some rather basic points with von Wright, but it was all very amiable. The next day, there was a symposium on the philosophical significance of higher-order languages, with Suppes (who talked about learning theory), Montague,[547] and Dana Scott;[548] the latter two had what seemed to be technically excellent papers {in math logic}, which however, seemed to me without any philosophical content whatever; and I came to sense the strange attitude that the ablest West Coast people seem to take re philosophy in general and philosophy of science in particular, and Carnap, who was in good shape, confirmed that impression with some pretty discouraging details. The third day brought a paper by Carnap, roughly on the justification of the basic principles of his inductive logic along the lines of De Finetti[549] and Shimony's coherence arguments; I thought it was very lucid, quite unpretentious and unsharp {in tone}, in relaxing contrast to the performance of Montague, for example (who seems to be a brilliant logician, but a tense, sharp, and difficult person). One evening, I went to a very pleasant and quite luxurious (yet unostentatious and informal) cocktail-buffet supper at Moody's[550] new and very elegant house; Mia Reichenbach drove me there and then took me back to the Helmers, where I was staying. Moody, I was told, is doing an excellent job as chairman and manages to cope with the clashes of opinions and the primadonna antics that seem to mark some of the deptl meetings.

Earlier in Dec., I had met Weingartner,[551] with whom I had previously had some correspondence; and I found him a very likeable person. He phoned to ask whether I would read them a paper, and when I declined – as I have done with all invitations so far – he suggested that I come to S.F. for just some informal talk and dinner at a Chinese

[546] Presumably Donald Kalish (1919–2000). Kalish was an American logician who studied psychology and philosophy at the University of California at Berkeley, where he also received his PhD in philosophy. He taught at Swarthmore College and UCLA. Kalish was an anti-war activist and acted as vice-chairman of the Peace Action Council.

[547] Richard M. Montague (1930–1971) was an American logician and philosopher. He studied philosophy and mathematics at the University of California at Berkeley, where he received his PhD in philosophy. He then taught at UCLA. He is best known for formulating the so-called Montague grammar.

[548] Dana Scott (1932–) is an American logician and mathematician. He studied at the University of California at Berkeley and received his PhD from the University of Princeton. He taught at several American and European universities.

[549] Bruno de Finetti (1906–1985) was an Italian statistician, known mainly for his work on "subjective probability".

[550] Ernest A. Moody (1903–1975) was an American philosopher and logician, working mainly on medieval philosophy. He was a professor and chair of the department at the University of California, Los Angeles. Moody was the president of the American Philosophical Association from 1963 to 1964.

[551] Presumably Rudolph H. Weingartner (1927–2020), a German-born philosopher. He studied at Columbia University, where he also received his PhD. He taught at San Francisco State College and the University of Pittsburgh. He also worked as a full-time administrator at several American universities.

restaurant with him and his colleagues, and that proved to be a very pleasant occasion. Weingartner seems to me perhaps the brightest and most stimulating man in the group, but perhaps that is too much of a snap judgment.

We had Pete here for the holidays', he came on a Princeton charter flight, a Constellation that took some 12–14 hours each way, but saved a lot of fare. He has begun to work much harder than he had done before, and on his initiative, which was of course very encouraging.

We are all well and still very glad we came here; it is really an unusually pleasant place. I just wish I did not so completely depend on transportation virtually for all purposes, but with Diane more than willing to drive me, that is a minor complaint.

Did I mention that we have been seeing a fair amount of the Gays[552] and the Gershoys[553] and have been enjoying both? And that Savage is an unusually nice bright man with lots of interesting ideas?

I hope all is well with you, Edith, and the boys! Love and best wishes to you all from Diane and me!

As ever yours,

4.1.143 Hempel to Nagel, May 29, 1964

[Typewritten, Hempel 1999, Box 28, Folder 2]

5-29-64

Dear Ernest:

I have to send you the distressing news that Ina Carnap committed suicide last Tuesday, May 26. She had been in an increasingly severe depression, and had only a few days before her death been put on antidepressant drugs; but these normally take effect only after two to six weeks, and of course even then, success is not certain. Electroshock had been mentioned as a possibility, but – understandably – Ina had resisted the idea; as Carnap told me, she had been very much afraid that if she were to be hospitalized for shock treatment or any other kind of treatment, she would lose control over her own life, and this she was determined not to do. About a year ago, her brother in Vienna had killed himself, and she was, it seems, afraid, that she was going downhill. She was entirely rational and to the last looked after Carnap, but she suffered from increasing indecision, and just to go to the market and shop for a meal was a matter that required long phone conversations, with requests for advice and extended subsequent discussion, with Maggie Helmer and apparently also with Mia Reichenbach. Carnap did everything he possibly could to keep her on an even keel, to reassure her, to have her hang on and not to give in to her suicidal impulses; he had persuaded (not forced) her to give him all her drugs for safekeeping, and had spent the morning of the fatal Tuesday talking with her; and then during half an hour or so before he looked for her again, she had hanged herself; he found her and called the police with oxygen equipment, but it was too late.

---

552 Presumably Peter Gay (1923–2015), a German-born historian who became an American citizen in 1946. He studied at the University of Denver and at Columbia University, where he also received his PhD. He taught at Columbia and Yale. He was an expert in intellectual history and devoted several works to Freud and the history of psychoanalysis.

553 Leo Gershoy (1897–1975) was a Russian-born historian. He studied at Cornell University and the University of Strasbourg. He taught at several prestigious American universities. He published widely on French history, particularly the French Revolution.

Olaf informed me within an hour, and Diane and I flew down to Los Angeles that afternoon. Carnap is deeply shaken, but also astonishingly in control of himself. His friends and his colleagues have rallied around him splendidly; the Helmers had him at their house for meals (but despite our joint urging, he did not stay overnight, but returned to his house), and they did all sorts of chores for him, apart of course from giving him the kind of support one needs at such a time. Mia Reichenbach was splendid, cooked, cleaned the house, and whatnot; Moody was around, and he will try to help in finding Carnap eventually a small apartment with maid service. We have urged Carnap to come to Stanford for a while, where he can have his privacy in a small apartment near our place and can have all meals with us and as much company as he would like; and we are hopeful that he will indeed come. He even was not entirely negative to the suggestion that Feigl had made in a letter just a few days before Ina's death, that Carnap consider coming to the Vienna Institute during October to December or mid-January, when Feigl will be there for a second round – he is flying to Vienna on the 31st. Carnap seemed in reasonably good physical shape, sat up without any exception when we were with him, walked without a cane, and also never showed that he was feeling sorry for himself: he wasn't. Also, I think that basically he knows that he cannot blame himself for what happened, for he clearly has been truly exemplary in his way with Ina during these difficult months of her depression. He wanted you to know what had happened, and I said I would write you.

There is much that goes through my mind on this occasion, and there would be much to talk about, but this message will have to do for now.

I have heard from Adolf about the conference he is planning, and I hope to attend it, too, of course.

Diane and I send our love to you, Edith, and the boys. We do hope you all are not too weary from a busy and demanding year, and especially that you will have a good rest in Vermont.

As ever,

PS. Carnap was evidently pleased with the publication of the Schilpp-volume,[554] and I think he would be cheered a bit if you said something about it. His address: 509,22d Street, Santa Monica; phone 213-GL 1 – 5218.

4.1.144 Nagel to Hempel, June 18, 1964

[Typewritten, Hempel 1999, Box 28, Folder 2]

25 Claremont Avenue
New York, N.Y., 10027
June 18, 1964.

Dear Peter,

I have been wanting to write you for a long time, but this has been an unusually hectic spring semester, and I just about managed to keep my nose half-way above the water. And then as soon as classes were over, I've had another attack of my ancient back-trouble which laid me flat – there has been a slow improvement, but I still feel uncomfortable and what is perhaps even worse I am under a constant threat that any move I make will send me off into another tail-spin.

Your letter with the terrible news about Ina arrived when sitting up was a misery, but I managed to write to Carnap anyway, and following your suggestion said something about

[554] Schilpp (1963).

the Schilpp volume. Ina's death shook Edith and me badly, and I still find it difficult to believe that the whole thing is not a nightmare. I did have an inkling of Ina's condition about a week before her end, for I met Dick Jeffrey[555] who had just returned from L.A. and told me about it – but I did not imagine that her state was as serious as it turned out to be. Poor Ina, she did not have a joyous life, not because external circumstances prevented it but because of her own temperament – at any rate, she showed an inclination to look at the seamiest side of experience ever since I first met her in Prague in 1934. And poor Carnap! He was so dependent on her that the adjustments he must make seem almost insuperable. He has not acknowledged the incoherent note I sent him, so that I know nothing about his state of mind beyond the information you sent me. If you have the chance, do let me know; and if there is something you think I might do to help, please tell me.

When I saw Dick Jeffries [Jeffrey], he said that you would be staying on in Stanford until August, so that I cannot look forward to seeing [you] before September. I've missed your absence very much, though I felt partly compensated by the thought that you were having a good year. I hope your writing plans have prospered, and am eager to hear about your accomplishments and even more eager to profit from reading them as soon as they are available.

Our plans for the summer are indeterminate. Bobby will finish school next week, so that we would be free then to go to Vermont. However, Sandy will start taking a course in German at City College next week, and Bobby also wants to audit a course (in English literature) given in the Columbia summer session; and while Sandy is old enough to be left on his own in the city, Bobby is not. Moreover, with my precarious back Vermont does not seem too inviting; and partly because of Edith's bad experience last summer up there (high temperature with some pain that put her in the hospital at Townshend for a week), as well as because she continues to be fatigued despite long daily rests, we may stay on the city until later in the summer, except for short trips to our place.

I will be on sabbatical leave next year (I got a grant from the NSF for half my salary to supplement the half Columbia will pay) – it is my last sabbatical before mandatory retirement. I want to work on the sequel to the Structure of Science, but do not expect to turn to it before the fall, since there is a mountain of obligations that I must first get out of the way and am counting on the summer to do this. But I work so slowly that I am not really sure that the summer will be long enough – writing is becoming more difficult (largely, I am afraid, because I have nothing worthwhile to say), and I have begun to wonder whether the time has not come for my being put in the pastures as a preparation for the glue factory.

I've seen a bit of John Cooley this spring, but not as much as I would have liked to. He is sailing at present, and we may visit him later on at his place in Connecticut or perhaps see him in Vermont. He decided that he did not want to teach next spring; and the year after that Charlie Parsons[556] will be coming to Columbia on a tenure appointment, so that we will have our domestic logician with no funds to ask John again. Sid Morgenbesser has been trying to write, but I cannot really tell how much he has been able to put down on paper – since he will be back in the fall on a full teaching schedule, he will probably not be able to get much done since his current leave is over. To my great dismay, both Gregory Vlastos and Stuart Hampshire came to the meeting of the Philosophy Club in May when it was my turn to

[555] Presumably Richard C. Jeffrey (1926–2002), an American logician and philosopher. He earned his MA from the University of Chicago and his PhD from Princeton under the supervision of Hempel. He taught at MIT, Stanford University, the City University of New York, the University of Pennsylvania, and Princeton. His most important work is *The Logic of Decision* (1965).

[556] Charles Parsons (1933–2024) was an American philosopher and logician. He studied at Harvard University and Cambridge and earned his PhD in philosophy from Harvard. He then taught at Cornell University, Columbia, and Harvard. He made important contributions to proof theory and the philosophy of mathematics.

read a paper (I managed to put together some criticisms of Popper and Feyerabend,[557] but it was a miserable performance) – I did not have much chance to talk to either of them, but I enjoy both of them immensely. I've also seen Abe Edel[558] since the death of May; he is finding it hard to pick up the pieces, but he has resilience and I am confident he'll pull through.

There is so much I would like to tell as well as ask you, but even the above mish-mash has tired me, and I must postpone the rest for another occasion. Meanwhile, Edith and I send our love to you and Diane (and the children, on the assumption they are with you), and all the best for the remainder of your California adventure.

As ever,
Ernest

4.1.145 Hempel to Nagel, June 20, 1964

[Typewritten, Hempel 1999, Box 28, Folder 2]

{Stanford (Behav Sci Center)}
June 20, 1964

Dear Ernest:

I have just received your long letter – thank you so much! I am terribly sorry to hear you have had such a tiring spring term (though I can add that *I* would not be able to stand up under your work load at all), and that your back has again given you trouble; that must be a wretched and discouraging business since there is no effective quick therapy for it, and the rest that is prescribed takes patience. Of course, this should be a time when you should allow yourself some relaxation; I wish it could be in some peaceful quiet setting outside New York!

We have been in touch with Carnap by phone and have also had some reports from the Helmers. He is as well as he can be expected under the circumstances and faces his situation with admirable composure and self-control. He had had some hope that a widowed daughter of his in Germany, whom he and Ina had seen in Mexico last fall, and whom both liked well, might be able to come over for a while; but last night Carnap told me on the phone that her job makes it impossible to do that, and that his former wife has now encouraged him to come to Germany for a while. He is giving this very serious consideration and thinks that he may fly over early in August; I do not know how long he would expect to stay, but I rather think it would be for a limited time. – His practical needs are looked after, in part, by a young colored girl who comes in for a few hours three times a week, cleans, and cooks him supper. On the other days, it seems, he has supper at the Helmers or some friend comes in and cooks for him. Mia Reichenbach has been very efficient and helpful, too. And Carnap told me

---

[557] Paul Karl Feyerabend (1924–1994) was an Austrian-American philosopher. He studied at the University of Vienna, where he received his PhD in philosophy. After his PhD, he went to the London School of Economics to study under the supervision of Popper. He taught at the University of Bristol and at the University of California at Berkeley. He is best known for his book *Against Method* (1975).

[558] Abraham Edel (1908–2007) was an American philosopher. He studied at McGill University and the University of Oxford, and received his PhD from Columbia University. He then taught at the City College of New York and at the CUNY Graduate School. An expert in moral philosophy, he wrote a monograph on ethical theory for the *International Encyclopedia of Unified Science—Science and the Structure of Ethics* (1961).

yesterday that his NSF grant has been renewed for three years, which is of course a great help.[559] He said to me, half choking up, that he wished Ina could have heard the news; she had been getting increasingly und unrealistically worried about finances. But Carnap agreed with my conjecture that even if the news of the grant had come earlier, it would have made no difference in the end. We will get more news, we hope, from the Helmers, who are planning to come here for a visit. We expect them tomorrow. Maggie has not been feeling altogether well, and the psychological demands made on her by Ina's pathetic long telephone conversations about what to buy for dinner and how to prepare it, have apparently contributed to her feeling somewhat low. So they announced their visit only tentatively, depending on how Maggie would feel; but we hope they'll make it; it would be a good change for both of them. Despite all the stress and strain they were extremely hospitable to Diane and me when we stayed there during our last visit to Los Angeles, at the time of Ina's death.

I am very glad you are having a sabbatical next year, and I hope very, very much that it will be a good and minimally strenuous year for you. I can empathize with the way you feel about the difficulty of writing, but you must emphatically have no right to complain that you have nothing worthwhile to say. *The Structure of Science* is a great achievement, and I need not remind you how very widely it is being read, cited, and discussed, and how stimulating it has proved already for current philosophical thinking. I do not know whether I mentioned how much impressed I was by your review of Watson's book;[560] I have often said to colleagues and students that your reviews are to me exemplars of what a review should be, giving the reader a clear idea of what the author tries to accomplish, and what he has to say, and then giving a really incisive and illuminating critical appraisal. I was therefore very glad to hear from Adolf that you had agreed to attend and participate in a conference on recent problems of explanation, and that you might deal with such issues as those raised by Feyerabend. So I promptly wrote Adolf that I would be glad to participate, too. I saw Paul Feyerabend at my so far only visit to Berkeley (I was invited to give the Howison Lecture, and while I had declined all other invitations, I did accept this one and was glad to see I as one of your successors on the list of speakers); and he was perfectly pleasant and nice to talk to. His piece in the third Minnesota vol.[561] seems to me to attribute to the covering-law models of explanation certain implications that are not implied in the least, such as his principle (5) on p.44. I will be very much interested in hearing you talk on the subject!

Congratulations on your NSF, Ernest![562] I just wish you could get even more time off than a year for research and writing; but the Carnap's experience is an encouraging example that the NSF may enable one to continue one's work without financial worries. And if anyone is entitled to that sort of support, then it's certainly you.

I have agreed to serve on Roger Buck's[563] committee that deals with NSF grants in the history and philosophy of science; I think it is for a period of two years. So far, I have been only a consultant for his group.

It is very good news that your Department is getting Charles Parsons; he is an extremely able man.

---

[559] Carnap got a grant from National Science Foundation in 1964 for 36 months to investigate "the theory of inductive probability."

[560] Nagel (1963c).

[561] Feyerabend (1962).

[562] There is no sign of Nagel receiving an NSF either in the 1965 Fifteenth Annual Report, or in the 1966 Sixteenth Annual Report, though Hempel is listed for both years as member of the Advisory Panel for History and Philosophy of Science.

[563] Roger Buck (1922–2002) was an American philosopher, specialized in history and philosophy of science. He studied at the University of Minnesota and Oxford University, and taught at the University College in Ghana, the University of Chicago, Duke University, Oberlin College, and Indiana University.

I did finish the work on my volume of collected essays.[564] It involved revising and annotating, and writing some postscripts to, a set of previously published articles, and then writing a new piece especially for this collection. It is called "Aspects of Scientific Explanation" and turned out to run to some 90,000 words.[565] The beginnings go back to my year in Oxford and before so I am not exactly a whirlwind when it comes to writing. But the whole ms went off last Monday to the Free Press. And for some time now I have been at work on the small book on the philosophy of the natural sciences that I am under contract to write for Prentice-Hall.[566] It is all familiar stuff, and yet I find the going painfully slow and the formulation of almost every idea difficult and tedious. So I won't be ready for the deadline of July 1st, and I am afraid not even by the time I get back to Princeton, but I hope to get it done during the fall term.

Diane and Toby are well and have enjoyed the stay here a great deal, as have I. And a few days ago, Peter arrived for the summer and for the common drive back; he is planning to take a very intensive German course at Stanford, and I hope he'll get something out of it.

Our love and good wishes to you all – it will be so good to see you again!

As ever,

4.1.146 Nagel to Hempel, October 27, 1964

[Typewritten, Hempel 1999, Box 28, Folder 2]

25 Claremont Avenue
New York, N.Y. 10027
Oct. 27, 1964

Dear Peter,

It was wonderful to see you and Carnap last Friday, though I am sorry that I had to rush off at such an exciting point in the discussion. Carnap seemed to me to have aged noticeably since I last saw him in 1960 – he has more of a stoop than I think he had, and there is a trembly quality to his writings that I did not notice before. But he is as alert as ever, and talking to him continues to be a joy – even though, alas, I find myself unable to agree with his approach to the justification of induction, and think his "apriorism" hair-raising. Have you been able to persuade him to settle down in Princeton? My guess is that the answer is no.

I had a note from John Cooley this morning, in which he replies to one I had sent him transmitting your invitation to renew our visits with you. He says he would be delighted to come down, but leaves the arrangements to us. He will be in Maine this week, but after that any day would be possible for him, and suggests that I consult you about possibilities. I talked to Sid M. over the phone, and he finds that Wednesdays are the best days for him, with Thursdays as second best. When you feel inclined to do so, could you suggest several dates from which we might choose?

I've finally written to Adolf suggesting a number of alternatives for the conference in Pittsburgh, but it is too soon to have had a reply.

The dinner bell is ringing, and I must stop. Love to you all,

Ernest.

[564] Hempel (1965a).

[565] Hempel (1965b).

[566] Hempel (1966a).

4.1.147 Hempel to Nagel, November 7, 1964

[Typewritten, Hempel 1999, Box 28, Folder 2]

November 7, 1964

Dear Ernest:

Thank you so much for your note. I would have answered sooner, but I would then have been unable to suggest some dates; for the painters were due to invade our house and repaint it from top to bottom, and we did not know how long this would take. Now they have started – what a mess it is! –, and after then, the carpenters are supposed to come in, take out the floors in the downstairs rooms and replace them by new ones; with the furniture in the house, that should be quite an operation, but we hope the results will be worth the inconvenience. We have not been told how long the carpenters will take, but we think we are not being overoptimistic in assuming that the whole job – which calls for the painters to return and do the baseboards – will be done by the end of this month. So let me suggest some dates in December – I am sorry to have to suggest so long a delay!

Wednesdays are slightly better for me than Thursdays, on which departmental meetings usually take place – though many Thursday afternoons stay clear. Grice is to read a paper here and has announced that he will come either Mo Dec. 7 or Wed Dec. 9; if he chose the former date, Dec. 9 would be splendid. I do not know when he will send final word, and knowing him, I suspect it may be fairly late. So I would suggest Wed. Dec. 2, Th. Dec. 3, Th. Dec 10; Wed. Dec. 16 as possibilities, and I hope that one of these will suit you, John, and Sid. If I should learn soon that Dec. 9 is free, I will let you know right away.

The rest of the informal discussion with Carnap, and supper at our house, continued to be very pleasant, and we all wished you could have stayed. Carnap stayed until Sunday, when we put him on a through plane at Newark. All in all, he seemed in very good shape and in good spirits. He hopes that the younger one of his daughters, Hanneliese, will come to L.A. perhaps in Jan. and will stay with him until he returns to Germany for another visit, perhaps in June. She is now learning how to drive; but even if she gets her license soon, being able to contend with L.A. traffic is quite a different matter. I suppose, however, that it might be sufficient if she can drive in the vicinity of Carnap's house; and of course just having somebody he likes around him will be a huge help. He promised to stop over in Princeton with his daughter on his way to Germany, or at least nearly promised; whether he would consider eventually settling down here, is an open question, and I think it is pretty unlikely.

Diane and I send you all many good wishes!

Cordially,

I am off to Washington tomorrow afternoon for my first NSF panel meeting.

4.1.148 Nagel to Hempel, November 16, 1964

[Handwritten, Hempel 1999, Box 28, Folder 2]

Nov. 16, 1964

Dear Peter,

I've heard from John Cooley. He suggests Dec. 2 for a visit with you, and this date is fine for both Sid and myself. Is this still convenient for you? And if so, what time would you like us to arrive?

In haste,

Ernest

{Yes}

4.1.149 Nagel to Hempel, January 30, 1965

[Typewritten, Hempel 1999, Box 28, Folder 2]

Jan. 30, 1965.

Dear Peter,

Thanks for your card of almost 2 weeks ago. I haven't acknowledged it because I wanted to get in touch with John first; but I did not talk to him until last night (he had just returned from a trip to Maine), and mentioned the dates you suggested for a visit. He thought Wed. Feb. 10th would be fine for him (I had previously talked to Sid M. who also found this a convenient one), and I promised to confirm it with you. If this is still open, and does not overburden you to have us, could you drop me a line? I also raised the question with John about having either Bob Nozick or Paul. B. present; but John is a dyed-in-the-wool conservative, and said that since everything went so magnificently without anyone else along, why should things be changed.

We are still waiting for a vacant bed for Edith. She called the surgeon's office this week, and the indications are that there will be an opening over this week-end or early next week. Meanwhile we are living in a kind of no-man's land, though Edith is standing the strain of the uncertainty remarkably well.

Love to you and Diane,
As ever,
Ernest

4.1.150 Nagel to Hempel, June 27, 1965

[Typewritten, Hempel 1999, Box 28, Folder 2]

June 27, 1965

Dear Peter,

I am terribly sorry that I will not be able to get to the conference with Carnap during Sept. 1–4.[567] If all goes well, we would still be in Vermont at that time, expecting to return to the city around the 15th; and it would be an imposition on Edith and the boys were I to insist that we leave Vermont before the end of August (as would be necessary for me to attend the Princeton meeting). I wish I could make a different decision, but I just can't. But please explain to Carnap when you see him the circumstances that prevent my being on hand. (By the way, I do not find in my collection of dittoed material from Carnap anything with the title "An Axiomatization of Inductive Logic", and have certainly not received what you refer to as section 19.[568] I do have a ms. entitled "Notes on Probab. and Induction", and a much

[567] Presumably the Conference on Inductive Logic at Princeton, 1965.

[568] The relevant letter of Hempel to Nagel seems to be missing. For Carnap's paper, there exists a dittoed manuscript, entitled "An Axiom System for Inductive Logic" that was announced for publication in *The Philosophy of Rudolf Carnap*. It is kept now in the Carnap Archive at Pittsburgh (https://digital.library.pitt.edu/islandora/object/pitt:31735061813139).

later one with the title "Remarks on Probab. and Content"[569] – the former has 18 sections, so that this *may* be the one to which you refer. I imagine the thing to do is to write Carnap directly, and will do so in the fall unless you advise me to the contrary.)

Edith was very much moved by your postcard, and asked me to thank you for your kindness – she is even a worse correspondent than I am even she feels at top form, and finds it difficult to write when she is as tired as she is at present. She is seeing her surgeon tomorrow, in the expectation that he will have no objection to our leaving for Vermont (hopefully next Wednesday, when Bobby will be finally through with school). Both boys will go with us, but Sandy has been talking of returning to the city after a while, in order to be able to use the Columbia Library – he wants to start reading for his senior essay for honors in mathematics: the subject is some recent development in Banach algebra, but the whole thing is way beyond my comprehension.

I've been attending a conference organized by Fordham University on "Images of Man", the theme vaguely centered around Teilhard de Chardin's[570] PHENOMENON OF MAN (which I forced myself to read through). Two of the other participants were Paul Weiss and John Smith.[571] I found it possible to carry on discussion with John, but Paul is a completely different kettle of fish. Indeed, I found myself in greater rapport with the Jesuit fathers (including Ernan McMullin[572] from Notre Dame, whom I think you know) than with my philosophical colleagues; I was quite surprised to find the sophistication, liberal outlook, and concern with carefully articulated arguments that the Jesuits present uniformly exhibited, and if I had to make a choice between joining the order and adopting an approach like Weiss's I certainly would feel more at home in the Church than in Paul's intellectual cosmos. Perhaps nothing came out of the Conference except a strong conviction by all the laymen present that the Church is moving rapidly and that perhaps even by the end of the present century its position on various intellectual and social issues will hardly be recognizable (e.g. *all* the Jesuits present made no bones over the inescapability of a liberal outlook on birthcontrol and artificial insemination).

Although I won't see you early in September, I expect to attend the meeting of the Advisory Council on the 24th (as I've written to Stuart) and am looking forward to some talk with you. Meanwhile, I hope all of you will have a restful and enjoyable summer, and Edith joins me in sending our love to you and Diane.

As ever,
Ernest

---

[569] "Notes on probability and induction" was published only posthumously, although it indeed has 18 sections; see Carnap (1973). The Stegmüller Nachlass has a manuscript from Carnap, entitled "Remarks on probability and content" from the year 1962, see here: https://www.uibk.ac.at/de/brenner-archiv/bestaende/stegmueller/.

[570] Pierre Teilhard de Chardin (1881–1955) was a French paleontologist and geologist. A Jesuit, he studied at the University of Paris and taught at the Catholic University of Paris. In his work, he tried to show the substantial compatibility between the theory of evolution and the Catholic faith. The book to which Nagel refers in his letter appeared posthumously in French in 1955 and was translated into English in 1959 under the title *The Phenomenon of Man*.

[571] John E. Smith (1921–2009) was an American philosopher. He studied at Columbia University and at the Union Theological Seminar; he then received his PhD in philosophy from Columbia University. He taught at Barnard College and at Yale University. His areas of research were the history of American philosophy and the philosophy of religion.

[572] Ernan McMullin (1924–2011) was an Irish philosopher and priest. He studied physics at Maynooth College and at the Dublin Institute for Advanced Studies, and received a PhD in philosophy from the University of Leuven. He taught for his entire career at the University of Notre Dame. McMullin wrote extensively on Newton and scientific realism.

4.1.151 Hempel to Nagel, July 23, 1965

[Typewritten, Hempel 1999, Box 28, Folder 2]

Friday, July 23, 1965.

Dear Ernest:

It was good to get your letter of June 27, which I appreciate all the more because when you wrote it you were already under the pressure of preparing your trip to Vermont. Diane and I do so hope that the journey went smoothly, and that you found your house in good shape and that you are enjoying your stay. Do you and Edith ever need a rest?

It would of course be foolish for you even to think of coming to the Carnap conference under the circumstances. The response to the invitation, by the way, has been amazingly favorable: So far, Martin,[573] Leblanc,[574] Kyburg, Achinstein,[575] Savage, Sid, Shimony, Salmon, and Jeffrey have accepted, and, apart from you, only Bohnert[576] has declined; he will then be abroad. Feigl is back in Mpls but not sure whether he can come; no definite word from the others. So I have just written Carnap and suggested Sept. 3–4 for the meeting; there was a slight pref. for that date. – We had a short note from Carnap, he is well and things have gone satisfactorily, but the weather hasn't been too good and we have a feeling that the second visit to Germany simply cannot be as overwhelmingly exciting as the first one had been. His daughter Hanneliese and her husband have decided, on amiable terms, to get a divorce; Hanneliese and her 12-year-old daughter will then accompany Carnap back to Los Angeles and will stay with him, at any rate for a year.

I have been working quite hard, but have made only slow progress. First I had to put a lecture ms into shape for publication in the third volume in the Pittsburgh PhSci series; it is a rather general piece on induction, largely expository.[577] Then I went back to my ms for Prentice-Hall, and it is slow sailing, though I enjoy it in a way. But the summer is going, and now I have to prepare a 1000-word summary of my comments on Dick Jeffrey's paper on decision and valuation; I have read it and found it tough sledding, and I simply have no idea as yet as to what I could say about it; I can only hope that a second reading will bring *some* ideas to my mind. Then, I am to tell Kerner[578] at Mich. State my topic in the Isenberg

---

[573] Richard Milton Martin (1916–1985) was an American logician. He earned his BA from Harvard University and his MA from Columbia University; he received his PhD from Yale University. He taught at several American universities. He wrote extensively on philosophic logic and the history of logic.

[574] Possibly Hugues Leblanc (1924–1999) was a Canadian logician. He studied at the University of Montreal and received his PhD in philosophy from Harvard University. He taught at Bryn Mawr College and Temple University. He was an expert in non-standard logics.

[575] Peter Achinstein (1935–) is an American philosopher who studied with Hempel at Harvard University (the year Hempel was a visiting professor there). He also received his PhD from the same university. Achinstein taught for most of his career at Johns Hopkins University. He has published influential works on the history and philosophy of science.

[576] Herbert G. Bohnert (1918–1984) was an American philosopher. He studied at the University of Chicago and received his PhD from the University of Pennsylvania. A student and then lifelong friend of Carnap, he worked at IBM before accepting a position at Michigan State University.

[577] See Hempel (1966b).

[578] George C. Kerner (1927–2001) was an Estonian-American philosopher. After an insecure life in Germany and France during the Second World War, he went to the United States and earned a PhD at Harvard in 1960. He was interested in moral philosophy and worked at Michigan State University.

Memorial series, in which you are going to speak, too.[579] I have settled my date, Nov. 5, but am still trying to decide on a subject; I have the feeling that with the other things on my mind I just cannot think up anything worth publishing. I think I will have to settle for something akin to my paper for the Pittsburgh volume as the subject of my lecture, and then to work in some more special stuff into the version to be published. I have some slight inklings that I might have some reactions to Nelson's views on projectibility and entrenchment. – What will your date be – and what are you going to talk about? Have you let them know?

Pete seems to be enjoying his work in Washington, which to judge by his accounts, is varied and interesting. He and the other interns (distributed over the offices of several Senators) have even been addressed by Humphrey,[580] and they are to meet the President before the summer is over. We hope to visit him for one weekend.

Tomorrow, Diane, Toby, and I are planning to take a day off and visit the World's Fair under the guidance of Diane's brother, who has a Press pass and can get us into many exhibits without the usual wait.

Perhaps Sid will come out for an evening; then we will be especially sorry you can't be here as well!

Your account of the conference on Images of Man sounded fascinating. Yes, I have met Ernan McMullin three times at last year's NSF Panel meetings and I like him; I am sorry that his term on the panel has expired.

Both Diane and I send you all our fond greetings and wish you a peaceful, restorative vacation.

Cordially,
Peter

4.1.152 *Hempel to Nagel, August 5, 1965*

[Typewritten, Hempel 1999, Box 28, Folder 2]

Thursday, August 5, 1965.

Dear Ernest:

I thought you might like to see how the plans for the Carnap conference stand; the response has been remarkably strong. It is nice that Feigl will very probably attend, too, and that Bar-Hillel has been added to the group, thanks to the fact that he met Carnap in London. I have had a brief letter from Carnap about his trip to London; it seems to have gone quite well, and he seems to have enjoyed it, though from Dick Jeffrey's account one get the impression that the atmosphere was rather highly charged, that the Popper-people were all antecedently convinced that Carnap, Jeffrey, and all others talking about induction were simply benighted, and so forth. Then there was a man by the name of Miller,[581] who was pushed into the debate after Jeffrey and Carnap had spoken and presented a new proof of inconsistency of inductive logic; and this was greeted with lively protest by Quine, Suppes,

[579] Hempel's paper was, after all, "On the Structure of Scientific Theories" (Hempel 1969a). We did not find any evidence about what topic Nagel talked about, after all.

[580] Hubert Humphrey (1911–1978) possibly. Humphrey was an American politician and statesman who served as the 38th vice president of the United States of America, under Lyndon B. Johnson's presidency.

[581] David Miller (1942–2024) was a British philosopher. He studied at the University of Cambridge and the London School of Economics, where he was strongly influenced by Karl Popper. He taught at the University of Warwick. He has made significant contributions to the theory of verisimilitude.

Jeffrey etc., who pointed out that the proof used some illicit quantification into a modal context. It seems that Miller, who is apparently a bright man, was eventually persuaded by Carnap that there might be a flaw in the inconsistency proof.

I hope that things will proceed in a less emotional and more enlightening manner here in September. I am very sorry you can't be here, but I am also very sure that you should have no regrets about it. And Sid will, I am sure, later give you a detailed report. He was here last night to pick up my set of Carnap's notes on the axiomatization of inductive logic, and we had him, Saul Kripke, and Dick and Edith Jeffrey for dinner! It was fun, everybody was in good shape, especially Dick and Sid.

It was a very pleasant surprise to receive a copy of *Meaning and Knowledge*,[582] and I am staggered at the thought of the time and work that must have gone into preparing the volume! The result, however, seems to me to justify the effort; this strikes me as a truly excellent collection of pieces, and I know that I will find it handy and helpful both for teaching purposes and in writing. I was delighted to receive, quite unexpectedly, in yesterday's mail, the check that had sent me on your and Dick Brandt's behalf! It is the first one, I think, I have ever received for any piece of mine that has been reprinted; and I appreciate the fact as well as the policy you are following!

I hope things are going smoothly with you, and that Edith is gaining some strength and getting a good rest!

Best wishes to you all from Diane and me!

As ever yours,

Peter

4.1.153 Nagel to Hempel, October 19, 1965

[Typewritten, Hempel 1999, Box 28, Folder 2]

25 Claremont Avenue
New York, N.Y. 10027
Oct. 19, 1965

Dear Peter,

Thank you most warmly for the copy of your book – I need not tell you how much I appreciate your sending it to me or how happy I am to have it.[583] It arrived less than a week ago, so that I have not had the time to read all the new material included in it. But I have managed to gulp down large chunks of the title essay, as well as the various "postscripts" and I am already heavily in your debt – a debt that will quickly increase, for I am going to use the book this semester in my graduate seminar in the philosophy of science (in fact, at the seminar meeting this morning I assigned various parts of the title essay to be reported on by students during the next five or six weeks). It must be a relief to you to have it out, but it certainly has been worth waiting for, and the publishers have done a fine job with its appearance.

It seems to me that you handle criticisms of your previously published analyses with admirable though characteristic generosity and clarity. I confess that I marvel at your patience in dealing with what seem to me either deliberate distortions or downright stupidities in Scriven, Feyerabend, etc. (I had assigned Feyerabend's long piece in the second

[582] See Nagel and Brandt (1965).

[583] Presumably Hempel (1965a).

Pittsburgh volume,[584] as a basis for discussion in my seminar until your book would be available; and I am glad to see that you dealt with his absurd remarks in connection with his "condition of consistency" in what I think is a definitive manner. By the way this piece seems to me to establish a new record of cheap philosophical writing – when I next write Adolf I'll speak my mind about it. Why Adolf tolerates so much "quatsch" is beyond me!) However, I'll postpone telling you about my enjoyment of the various plums in the book until I've had a chance to read all that isn't yet familiar.

I've had a brief note from John Cooley a couple of weeks ago, still from Maine, expressing the hope that we may be able to visit you after his return. I've been too busy to write him, so that I do not know what his own plans are. My own schedule this semester seems an unusually crowded one; and in addition I've been having a perfectly miserable time with my back, despite my spending most of the time I do not have to be on the campus in a horizontal position. So although I've managed to keep all my obligations thus far, I curse my fate. On the other hand, Edith is in reasonably good condition, Bobby (who is living at home because there is no room available in the Columbia dorms) is working hard but is happy, and all the news from Sandy is good. Perhaps you'll come down to the next Philosophy Club meeting – I'll look forward to it. For the moment, love to you and Diane.

Ernest

4.1.154 Hempel to Nagel, November 2, 1965/1

[Typewritten, Hempel 1999, Box 28, Folder 2]

November 2, 1965

Dear Ernest,

It was terribly nice of you to write me despite the many demands that are being made on your time and energy, and despite the trouble your back has been giving you again. But I must say I was truly delighted to get your letter, and very pleased to hear that your first impression of the book is favorable. I find – and I am sure you know the feeling – that as soon as I have got something published, there are all sorts of points that are not quite right, things that should have been added, and others that should have been cut out. But one just has to stop somewhere in preparing a manuscript and then just take the plunge. One thing of a somewhat minor sort does dismay me, namely that there are quite a number of misprints in the book, although I did my best reading proofs, and although I had an undergraduate student to help me by reading the page proofs independently of me. Most of the misprints are trivial, but here is one more serious one, which must actually be an oversight by me: on p. 405, 'S_2' should be replaced by 'S_3' on lines 4,6,7 from bottom.

The Hanson-Feyerabend (and I suppose even Ryle-) idea of terms, including observation terms, becoming theory-laden to such an extent that they change their meaning when the theory is changed – that idea I think needs full discussion, and I have said nothing about it here. I may touch on it in the printed version of my paper for the Arnold Isenberg series, though I do not intend to discuss it in my oral presentation, which will be rather allgemeinverständlich. I expect to leave Friday morning and to get back Sunday afternoon. Thus, alas, I will be missing your visit here on the occasion of the Osgood evaluation meeting, to which I was belatedly invited. Herbert Feigl is to arrive here Thursday late pm, and I expect to see him that evening at the Oppenheims, where he is to have supper; I am glad that Diane and I will at least catch a glimpse of him. – The very evening of the next

584 Feyerabend (1965).

Philos. Club meeting, we will have again a departmental seminar, with Paul Ziff[585] as speaker, and I feel under obligation to be here. I am all the more eager for an informal get-together here with you, Sid, and John; I'll suggest some dates after my return from Michigan. Right now, Diane is somewhat under the weather, and Toby is in bed with some not clearly diagnosed trouble; our – very competent – physician, to whom she was taken today, said that the most likely diagnosis was appendicitis, although some of the typical symptoms are missing. (*My* diagnosis for her had been sinusitis; so an M.D. does not yet seem within reach for me.) Pete is tryingly non-corresponding, but his latest letter, two or so weeks ago, sounded rather better than his first impressions, which came by phone. Diane and I are so glad you have good news from Sandy, and that Bobby seems to be happy at Columbia. – Love from both of us to you and Edith, and best wishes to you all!

As ever,
Peter

4.1.155 Hempel to Nagel, November 2, 1965/2

[Typewritten, Hempel 1999, Box 28, Folder 2]

2.
November 2, 1965

Dear Ernest,

I must tell you one more thing: Last Friday, the Scientific American came in the mail; I did not at the moment recall my conversation with you about Walter Kaufmann's *Hegel*,[586] and then I came upon your review and read it eagerly.[587] I really sat there shaking my head, and I told Diane: I just do not understand how you can span so many fields! I do not know the first thing about Hegel, and here you give this hugely informative and balanced account and appraisal and weave into it also an evaluation of Walter's book – I am amazed, and truly full of admiration. My impression is that you are being very fair to Walter Kaufmann in the reservations you express, and that you acknowledge very handsomely the features of the book that you find meritorious. He has not said anything to me about the review, but I should think that you have given him no reason at all to feel that he has been treated harshly, let alone unfairly!

Yours,
Peter

4.1.156 Hempel to Nagel, January 23, 1966

January 23, 1966

Dear Ernest:

I so enjoyed seeing you the day before yesterday; but I was sorry there was so little time really to talk with you. So I am all the more eager to suggest some dates for a get-together

[585] Paul Ziff (1920–2003) was an American philosopher and artist. He worked at Harvard, Oxford, Princeton, and later at the University of Pennsylvania.

[586] See Kaufmann (1965).

[587] Nagel (1965).

here. And Diane is much looking forwards to seeing you, Sid, and John again. Her schedule makes Fridays and Sundays the most convenient possibilities; would any of these dates suit you: Fri 2/11, Sun 2/13; Fri 2/18, Sun 2/20? I very much hope that John can make it, too; but if not, let's not put it off just for that reason!

Many good wishes to you all!

4.1.157 Nagel to Hempel, May 21, 1966

[Handwritten, Hempel 1999, Box 28, Folder 2]

May 21, 1966

Dear Peter,

The book I mentioned to you last night is:

C.E.M. HANSEL, ESP: A scientific evaluation (Scribner's)[588]

It was wonderful to talk to you yesterday, I wish I could do it more often. If I don't see you before the fall, all the best to you and Diane.

Affectionately,

Ernest

{+ Carl Boyer, Rainbow}

4.1.158 Nagel to Hempel, July 5, 1966

[Typewritten, Hempel 1999, Box 28, Folder 2]

South Wardsboro, Vt. 05355
July 5, 1966

Dear Peter,

We've been here for a little more than a week, though we made an earlier trip to open the place before going to Cambridge for Sandy's commencement. Just Edith and I are here now, and it feels very lonely indeed. Sandy is in Europe, on a nine week grand tour (he left June 25), and Bobby returned to New York yesterday to take a summer session course in Greek. The Hooks, whose house is a five minutes walk from ours, are also away (in Italy), but expect to be back early in August. So far a while we are rattling around by ourselves, and we haven't got quite used to it.

But the real reason for my writing is not to tell you these trivia. You mentioned to me last time we met that you will be on leave the year after the coming one, and that you were going to apply to the NSF for a grant to supplement your Princeton half-salary. And it has occurred to me that perhaps instead of the latter you might be interested to come as a visitor for either a half year or the whole year to the Rockefeller. I am unable at the moment to specify details either about stipend or other matters (e.g., whether there would be expectation that you would use an office that would be provided with any regularity); but I am confident that an arrangement entirely satisfactory to you could be worked out, and I need not tell you how happy I would be if you decided that the question is worth your consideration. Please let me know at your convenience what you think.

---

[588] Hansel (1966).

Except for missing the boys, Edith is feeling pretty well – though she was disappointed by what the X-rays showed when she had them taken two weeks ago: the wound from her last operation (now a year and a half old) still continues to drain. But the surgeon doesn't think there has been any deterioration, which is something to be thankful for. She will be giving a course next fall in electricity and magnetism, and is busy preparing herself for it. As for myself, I continue to feel both tired and anxious about the bed I made for myself, but am hoping that I'll adjust myself eventually to the new situation.

Before I left the city I heard the news about Donald Davidson becoming your colleague again – it will be almost like old times at Queens! As far as I know there have been no developments at Columbia, and the department will be terribly short-handed during the coming year. Did you see the announcement in the papers last month of the death of my friend Jim Newman? Although he had been complaining during the past two years about circulatory spams, the doctors assured him there was nothing seriously wrong, so that the news was an unexpected blow. I went down to Washington for the services. Denis Flanagan,[589] the editor of SCIENTIFIC AMERICAN, was also there, and he didn't know just what would be done to replace Jim – it will not be an easy task, for Jim had rare qualities. One final item. Do you know Marshall Cohen,[590] who used to be at Harvard and subsequently at Chicago? Harry Frankfurt[591] and I asked him whether he would be interested in coming to the Rockefeller, and he was eager to make the move. If things turn out as anticipated, he'll join as of next January – but I am keeping my fingers crossed that the appointment will be a reasonably good one. On this and other accounts, pray for me!

Love to you and Diane,

Ernest

4.1.159 Hempel to Nagel, July 8, 1966

[Typewritten, Hempel 1999, Box 28, Folder 2]

July 8, 1966

Dear Ernest:

It was so good to get your letter – thank you very much! I had tried a number of times to reach you by phone and finally concluded that you must have left town. Recently somebody – I forgot who – mentioned having seen you and Edith at the time of Sandy's graduation at Harvard, and he mentioned that Sandy had graduated summa cum, which is really remarkable! Congratulations to him – he surely has earned his trip to Europe, and you and Edith must be very proud of him. I am terribly sorry his application to Princeton was unsuccessful, but of course he will do very well also at Columbia.

Our Pete, though not resting on the same kind of laurels, is in Europe, too, on money he earned last summer. It turned out, rather in the last minute, that there would be a charter flight he could join, and he decided to take advantage of it, planning to spend most of his time in England. He left four weeks ago today, and we have had only one card from him, which came about three weeks ago from Paris; we don't even have his address and don't know

[589] Denis Flanagan (1919–2005) was the founding editor of the magazine *Scientific American*.

[590] Marshall Cohen (1929–) is an American philosopher and legal thinker. He studied at Dartmouth College and at the Universities of Harvard and Oxford. He then taught at several American universities, including the University of Chicago.

[591] Harry Frankfurt (1929–2023) was an American philosopher. He studied at Johns Hopkins University, where he also received his PhD. He taught at several American universities, including Rockefeller University. He wrote extensively on free will and moral responsibility.

where he is, and so we feel a combination of annoyance and worry. – Toby has graduated from high school and has accepted an offer of admission from Brandeis, with which she was very pleased; Brandeis had been her favorite.

It is good to hear that Edith's condition is, at the very least, stable; but I hope very much that her wound will stop draining; if only she could relax completely, both physically and mentally, for a number of weeks, I think it could do the trick. But here she goes again, preparing her classes for next year, instead of lying in a hammock, reading detective stories! It should be fun for both of you when the Hooks come back and tell you all about their travels; I recall Sidney mentioning, when I last saw him at the N.Y. Philos. Club meeting, that he was headed for the fabled villa on Lake Como.

You don't say anything about having had another checkup; so I suppose you haven't! But what I am most sorry about is that you continue to feel worried about your change to Rockefeller U. Somehow – and I don't believe irrationally so – I feel quite sure that a few weeks after moving into your new office and getting acquainted with your new surroundings you will settle in very well and will feel very comfortable there. I think it just impossible that there could be nearly as much work for you as there was at Columbia (not counting, of course, your own research and writing); and even your administrative duties will not, I am entirely confident, prove very burdensome after all, how many appointments can you expect to be in store for the Department of Philosophy? Getting Marshall Cohen would be a very good move; I know him slightly and I have heard him lecture and found him very bright; he was on several occasions discussed as a potential candidate for Princeton.

It is terribly nice of you to suggest the possibility of my coming to the Rockefeller as a visitor during part or all of my next leave, which I expect to get in 1967–68. I would very much like to talk some more with you about the possibilities, though, as matters look at present, it does not seem likely that it would work out for my next leave. Next summer, as I may have mentioned to you, I have agreed to teach at the summer institute to be held on the Stanford campus under the aegis of the Council for Philosophical Studies; and in September, after Toby has settled in for her sophomore year, Diane and I hope to go off to Europe for a few months, possibly a semester; during that time, I hope to get some change of pace and of intellectual scenery by not planning to write, but to read and to talk to people; but then, I want to settle down and to get some sustained writing done on a project that has still to be crystallized. And I have the notion that for that purpose, I should probably simply stay put in Princeton, since I find that I need a good deal of just uninterrupted quiet time if I am to get any writing done at all – and I can't say that I find my productivity increasing nor the ease with which I formulate things getting any greater! But at any rate, I would like to talk some more with you about it. And it occurred to me also that if there is some arrangement under which I could perhaps read a paper or give a talk or two at Rockefeller, that would interest me, too. In any event, I look forward to visiting you in your new quarters!

Maybe we will even have a meeting of the Nominating Committee of the Ph of Sci Assoc in your office! As you presumably know, Paul Benacerraf has agreed to act as chairman and use his secretarial help in this connection; and you, he, and I are to be the committee. I have a notion that one meeting should suffice to arrive at agreement on the nominations, and that this will be a pleasanter and more efficient way than correspondence. Another meeting place, highly to be recommended, would be Princeton! And at any rate, Diane and I expect you and Sid to come out as of old, of course.

I was quite shocked to read of James Newman's death, and I thought of the loss it must mean to you. I met Newman only once, I think; that must have been in Washington or at any rate in that area; I believe he came to the hotel where you and I were staying and wanted to talk to you. How sad – he was still a relatively young, and at any rate very vigorous, man.

Many thanks also for your earlier card with the title of the book critically examining ESP. It was not in the Princeton Library, and I immediately put in a request that it be purchased; but it has not come yet.

I am sweating away on my ms for the Isenberg volume,[592] which I had mistakenly thought was all but finished. And then I have several more commitments. – I hope *your* calendar is not too crowded.

Diane and I wish you both a very good summer.

As ever

P.

4.1.160 Hempel to Nagel, September 2, 1966

[Typewritten, Hempel 1999, Box 28, Folder 2]

September 2, 1966

Dear Ernest:

I do not know whether you are still in Vermont, but I suppose the letter will be forwarded if you have already returned to New York. I very much hope your summer has been restful and pleasant for you and for Edith; you both certainly needed it. Are you and Edith feeling all right? And are you looking forward with a certain curiosity to Rockefeller University? I hope to pay you a visit there soon!

The news of Frank's death was very sad; but I had heard from Paul Feyerabend at the Minnesota workshop last May that both he and Hania were quite senile and very weak and pathetic, and how they had begged Feyerabend and a friend to take them for a ride through the streets; Frank asking Paul whether he knew a philosopher called Feyerabend. And now poor Hania is all alone – it is dreadful to contemplate.

I thought I might send you a copy of my reply to Morris, which I am also sending to Carnap and Feigl. I suspect that you will feel inclined to agree with my hesitations about efforts to revitalize the Institute[593] beyond its mainly administrative role. And I must say that while I like Bob Cohen personally very much and think him extremely bright, I doubt very seriously that he would be able to revive the Institute in a way that was significantly related to its original purposes.

We are reasonably well, though Diane and I feel rather tired. We were unable to get away during the summer; now Toby's departure and all the requisite preparations are approaching, and she is understandably anxious even though she is also excitedly looking forward to the experience. Pete is with us for a visit of perhaps two weeks, he returned a few days ago from his charter flight-visit to Europe, which he greatly enjoyed. He even managed to visit Eva's sister Lux and her husband, who received him very nicely and showed him something of Munich. But about his career, there are still considerable uncertainties, which is a bit worrisome.

I recently had occasion to look through notes and letters dating back to the time of Eva's death, and I came to realize again how much help – in more ways than one – you and Edith gave me.

Warmest good wishes for a good next academic year for each and every one of the four Nagels!

As ever,

Peter

---

[592] Hempel (1969a).

[593] By the late 1950s, the Institute stopped any public activities, and after Frank's death in 1966, various people (including Carnap, Morris, Nagel, Feigl, Holton, and others) debated the future of the Institute. For some time, it was supposed to be integrated into Feigl's institute in Minnesota, but it had ties through Robert S. Cohen to the Boston Colloquium for the Philosophy of Science as well.

4.1.161 Nagel to Hempel, September 3, 1966

[Typewritten, Hempel 1999, Box 28, Folder 2]

So. Wardsboro, Vermont 05355
Sept. 3, 1966

Dear Peter,

Your most welcome letter arrived today, and I hasten to tell you what I wrote to Morris – but I did not have the presence of mind to make carbons, and am compelled to give you only the gist of my reply to him. My main point was that there are already in existence a number of "institutes" and "centers" concerned with the philosophy of science, that I do not think there is a need in this country for another one, that it is unlikely there is sufficient interest or enthusiasm among younger men to warrant "revitalizing" the IUS along the specific ideas of its founders (especially Frank and Neurath), and that what might be done is to have one of the existing "centers" assume the *name* of the Institute. The latter would be an act of piety, and help keep green the memory of Frank and Neurath; but although this would certainly be appropriate, the question is one of sentiment (which I happen to share) for which I cannot give any cogent reasons and to which I do not attach much significance. On the other matters raised by Morris, I think I am in even greater agreement with you, except in one case: I expressed to him a jaundiced view on a prize-essay contest; I did not think there is much need for a volume of essays on the Unity of Science, especially in the light of the talks to be given this fall at Johns Hopkins (which may be published) on the heritage of logical positivism;[594] I strongly approved of the Institute having a role in the Memorial meeting for Frank, as well as of the suggestion that authors of the Encyc. monographs be given an opportunity to up-date the latter. (Incidentally, I have no idea whether my own monograph is still in print, for I haven't received any royalties for at least a half dozen years.) The point on which I may differ from you is on the Neurath volume. It has been my understanding all along that the IUS was committed to giving some financial help in the publication of the volume; and on the assumption that there has been this commitment, I took the view that the obligation should be kept if some moneys are available.

Earlier in the summer Morris sent me a draft of the bibliography monograph for the Encyc, and I finally sent some suggestions and queries to Herbert Feigl about two weeks ago. (I imagine you must also have seen the draft.) Quite frankly, I do not know what this monograph is good for, especially in its present form, for I do not understand what principle (if any) controls what is included in the listings or what use an unannotated catalogue might have for the probable audience of the monograph.

About a month ago we had a brief visit from an old friend of yours at Yale (her husband is at the Medical School at Dallas, but cannot recall the name),[595] and she told us how hard you have been working this summer. And so I can very well imagine that you feel tired, and hope that both you and Diane will get some relaxation and refreshment before the academic year begins. Are you going to give one of the talks at {at Johns Hopkins} on the heritage of logical positivism? I haven't seen the list of expected speakers, although I am supposed to discuss logical positivism and apriori knowledge. I've been trying to think of what to say during most of the summer, but have nothing to show for my pains thus far. Part of my difficulty comes from my usual trouble in getting anything on paper; but the difficulty has been aggravated greatly this year by my continuing depression associated with my leaving Columbia, and with pathological apprehensions as to what awaits me at the Rockefeller. But I won't burden you any further with my miseries!

---

[594] It appeared as Achinstein and Barker (1969).

[595] A pencil marginal note says "Donald Seldin".

The rest of the family is in pretty good shape. Edith finished her preparation for the new course she will be giving at City College, and has enjoyed doing the work as well as taking care of her flower garden. Bobby's experience of being alone in the city for six weeks (he is here now) was a valuable one for him, for it convinced him that he can take care of himself, and has given him some needed self-confidence. Sandy returned from Europe a week ago, and he thoroughly enjoyed his grand tour, even though the sight-seeing was a bit strenuous and perhaps somewhat too long. We shall miss both of them this fall, for both of them will be living away from home; and I imagine you will have similar regrets when Toby leaves for Brandeis and Pete for points south.

We expect to be back in the city by Sept. 13th, and I hope to see you before long. Meanwhile, all good wishes and warmest regards to you and Diane from all of us.

As ever,

Ernest

P.S. Colodny[596] sent me a copy of his Introduction to Vol. III in the Pittsburgh series,[597] and I am eager to read your discussion of the justification of induction. I've sent away for the volume, but have not received it yet and will probably not see it until my return to New York.

E.

4.1.162 Hempel to Nagel, September 7, 1966

[Typewritten, Hempel 1999, Box 28, Folder 2]

September 7, 1966

Dear Ernest:

Your letter just arrived; thank you so much! I only wish I could successfully inoculate you with my quite firm conviction that before long, you will be very pleased with the change from Columbia to Rockefeller! Since there is no way of doing that, I will just have to sit back and look forward to the time when I can say to you: "I told you so"! I hope that will be very soon. You say nothing of your health. I hope that means you have been feeling reasonably well.

Re Johns Hopkins: Yes, I am to read a paper in the series on December 15, with a kind of seminar the next day. I was asked to speak on logical empiricism and the social sciences; that is why I have been reading or re-reading a fair dose of Neurath recently. My paper is not yet in shape, but I have a broad idea of what I am going to say. Reading Neurath, I was surprised to find that even in his *Empirische Soziologie*, he cited virtually no specific evidence in support of his sociological and anthropological pronouncements. – Achinstein and Barker[598] wrote me recently – probably you have received the same letter, saying that I would receive a full list of speakers and topics as soon as all plans were completed. I was surprised to hear

[596] Robert G. Colodny (1915–1997) was an American philosopher. He studied at Columbia University and at the University of Chicago. He fought in the Spanish Civil War, as he had joined the Abraham Lincoln Brigade; after that, he joined the US Army and fought in World War II. After the war, he received a PhD in history and philosophy from the University of California at Berkeley. He taught at San Francisco State College, the University of Kansas, and the University of Pittsburgh.

[597] Colodny (1966). Hempel contributed to the volume with the essay "Recent Problems of Induction" (1966b).

[598] Stephen Francis Barker (1927–2019) was an American philosopher of mathematics. He studied at Swarthmore College and Harvard University, where he also received his PhD. He then taught at several American universities but spent most of his career at Johns Hopkins University.

that the arrangements were not yet complete, for I was approached by them more than a year ago. I had meant to mention the series to Morris, but felt that I should wait for fuller details. I quite agree with you that the Johns Hopkins series militates further against the idea of a volume of essays on the Unity of Science.

It would be nice, of course, if one of the existing centers were to assume the name of the institute, but I frankly doubt that any of the present organizations would be willing to do so, because – among other reasons – it might seem to imply a commitment to a doctrine to which the group might be reluctant to subscribe in such an explicit fashion. And I think you are entirely right that we should support publication of the Neurath volume if there was an earlier commitment or even near-commitment to this effect. Thanks lots for briefing me on your response to Morris's letter!

Re royalties for your monograph: I am glad you mentioned the matter; for I received some months ago a sizeable check from Morris, who indicated that he had taken charge of these payments. I had not had a check in a very long time, just like you; and I am sure that your monograph is selling well (note Morris's general remark that the sales have continued satisfactorily; and yours must surely be among the bestsellers in the group); so I think you shld ask Morris about this; there must have been a slip somewhere.

The old friend who visited you must be Muriel Seldin, wife of Donald Seldin,[599] who had visited us when Donald was in New York to discuss an offer of a professorship in New York (he had declined what seems to have been an attractive offer from Harvard). We had no idea Muriel was going to see you, and I am quite surprised to hear that she did.

I am enclosing a reprint of my piece on induction since you so nicely express an interest. Quite frankly, I had not done so before because there is very little that is new in the piece, it's rather more like a survey article. But the Colodny volume in which it appeared seems to me exceptionally substantial; you will be particularly interested in a long piece by Shapere which critically examines Feyerabend,[600] with some side remarks on Kuhn. I would enjoy talking with you about the piece when I see you next time.

It was good to hear that the summer was satisfactory for Edith and for the boys. Pete returned from Europe a few days ago and it is clear that he enjoyed his summer hugely; the worrisome thing is that the direction of his interests is still unclear, und that he has considerable doubts about grad. work in English, though he does intend to continue this coming academic year.

You have no doubt read the sad news of Buz Henry's[601] sudden death in his summer place in the Adirondacks; he was a very fine person. His obituary in the Times revealed an amazingly rich and varied background of experiences and competencies, which he never mentioned. I am sure you recall him, he was on our deptl. Advisory Council and lived on the outskirts of Princeton.

Greetings, and all the best to you all – and have a good trip home!

As ever,

---

[599] Donald W. Seldin (1920–2018) was an American physician and nephrologist. He received a bachelor's degree from New York University and a medical degree from Yale University School of Medicine. He worked at Yale before being appointed professor in the Department of Medicine at the University of Texas Southwestern Medical School.

[600] Shapere (1966).

[601] Presumably Barklie McKee Henry (1902–1966), known as Buz Henry, a trustee of The Rockefeller University. He graduated from Harvard in 1924 and acted as a worker and leader of various community services and emergency committees in New York. He was also Governor of the Society of The New York Hospital and later its vice president and president. See the full obituary here: https://digitalcommons.rockefeller.edu/cgi/viewcontent.cgi?article=1006&context=occasional-papers-rockefeller-institute (accessed on September 24, 2024).

4.1.163 Hempel to Nagel, March 9, 1967

[Typewritten, Hempel 1999, Box 28, Folder 2]

March 9, 1967

Dear Ernest:

Unfortunately, I will once more be unable to come to the meeting of the Philos Club; this time because I have to participate in a two-man presentation (Gillispie and I) to the Trustees on our Program in the History and Philosophy of Science next Friday evening.

It was such fun seeing you, Sid, and John the other day, and we hope you will soon come again! Recently, Van Quine was here to read a paper in our colloquium, and that was a very pleasant and lively occasion.

I telephoned with Carnap yesterday, and learned from him that he will be flying to Hawaii this Sunday with his daughter and with Herbert Feigl; they are going to give some sort of seminar there. And [e]arly in July, Carnap will be going to Europe for some months! He sounded quite chipper. His eye is gradually getting worse, but no surgery seems indicated this year, he said.

Enclosed is a copy of my reaction to the Morris-Szekely piece and to Morris' latest letter.[602]

This weekend, I have to be in Washington on the usual NSF business, with meetings scheduled for Sunday pm and for Monday. This time, Diane will come along, and we will go already on Saturday in the hope of seeing something of a museum or two.

I hope that you are all well, and that things are looking more cheerful to you every day! My very best to you all!

As ever,

Peter

4.1.164 Nagel to Hempel, March 11, 1967

[Typewritten, Hempel 1999, Box 28, Folder 2]

25 Claremont Avenue
New York, N.Y. 10027
March 11, 1967

Dear Peter,

There must be some sort of telepathic influence! I've been thinking of you repeatedly during the past week, and even planned to call you for a chat, when your letter with its

[602] In 1966, the logician David L. Székely published a 66 pages long booklet in his monograph series ("Association for Unification and Automation in Science") under the title "Otto Neurath and the Unity of Science Movement." It contained some pages from Charles Morris and a reprint of his 1962 article, "On the History of the International Encyclopedia of Unified Science," along with some papers and remarks of Székely on Neurath and on his own ideas of unification. See Székely and Morris (1966). In 1967, Morris sent out a circular letter to the Institute for the Unity of Science, because Székely wanted to republish it with the help of the Institute to distribute it in America (Morris to the Institute, 20 February 1967, RC 102-48-09). In return, Carnap answered that he was happy to have some of Morris's papers in one volume, though Székely's papers were quite weak; thus he did not support the republication of the volume with the help of the Institute.

enclosure arrived earlier today. Now I'll save the call for another time, to tide over the long interval before you come to New York again.

Thank you for the copy of your reply to Morris. I wrote him more than a week ago, but unfortunately did not make a carbon. However, my reply as I recall it was almost identical with yours, down to my appreciation of Neurath's elephant signature. Needless to say, I too found the Szekely piece just jabberwocky, and strongly urged that the Institute support neither its publication nor any lecture trip by him.

It was good to get your news about Carnap and Feigl, for I rarely hear from either of them. I had a letter from Carnap about two months ago, acknowledging one I sent him last summer in which I noted what I thought were a couple of small slips in his PHILOS. FOUND. OF PHYSICS[603] – he complained about his eyesight then, but seemed in amazingly good spirits.

I see Sid Morgenbesser once in a while (usually at Tom Kuhn's seminar at the Rockefeller, which will be over in a couple of weeks and has not been very successful), who continues to have about six dozen irons in the fire, so that there is little time for a leisurely talk; but I haven't heard from John since our visit to Princeton. A week ago I heard Hilary Putnam read a paper on OTHER MINDS,[604] it was sensible, though as usual with him he tends to be slap-dash – but he is lively, and I enjoyed the affair. And this week Harmon[605] gave a paper at Columbia on QUINE AND THE DEATH OF MEANING – it was just a restatement of Van's views, presented with an air of finality with no indication that there might be difficulties in Van's claims. I wish I could discuss with you Van's position some time, especially to learn how far you go along with him.

I continue to go regularly to my office at the Rockefeller, but am looking forward to the end of this adventure. It has been a traumatic experience for me, and I have not been able thus far to pull out completely from the tail spin into which it sent me. My leaving it has, perhaps, been the only thing I've done for the place, for as a consequence quite a number of new appointments have been made. In addition to Joel Feinberg, Bob Nozick, and Margaret Wilson,[606] it seems that Fodor[607] will be coming, and there is talk of asking Grice; and as visitors for next year, there will be Shoemaker,[608] Pears (fall),[609] Sid Morgenbesser (Spring).

This forenoon I attended the meeting of the Woodrow Wilson Dissertation Selection Committee – all of the applicants from Princeton were voted grants. But the entire

[603] Carnap (1966).

[604] The paper was later published as Putnam (1972).

[605] Nagel presumably meant Gilbert Harman (1938–2021), an American philosopher who published extensively in the fields of epistemology and moral philosophy. He studied at Swarthmore College and received a PhD from Harvard under the supervision of Quine. He taught at Princeton University for his entire career. See his paper on Quine and meaning as Harman (1967).

[606] Margaret Wilson (1939–1998) was an American philosopher who earned his PhD from Harvard University in 1965. She worked at Oxford, Columbia, at the Rockefeller Institute between 1967–1970, and became the first female member of the philosophy faculty at Princeton in 1970. She worked on early modern philosophy, philosophy of religion and mind.

[607] Jerry A. Fodor (1935–2017) was an American philosopher and cognitive scientist. He studied at Columbia University and earned his PhD in philosophy from Princeton University. He then taught at MIT, CUNY, and Rutgers University.

[608] Sidney S. Shoemaker (1931–2022) was an American philosopher who worked in the field of the philosophy of mind. He studied at Reed College and then went on to earn a PhD in philosophy from Cornell University. He taught at Ohio State University and at Cornell.

[609] David Pears (1921–2009) was a British philosopher. He studied at Oxford, where he also worked for most of his career. A specialist on Wittgenstein, he also wrote influential works on the philosophy of language and the philosophy of mind.

W.W. program is folding up: the Ford Foundation has declined to continue the grant, so that this is the final year during which Wilson Fellowships will be awarded. The rumor is (though no one really knows) that the Ford Foundation will be making sizable grants for graduate fellowships to ten universities; but my feeling is that this will probably be giving to them that hath, and that the present program is preferable.

I hope that all is well with you, and that you and Diane will enjoy the trip to Washington. My love to you both,

Always yours,
Ernest

4.1.165 Hempel to Nagel, August 8, 1968

[Typewritten, Nagel 2020, Box 4, Folder 4]

Princeton, August 8, 1968

Dear Ernest:

I am very sorry I did not see you before you and Edith went off to Vermont. I recall that you were going off to lecture or perform some other professional function some place – was it at the meeting on biological explanation? I hope you found it worth your while!

Diane and I actually decided to go on a three-week vacation in Europe on one of these 21-day excursion deals, which offer an enormously reduced fare; and we are very glad we did. It was a very successful vacation; one week in Amsterdam, with side trips to Haarlem, The Hague, and Delft; then to London, with side trips to Oxford, Chichester, and Bristol. We walked and walked, and that was great fun; we saw some splendid exhibits and beautiful galleries, went to the ballet in Amsterdam and to two plays in London, and lived on a houseboat on the Thames in a choice section of London! I have no doubt that this kind of vacation is much more satisfying for Diane than it would have been for her to come with me to the Pacific Division meeting in Hawaii, for which I expect to leave about August 26.[610] On the way there, I hope to stop over in L.A. to visit Carnap and perhaps briefly the Helmers; on the way back, Michael Scriven invited me to stay at the apartment he keeps in San Francisco; and I also hope to go to Stanford to see the Goheens and, if they are still there, the Vlastoi. I accepted an invitation to stay on in Honolulu for a few days and to give a lecture and two seminars at the university. The fee for this is fairly small, and it might cost me more to stay in a hotel, but Abe Kaplan has very nicely asked me to stay at his house. He is in charge of the 1969 East-West Philosophers' conference, according to Negley,[611] the chairman at the U of Hawaii; and it seems to me someone told me that Abe Kaplan was leaving Michigan to go to Hawaii for good; but I am not sure about this.

In Europe, by the way, I did not make contact with any of the philosophers, except that we saw the Hampshires in London (he will be on leave there this coming year).

Diane has accepted, at least temporarily, a second editorial assignment; she is in full charge of the quarterly bulletin of the Princeton Graduate Alumni Association and has to get her first issue out this September; so she is faced with an awful lot of work – all the more so

[610] The 42nd annual meeting of the Pacific Division of the American Philosophical Association was held in Honolulu, Hawaii, from August 29 to September 2, 1968. Hempel delivered an invited address entitled "On the Limits of Scientific Understanding", and a similarly titled paper was published as Hempel (1973).

[611] Winfield Eugene Nagley (1918–1996) was an American philosopher who was professor of philosophy at the University of Hawaii and its chairman.

because she also has assumed considerable responsibilities on the committee that runs the Princeton Adult School program; and she has come up with some excellent new ideas, some of which were tried out in the past year and worked out beautifully.

Toby and her husband are spending the summer in a house they have rent-free for July and August in Brooklyn. To our great relief, we heard not long ago that Sam has found a job at Boston U. for the coming academic year; and this means that Toby can go on at Brandeis. She visited us twice before we went to Europe, and we expect her to come out soon again. Pete is taking a summer course in German at Harvard and is enjoying life in Cambridge a great deal.

I hope that you and Edith are getting some rest and relaxation this summer! Have the boys spent some time with you? I recall that neither intended to be in Vt. for the whole summer.

It will be good to swap news with you! How are you feeling? How is your back etc. behaving?

Love to you all from Diane and me!

As ever,

Peter

4.1.166 Nagel to Hempel, August 21, 1968

[Typewritten, Hempel 1999, Box 28, Folder 2]

South Wardsboro, Vermont 05355
August 21, 1968

Dear Peter,

I am back again in Vermont, hopefully for the remainder of the rapidly passing summer. On Monday of this week most of those serving with me as jurors were excused from further service (during the six days I was in the court building I wasn't even put on a panel of jurors to go to a court room), and I took the first opportunity to return here by an early bus yesterday. I had hoped to call you Monday evening, but there was no chance to do so since I promised to go out with Bobby and got back late. And so the plan of visiting you and Diane in Princeton before you left for Hawaii has unfortunately come to nothing. It is such a long time since I've seen you and had a decent talk with you! Perhaps after both of us get settled for the fall we can arrange for a get-together. I'll be looking forward to it.

Your letter of two-weeks ago was waiting for me here. Your description of your European journey sounds most inviting – Edith and I had been talking of doing something similar, but our energy to undertake even a brief trip evaporated into idle dreaming. Even our annual intention to visit Quebec has come to naught. Once we are here in Vermont it is so tempting to remain that I am afraid we have become permanent sticks-in-the-mud.

I don't recall whether I mentioned to you on the phone last week that we saw Chas. Stevenson and his wife a short while ago. He had had no news of John Cooley, and neither have I. But I asked Steve whether Abe Kaplan had left Michigan for good, and his answer was No – Abe was on leave at Hawaii to plan the 1969 East-West Conference. I tried to reach Sid Morgenbesser several times while I was in the city, but though he is repeatedly in town I had no luck. I had a rumor that he has proposed marriage to a girl (whom I did not meet), but that for reasons not told me nothing came of it in the end. I suppose I ought to be resigned to the fact that Sid will never settle down, though I am convinced that he both needs and would eventually enjoy a stable married life.

The latest news about Russian and Polish troop movements into Czechoslovakia is pretty depressing, especially after the hopes that had been raised concerning the amicable settlement of differences between Moscow and Prague. Although Sidney Hook did not take much stock in the apparently promising news about the Russo-Czech conference, he too is quite upset by the latest turn of events. Will we ever see the day without opening the newspaper without dread?

But enough of these gloomy thoughts. Have a good trip to points west, if you think of it remember me to Carnap, and love to you and Diane from the two of us.

Ever yours,
Ernest

4.1.167 Nagel to Hempel, October 26, 1968

[Handwritten, Hempel 1999, Box 28, Folder 2]

25 Claremont Ave
New York, NY 10027
Oct 26. 1968

Dear Peter,

I talked to John Cooley on the phone last night. He had returned only recently from his final sail of the season, and sounded quite happy and in good form.

I mentioned that I had seen you a week ago, and that you raised the question whether he, Sid, and I might pay you a visit. John jumped at the idea and asked me to find out from you what dates would be possible for you – perhaps in November if you would have the time for a visit from us. Could you send me a list of dates convenient to you? I'll then communicate with John and Sid, and let you know. (Mondays are not possible for me, and Tuesdays also raise a bit of a problem though not an impossible one.)

It was a great joy to me to talk with you the other day.

Love to you and Diane,
Ernest

4.1.168 Nagel to Hempel, November 2, 1968

[Handwritten, Hempel 1999, Box 28, Folder 2]

25 Claremont Ave
New York, NY. 10027
Nov. 2. 1968

Dear Peter,

I was able to get in touch with John and Sid fairly quickly, and both of them are eager to see you soon. They gave as first choice among the dates you proposed Wed. Nov. 20, and needless to say this is my preference also. Unless you veto this, I'll assume you'll expect us on that day in your office around 3 pm.

It seems unlikely that I'll be able to attend the Amer. Philos. Society meeting two weeks hence. Edith teaches until 5:30 on Thursdays, so that we couldn't get to Philadelphia until late that evening and so miss most of the scheduled events.

I had a chance to look at the *Jr. of Philos.* review copy of Kyburg's book[612] on the philosophy of science. He is surely an able man, but what poor judgment he shows in writing his books! I don't believe it will have much of an audience.

However this may be I'm looking forward to the visit to Princeton. Meanwhile, love to you and Diane.

As ever,
Ernest

4.1.169 Hempel to Nagel, August 10, 1969

[Handwritten, Nagel 2020, Box 3, Folder 4]

8-10-69

Dear Ernest,

From time to time, I have decided to write you and then put it off again – and gradually, the summer will be running out!

I hope you and Edith – and perhaps the boys part-time –are having a relaxing and refreshing change of pace, air, and scenery in your summer retreat and that you and Edith have been feeling well. I suppose one of these days you'll be seeing Cobarrubias,[613] who came out here for a few hours last week to chat with me, and who said he hoped to visit you in connection with a trip he was planning to Albany. He seemed a lively, likeable, and enterprising man, who seems to have read a good deal of the English and American literature in epistemology and the philosophy of math; but I found it impossible to form an impression of his competence in these fields, or of the extent to which he had understood the many books and articles he referred to. I certainly wish he will find an opportunity to get his PhD in this country; but given his current professional (?) position, it would no doubt be slightly [n/a] for him and for his teachers.

Perhaps six weeks ago, I received from Stegmüller[614] a fat volume he had just published, *Wissenschaftliche Erklärung und Begründung*.[615] I was very much touched to find it dedicated to me; all the more so because in a letter announcing the book (without mention of the dedication) he told me that much of it had been written during a terrifying illness, of the likes of which I had never heard before. Its main symptom seems to have been an incredible restlessness, making him pace his room day and night and making sleep impossible for days on end, even with the strongest available drugs. There must have been much anxiety associated with it all, and he sometimes despaired of ever recovering. He had a stay in a sanatorium, and finally, his condition improved, and he is back teaching. It must have been a fearful ordeal.

---

[612] See Kyburg (1968).

[613] Juan J. Gonzalez Cobarrubias was a professor of philosophy at the University of Puerto Rico who eventually had a PhD from Columbia University and taught at the University of Mexico and other places. He published about ethics, planning theory, and language.

[614] Wolfgang Stegmüller (1923–1991) was an Austrian philosopher. He studied at the University of Innsbruck, where he also received his Phd. He worked at Oxford, the University of Innsbruck, and the University of Munich. A brief account of Stegmüller's thought is provided in the obituary that Hempel wrote for *Erkenntnis* in 1992, entitled "In Memoriam Wolfgang Stegmüller" (Hempel 1992).

[615] Stegmüller (1969).

The book is done in his amazingly lucid and thorough manner; much of it consists of detailed reports on recent Anglo-American literature in the field, and some of it of criticism and further suggestions, some of the latter – and the very end – based in part on ideas of a man called Käsbauer,[616] which look quite interesting. I wonder whether J.Phil. will review the book – of course, it isn't easy to find a person who is both able and willing to do the job.[617]

No doubt you have heard that Donald Davidson – on leave this year at the Behav. Sci. Center – is leaving us for "Rocky U",[618] as the place seems to be generally referred to. I am very sorry about it, and the more so because it is very likely that John Wallace[619] will follow him. But my main worry is now that Gil Harman and Tom Nagel[620] stay; both are very original, productive, and congenial people; good friends of longstanding, but of quite different philosophical bent.

Diane went on a four-week trip to Israel with Toby; it was her graduation present. She graduated after 3 years, and even with honors; but she didn't even go to her commencement, though that isn't very unusual nowadays, and we are awfully glad she did complete her undergraduate studies. What will come next is unclear. She says she is tired of going to classes, and that's understandable. She and Diane had a most interesting time in Israel, saw the Bar-Hillels (who were *extremely* nice + helpful), and Poznanski,[621] whom we like very much, as well as some relatives of Diane's. They also visited Masadah, attended the Knesset, etc. But now Toby is back in Boston, where Diane is visiting her this weekend ("meeting her", to be more exact). Pete is travelling in Europe: England, France, Spain, Holland!

I have been lucky enough to have the use of an air-conditioned office on the floor below mine and have been holding my nose to the grindstone. It keeps my conscience moderately manageable, but I have struck very few sporks (what can one expect of a nose?). It is all aimed at putting down notes and trying to get some more ideas for the Carus lectures. The work has been somewhat overshadowed by the expectation that I might have to have my cataract removed this fall. But when I recently saw the ophthalmologist, he said he cld detect no chance since six months ago, and while the operation cld be done any time, it had not become more urgent than before. So I am to see him if I notice any deterioration, and at any rate in six months.

Carnap is doing well with his eye glasses (but he had no contact lenses yet when I last spoke to him by phone); he takes walks, reads, writes, and seems in good shape.

---

[616] Max Käsbauer (1927–2024) was a German philosopher who obtained his PhD from the University of Würzburg in 1954. He was assistant to Stegmüller between 1960 and 1965 in Munich. He worked on metamathematics, logic, and philosophy of science. In 1981, he changed his name to Max Drömmer.

[617] Eventually Feigl (1970) will review the volume in *The Journal of Philosophy*.

[618] Rockefeller University, where Davidson was professor of philosophy from 1970 to 1976.

[619] John Wallace is an American philosopher, specialized in the philosophy of language. He studied at Yale, received his PhD from Stanford University, and then spent most of his career at the University of Minnesota.

[620] Thomas Nagel (1937–) is an American philosopher born in Yugoslavia. He studied at Cornell University and Oxford; he received his PhD in philosophy from Harvard University. He then taught at University of California at Berkeley, Princeton University, and New York University. A prolific writer, he is best known for his influential essay in the philosophy of mind "What Is It to Be Like a Bat?" (Nagel 1974).

[621] Edward Poznański (1901–1974) was a Polish logician and philosopher of science. He immigrated to Palestine in 1939 and was appointed academic secretary of the Hebrew University in 1946.

The Hopkins volume, *The Legacy of Log. Pos.*,[622] has just come out. It is quite interesting, but not well balanced. Scriven writes on Log. Pos. and the Behavioral Sciences, and I have a piece called "Log Pos and the Social Sciences"[623] (I have so far only glanced at Scriven's, and found no significant overlap); but various other facets are missing. I still hope you'll publish your lecture – I bet you, it would have been *very* useful! And it will be, no matter where you now publish it.

If you shld see John Cooley, please do give him my warmest greetings.

Love and best wishes to you and Edith, and lot of luck to the boys, one of whom we look forward to seeing in Princeton soon!

As ever,

Peter

P.S. Some six weeks ago, Adolf Grünbaum wrote me that Henry Mehlberg's wife Janina had died on May 25 of a heart ailment; that Henry was hit very hard, but that at least he was being looked after by the widowed sister of Janina. I wrote Henry, and recently received a letter which showed that he evidently values expressions of sympathy very much. He mentions also that he is embarked on various work projects, some of which he hopes to see published soon. I have his letter in the office, but I think he said one of the planned publications was to include (part of?) his earlier "La théorie causale du temps".[624] It *would* be great if that came out in English at last.

P.

4.1.170 Nagel to Hempel, August 13, 1969

[Typewritten, Hempel 1999, Box 28, Folder 2]

So. Wardsboro, Vt. 05355
August 13, 1969

Dear Peter,

I too had intended to write you long before this – I hesitated to do so lest you might find correspondence during the summer an unwanted distraction – but your welcome letter gives me an excuse for ignoring that consideration. It is relatively good news that there is no urgency for the cataract operation (what a pity that this is hanging over you!), and that your and Diane's relations with Toby have apparently improved. The last time we talked, at the N.Y. Philosophy Club, you seemed overly burdened with worries (again I resisted the temptation to call you for fear of intrusion), and it is a relief to know that your summer has been a fairly good one.

Cobarrubias visited us with his (attractive) wife last week, and mentioned how much he enjoyed meeting you. But I had very little chance to talk to him, for by coincidence he arrived when Isaac Levi, his wife Judy, Arthur Danto (who is spending the summer in Bennington) *cum* wife and daughter, also were here. Edith managed to feed guests all right, but the occasion was not one for philosophical shop-talk. I like the Levis very much, and though his book GAMBLING WITH TRUTH[625] is badly written I think Isaac has much on the ball. Arthur Danto had received what he called an "attractive offer" from Yale, but finally turned it down, I gather a good boost in salary at Columbia made it easier for him to reach this decision. Arthur is now enjoying quite a professional reputation, he does not conceal his

[622] Achinstein and Barker (1969).
[623] Scriven (1969) and Hempel (1969b).
[624] Mehlberg (1935).
[625] Levi (1967).

sense of his own importance, but while he writes easily and with a certain amount of grace I continue to think of him as an intellectual lightweight.

I learned about Janina Mehlberg's death only two weeks ago, and wrote Henry – it is perhaps too soon for a reply. I met her in New York long before he left Poland, and thought she was a remarkabl[y] strong and able person who looked after Henry with a sure hand. From what I heard, he has been badly broken up by her death, and I hope he is on even keel again.

I did not know of Stegmuller's new book, your comments on it make me want to get hold of it, and I will write Stechert-Hafner for a copy. It is a strange illness from which he suffered, and almost unbelievable that he could write a solid book under those conditions.

Harry Frankfurt wrote me earlier in the summer about some of the changes at Rockefeller, including the addition of Donald Davidson and the probable addition of John Wallace to their staff. There will be ten graduate students in philosophy this year, and this may make the place livelier. However, although I never got to know Donald very well, I had the impression that he liked teaching and enjoyed influencing students; and it will be interesting to see whether Rockefeller will be to his liking for many years.

I saw John Cooley in the city before we left for the summer, but haven't heard from him since. Chas. Stevenson (whom we visited briefly a couple of weeks ago) had a note from him, saying that he might drop in on Steve between cruises. Steve wasn't sure whether Nelson Goodman spent any time in Vermont this year, though Kate has been here most of the time. Have you read Nelson's LANGUAGES OF ART?[626] I finally finished it the other day, and have mixed feelings about it. Some of it is obviously clever, but much of it seemed to me dull (perhaps because I do not appreciate the problems he is dealing with), and I am frankly puzzled by Nelson's attempt in the final chapter to assimilate art to science and to minimize the notions of truth in science.

Two weeks ago we went to fetch Sandy in Montreal where he spent July attending what he found an exciting Seminar in mathematics. He is with us at the moment, but will be leaving in a few days to go to England for a month on a bargain rate flight. Bobby is in Woods Hole, working for your colleague in the physics department at Princeton (Prof. G. Reynolds)[627] on some experiments dealing with bio-luminescence. He will be through this Friday, and we are going to pick up his belongings (especially his bike). But he will have to return to New York: He was reclassified as 1A last month, despite his submitting a letter from his eye doctor that Bobby lacks binocular vision (allegedly a sufficient ground for disqualification for service in the armed forces); and last week-end he was notified to appear for a physical examination in New York next Tuesday. We are moderately confident that his defective vision will exclude him from the draft, but are nonetheless anxious. I have no idea what Bobby will do if he should be drafted; and the uncertainty has generated a melancholy mood that controls our waking hours.

Otherwise, Edith is fine and enjoying the combination of doing some physics and working in her garden. It would be wonderful if you and Diane could visit us some time: is there the remotest possibility of this? But in any case, our love to you both,

Always yours,
Ernest

---

[626] Goodman (1968).

[627] George T. Reynolds (1917–2005) was an American physicist, working mainly on particle and biophysics, and environmental science. He obtained his PhD from Princeton in 1943, and worked there for the most part of his career.

4.1.171 *Hempel to Nagel, October 30, 1969*

[Handwritten, Hempel 1999, Box 28, Folder 2]

10–30–69

Dear Ernest,

I had a letter from Herbert Feigl today saying that Dr. Franz Oppacher[628] (who is in Europe, I think) will, at Feigl's suggestion, write a review of Stegmüller's book, & that Dick Rudner has agreed to publish it in *Ph. Sci.*

He adds that he would be "glad to write a *brief* review of St's book before Feb 28 ('70) and send it to the *Journal of Philosophy*", *if* I (C.G.H.) "have heard affirmatively from Ernest Nagel". So if the review has not been assigned otherwise, perhaps the editors of *J Phil* would like to get in touch with Herbert?[629]

I am very much impressed with your *Festschrift*,[630] of which I received a copy two days ago. It is marvelous to see the wide range of well-known scholars joining to honor you – as you ought to be! The intellectual density of the volume seems to me remarkably high – you even have one of those rare Nozicks, which is great!

All the best to you all!

As ever,

Peter

4.1.172 *Nagel to Hempel, December 19, 1971*

[Typewritten, Hempel 1999, Box 28, Folder 2]

25 Claremont Avenue
New York, N.Y. 10027
Dec. 19, 1971

Dear Peter,

Although I have not written until now, and feel ashamed for not having done so, I've thought of you often, and have been keeping my fingers crossed that all is well with you and all three of you are thriving. For reasons not quite clear to me, this has been a rather busy semester, and I've not had a single day free from sort of chore. Some of this felt (as distinct from actual) burden is doubtless the consequence of my teaching commitments – two courses at Columbia, one at the Graduate Center (CUNY), and in addition I foolishly made and kept a promise to attend (nominally as an auditor) a seminar on Philos. Questions of Modern Physics which is run jointly by Sid Morgenbesser and Gerald Feinberg.[631] (Half of the latter has been on conventionalism, for which Sid has been primarily responsible, the other half on measurement in quantum mechanics for which Gary is the leader. The first half

---

628 Franz Oppacher was a philosopher, working at Washington University, and then at Concordia University in Montreal. The mentioned review in *Philosophy of Science* was seemingly never published.

629 See Feigl (1970).

630 See Morgenbesser et al. (1969).

631 Gerald Feinberg (1933–1992) was an American physicist who earned his BA, MA, and PhD from Columbia University, where he also spent most of his career as a professor of physics.

has been a sheer waste of time, partly because the students are not very good, but in good measure because Sid is as disorganized and undisciplined as ever. But once I started attending, I could not drop out, without wounding Sid.) There have also been numerous consultations with students, and seemingly endless (and fruitless) committee meetings. It seems likely at the moment that I will be reappointed for another year at Columbia (my final fling, so to speak), but I am really fed up with the Graduate Center, and will not continue with it next year even if the opportunity to do so is offered me. Although I manage to do a minimum amount of reading, I've not done a solid piece of work on anything of my own for ever so long; and as you can imagine, this state of affairs is hardly conducive to a cheerful frame of mind.

But let me return to other things. Edith has been teaching as usual, and is in pretty good shape. But she has not yet gone to see her surgeon, to find out what the operation succeeded in accomplishing – on the principle that no news is good news. But I'll continue trying to persuade her to make the visit after the coming holidays. The boys also have been working hard, and we are looking forward to see them this week. Sandy is now looking for a job, for his 2 year appointment at Madison comes to an end next June, and though one of the senior men in the department is trying to have him reappointed, the chances are small because of shortage of funds. There are a number of alternative possibilities (including spending a year in Zurich), but he will not know what's what for some time, and meanwhile he feels far from relaxed. Bobby is now thinking hard about a problem in solid state physics which was suggested to him by Schnatterly[632] as a PhD topic, but the going has been slow thus far and he is often discouraged.

I saw John Cooley this past week. He came down from Bakersville for a day or so, had lunch with several of us here at Columbia, and later that evening he had dinner with Edith and the two Frankels.[633] He seems in excellent shape, but regrets that he will not be seeing you during the semester ahead.

As you know, APA is meeting in New York just after Xmas,[634] and there are a number of resolutions to be presented at the Business Meeting on the question of the Association passing on political matters as a body. I myself am against this increasing practice of committing a professional association on matters external to its main purpose – the practice is extremely divisive, and to my knowledge accomplishes nothing except the venting of emotions. I anticipate the meeting will be unpleasant.

I've started reading the just published Carnap & Jeffrey, Studies in Inductive Logic and Probability, Vol. 1.[635] The forty odd pages I found time to read thus far contain nothing new but perhaps the real meat will come later, maybe not until the next volume.

But I must put an end to this chatter, or this will not reach you before the coming new year becomes quite old. I do hope you are well, and that your stay in London is both pleasant and intellectually profitable. In any case, our love to you both.

As ever,
Ernest

---

[632] Presumably Stephen Eugene Schnatterly (1938–2019), who was a physicist, with a PhD from the University of Illinois in 1965. He joined the Physics Faculty at Princeton the same year.

[633] Presumably Charles Frankel (1917–1979). Frankel studied at Columbia University, where he also received his PhD in philosophy and where he taught for most of his career. A political philosopher by training, he was a public figure, serving as Assistant Secretary of State for Education and Cultural Affairs and helping to found the National Humanities Center, which he directed until his death.

[634] The sixty-eight annual meeting of the American Philosophical Association Eastern Division was held in New York, from 27 to 29 December 1971.

[635] Carnap and Jeffrey (1971). The second volume was published as Jeffrey (1980).

4.1.173 Hempel to Nagel, February 21, 1972

[Typewritten, Hempel 1999, Box 28, Folder 2]

6 Rumbold Road
London S. W. 6
February 21, 1972

Dear Ernest:

Recently, I have been reading a fair amount in *The Structure of Science* again and have marvelled anew at it; what a substantial work this is, how carefully argued, how lucidly presented. My copies (I have two, but only one of them here) are both full of underscorings and annotations and cross-references; the cover which came with one of them has gone to pieces, but I saved the excellent foto of you on the back and put it inside the book and – though this sounds sentimental – it cheers and reassures me to look at it from time to time.

Many thanks for your letter last December with all sorts of interesting news. Diane and I were glad to hear that Edith is pretty well; we hope that this has by now been confirmed by a verdict of her physician!

What a shame that Sandy should have to worry about his job; we hope that it will all have a happy ending in one form or another. If he could get a year in Zurich, that might be very good for him in many ways; it could be broadening not only professionally, but also in various other ways. The one hitch about it that occurs to me is just that it is harder to place oneself in the U.S., while one is abroad. We are keeping ou[r] fingers crossed! Please give both boys our best wishes.

I wish I could have a chat with you and hear what went on at the APA meeting!

Diane and I are well. Toby has gone back to New York and writes us frequent and very contented letters about a new job she has, principally as a secretary, but also as a copy writer and occasional illustrator of materials issued by a Jewish Youth organization. She had one bad shock, though: on a supposedly safe Street in Bklyn, she was mugged and pushed to the ground; fortunately, apart from the loss of her handbag, she suffered no harm, and she seems to have gotten over the awful experience all right. We have a new type of correspondence, namely by tape cassette; she and Pete and we each have a small tape recorder and thus can produce and play back spoken letters. But we can use this method only occasionally.

I have been invited to give a lecture and two seminars at the Hebrew Univ, and we expect to visit Israel from April 20 to about May 3.

My health seems to be all right, but I am feeling less peppy than I used to, and trying to work something out and write it down is sometimes a discouraging struggle.

Love to you all from Diane and me!

As ever,
Peter

4.1.174 Hempel to Nagel, June 14, 1973

[Typewritten, Hempel 1999, Box 28, Folder 2]

June 14, 1973

Dear Ernest:

I sure do hate to trouble you with this, and I am all the more grateful to you for your willingness to take on the chore!

I enclose the green sheet for the letter of reference; the address to which it is to be mailed is on the back of it. Then, there is a copy of my application, to which I have attached at the end two pages from my list of publications, on which the items referred to in my proposal are listed.

You will, I fear, take a rather dim view of my project; but as you know, I have for some time been intrigued with the subject, and I would like to become clearer in my own mind about the basic issues that are raised afresh by Kuhn's onslaught.

I am afraid that readers other than the referees named by me (you, Herbert Feigl and Gregory Vlastos), even if favorably disposed, will think that this application should properly go to the NSF. Well, if I am turned down, I hope I'll still have a chance to try that: that's an advantage of having such a very early deadline.

Did you have a good trip back to Brattleboro? During your brief return visit to New York, you probably just hit the record heat wave we have had; yesterday, things started to improve, and today, we have splendid dry and sunny weather.

We had a little party Tuesday night in honor of Doug Quine's[636] graduation; Van and Marge Quine and Doug's sister were here, and Willis Doney, the Jeffreys, Morty and Lucia, Paul Oppenheim, and Paul Benacerraf and the Quine's old friend Nini Borgerhoff,[637] who is a Dean here. Despite heat and a threat of rain, it was very animated and pleasant, and we were able to sit outside for our buffet dinner.

I hope you have good news from the younger generation, that Edith is enjoying the country, and that you will all have a very good summer. Warm good wishes from both of us, and many thanks again for your help!

As ever,
Peter

4.1.175 Nagel to Hempel, June 19, 1973

[Typewritten, Hempel 1999, Box 28, Folder 2]

So. Wardsboro, Vt. 05355
June 19, 1973

Dear Peter,

Your letter and enclosure came last night, and I am sending off my report to the NEH with this mail. In my opinion your project is an excellent one, and is most persuasively written. Needless to say, I have endorsed it in the strongest terms possible, and am confident that the Endowment will act in the same way. (I've looked through my files of correspondence with NEH, and unless I am mistaken the committee on Research Grants will meet sometimes in September.)

We opened up the summer house a week ago, but Edith and I both returned to New York on Thursday and did not get back to Vt. until Sunday. There was another informal meeting of the panel on the NEH interdisciplinary institutes (all but two of the members were present), and it is becoming clearer to both Dick Brandt and myself that the needs of philosophers are in general different from the more historically oriented humanities, and that we shall have to

---

[636] Douglas B. Quine, son of W.V.O.

[637] Nini Borgerhoff was Cornelia C. Newlin Borgerhoff (1920–2011), who graduated from Vassar College (where she took a secret cryptography course: https://www.vassar.edu/vq/issues/2011/01/a-look-back/for-school-for-country.html). Her husband was E.B.O. Borgerhoff (1908–1968), a professor of modern languages at Princeton.

write what will perhaps be a minority report. All of this takes much time that I can ill afford, not only the writing but the worrying about matters about which I know just about nothing. If there were a way of my coming *dis*involved, I would take it.

Edith is in good form and has been busy with her garden (both flower and vegetable). But the house is empty without the boys, and we miss them a good deal (at least frequently). If you should find a way to come to New England, please be sure to pay us a visit. (Our telephone is: 802-896-2541). For the moment, love to you and Diane from both of us.

Ernest

4.1.176 Hempel to Nagel, June 24, 1973

[Typewritten, Hempel 1999, Box 28, Folder 2]

June 24, 1973

Dear Ernest:

Thank you so much for your very nice letter and for your help and encouragement!

I found your letter waiting for me on my return from a week's stint at the Summer Institute at the Catholic Univ., which was very effectively organized and directed by Prof. William Wallace[638] (two years ago, when I participated, Seeger[639] had been in charge). All in all, it was quite a stimulating and interesting experience even though it was hard, if not impossible, to find a suitable common denominator for the discussion, given the great diversity of backgrounds of the participants: from astron. through physics and biology to sociology and the history (and even philosophy) of science; with only a few people having had much philosophical training. But it was lively, they all seemed interested, and Diane and I found especially some of the younger participants most likeable and sympathetic. Several of them will be going back for a terminal year of teaching; there are problems all over the place. – The priests in the audience – except for Wallace himself, whom I find a pleasant and intelligent man – were the most inscrutable members of the audience; when they discussed, their reactions seemed anxious and intense, and I could feel no human touch at all. Well, they were only a very small sample of priests, let me hasten to add!

I was matched up with John Heilbron,[640] historian of science from Berkeley, who gave a series of very interesting lectures on science and the history of science and the scientific establishments in Western Europe and the US, around the turn of the century.

Diane and I had some opportunities to see museums, and on the last day, also to stroll a bit in Georgetown, which was very pleasant.

---

[638] William A. Wallace (1918–2015) was an American clergyman and philosopher of science. He earned a degree in electrical engineering from Manhattan College and an M.S. in physics from the Catholic University of America; he also studied theology at the Dominican House of Studies in Washington. Wallace taught at several institutions and was an influential voice in Aristotelian-Thomistic philosophy of science.

[639] Raymond J. Seeger (1906–1992) was an American physicist. He earned his BA in physics from Rutgers University and his PhD in theoretical physics from Yale University. He then taught at several American Universities. Seeger served as Director of the Summer Institute for the History, Philosophy, and Sociology of Science at The Catholic University of America in 1971.

[640] John Lewis Heilbron (1934–2023) was an American historian of science. He earned his BA, MA, and PhD from the University of California at Berkeley, where he also spent most of his career. A student of Kuhn, he wrote extensively on Galileo Galilei and the history of physics.

What a pity that the plan for NEH interdisciplinary institutes does not seem to fit the needs of philosophers! [n/a] well imagine how frustrating this must be to you, and I am very sorry about it!

Diane and I are glad to hear that Edith is so active in the garden, and we wish you both a bumper crop of vegetables, flowers, and summer joys!

As ever yours,

Peter

4.1.177 Nagel to Hempel, August 14, 1973

[Typewritten, Hempel 1999, Box 28, Folder 2]

So. Wardsboro, Vt. 05355.
Aug. 14, 1973.

Dear Peter,

This is a hasty note to tell you that I returned last night from Boston, from the meeting of the philosophy panel for selecting recipients of NEH Senior Fellowships. I thought you might be interested to know that the panel selected 18 of the applicants for awards, and that you are one of them. This is not a final disposition of the applications (our panel had 91 to consider, and the total number of Senior Applications was approximately 900), for our recommendations go to another panel (which goes over the recommendations of the other panels like the one on which I served), and then finally to the Council of NEH. We were told yesterday that about 100 Senior Fellowships will be awarded, and that the list of them will not be known until the first week in November. So there is still some distance to go before the matter is settled. But it is good to know that at least the first outcome of the selection process is what I had expected it to be.

May Brodbeck[641] also was on the panel, so I saw her again after many years. Like the rest of us, she has aged. But did you know that is now Dean of the Graduate School at Minnesota?

We are well here. Sandy and Yvonne[642] are with us at the moment (I drove back with them from Boston), but they will have to return to Madison in about a week. We expect Bobby (who is working away at Princeton) to come up next Friday or Saturday for a short holiday.

Hope you and Diane are having a reasonably comfortable summer – it was almost unbearably hot here last week, so that I imagine it was even less pleasant in Princeton. Love to you both from both of us.

Ernest

[641] May Brodbeck (1917–1983) was an American philosopher. She studied chemistry at New York University, and then earned both her MA and her PhD from the University of Iowa. She taught at the University of Minnesota from 1947 to 1974—and she was Dean of the Graduate School for 2 years from 1972 to 1974—before returning to the University of Iowa. While her first publications were on Dewey's logical thought, she then established herself as a specialist in the philosophy of social sciences.

[642] Yvonne Nagel is a mathematician who retired from the University of Wisconsin in 2011. She is the wife of Alexander Nagel.

4.1.178 de Bary to Hempel, October 22, 1973

[Typewritten, Hempel 1999, Box 28, Folder 2]

October 22, 1973

Dear Professor and Mrs. Hempel:

On Monday evening, November 19 Ernest Nagel will deliver a university lecture in the Low Rotunda. Some of his friends are entertaining him at dinner beforehand, and we should very much like to have you join us at 6:00 p.m. in the President's Room of the Faculty Club. Please let us know on the enclosed card.

Sincerely,

Wm. Theodore de Bary[643]

4.1.179 Nagel to Hempel, February 11, 1975

[Handwritten, Hempel 1999, Box 28, Folder 2]

Jerusalem, Febr 11, 1975

Dear Peter,

Although I've thought of you almost daily since I got here, I've delayed writing until Edith joined me (which happened last week). With the help of one of the Department secretaries, I found a furnished apartment in the Beit Hakerem section of the city (about 10–15 minutes walk from the Campus), and we found it quite pleasant, except for the inadequate heating (so that we huddle over a small electric heater during the day). By the way, Mrs. Gatt was of no help whatsoever, not only to me but other visitors from abroad. There have been a few beautiful days, but the temperatures are still low, and snow fell yesterday and this morning. We are looking forward to real spring and warmer weather.

Jerusalem is a fascinating and in good part a beautiful city. I tramped some of the streets before Edith came, and we went together once to the old city and the Temple Mount – we'll go back and to other parts when the weather improves. I've met immeasurable former students, and I almost feel embarrassed by their warm hospitality. There have been others too who have been exceedingly kind – eg. Elkana[644] whom I had never met before, a Dr. Amit[645] in physics, and yesterday Sambursky[646] who, like his wife, is a real charmer.

[643] William Theodore de Bary (1919–2017) was an American sinologist and scholar of East Asian philosophy. He spent almost his entire career at Columbia University, where he received his MA, BA, and PhD. He was an influential figure in the field of Confucian intellectual history.

[644] Presumably Yehuda Elkana (1934–2012), a Yugoslavian-born philosopher and historian of science who immigrated to Israel in 1948. He studied at the Hebrew University of Jerusalem and obtained his PhD from Brandeis University. He taught at several universities, in the USA, Europe, and Israel. He is best known for his work on the anthropology of knowledge.

[645] Presumably Daniel J. Amit (1938–2007) an Israeli physicist who was born in Poland and emigrated to Tel Aviv in 1940 after the Nazi occupation. He received his BA in physics in 1961 at the Hebrew University of Jerusalem and did a PhD at Brandeis University in 1963.

[646] Shmuel Sambursky (1900–1990) was a German-Palestinian-Israeli physicist. He joined the physics faculty at the Hebrew University in Jerusalem in 1928. He was a popular lecturer and received in 1968 the Israel Prize for Humanities.

There have been innuendos about the precarious prospects for the Dept of the History and Philosophy of Science, but nothing very definite. [N/A] who has impressed me as a very capable man, is still acting chairman of the department.

I've seen Bar-Hillel a couple of times (have him your regards) as well as his daughter, and while he was in a better shape than I anticipated it is clear that he is pretty much out of things. There is no doubt someone is needed to give direction and prestige to the department of philosophy, and in whom people have confidence as they did in Bar-Hillel. Except for Eddie Zemach,[647] the current chairman of the philosophy department and Yehuda Melzer[648] (a former Columbia student and now a lecturer) I still don't have a good sense of what the philosophers here are like. Sanbursky has a rather low opinion of what he calls "the German Mafia" in it (apparently Rotenstreich[649] and perhaps Yovel[650]), and while he likes Elkana he has genuine doubts about him as a philosopher (as desired from a historian and sociologist of science).

My classes keep me quite busy, in part because of a lifelong habit of being nervous over my own shortcomings. The undergraduate course in the philosophy of biology gave me less trouble, for I brought along a fair set of notes; but I have nothing comparable for the seminar in the philosophy of science, and since I have few books with me preparation does not come easy. I've been using Popper's *Objective Knowledge*[651] as a part of text, and have been lecturing rather than having students report – this is partly because the class is rather mixed and I haven't dared to place the burden on its members to present any material. I'm afraid also that Popper may not be a good choice for some of the group for seemingly they have had a good deal of Sir Karl in Bar-Hillel's courses.

Sid Morgenbesser had played with the idea of coming here permanently, but told me before I left NY. that he finally decided against it, though I am not sure why. Just about everyone I've talked to is some variety of "dove" on the "Israeli-Arab" question, but that doesn't seem to be the attitude of the present government. On the other hand, I haven't yet met anyone who is optimistic about the future, so that despite an external cheerfulness there is a deep anxiety – it is sad and a bit frightening, though I hope nothing like another war will take place.

I have no typewriter, and feel very apologetic about the handwriting I have to inflict on you. I'll write again, and meanwhile our love to you and Diane.

As ever,
Ernest

---

[647] Eddy M. Zemach (1935–2021) was an Israeli philosopher who was an expert in aesthetics and metaphysics. After receiving his PhD from Yale University, he taught at several American universities and at the Hebrew University of Jerusalem.

[648] Yehuda Melzer is an Israeli philosopher and editor who received his PhD from Columbia University and taught at the Hebrew University of Jerusalem and at Tel Aviv University. He worked and wrote on the concept of just war.

[649] Possibly Nathan Rotenstreich (1914–1993), an Israeli professor of philosophy. He studied philosophy at the Hebrew University of Jerusalem and received his PhD in 1938. Rotenstreich served as the Dean of the Faculty of Humanities (1958–1962) and was Rector of the University between 1965 and 1969.

[650] Presumably Yirmiyahu Yovel (1935–2018), an Israeli philosopher and public intellectual. He studied philosophy and economics at the Hebrew University of Jerusalem, later at Sorbonne and Princeton, and had a PhD from the Hebrew University (1968)—his adviser was Nathan Rothenstreich. Yovel worked mainly on existentialist philosophy, Kant's philosophy, and rationalism.

[651] Popper (1972).

4.1.180 Hempel to Nagel, June 2, 1975

[Typewritten, Hempel 1999, Box 28, Folder 2]

Princeton, June 2, 1975

Dear Edith, dear Ernest:

Thank you so much for your several missives from Israel and Greece; we were truly delighted to receive your good news, especially since, as you can imagine, we were thinking of you a good deal, and with some concern, at critical times of your stay in Israel. We were very sorry to hear of the difficulties at the start and the unhelpfulness of Mrs. Gatt and then the cold first months; in fact, when I read your account, Ernest, I was reminded of my own experience in March and even April of 1974, when I often wished I had brought a warm sweater along – I should have remembered in time and should have forewarned you! I very much hope that the balance of your impressions and appraisals is definitely positive, all the same as it has been for us. We are very eager indeed to exchange experiences and opinions with you and also to learn more about the recent history and the prospects of the Dept. of Hist. and Phil. of Science. At the end of this week, we are expecting Chaim Gaifman[652] here in Princeton; he is to give an informal talk or two and also wants to discuss with us the first part of a piece he has been working on for ERKENNTNIS. And this week, Saul Kripke will give three lectures on Truth, which will probably draw large audiences despite the fact that they come so uncomfortably late in the academic year.

I am very sorry not to have written you sooner, and in fact to have written you only one letter, which must have crossed Ernest's first one in the mails. I wrote from Berkeley, where I had a very satisfactory winter quarter, being pleasantly ensconced in the Faculty Club right in the middle of the campus, and having very eager, enthusiastic, and inquisitive students – the experience gave me quite a psychological boost. Diane, being tied down by her editorial job here, managed at least a bit better than a week of unpaid leave, and we had a very enjoyable time, going to S.F. visiting the Goheens, calling on the Bouwsmas,[653] etc.

While in Berkeley, I received an invitation to give some lectures at the U. of Mexico, and Diane fortunately was able to come along. I went with apprehensions concerning the audiences' command of English and their interest in my field, but it turned out that I need not have worried: there was a lot of interest and many lively discussions; I gave three evening lectures and in addition two morning seminars (for invited faculty and grad. students) in three days! But the Univ. paid for Diane's and my trip, put us up in the hotel where the Shah of Iran was staying, paid all our food bills there, and assigned us to the very hospitable care of a professor and his wife, who drove us about and were excellent company; so we saw a lot of Mexico City as well as the nearby Aztec pyramids and other places. On the way back, we stopped – a first for both of us – in Austin to visit Pete, who had just finished

[652] Haim Gaifman (1934–) is an Israeli logician, mathematician, and philosopher. He studied at the Hebrew University of Jerusalem and received his PhD in mathematics from the University of California, at Berkeley. He taught at several American universities and at the Hebrew University of Jerusalem. He has made important contributions to several fields, from mathematical logic to probability theory and the philosophy of language.

[653] Oets Kolk Bouwsma (1898–1978) was an American philosopher. He studied at Calvin College and at the University of Michigan, where he also received his PhD. He then taught philosophy at the University of Nebraska and the University of Texas at Austin from 1965. While at the University of Nebraska, he took a leave of absence that allowed him to travel to the UK, where he met Wittgenstein. His notes of the conversations he had with Wittgenstein were published as *Wittgenstein: Conversations, 1949–1951* (1986).

his dissertation and hopes to defend it later this week, then on to St. Louis, where I was tickled to receive an honorary Doctor of Science from Washington Univ.

I hope this greeting will still reach you in Israel; it is to bring you our best wishes for a pleasant smooth homeward [...][654] I have a hunch that you will be going off to Vt. almost immediately; but it would be great if you should have time for a brief phone call at least. Diane and I expect to be in Evanston, Illinois from 6/13 to 6/16, but then we plan to stay put. I intend to go to the London, Ont. meeting to hear Stegmüller, Kuhn, and Sneed[655] on Stegmüller's Sneedification of Kuhn.

Sad news: two or three days ago, I heard that Alice Mandelbaum had died in May. When I saw her last, perhaps a year and a half ago, she had some difficulty walking, and the trouble seems to have gotten rapidly worse, leading even to inability to recognize friends, so that her death was merciful.

I called up Herbert quite recently, and he sounded more cheerful than he had in some time, though there seems to have been no improvement in his condition.

We hope that all the Nagels, big and small, are well, and we send your love to you all.

As ever,

4.1.181 Nagel to Larry Laudan, May 29 1979

[Typewritten, Hempel 1999, Box 28, Folder 2]

South Wardsboro, Vermont 05355
May 29, 1979
Professor Larry Laudan
Department of Philosophy, 620 Mervis Hall
University of Pittsburgh
Pittsburgh, Pa. 15260

Dear Professor Laudan,

Your letter of May 4 was forwarded from Columbia to my home, and reached me two weeks after that date when I was in the midst of packing in preparation for leaving the city for the summer. After arriving here (address as above) it has taken a number of days to get settled for the season, and this is the first opportunity I've had to write you.

I am distressed that you find the footnote in the recent collection of my essays, which comments on some theses presented in your book,[656] "a highly misleading caricature" of your position. I read your book with care, and wrote those comments in good faith. It is of course possible that because of the brevity with which I made them I did not state what I wanted to say as clearly as I thought I did; and I may have been guilty of poor judgment in dealing with your views so briefly. But I do not think I can rightly be charged with caricaturing or misrepresenting them; and I continue to believe, in the face of your rejoinders, that your book contains no answers to the questions I raised in my footnote.

---

[654] A small part of the letter is missing, presumably half of a sentence.

[655] Joseph D. Sneed (1938–2020) was an American physicist and philosopher. He earned his BA from Rice University and his MS in physics from the University of Illinois; he received his PhD in philosophy from Stanford University. He then taught at several American universities. His most important work is *The Logical Structure of Mathematical Physics* (1971). See also Stegmüller (1975).

[656] See Nagel (1979, note 3, pp. 317–318), and Laudan (1977).

Before turning to details, I must make a preliminary observation. My footnote was not intended to be an evaluation of all the materials presented in your book, and the footnote would indeed be unfair to you were it construed as a review of your essay. In the Introduction to my own book I refer to some attempts by "the new philosophy of science" to explicate the notion of rationality in science; and since in my opinion the older of these analysis all contain serious flaws, I thought it would be appropriate to deal briefly with your own explication, the most recent one with which I was familiar. I have much sympathy with your proposal to explicate the notion in terms of procedures for solving the problems that initiate scientific inquiry, or in your language in terms of scientific progress. On the other hand, your general criterion of progress in science, and in particular your proposed "measure" of the "problem-solving effectiveness" of a theory, raise what I think are fundamental difficulties; and to call attention to them was the primary purpose of my writing the footnote.

I will now address myself to the specific points you make in your letter:

1. In reply to the query in my footnote as to how the number and importance of empirical problems is to be assessed, you say (as does your book) that the importance of a solution to a problem is the greater the more generally we characterize the problem. But neither this paragraph of your letter nor any passage in your book answers my question whether Newtonian theory solves *one* problem or *nine*. For it does not follow from your explicitly stated thesis that the importance or weight of a problem is a function of its generality, that the different solutions of ostensible instances of a general problem are to count as just one solution of a single problem (rather than as many solutions of that many different problems). (I will ignore the question as not relevant to the issue under discussion whether the determinations of the orbits of the nine planets are indeed just special cases of a single general problem, and whether each such determination does not involve assumptions and techniques distinctive in each case that are not provided by Newtonian theory.)

   Nor do I find either in your letter or in your book an answer to my further query whether or not the problem of determining planetary orbits is more important than the problem of accounting for the behavior of the tides. Neither of these problems is more general than the other (in the sense of generality specified on page 35 of your book), so that the query is not answered by referring to the weighting of problems in terms of their relative generality. But neither do I find an answer to my question in your discussion on pages 33–40 of other "methods of differential weighting."
2. In reply to my question whether it is the sheer number or the variety of solved problems that counts in assessing a theory, you claim that you have answered it, and refer me to pages 31–40 of your book. I have reread those pages, but find that nowhere in them is there even an explicit mention of variety as a feature of problems, to say nothing of a discussion of the relevance of variety in assessing the problems solved by a theory. Accordingly, those pages contain no answer to my query. Nor is the "unambiguous answer" to my puzzle which you give in your letter (that "problem weight – which includes but is not reducible to variety – is decisive") of any help. Problem weight may be decisive; but since in your explication of this concept variety is not stated to be a factor to be considered, I fail to see how your "unambiguous answer" could possibly be an answer to my query.
3. In reply to my question how one is to decide whether or not quantum mechanics generates the unanswered conceptual problem which Einstein raised, you say that the decision involves ascertaining, among other things, "whether the quantum theory of measurement is consistent with the best established epistemological and methodological theories." But is this really an answer to my question? It would doubtless be counted as an answer by most readers if there were established and readily identifiable epistemological or methodological doctrines such that it can be determined whether or not a conceptual problem allegedly generated by a theory is consonant with them. However, I do not believe there are such doctrines; and if I am correct in this claim, you provide no answer

to my question, no more than the suggestion to use perpetual motion machines provides a solution to our current energy problem. Indeed, to give an example additional to the Einstein-Bohr debate in support of my claim, you yourself point out in your book that the notion of action-at-a-distance was regarded by Leibniz and some of his contemporaries as a grave conceptual problem for Newtonian theory, but that it was not so regarded either by Cotes or by Kant. Accordingly, while I admit that something very much like what you say in your letter under point 3) is also stated in your book, I do not think I was caricaturing your views when I declared that the book gives no hints on how to resolve my query about conceptual problems.

I regret that on these matters I am compelled to be stiff-necked. But there are other things in your book (as well as in your article "The Philosophy of Progress"[657] which you were good enough to send me and for which I thank you most warmly) which I would like to discuss with you; and I hope, as you do, that there will be an opportunity to do so in the not too distant future. Meanwhile, my best wishes to you.

Sincerely yours,
Ernest Nagel
P.S. I am sending copies of this letter to Adolf Grünbaum and Peter Hempel.
E.N.

4.1.182 Nagel to Hempel, May 30, 1979

[Typewritten, Hempel 1999, Box 28, Folder 2]

South Wardsboro, Vt. 05355
May 30, 1979

Dear Peter,

I don't know whether Larry Laudan has discussed with you the contents of the footnote in my TELEOLOGY[658] etc. in which I criticized certain views he expressed in his own book. In any case, he wrote me a letter, charging me with publishing "a highly misleading caricature" of his views and replying to the various points in my criticism. I am very much perturbed by this charge, for if it is true I am something of an intellectual blackguard; and I would hate to think that friends whose opinions of my character I value (above all, yours) subscribe to this accusation. For this reason I am enclosing a copy of my letter to Laudan, in which I try to disprove the charge by examining the details of his reply to my criticism. You will judge for yourself whether or not my attempted disproof is successful.

We've been here in Vermont for a week (it has been raining cats and dogs since our arrival), and are almost settled for the season. My back continues to be a bother, and gets in the way of physical as well as sedentary activities; but in other respects I really ought to count my blessings.

This note may not reach you before you leave for Australia. In any event, I hope both you and Diane are in excellent shape, and that you will have a most enjoyable time in your travels.

Edith joins me in sending our love to you both.
Ernest

[657] Laudan (1978).
[658] Nagel (1979).

4.1.183 Nagel to Hempel, June 14, 1979

[Handwritten, Hempel 1999, Box 28, Folder 2]

June 14, 1979

Dear Peter,

Warm congratulations! Yesterday's *NY TIMES* (which arrived this morning) carried the news of your receiving an honorary degree from Princeton. It was a much deserved recognition of the work you've done, and both Edith & I are happy over it. Yesterday we went to Bennington to visit Nora Stevenson.[659] She seems to be in excellent shape, with boundless energy and busy with fixing up the house she had bought in North Bennington as well as with her mum. There will be a memorial service for Steve on June 30 in Wilmington, Vt, and all eight children (4 from Nora's previous marriage) are expected to be there. I'm returning to NY next Sunday to see my urologist the following day, and will after fly to the mid-west to visit with Sandy et al and Bobby. Love to you and Diane from both of us.

Ernest

4.1.184 Nagel to Hempel, September 13, 1979

[Typewritten, Hempel 1999, Box 28, Folder 2]

So. Wardsboro, Vt. 05355
Sept. 13, 1979

Dear Peter,

Thank you most warmly for the Xerox copy of your *Nachwort*.[660] It arrived a week ago, but I did not have the chance to read it until yesterday. I also hope that it was not too great a bother to make the copy. In any case, I think the Addendum is a splendid piece, incisive and typically clear, and I very much enjoyed reading it. I think you ought to publish an English version, for relatively few American philosophers can manage German. There are some parts of the *Nachwort* I cannot fully appreciate – those dealing with Coffa[661] and Stegmüller, for I do not recall in any detail their views, and their papers are not available to me here; but for the rest, I think I am in complete (or almost complete) agreement with your comments on Salmon.

One of the places about which I *may* have some reservations is your comments in 3.7.3 on Salmon's requirement of an "absolute" homogeneity for reference classes, especially in the light of what you say on pg.115. As I understand the matter, what you say on this page summarizes the conclusions of Abraham Wald's paper[662] in Karl Menger's[663] *Ergebnisse*

---

[659] Nora Stevenson was the second wife of Charles Stevenson (they married in 1965).

[660] Hempel (1977).

[661] J. Alberto Coffa (1935–1984) was an Argentinean philosopher who was educated in Buenos Aires, and later at Notre Dame, and earned a doctorate from the University of Pittsburgh in 1972.

[662] For example, Wald (1936).

[663] Karl Menger (1902–1985) was an Austrian-American mathematician. The son of Carl Menger, he studied at the University of Vienna, where he earned his PhD. He was a member of the original Vienna circle. He then taught at several universities both in Europe and in the USA. In 1931 Menger founded the "Ergebnisse eines Mathematischen Kolloquiums", to which Nagel refers in his letter.

etc. in which he shows that von Mises' original definition of randomness is inconsistent, so that there can be no *Kollektives*. Why does this argument not also show that Salmon's requirement of homogeneous reference classes is similarly self-contradictory? If I am right in this, it seems to me that your criticism in 3.7.3 does not go far enough.

This is one of the things I had in mind (but failed to make clear) when we last talked over the phone. As I recall, I mentioned that one of my difficulties with Salmon's account of statistical explanations is that he *formally* accepts the requirement of total evidence, but then seems to forget this principle when he specifies what he thinks is the correct structure of statistical explanation. He forgets it a) when he gives alleged counter-examples to your requirements for such explanations – for they are counter-examples only if one *already knows* that some item in the explanans is statistically irrelevant so that it is strictly speaking not necessary to mention statistical irrelevance as an *additional* requirement; and also when b) he introduces absolute homogeneity as a requirement, although familiarity with Wald's results is presumably part of the total evidence. Since Wald also showed that a consistent notion of randomness requires relativization to a denumerably infinite number of place-selections, then as you point out so clearly the problem remains of specifying *which* set of place-selections go into the definition of homogeneity.

I don't want to weary you by pursuing the matter any further. But I would love to discuss with you face to face Salmon's approach to explanations – including his view (with which you now agree) that the explanandum need not have a high probability relative to the explanans. I confess that I continue to be unhappy with the idea that a necessary condition for an occurrence can count as an explanation for the latter.

When I talked to Adolf last week, he gave me the sad news about Dick Rudner's death, of which I was unaware. I saw him in St. Louis a year ago last spring, and he seemed to be in excellent health and fine form. What a pity he is gone!

Thank you again for your Addendum, it is an unusually stimulating discussion.

Please remember me to Adolf, and tell him that I will now turn to the psycho-analytic paper he sent me.[664]

And my warmest love to you and Diane.

Ernest

4.1.185 Nagel to Hempel, October 25, 1979

[Typewritten, Hempel 1999, Box 28, Folder 2]

25 Claremont Avenue
Oct. 25, 1979

Dear Peter,

Although I had intended to express my warm thanks to you (and also my chagrin that you were bothered by Mrs. Hackett to do this extra chore) for sending the Hackett Pub. Co. such a fine endorsement of the new printing of my book, it slipped my mind when we talked last night. I had no idea that you were being harassed by her until she sent me a copy of your letter to her which reached me last week, and as I wrote to her I wish she hadn't imposed that task on you. But this is now water under the bridge, and I can only tell you how deeply appreciative I am of your kindness.

I forget whether I mentioned to you that last December Harcourt Brace & Jovanovich wrote me that they were letting the book go out of print because at the last count less than 500 copies were sold in the year just ending. I was rather angry with the firm, for they made plenty of money on the Cohen & Nagel text which it published in 1934; and it seemed to me

[664] Possibly Grünbaum (1979).

that the sales of STRUCTURE OF SCIENCE (though admittedly modest) were good enough to merit keeping the book in print. And so I was glad when Wm. Hackett wrote me, saying that he was interested in giving the book a longer lease of life. The new printing will be in paper as well as hard cover, and copies are supposed to be available this month.

Many, many thanks again. And it was wonderful talking to you yesterday. Love to you and Diane.

Ernest

4.1.186 Hempel to Nagel, July 5, 1980

[Handwritten, Nagel 2020, Box 3, Folder 4]

London, 7-5-80

Dear Edith, dear Ernest:

Greetings from London, which is the first stop on our return journey, and where we are planning to stay through the conference on Popper's philosophy, to be held at LSE 7/14-16. Our visits to Munich + Vienna were interesting, but our enjoyment was dampened by the death of Diane's father; Diane flew to Miami for the funeral. Her father was 92 and had a peaceful death, but still it was a severe blow. – Both of us took turns with various ailments on the Continent, but we are all right now. At my Vienna lecture, I saw an acquaintance who had been a fellow student 50 yrs ago in Vienna, and I also met Paul Neurath[665] after many years; he is now teaching (with some time limit) at the U. of Vienna, of which he pressed a very dim view. Also saw Lazarsfeld's first wife, Marie Jahoda,[666] who now teaches at U. of Sussex, and whom I found very sympathetic. – As a city, we liked Munich much better than Vienna. Eva's sister Lux and her husband were extremely nice + hospitable, and for 4 weeks we had their aptment to ourselves while they were vacationing. – Among speakers scheduled for Popper conf:[667] Musgrave,[668] Adolf Grünbaum, Wes Salmon, Shimony, John Stachel,[669]

[665] Paul Neurath (1911–2001) was Otto Neurath's son. After studying and working in Vienna, Neurath was deported to Dachau and Buchenwald concentration camps between March 30, 1938, and May 27, 1939, when he was released to Sweden. He went to the States and studied sociology and statistics at Columbia University and taught statistics at City College, New York, between 1943 and 1946. He defended his PhD thesis on the social life in the Dachau and Buchenwald camps in 1943. His book appeared only later as *The Society of Terror: Inside the Dachau and Buchenwald Concentration Camps*.

[666] Marie Jahoda (1907–2001) was an Austrian-born social scientist and social psychologist. She studied at the University of Vienna, where she met Paul Lazarsfeld, whom she married in 1927. After divorcing Lazarsfeld, who moved to the USA, she remained in Vienna, where she was arrested for political reasons in 1936. Once released, she moved to Great Britain and, after the end of WWII, to the USA, where she held positions at several universities. She worked extensively on the topics of unemployment and mental health and made influential contributions to these fields.

[667] There was a conference in honor of Popper in London, but the details are unknown.

[668] Alan Musgrave (1940–) is an English-born New Zealand philosopher and historian of science. He studied under Popper at the London School of Economics, where he also received his PhD. He has taught for most of his career at the University of Otago. Though he has written several books in epistemology and philosophy, he is probably best known as the editor, together with Imre Lakatos, of the hugely influential collection of essays *Criticism and the Growth of Knowledge* (1970).

[669] John Stachel (1928–2025) was an American physicist, historian and philosopher of science. He earned his BS from The City College of New York and both his MS and PhD from the Stevens Institute of Technology. He then taught at Lehigh University, the University of Pittsburgh, and Boston University. He served as the Director of the Center for Einstein Studies and was also the founding editor of the *Collected Papers of Albert Einstein*.

John Worrall[670] (editor of BJPS) and others. (I am *not* a speaker, and glad of it; I had begun to feel rather exhausted). We hope you are well, and we wish you a very pleasant + restorative summer in B'boro.

As ever – Peter

4.1.187 Hempel to Nagel, January 2, 1982

[Handwritten, Nagel 2020, Box 3, Folder 4]

Jan. 2, 1982

Dear Ernest:

Diane and I were deeply distressed to hear that you had to undergo such serious and complex surgery, and we are relieved at learning from Edith that you now are out of the intensive care unit and recovering in your own room. We have been sending you and Edith our affectionate thoughts and good wishes, and we hope that every day will bring you further progress in your recovery.

It was wonderful that Sandy and Robert could be there in the time of greatest need, and it is good to know that Sandy will be rather close by during the coming term. I greatly enjoyed speaking with him on the phone and I was again struck by the uncanny similarity in certain Gestalt qualities between his and your ways of thinking and talking. Well, that won't be news to you!

You will have heard, I suppose, that the Philadelphia meeting went rather smoothly and that a conciliatory tone seems to have prevailed; I am very glad about that.

I was astonished and happy to hear from Edith a short while ago that you had phoned her from the hospital! I know that must be quite a big effort for you, and I will not try to call you for a while.

All the very best and much love to you from Diane and from me. If good wishes could cure you, you would be heading home by now; but it will come soon, I am quite sure. Do try to take it easy!

As ever yours,

Peter

4.1.188 Hempel to Nagel, September 28, 1985

[Handwritten, Nagel 2020, Bos 3, Folder 4]

London, September 28, 1985.

Dear Edith:

Diane and I are deeply saddened by the news of Ernest's death, which reached us today in London.

---

[670] John Worrall (1946–) is a British philosopher of science. He studied at the London School of Economics, where he obtained his PhD under the supervision of Imre Lakatos, and then worked and taught there for the rest of his career. He is best known for his article "Structural Realism: the Best of Both Worlds" (1989), in which he made a case for structural realism in the debate between realism and anti-realism in the philosophy of science.

Ernest was one of my oldest and staunchest friends. I met him shortly after coming to the United States in 1937 and he soon became a wise and sympathetic counselor to me in an unfamiliar social and academic environment. He put me in touch with stimulating thinkers and lively professional groups and he helped me find my first academic teaching position. Later, in the difficult years after Eva's death, you and Ernest gave me warm human support as well as material help in engaging Mrs. Richmond to look after little Peter Andrew. Ernest would not accept a word of thanks, regarding it all as a matter of course. He was an exemplar of dedication to the highest mood and intellectual standards, and my admiration for him continued to grow over the years. He has exerted a strong and very salutary influence on my thinking and he contributed to keeping me from espousing a too formalistic approach to philosophical issues.

Dear Edith, we mourn with you the great loss that you and Sandy and Robert and the other members of your family have suffered. We send you all our deepest sympathy and most affectionate thoughts.

As ever,

Peter and Diane.

References

Achinstein, Peter, and S. F. Barker, eds. 1969. *The Legacy of Logical Positivism*. Baltimore: Johns Hopkins University Press.

Anastasi, Anne. 1937. *Differential Psychology*. Oxford: Macmillan Co.

Austin, John L. 1966. Three Ways of Spilling Ink. *The Philosophical Review* 75 (4): 427–440.

Ayer, Alfred J., C. H. Whiteley, and Max Black. 1936. Truth by Convention: A Symposium by A.J. Ayer, C.H. Whiteley, M. Black. *Analysis* 4 (2/3): 17–32.

Ayer, A. J. 1962. Names and Descriptions. *Logique et Analyse* 5 (20): 199–204.

Baumrin, Bernard, ed. 1963a. *Philosophy of Science: The Delaware Seminar, Volume 1, 1961–1962*. New York/London: Interscience Publishers.

Baumrin, Bernard, ed. 1963b. *Philosophy of Science: The Delaware Seminar, Volume 2, 1962–1963*. New York/London: Interscience Publishers.

Bergmann, Gustav. 1940a. On Physicalistic Models of Non-Physical Terms. *Philosophy of Science* 7 (2): 151–158.

Bergmann, Gustav. 1940b. On Some Methodological Problems of Psychology. *Philosophy of Science* 7 (2): 205–219.

Berlin, Isaiah. 1954. *Historical Inevitability*. London: Oxford University Press.

Braithwaite, Richard. 1953. *Scientific Explanation: A Study of the Function of Theory, Probability and Law in Science*. Cambridge: Cambridge University Press.

Bridgman, Percy W. 1959. *The Way Things Are*. Cambridge, MA: Harvard University Press.

Bronstein, Daniel J., Yervant H. Krikorian, and Philip P. Wiener, eds. 1948. *Basic Problems in Philosophy*. New York: Prentice-Hall.

Burke, Christopher, and Adam Tamas Tuboly. 2025. *Otto Neurath in Britain*. Cambridge: Cambridge University Press.

Carnap, Rudolf. 1929. *Abriss der Logistik*. Wien: Springer.

Carnap, Rudolf. 1934. *Logische Syntax der Sprache*. Wien: Springer.

Carnap, Rudolf. 1936. Testability and Meaning. *Philosophy of Science* 3 (4): 419–471.

Carnap, Rudolf. 1937a. Testability and Meaning – Continued. *Philosophy of Science* 4 (1): 1–40.

Carnap, Rudolf. 1937b. *Logical Syntax of Language*. Trans. Amethe Smeaton. London: Routledge & Kegan Paul.

Carnap, Rudolf. 1939. *Foundations of Logic and Mathematics*. Chicago: University of Chicago Press.

Carnap, Rudolf. 1942. *Introduction to Semantics*. Cambridge, MA: Harvard University Press.
Carnap, Rudolf. 1943. *Formalization of Logic*. Cambridge, MA: Harvard University Press.
Carnap, Rudolf. 1946. Modalities and Quantification. *Journal of Symbolic Logic* 11 (2): 33–64.
Carnap, Rudolf. 1947a. *Meaning and Necessity: A Study in Semantics and Modal Logic*. Chicago: University of Chicago Press.
Carnap, Rudolf. 1947b. On the Application of Inductive Logic. *Philosophy and Phenomenological Research* 8 (1): 133–148.
Carnap, Rudolf. 1948. Reply to Nelson Goodman. *Philosophy and Phenomenological Research* 8 (3): 461–462.
Carnap, Rudolf. 1950. *Foundations of Probability*. Chicago: University of Chicago Press.
Carnap, Rudolf. 1952. *The Continuum of Inductive Methods*. Chicago: University of Chicago Press.
Carnap, Rudolf. 1955. Meaning and Synonymy in Natural Languages. *Philosophical Studies* 6 (3): 33–47.
Carnap, Rudolf. 1966. *Philosophical Foundations of Physics. An Introduction to the Philosophy of Science*. New York: Basic Books.
Carnap, Rudolf. 1973. Notes on Probability and Induction. *Synthese* 25: 269–298.
Carnap, Rudolf, and Friedrich Bachmann. 1936. Über Extremalaxiome. *Erkenntnis* 6 (1): 166–188.
Carnap, Rudolf, and Friedrich Bachmann. 1981. On Extremal Axioms. Trans. H.G. Bohnert. *History and Philosophy of Logic* 1981 (1–2): 67–85.
Carnap, Rudolf, and Richard Jeffrey, eds. 1971. *Studies in Inductive Logic and Probability: Volume I*. Berkeley: University of California Press.
Cassidy, Harold G. 1962. *The Sciences and the Arts. A New Alliance*. New York: Harper.
Cohen, Morris R., and Ernest Nagel. 1934. *An Introduction to Logic and Scientific Method*. New York: Harcourt Brace.
Cooley, John. 1959. On Mr. Toulmin's Revolution in Logic. *The Journal of Philosophy* 56 (7): 297–319.
Colodny, Robert G. 1966. Introduction. In *Mind and Cosmos*, ed. R. G. Colodny, xi-xviii. Pittsburgh: University of Pittsburgh Press.
Creath, Richard, ed. 1990. *Dear Carnap, Dear Van: The Quine-Carnap Correspondence and Related Work*. Berkeley: University of California Press.
Danto, Arthur. 1962. Narrative Sentences. *History and Theory* 2 (2): 146–179.
Danto, Arthur. 1965. *Analytical Philosophy of History*. Cambridge: Cambridge University Press.
Davey, Jocelyn. 1960. *A Touch of Stagefright*. London: Chatto & Windus.
de Santillana, Giorgio, and Edgar Zilsel. 1941. *The Development of Rationalism and Empiricism*. Chicago: University of Chicago Press.
Dray, William H. 1957. *Laws in Explanation in History*. Oxford: Oxford University Press.
Feigl, Herbert. 1970. Review of Stegmüller's Erkläung und Begründung. *The Journal of Philosophy* 67 (8): 248–250.
Feigl, Herbert, and May Brodbeck, eds. 1953. *Readings in the Philosophy of Science*. New York: Appleton-Century-Crofts.
Feigl, Herbert, and Grover Maxwell, eds. 1962. *Scientific Explanation, Space and Time*. Minneapolis: University Minnesota Press.
Feigl, Herbert, and Michael Scriven, eds. 1956. *Foundations of Science and the Concepts of Psychology and Psychoanalysis*. Minneapolis: University of Minnesota Press.
Feller, William. 1950. *An Introduction to Probability Theory and its Applications*. New York: John Wiley and Sons Inc.
Feyerabend, Paul. 1962. Explanation, Reduction, and Empiricism. In *Scientific Explanation, Space and Time*, ed. Herbert Feigl and Grower Maxwell, 28–97. Minneapolis: University of Minnesota Press.
Feyerabend, Paul. 1965. Problems of Empiricism. In *Beyond the Edge of Certainty: Essays in Contemporary Science and Philosophy*, ed. Robert G. Colodny, 145–260. Pittsburgh: University of Pittsburgh Press.

Frost-Arnold, Greg. 2013. *Carnap, Tarski, and Quine at Harvard: Conversations on Logic, Mathematics and Science*. Chicago: Open Court.

Gardiner, Patrick, ed. 1959. *Theories of History: Readings from Classical and Contemporary Sources*. Glencoe: Free Press.

Gellner, Ernest. 1959. *Words and Things: A Critical Account of Linguistic Philosophy and a Study in Ideology*. London: Victor Gollancz Ltd.

Gödel, Kurt. 1934. *On Undecidable Propositions of Formal Mathematical Systems. Lecture Notes by Kleene and Rosser.* Princeton: Princeton University Press. Reprinted in *Kurt Gödel: Collected Works, Volume 1*, eds. S. Feferman, et al., 338–371. Oxford: Oxford University Press.

Gödel, Kurt. 1995. Is Mathematics Syntax of Language? In *Collected Works, Volume III, Unpublished Essays and Lectures*, ed. Solomon Feferman et al., 334–355. Oxford: Oxford University Press.

Goenner, Hubert F. M. 2014. On the History of Unified Field Theories. Part II. (ca. 1930 – ca. 1965). *Living Reviews in Relativity* 17 (5): 7–241.

Goodman, Nelson. 1947. On Infirmities of Confirmation Theory. *Philosophy and Phenomenological Research* 8 (1): 149–151.

Goodman, Nelson. 1955. *Fact, Fiction, and Forecast.* Cambridge, MA: Harvard University Press.

Goodman, Nelson. 1968. *Languages of Art: An Approach to a Theory of Symbols.* Indianapolis: Bobbs-Merrill.

Gregg, John R., and F. T. C. Harris, eds. 1964. *Form and Strategy in Science: Studies Dedicated to Joseph Henry Woodger on the Occasion of his Seventieth Birthday.* Dordrecht: D. Reidel.

Grelling, Kurt, and Paul Oppenheim. 1937/38. Der Gestaltbegriff im Lichte der neuen Logik. *Erkenntnis* 7: 211–225.

Gronda, Roberto, and Adam Tamas Tuboly, eds. 2026. *Ernest Nagel on Science and Philosophy: Volume 2. Selection of (Un)published Writings.* Cham: Springer, forthcoming.

Grünbaum, Adolf. 1963. *Philosophical Problems of Space and Time.* New York: Alfred Knopf.

Grünbaum, Adolf. 1979. Is Freudian Psychoanalytic Theory Pseudo-Scientific by Karl Popper's Criterion of Demarcation? *American Philosophical Quarterly* 16 (2): 131–141.

Gunther, John. 1939. *Inside Asia.* New York: Harper & Brothers.

Hansel, C. E. M. 1966. *ESP: A Scientific Evaluation.* Introduction by Edward G. Boring. London: Mac Gibbons and Kee Ltd.

Hanson, Norwood Russell. 1963. *The Concept of the Positron: A Philosophical Analysis.* New York: Cambridge University Press.

Harman, Gilbert. 1967. Quine on Meaning and Existence, I. The Death of Meaning. *Review of Metaphysics* 21 (1): 124–151.

Hempel, Carl G. 1934. *Beitrage zur logischen analyse des wahrscheinlichkeitsbegriffs.* Jena: Universitats-buchdruckerei G. Neuenhahn.

Hempel, Carl G. 1935a. Some Remarks on 'Facts' and Propositions. *Analysis* 2 (6): 93–96.

Hempel, Carl G. 1935b. On the Logical Positivists' Theory of Truth. *Analysis* 2 (4): 49–59.

Hempel, Carl G. 1936. Eine rein topologische Form nichtaristotelischer Logik. *Erkenntnis* 6: 436–442.

Hempel, Carl G. 1937a. A Purely Topological Form of Non-Aristotelian Logic. *The Journal of Symbolic Logic* 2 (3): 97–112.

Hempel, Carl G. 1937b. Karl Popper, Logik der Forschung. *Deutsche Literaturzeitung* 58: 309–314.

Hempel, Carl G. 1939. Vagueness in Logic. *Philosophy of Science* 6:163–180.

Hempel, Carl G. 1945a. Studies in the Logic of Confirmation (I). *Mind* 54 (213): 1–26.

Hempel, Carl G. 1945b. Studies in the Logic of Confirmation (II). *Mind* 54 (214): 97–121.

Hempel, Carl G. 1945c. Geometry and Empirical Science. *The American Mathematical Monthly* 52 (1): 7–17.

Hempel, Carl G. 1950a. Problems and Changes in the Empiricist Criterion of Meaning. *Revue Internationale de Philosophie* 11: 41–63.

Hempel, Carl G. 1950b. A Note on Semantic Realism. *Philosophy of Science* 17 (2): 169–173.

Hempel, Carl G. 1951a. The Concept of Cognitive Significance: A Reconsideration. *Proceedings of the American Academy of Arts and Sciences* 80 (1): 61–77.

Hempel, Carl G. 1951b. General System Theory: A New Approach to the Unity of Science. *Human Biology* 23: 323–322.

Hempel, Carl G. 1952. *Fundamentals of Concept Formation in Empirical Science*. Chicago: University of Chicago Press.

Hempel, Carl G. 1955. Review of Reichenbach's Nomological Statements and Admissible Operations. *The Journal of Symbolic Logic* 20 (1): 50–54.

Hempel, Carl G. 1956. Review of Gregg's, The Language of Taxonomy. *The Journal of Symbolic Logic* 21 (4): 396–397.

Hempel, Carl G. 1960. Science and Human Values. In *Social Control in a Free Society*, ed. Robert E. Spiller, 39–64. Philadelphia: University of Pennsylvania Press.

Hempel, Carl G. 1961. Introduction to Problems of Taxonomy. In *Field Studies in the Mental Disorders*, ed. J. Zubin, 3–32. New York: Grune & Stratton.

Hempel, Carl G. 1962. Deductive-Nomological vs. Statistical Explanation. In *Scientific Explanation, Space & Time*, ed. Herbert Feigl and Gordon Maxwell, 98–169. Minneapolis: University of Minnesota Press.

Hempel, Carl G. 1963a. Explanation and Prediction by Covering Laws. In *Philosophy of Science: The Delaware Seminar*, ed. B. Baumrin, 107–133. New York: Interscience.

Hempel, Carl G. 1963b. Implications of Carnap's Work for the Philosophy of Science. In *The Philosophy of Rudolf Carnap*, ed. P. A. Schilpp, 685–709. La Salle: Open Court.

Hempel, Carl G. 1965a. *Aspects of Scientific Explanation*. New York: Free Press.

Hempel, Carl G. 1965b. Aspects of Scientific Explanation. In *Aspects of Scientific Explanation*, 331–496. New York: The Free Press.

Hempel, Carl G. 1966a. *Philosophy of Natural Science*. Englewood Cliffs: Prentice-Hall.

Hempel, Carl G. 1966b. Recent Problems of Induction. In *Mind and Cosmos*, ed. R. G. Colodny, 112–134. Pittsburgh: University of Pittsburgh Press.

Hempel, Carl G. 1969a. On the Structure of Scientific Theories. In *The Isenberg Memorial Lectures Series 1965–1966*, 11–38. East Lansing: Michigan State University.

Hempel, Carl G. 1969b. Logical Positivism and the Social Sciences. In *The Legacy of Logical Positivism*, ed. P. Achinstein and S. F. Barker, 163–194. Baltimore: Johns Hopkins University Press.

Hempel, Carl G. 1973. Science Unlimited? *Annals of the Japan Association for Philosophy of Science* 4 (3): 187–202.

Hempel, Carl G. 1977. Nachwort 1976. In *Aspekte wissenschaftlicher Erklärung*, 98–123. Berlin: Walter de Gruyter.

Hempel, Carl G. 1992. In Memoriam Wolfgang Stegmüller. *Erkenntnis* 36: 5–6.

Hempel, Carl G. 1999. Carl Gustav Hempel Papers, 1903–1997, ASP.1999.01. Archives of Scientific Philosophy, Archives & Special Collections, University of Pittsburgh Library System.

Hempel, Carl G. 2000. Prologue: An Intellectual Autobiography: Carl G. Hempel. In *Science, Explanation, and Rationality: Aspects of the Philosophy of Carl G. Hempel*, ed. James H. Fetzer, 3–35. Oxford: Oxford University Press.

Hempel, Carl G., and Paul Oppenheim. 1936. *Der Typusbegriff im Lichte der neuen Logik. Wissenschaftstheoretische Untersuchungen zur Konstitutionsforschung und Psychologie*. Leiden: A.W. Sijthoff's Uitgeversmaatschappij.

Hempel, Carl G., and Paul Oppenheim. 1945. A Definition of 'Degree of Confirmation'. *Philosophy of Science* 12 (2): 98–115.

Hesse, Mary B. 1963. *Models and Analogies in Science*. London: Sheed & Ward.

Hook, Sidney. 1955. *Marx and the Marxists: The Ambiguous Legacy*. New York: Van Nostrand Reinhold Company.

Hook, Sidney, ed. 1958. *Determinism and Freedom in The Age of Modern Science*. New York: Collier Books.

Hook, Sidney, ed. 1959. *Psychoanalysis, Scientific Method, and Philosophy*. New York: New York University Press.
Hosiasson-Lindenbaum, Janina. 1940. On Confirmation. *The Journal of Symbolic Logic* 5 (4): 33–148.
Hospers, John. 1946. On Explanation. *The Journal of Philosophy* 43 (13): 337–356.
Hull, Clark L. 1937/1987. Logical Positivism as a Constructive Methodology in the Social Sciences. In *Unified Science*, ed. Brian McGuinness, 159–162. Dordrecht: D. Reidel.
Hutten, Ernest H. 1960. Physics and Biology. *The British Journal for the Philosophy of Science* 11 (42): 101–108.
Jeffrey, Richard, ed. 1980. *Studies in Inductive Logic and Probability: Volume II*. Berkeley/Los Angeles: University of California Press.
Jeffreys, Harold. 1931. *Scientific Inference*. Cambridge: Cambridge University Press.
Jeffreys, Harold. 1939. *Theory of Probability*. Oxford: Clarendon Press.
Joad, C. E. M. 1941. Bertrand Russell's New Book. *The New Statesman and Nation* 21: 218.
Kaufmann, Walter. 1965. *Hegel: Reinterpretation, Texts, and Commentary*. New York: Doubleday & Company.
Kneale, William. 1949. *Probability and Induction*. Oxford: Clarendon Press.
Krikorian, Yervant H., ed. 1944. *Naturalism and the Human Spirit*. New York: Columbia University Press.
Kuhn, Thomas. 1962. *The Structure of Scientific Revolutions*. Chicago: The University of Chicago Press.
Kyburg, H. E. 1968. *Philosophy of Science: A Formal Approach*. New York: The Macmillan Company.
Kyburg, H. E., and Ernest Nagel, eds. 1963. *Induction: Some Current Issues*. Middletown: Wesleyan University Press.
Laudan, Larry. 1977. *Progress and Its Problems*. Berkeley: University of California Press.
Laudan, Larry. 1978. The Philosophy of Progress.... *PSA: Proceedings of the Biennial Meeting of the Philosophy of Science Association* 2: 530–547.
Lazarsfeld, Paul F., and Allen H. Barton. 1951. Qualitative Measurement in the Social Sciences: Classification, Typologies, and Indices. In *The Policy Sciences*, ed. Daniel Lerner and Harold D. Lasswell, 155–192. Stanford: Stanford University Press.
Lazarsfeld, Paul F., and Morris Rosenberg, eds. 1955. *The Language of Social Research: A Reader in the Methodology of Social Research*. Glencoe: The Free Press.
Leighton, Joseph A. 1939. History as the Struggle for Social Values. *Philosophical Review* 48 (2): 118–154.
Levi, Isaac. 1967. *Gambling with Truth: An Essay on Induction and the Aims of Science*. New York: Alfred A. Knopf.
Lewis, C. I. 1934. Experience and Meaning. *The Philosophical Review* 43: 125–146.
Lynd, Robert S., and Helen M. Lynd. 1929. *Middletown: A Study in Modern American Culture*. New York: Harcourt, Brace and Company.
Lynd, Robert S., and Helen M. Lynd. 1937. *Middletown in Transition: A Study in Cultural Conflicts*. New York: Harcourt, Brace and Company.
Mazurkiewicz, Stefan. 1934. Über die Grundlagen der Wahrscheinlichkeitsrechnung. *Monatshefte für Mathematik und Physik* 41: 343–352.
McGill, Vivian J. 1936. An Evaluation of Logical Positivism. *Science & Society* 1 (1): 45–80.
McGill, Vivian J. 1937. Logical Positivism and the Unity of Science. *Science & Society* 1 (4): 550–561.
McGill, Vivian J. 1939. Concerning the Laws of Contradiction and Excluded Middle. *Philosophy of Science* 6: 196–211.
Mehlberg, Henry. 1935. Essai sur la théorie causale du temps. *Studia Philosophica* 1:119–260; 2: 111–231.
Merton, Robert K. 1949. Manifest and Latent Functions. In *Social Theory and Social Structure: Toward the Codification of Theory and Research*, 21–81. Glencou: The Free Press.

Morgenbesser, Sidney, Patrick Suppes, and Morton White, eds. 1969. *Philosophy, Science, and Method: Essays in Honor of Ernest Nagel*. New York: St. Martin's Press.
Morris, Charles. 1938. *Foundations of the Theory of Signs*. Chicago: University of Chicago Press.
Myhill, John. 1961. Review of Gödel's Proof. *The Journal of Philosophy* 58 (8): 209–218.
N/A. 1936–37. Abstracts of Papers to be Read at the Thirty-Seventh Annual Meeting of the Eastern Division of the American Philosophical Association, Princeton University, December 28, 29, and 30, 1937. *The Journal of Philosophy* 34 (25): 673–687, 680–681.
Nagel, Ernest. 1933. A Frequency Theory of Probability. *The Journal of Philosophy* 30 (20): 533–554.
Nagel, Ernest. 1934. The Eighth International Congress of Philosophy. *The Journal of Philosophy* 31 (22): 589–601.
Nagel, Ernest. 1935a. Review of Logische Syntax der Sprache. *The Journal of Philosophy* 32 (2): 49–52.
Nagel, Ernest. 1935b. Review of Reichenbach's Wahrscheinlichkeitslehre. *The Journal of Philosophy* 32 (14): 389–392.
Nagel, Ernest. 1936a. Impressions and Appraisals of Analytic Philosophy in Europe, Part I. *The Journal of Philosophy* 33 (1): 5–24.
Nagel, Ernest. 1936b. Impressions and Appraisals of Analytic Philosophy in Europe, Part II. *The Journal of Philosophy* 33 (2): 29–53.
Nagel, Ernest. 1936c. Reduction and Autonomy in the Sciences. In *Actes du huitième International de Philosophie*, 181–186. Paris/Prague: Orbis.
Nagel, Ernest. 1936d. Review of Hempel and Oppenheim's Der Typusbegriff im Lichte der neuen Logik. *The Journal of Philosophy* 33 (22): 611.
Nagel, Ernest. 1936e. Critical Notices. *Mind* 45 (180): 501–514.
Nagel, Ernest. 1936f. The Meaning of Probability. *Journal of the American Statistical Association* 31 (193): 10–26.
Nagel, Ernest. 1937. Review of Carnap's Logical Syntax of Language. *The Journal of Philosophy* 34 (11): 303–304.
Nagel, Ernest. 1938a. Some Theses in the Philosophy of Logic. *Philosophy of Science* 5 (1): 46–51.
Nagel, Ernest. 1938b. Review of Reichenbach's Experience and Prediction. *The Journal of Philosophy* 35 (10): 270–272.
Nagel, Ernest. 1939. *Principles of the Theory of Probability*. Chicago: The University of Chicago Press.
Nagel, Ernest. 1939/1954. Probability and the Theory of Knowledge. In *Sovereign Reason and Other Studies in the Philosophy of Science*, 225–265. Glencoe: The Free Press.
Nagel, Ernest. 1942. Review of Carnap's Introduction to Semantics. *Journal of Philosophy* 39 (17): 468–473.
Nagel, Ernest. 1943a. Review of Carnap's Formalization of Logic. *The Journal of Philosophy* 40 (12): 332–334.
Nagel, Ernest. 1943b. Review of Susanne K. Langer's Philosophy in a New Key. *The Journal of Philosophy* 40 (12): 323–329.
Nagel, Ernest. 1944a. Truth and Knowledge of the Truth. *Philosophy and Phenomenological Research* 5 (1): 50–68.
Nagel, Ernest. 1944b. Rejoinder to Mr. Kaufmann's Reply. *Philosophy and Phenomenological Research* 5 (1): 75–79.
Nagel, Ernest. 1944c. Review of the Philosophy of G. E. Moore. *Mind* 53 (209): 60–75.
Nagel, Ernest. 1944d. Russell's Philosophy of Science. In *The Philosophy of Bertrand Russell*, ed. P. A. Schilpp, 317–349. Evanston: Open Court.
Nagel, Ernest. 1949. The Meaning of Reduction in the Natural Sciences. In *Science and Civilization*, ed. R. C. Stouffer, 99–135. Madison: University of Wisconsin Press.
Nagel, Ernest. 1950a. Dewey's Theory of Natural Science. In *John Dewey, Philosopher of Science and Freedom: A Symposium*, ed. Sidney Hook, 231–248. Westport: Greenwood Press.

Nagel, Ernest. 1950b. Memorial Address for Felix Kaufmann. *Philosophy and Phenomenological Research* 10 (3): 464–468.
Nagel, Ernest. 1950c. Science and Semantic Realism. *Philosophy of Science* 17 (2): 174–181.
Nagel, Ernest. 1950d. Review of Kneale's Probability and Induction. *The Journal of Philosophy* 47 (19): 545–551.
Nagel, Ernest. 1950e. Review of Reichenbach's The Theory of Probability. *The Journal of Philosophy* 47 (19): 551–555.
Nagel, Ernest. 1951. Reflections on the Casual Character of Modern Physical Theory. In *Freedom and Reason: Studies in Philosophy and Jewish Culture: In Memory of Morris Raphael Cohen*, ed. Salo W. Baron, Ernest Nagel, and Koppel S. Pinson, 244–268. New York: The Free Press.
Nagel, Ernest. 1953. Teleological Explanation and Teleological Systems. In *Vision and Action: Essays in Honour of Horace M. Kallen on His 70th Birthday*, ed. Sidney Ratner, 537–558. New Brunswick: Rutgers University Press.
Nagel, Ernest. 1954. Review of Toulmin's Philosophy of Science. *Mind* 63 (251): 413–412.
Nagel, Ernest. 1954/1956a. Naturalism Reconsidered. In *Logic Without Metaphysics and Other Studies in the Philosophy of Science*, 3–18. New York: The Free Press.
Nagel, Ernest. 1954/1956b. Symbolism and Science. In *Logic Without Metaphysics and Other Essays in the Philosophy of Science*, 103–141. Glencoe, Ill.: The Free Press.
Nagel, Ernest. 1957a. Determinism and Development. In *The Concept of Development: An Issue in the Study of Human Behavior*, ed. Dale B. Harris, 15–24. Minneapolis: University of Minnesota Press.
Nagel, Ernest. 1957b. A Budget of Problems in the Philosophy of Science. *The Philosophical Review* 66 (2): 205–225.
Nagel, Ernest. 1961. *The Structure of Science: Problems in the Logic of Scientific Explanation*. New York: Harcourt, Brace & World, Inc.
Nagel, Ernest. 1963a. Introduction. In *Induction: Some Current Issues*, ed. H. E. Kyburg and E. Nagel, xi–xxi. Middletown: Wesleyan University Press.
Nagel, Ernest. 1963b. Assumptions in Economic Theory. *The American Economic Review* 53 (2): 211–219.
Nagel, Ernest. 1963c. Review of W. H. Watson's Understanding Physics Today. *Scientific American* 208 (October): 145.
Nagel, Ernest. 1963d. Carnap's Theory of Induction. In *The Philosophy of Rudolf Carnap*, ed. P. A. Schilpp, 785–826. La Salle: Open Court.
Nagel, Ernest. 1965. Review of Kaufmann's Hegel. *Scientific American* 213 (5): 133–136.
Nagel, Ernest. 1979. *Teleology Revisited and Other Essays in the Philosophy and History of Science*. New York: Columbia University Press.
Nagel, Ernest. 2020. Ernest Nagel Papers, 1925–1985, ASP.2020.01. Archives of Scientific Philosophy, Archives & Special Collections, University of Pittsburgh Library System.
Nagel, Ernest, and Richard B. Brandt, eds. 1965. *Meaning and Knowledge: Systematic Readings in Epistemology*. New York: Harcourt, Brace & World, Inc.
Nagel, Ernest, and James R. Newman. 1958. *Gödel's Proof*. New York: New York University Press.
Nagel, Ernest, and James R. Newman. 1961. Putnam's Review of Gödel's Proof. *Philosophy of Science* 28 (2): 209–211.
Nagel, Ernest, Patrick Suppes, and Alfred Tarski, eds. 1962. *Logic, Methodology and Philosophy of Science*. Stanford: Stanford University Press.
Nagel, Thomas. 1974. What Is It Like to Be a Bat? *The Philosophical Review* 83 (4): 435–450.
Neurath, Otto. 1944. *Foundations of the Social Sciences*. Chicago: University of Chicago Press.
Newman, James R. 1950. BOOKS: The 50 Years and the Popularization of Science: A Listing of 20 Good Examples. *Scientific American* 183 (3): 97–107.
Oppenheim, Felix E. 1944. Outline of a Logical Analysis of Law. *Philosophy of Science* 11 (3): 142–160.

Pap, Arthur. 1958. *Semantics and Necessary Truth: An Inquiry Into the Foundations of Analytic Philosophy*. New Haven: Yale University Press.
Popper, Karl. 1972. *Objective Knowledge: An Evolutionary Approach*. New York: Oxford University Press.
Price, H. H. 1953. *Thinking and Experience*. London: Hutchinson.
Putnam, Hilary. 1960. Review of Gödel's Proof. *Philosophy of Science* 27 (2): 205–207.
Putnam, Hilary. 1972. Other Minds. In *Logic and Art: Essays in Honor of Nelson Goodman*, ed. Richard Rudner and Israel Scheffler, 78–99. Bobbs-Merrill: Indianapolis.
Quine, W.V.O. 1950. *Methods of Logic*. New York: Henry Holt & Company.
Quine, W. V. O. 1951. Two Dogmas of Empiricism. *The Philosophical Review* 60 (1): 20–43.
Reichenbach, Hans. 1935/1978. Induction and Probability: Remarks on Karl Popper's The Logic of Scientific Discovery. In *Selected Writings: 1909-1953*, ed. Maria Reichenbach and Robert S. Cohen, 372–387. Dordrecht: D. Reidel.
Reichenbach, Hans. 1935a. *Wahrscheinlichkeitslehre: Eine Untersuchung über die logischen und mathematischen Grundlagen der Wahrscheinlichkeitsrechnung*. Leiden: A. W. Sijthoff.
Reichenbach, Hans. 1935b. Über Induktion und Wahrscheinlichkeit. Bemerkungen zu Karl Poppers Logik der Forschung. *Erkenntnis* 5 (4): 267–284.
Reichenbach, Hans. 1936. Logistic Empiricism in Germany and the Present State of Its Problems. *The Journal of Philosophy* 33 (6): 141–160.
Reichenbach, Hans. 1938a. On Probability and Induction. *Philosophy of Science* 5 (1): 21–45.
Reichenbach, Hans. 1938b. *Experience and Prediction: An Analysis of the Foundations of the Structure of Knowledge*. Chicago: The University of Chicago Press.
Reichenbach, Hans. 1949. *The Theory of Probability: An Inquiry into the Logical and Mathematical Foundations of the Calculus of Probability*. Berkeley: University of California Press.
Reisch, George, and Adam Tamas Tuboly. 2021. Philipp Frank: A Crusader for Scientific Philosophy. In *Ph. Frank: The Humanistic Background of Science*, ed. G. Reisch and A. T. Tuboly, 1–67. Albany: SUNY Press.
Scheffler, Israel. 1963. *The Anatomy of Inquiry: Philosophical Studies in the Theory of Science*. New York: Knopf.
Schilpp, Paul A., ed. 1963. *The Philosophy of Rudolf Carnap*. La Salle: Open Court.
Schlauch, Margaret. 1942. *The Gift of Tongues*. New York: Viking-Modern Age.
Schlick, Moritz. 1936. Meaning and Verification. *The Philosophical Review* 45 (4): 339–369.
Scholz, Heinrich. 1938. Die mathematische Logik und die Metaphysik. *Philosophisches Jahrbuch der Görres-Gesellschaft* 51: 257–291.
Scriven, Michael. 1969. Logical Positivism and the Behavioral Sciences. In *The Legacy of Logical Positivism: Studies in the Philosophy of Science*, ed. P. Achinstein and S. F. Barker, 195–209. Baltimore: Johns Hopkins University Press.
Shapere, Dudley. 1964. The Structure of Scientific Revolutions. *Philosophical Review* 73 (3): 383–394.
Shapere, Dudley. 1966. Meaning and Scientific Change. In *Mind and Cosmos*, ed. R. G. Colodny, 41–85. Pittsburgh: University of Pittsburgh Press.
Sklenicka, Carol. 2019. *Alice Adams: Portrait of a Writer*. New York: Scribner.
Sommerhoff, Gerd. 1950. *Analytical Biology*. Oxford: Oxford University Press.
Stadler, Friedrich. 2015. *The Vienna Circle: Studies in Origins, Development, and Influence of Logical Empiricism*. Dordrecht: Springer.
Stegmüller, Wolfgang. 1969. *Wissenschaftliche Erklärung und Begründung*. Berlin/New York/ Heidelberg: Springer.
Stegmüller, Wolfgang. 1975. Structure and Dynamics of Theories: Some Reflections on J. D. Sneed and T. S. Kuhn. *Erkenntnis* 9 (1): 75–100.
Stevens, S. S. 1946. On the Theory of Scales of Measurement. *Science* 103: 677–680.
Stouffer, R. C., ed. 1949. *Science and Civilization*. Madison: University of Wisconsin Press.
Suppes, Patrick. 1960. A Comparison of the Meaning and Uses of Models in Mathematics and the Empirical Sciences. *Synthese* 12: 287–301.

Székely, D. L., and Charles Morris. 1966. *Otto Neurath and the Unity of Science Movement.* Jerusalem: Mimeographed monograph of the Association for Unification and Automation in Science.

Taylor, Richard. 1963. *Metaphysics*. New York: Prentice-Hall.

Toulmin, Stephen. 1958. *The Uses of Argument*. New York: Cambridge University Press.

Turner, Joseph. 1955. Maxwell on the Method of Physical Analogy. *British Journal for the Philosophy of Science* 6 (23): 226–238.

Urmson, J. O. 1956. *Philosophical Analysis: Its Development between the Two World Wars.* Oxford: Oxford University Press.

Verhaegh, Sander. 2020. Coming to America: Carnap, Reichenbach and the Great Intellectual Migration. Part II: Hans Reichenbach. *Journal for the History of Analytical Philosophy* 8 (11): 1–23.

Vivas, Eliseo, Howard Dykema Roelofs, and Philip Blair Rice. 1939. The New Encyclopedists: A Symposium. *The Kenyon Review* 1 (2): 159–182.

von Mises, Richard. 1939. *Probability, Statistics, and Truth*. New York: The Macmillan Company.

von Mises, Richard. 1951. *Positivism: A Study in Human Understanding*. Cambridge, MA: Harvard University Press.

von Wright, Georg Henrik. 1965. Paradoxes of Confirmation. *Theoria* 31 (3): 255–274.

Wald, Abraham. 1936. Die Widerspruchsfreiheit des Kollektivbegriffes der Wahrscheinlichkeitsrechnung. *Ergebnisse eines Mathematischen Kolloquiums* 8: 38–72.

Wang, Hao. 1952. Truth Definitions and Consistency Proofs. *Transactions of the American Mathematical Society* 73: 243–275.

White, Morton. 1950. The Analytic and the Synthetic: An Untenable Dualism. In *John Dewey, Philosopher of Science and Freedom: A Symposium*, ed. Sidney Hook, 316–330. Westport: Greenwood Press.

White, Morton, ed. 1953. *Academic Freedom, Logic, and Religion*. Philadelphia: University of Pennsylvania Press.

White, Morton. 1999. *A Philosopher's Story*. Universiry Park: The Pennsylvania State University Press.

Wiener, Philip P., ed. 1953. *Readings in Philosophy of Science: Introduction to the Foundations and Cultural Aspects of the Sciences*. New York: Charles Scribner's Sons.

Wild, John. 1960–1961. The Exploration of the Life-World. *Proceedings and Addresses of the American Philosophical Association* 34: 5–23.

Wisdom, John O. 1952. *Foundations of Inference in Natural Science*. London: Methuen & Co. Ltd.

Wohlstetter, Albert. 1940. Review of Woodger's The Technique of Theory Construction. *Journal of Symbolic Logic* 5 (1): 23–24.

Wohlstetter, Albert. 1959. The Delicate Balance of Terror. *Foreign Affairs* 37 (2): 211–234.

Woodger, Joseph H. 1937. *The Axiomatic Method in Biology*. Cambridge: Cambridge University Press.

Woodger, Joseph H. 1939. *The Technique of Theory Construction*. Chicago: University of Chicago Press.

Index

R. Gronda, A. T. Tuboly (eds.), *Ernest Nagel on Science and Philosophy*, Vienna Circle Institute Library 14, https://doi.org/10.1007/978-3-031-99458-6

MIX
Papier aus verantwortungsvollen Quellen
Paper from responsible sources
FSC® C105338

If you have any concerns about our products,
you can contact us on
ProductSafety@springernature.com

In case Publisher is established outside the EU,
the EU authorized representative is:
Springer Nature Customer Service Center GmbH
Europaplatz 3, 69115 Heidelberg, Germany

Printed by Libri Plureos GmbH
in Hamburg, Germany